# READINGS IN
# INPUT-OUTPUT ANALYSIS

# READINGS IN INPUT-OUTPUT ANALYSIS

Theory and Applications

Edited by
IRA SOHN
Montclair State College

New York   Oxford
OXFORD UNIVERSITY PRESS
1986

Oxford University Press

Oxford New York Toronto
Delhi Bombay Calcutta Madras Karachi
Petaling Jaya Singapore Hong Kong Tokyo
Nairobi Dar es Salaam Cape Town
Melbourne Auckland

and associated companies in
Beirut Berlin Ibadan Nicosia

Published by Oxford University Press, Inc.,
200 Madison Avenue, New York, New York 10016

Oxford is a registered trademark of Oxford University Press

Library of Congress Cataloging-in-Publication Data
Main entry under title:

Readings in input-output analysis.

   Includes index.
   1. Input-output analysis—Addresses, essays,
lectures. I. Sohn, Ira.
HB142.R39   1986          339.2'3          85-15351
ISBN 0-19-503674-3

Printing (last digit): 9 8 7 6 5 4 3 2 1

Printed in the United States of America
on acid-free paper

To my parents

# Foreword

Input-Output has been steadily gaining recognition as a useful and, in some instances, indispensible tool of economic inquiry. Practical application of this relatively new analytical technique is advancing in the United States more rapidly than its adoption as a subject of academic instruction, hence the absence of easily accessible teaching material.

The purpose of this volume is to provide students and researchers with a single source for some of the best work that has been done in input-output economics over the past several decades. The twenty-eight articles collected here for the first time in one volume appeared originally in a great variety of economic journals and conference proceedings volumes, some of which are now out of print. They were selected from among many papers published annually on input-output analysis for their importance either as theoretical contributions to understanding the basic properties of static and dynamic input-output models, or as representative examples of practical applications of that flexible tool in areas as diverse as the study of the economic structure of metropolitan areas and analysis of the competitive position of Japan.

Dr. Ira Sohn, who made the selections and wrote the introductory chapter, was closely associated with me for a number of years as a member of the Senior Research staff of the Institute for Economic Analysis at New York University.

Wassily Leontief

# Acknowledgments

As any fledgling student of input-output analysis is apt to quickly learn, the indirect effects stemming from a policy change are often as powerful as the direct effects of that stimulus. In attempting to apportion credit to those who have been instrumental in the publication of this volume, the contributions fall into two categories: direct and indirect.

I would like to acknowledge my debt to the scores of graduate students in the input-output course at New York University, whose repeated and amplified complaints over a period of seven "lean" years provided the direct incentive to conceive this volume. Regrettably, Stendhal's dictum, "Il n'y a pas de force que la force" ("There is no force but force"), seems all too accurate in this instance. In atonement for past transgressions I hope that this book will provide future students and practitioners of input-output analysis seven (and even fourteen) "bountiful" years of use!

I would like to express my thanks to Professor Wassily Leontief, who permitted me to climb mountains—enabling me to both broaden and deepen my understanding of the ever-evolving economic landscape, which is shaped by technological, social, political, environmental and financial forces. I hope the selections in this volume adequately illustrate this panorama from the summit.

And last, and needless to say, not least, I am indebted to those responsible for overseeing this volume of readings as it progressed from its pre-manuscript form to bound books. In particular, I would like to thank Herbert J. Addison, Vice-President and Executive Editor, Ellen B. Fuchs, Managing Editor, and Wendy Warren Keebler, Associate Editor, all of Oxford University Press, and Mona Wong, previously at the Institute for Economic Analysis, and now also at Oxford.

# Editor's Note

Since all the articles in this volume were either published previously or presented at scientific meetings, I would like to clarify at the outset the editorial changes that were introduced into the text.

Differences in American and British spelling were retained throughout, but punctuation has been changed to conform to American style.

All figures, tables and equations were renumbered, and some figures and tables were redrawn.

All footnotes were placed at the end of each contribution.

Text references were standardized throughout the volume.

In addition, when typographical or syntactical errors were discovered in the different articles, they were corrected.

I hope these modest changes have improved the coherency of the volume and, as a result, have enhanced the fluency of the book.

I.S.

# Contents

## Part IV: Economic Analysis

# READINGS IN
# INPUT-OUTPUT ANALYSIS

# Introduction

IRA SOHN

When I was first asked by Professor Leontief to begin reviewing the literature on input-output analysis with a view toward selecting a small subset of that literature for a volume of readings, my initial reaction was, "It's about time!" On further reflection, and with the hindsight of a preliminary survey (as opposed to a census) of the input-output literature, I concluded that "it couldn't be done." With the body of input-output literature now more than a few thousand books and articles strong, I was confronted with the following issues during the nominating process: Which criteria would have to be invoked so that only thirty or so of the articles could be selected for inclusion in this volume? And what compromises would have to be struck in order to insure that this collection (well under 1 percent of the estimated population of articles) exhibits a balance concerning the choice of topics, uniformity regarding the high quality of the papers, and relevance with respect to the subjects addressed. Whether or not this unenviable selection task was successfully accomplished only its readers can judge.

## Objectives of the Volume

The purpose of this collection is two pronged: first, as an educational tool for students of economic modelling in general and input-output analysis in particular; and second, as a reference guide for practitioners of input-output analysis.

For the past seven years I have assisted Professor Leontief in teaching a graduate-level course in input-output analysis at New York University. Each year, every new crop of students would voice the same collective complaint a few weeks into the semester: "Isn't there a self-contained book on the subject that covers the topics in the course syllabus and also provides both background and supplementary reading materials?" Perennially, my response would be, "Not yet." This volume provides a first step toward equipping students with adequate course materials which are "under one roof." A companion volume by Leontief, consisting of a collection of some of his past and more recent articles, has also been published.[1] Together, they should provide the student with the necessary reading materials from which he or she can go on to "do" input-output analysis either by extending its methodological frontiers or by adding to its empirical applications.

To carry out the second purpose for assembling this anthology (i.e., as a

3

reference guide for practitioners of input-output analysis), the subjects which comprise Parts II and III of the volume were chosen to address most of the problems that users of input-output analysis are likely to confront: issues related to classification, aggregation, by-products, changes in coefficients, and projections for national and regional economies. In addition, some examples of the uses of the input-output approach as a tool of economic analysis are presented in the last part of this anthology. While many of the so-called state-of-the-art papers can be found in recent journals, the articles selected for this volume present, in some cases for the first time, the cause and nature of the problem studied and *a* (rather than *the*) solution using the context of the input-output approach. For those with an interest in the history of economic analysis, this collection should be of special value since, to some extent, it traces out the evolution of a relatively new and specialized methodology as it has developed over the past forty years.

**Restrictions on the Selection Process**

A few words are appropriate concerning the self-imposed restrictions placed on the selection process. First of all, this collection is limited to articles that first appeared in, or were later translated into, English. While many studies using input-output analysis are undertaken and published by statistical offices and research centers in many non-English-speaking countries, the contents of these studies often find their way into an English language publication, either directly or through some intermediary, such as an international forum. (Appropriately, I believe it was Max Born who once remarked that it was not English—but *broken* English—that was the real international language of science!)

Second, even though Leontief has written at some time in his long career on almost all of the above subjects, his contributions were deliberately omitted from this volume. Leontief's name has been so strongly associated with input-output analysis that the two have been used interchangeably. While it may be accurate to say that Leontief is synonymous with input-output analysis, the selections that comprise this volume should go far to dispel the notion that input-output analysis is Leontief and only Leontief.

Third, I have attempted to strike a balance between some of the now "classic" literature in input-output analysis and the more recent developments in the field. Some may be of the opinion, after scanning the table of contents, that I have had an excessive bias for the "classics" and have not, therefore, sufficiently discounted the past. (More about this below.) Also, I have sought to provide a reasonable complement of input-output studies which, on the one hand, articulate methodological issues and, on the other, exemplify empirical ones which, as might be expected, lend themselves more to economic policy analysis.

In addition, the selections in this reader have been restricted to articles that have appeared in journals and periodicals or have been presented at scientific conferences (some of which were subsequently published in a volume of proceedings.)[2] Even after considering all of the above restrictions—most of which are nonintersecting—the number of remaining candidates from which a final selection was made depended ultimately on space considerations. Therefore, given the various (and very different) constraints within which I permitted myself to roam in the selection process, I believe (or, at the very least, hope) that this brief anthology captures the spirit and the body of this still fresh and innovative approach to describing and analyzing some of the issues with which economic science concerns itself.

**Organization of the Volume**

The articles selected were subdivided into three general groups: national and regional studies, studies on the coefficients matrix, and economic analysis using the input-output methodology. Preceding these groups of articles is a short survey—a brief progress report on the technique—by the recent Nobel laureate, Professor Sir Richard Stone of Cambridge University who has been a leading proponent of the input-output approach in the United Kingdom for over thirty years.

Leontief's pioneering study, *The Structure of the American Economy 1919–1929*, published in 1941, reconstructed the flows between industries and to the component parts of final demand for the U.S. economy for 1919–1929. With this basic framework in place, the seeds bearing the fundamental questions that input-output analysis was designed to answer began to sprout. Subsequent studies addressed questions of resource availabilities and sectoral capacity constraints for a given technological base and levels of final deliveries. Since the number of unknown variables always exceeds the number of independent equations in a so-called open input-output system, these questions can always be turned on their head. For example, given resource availability, sectoral capacity levels, and a technological base, for a given *structure* of final demand what *levels* of final demand can be attained? More generally, the introduction of the use of input-output analysis into statistical offices in many different countries afforded those designing national economic policy alternatives a real laboratory for assessing and analyzing these alternative policies. For the first time, detailed, quantitative projections at the sectoral level—computed in an internally consistent way, side by side with the national accounts—could be used to formulate macro- and microeconomic policies that were feasible within the set of national constraints (i.e., a given technological base, available labor force, sectoral capacity levels, available foreign exchange, etc.). One of the first countries to use input-output analysis in this context was Norway in the early 1950s when it was used as an analytical tool to aid in the reconstruction of the Norwegian economy in the postwar era. The important role that input-output analysis continues to play in the formulation of Norwegian economic policy is demonstrated in the article by Bjerkholt.

To date about 90 countries have constructed input-output tables. These countries include the Socialist-oriented economies, the Western market-oriented countries, and an increasing number of developing countries. Included in the section on national and regional studies in this volume are contributions that describe the current use of input-output analysis in some of these countries (Bjerkholt, Buckler et al., Miernyk, Polenske, and Bulmer-Thomas).

As the use of input-output analysis became more and more institutionalized throughout the world, it was inevitable that some standardization of conventions would become necessary. The article by Aidenoff describes the basis for the integration of input-output analysis into the system of national accounts (SNA) adopted by the Statistical Commission of the United Nations. This was an indispensable contribution to implement the global modelling efforts begun by Leontief, Klein, and Tinbergen, and their associates, and Almon, Bottomley, and others during the 1970s and up to the present time. The first section ends with an article by Stone on the incorporation of demographic factors into the input-output framework. Those working with input-output models are aware that for a given technological base [consistent with the Simon-Hawkins conditions (see below)] and a *structure* of final demand, *unique* output levels—rather than output ratios—can be determined only after the relevant demographic information is incorporated, either directly or indirectly, through *levels* of final demand. While the role played by technological change and/or changes in consumer preferences over time (in response to relative price changes among other things) is central to any discussion of long-term economic change, most growth in sectoral output levels over time can be attributed to population dynamics. One does not need to be a prophet to predict that the "funeral business" will be a high-growth sector of the U.S. economy in the year 2020–2030 if current actuarial figures are to be relied upon. Also, due to the "echo" of the postwar "baby boom" in the United States and Europe, the late 1980s and the last decade of this century should prove to be "boom years" for the primary and secondary education sectors. Again, this is nothing more than the normal course of population dynamics making itself felt.

Ultimately, if the economy is about anything, it is about people—how they live, what they eat, how they earn their income, what they buy, how they spend their leisure time, etc. The input-output table fills in the details to provide answers to these questions by linking the consumption and distribution activities with the production ones. The article by Stone links these demographic aspects with the economy, and, in that way, enables users of input-output analysis to incorporate explicitly the single most important factor that determines the *level* of sector inputs, sector outputs, and final output—people.

The second set of articles is devoted to studies of the coefficients matrix, the heart of the input-output model. Those who are familiar with the mechanics of input-output analysis are, in some cases, often painfully aware of the reason for an unsuccessful "run" of their model: "Something was amiss in the coefficients matrix." Whether the disorder was diagnosed as a matrix with "rank problems" or one with "a bad condition number," or one in which information was erroneously stored, the outcome is the same: the results of such a "run" are either nonsense and/or they cannot be guaranteed mathematically. Consequently, it should come as no surprise to input-output practitioners that so much attention was devoted to enumerating the formal properties of a "sound" coefficients matrix and its relationship with the bill of final demand. Around the early 1950s, these issues captured the interest (and time) of no fewer than five future Nobel laureates in economics: Kenneth Arrow, Gerard Debreu, Tjalling Koopmans, Paul Samuelson, and Herbert Simon. Their pioneering work on the substitution theorem and the properties that the coefficients matrix must exhibit are reprinted below.

By the mid 1950s with these concerns satisfactorily addressed, the focus of interest in the coefficients matrix shifted to the problem of aggregation and the potential biases that this problem could introduce into the analysis. Economists were finally endowed with (or more precisely, were given the necessary tools to construct) a framework that could exploit the rich databases of technical information that government surveys and censuses provided. However, with large-scale computing hardware still in its infancy at that time, this detailed information could be made operational only in aggregated form. The responses to this obstacle spawned a vigorous flurry of technical articles on aggregation, one of which is reprinted in this volume (Fisher). The problem of aggregation is, in the end, very much like a chronic skin ailment—it can be diagnosed and even treated but never really cured!

In the 1960s and 1970s, once the data on technology introduced after World War II became widely available throughout the Western economies, much attention was directed toward: first, assessing the economy-wide economic impact of changing technologies (Sevaldson and Vaccara); second, methods, pioneered by Stone, to update and project input-output coefficients (Lynch); and third, refinements, extensions, and applications of the dynamic model introduced by Leontief in the mid 1950s (Almon, Fisher and Chilton, and Johansen). The question of incorporating, within the input-output framework, the treatment of by-products or secondary products in the production process, or (in some cases) the equivalent of producing a single product using alternative methods of production also captured the interest of input-output theoreticians and practitioners in the 1960s (Almon).

The last section of this volume is devoted to articles that apply the input-output methodology to empirical problems requiring economic analysis. Linear growth models (as they are conventionally defined) had their genesis in J. von Neumann's pathbreaking article that was first published in 1937 and which appeared in translation in 1946 as "A Model of General Economic Equilibrium."[3] (P. Sraffa's, *The Production of Commodities by Means of Commodities*, though conceived in the 1930s, did not appear in print until 1960.) While the theoretical underpinnings of Leontief's input-output system were shown to be a special case of the more general von Neumann construct, the real strength of the input-output methodology lay not in its theoretical foundation but rather in its practical uses as an implement of economic analysis. In some cases input-output analysis has been used as just one more instrument in the economist's toolkit (e.g., for projecting GNP or sector output levels). For other uses, it occupies a unique position as an analytical device that has helped economists and policy makers to describe and quantify the tradeoffs that are inherent in any set of feasible policy alternatives. For example, the empirical work undertaken in response to its pioneering use in the description and analysis of U.S. imports and exports in the 1950s (the so-called Leontief Paradox) generated an extraordinary influence on advancing the measurement of factor intensity in production in general, and of our understanding of the changing position of the national economy within the wider framework of the world economy in particular. The articles by Heller and Riedel exemplify these advances in our understanding: the former provides evidence of the changing role played by the most developed Asian country—Japan—and the latter demonstrates the economic rationale for the heavy reliance of Taiwan on imported materials to fuel the development and industrialization process that began in the early 1960s and is continuing today.

The practical demonstration of the input-output methodology for the description and analysis of regional problems has provided regional economists and other regional scientists the ability to focus their attention on economic activities within a well-defined spatial environment, while at the same time continuing to track the larger and often more dominating economic activities that surround the regional economy. The article by Hirsch, written over 25 years ago, reports on one of the first regional input-output tables produced which had as one of its objectives "to develop more adequate methods for a better understanding of the economic fabric of a metropolitan area." Regional input-output analysis had its genesis in the work of Isard in the early 1950s; today in the United States, it is firmly institutionalized in governmental and quasigovernmental agencies such as the Port Authority of New York and New Jersey, aside from the studies undertaken in many university and private research centers.

One of the first major studies involving international comparisons of the structure of production which utilized the tool of input-output analysis was carried out by Chenery and Watanabe in 1958 (*Econometrica*, Vol. 26, no. 4, October 1958, pp. 487–521). The objective of this type of analysis was, in the words of the authors of this now "classic" article, to "shed light on the bases for international trade, the mechanism of economic growth and other economic problems whose understanding requires an empirical knowledge of the nature of interdependence." Two subsequent studies, both of which appeared in 1965 and are reprinted in this volume, extended the work of Chenery and Watanabe to gain new insight into the industrialization and development process. The paper by Simpson and Tsukui demonstrated "that the economic systems of Japan and the United States, although superficially dissimilar, contain almost identical patterns of industries which are strongly inter-related." Their findings suggest "that there are certain fundamental elements which may be found in the productive structure of modern economic systems which are purely technical in character." This latter suggestion is precisely the objective of the paper by Yan and Ames. They used input-output data as the basis for "testing empirically a variety of assertions made about certain tendencies in the economic history of industrializing economies." The article by Weisskoff and Wolff is a lineal descendant of the work of Chenery and Watanabe. In this case, the authors set out "to measure the net effects of the industrialization process on the structure of import flows and to examine the resulting changes in import dependence" for Puerto Rico.

If the problems of the debt crisis in some of the developing countries and trade protectionism in the developed countries emerged as the central economic issues of the first half of the 1980s, then the energy crisis and questions about the environment could qualify as the major subjects of concern to economists in the 1970s. (The accelerated rates of inflation which plagued the western developed economies throughout the past decade were, to a large extent, attributable to the slow and painful adjustment process dictated by the new regime of energy prices, and, to a lesser extent, to the compliance of the private and public sectors with recently enacted environmental and other social legislation. The coinciding downturn in private investment and poor performance in labor productivity growth over this same period can be, at least in part, imputed to the poor business environment that resulted from the energy and environmental issues that awaited resolution). Appropriately, the article by Anne Carter confronts both of these issues and assesses their effects on future economic growth in the United States. Using a closed dynamic input-output model, she "appraises the effects of specific pollution abatement and new energy technologies on the rate of growth" of the U.S. economy. Carter's computations show that "the effects of increased pollution abatement and energy costs" could be offset by increased energy conservation or by increased saving rates.

The article by Robert Costanza addresses the issue of embodied energy in the spirit of Georgescu-Roegen. Using the pioneering work of Hannon, Herendeen, and Bullard on the description and analysis of energy use and production in the U.S. economy, Costanza presents the case "for the interdependence of the currently defined primary factors" (labor, capital, and natural resources). In this article, "the earlier input-output analyses of energy-economy linkages are extended by incorporating the energy costs of labor and government services and solar energy inputs." Costanza maintains that "the required perspective is an ecological or 'systems' view that considers humans to be a part of, and not apart from, their environment."

In concluding these introductory remarks, I thought I should share some of my own personal impressions regarding the usefulness and suitability of viewing economic facts and assumptions through the prism of the input-output technique, having been first a student, and more recently, a practitioner of input-output analysis. The aspects of research that were, and continue to be, most satisfying to me began with the daily "shredding" of newspapers and scientific articles in search of what proved to be bounties of economic facts and assumptions. These were later transformed into arithmetical arguments enabling them to be incorporated into the methodological framework. With these data in place, and the alternative hypothetical changes that these data are likely to undergo in the future stored, a description of the alternative future course(s) of the economy, based on the above data and their assumed changes, can be formulated using the input-output technique. This versatility, or flexibility, enabling one to look at alternative future states of the economy simultaneously in a detailed and internally consistent way, attests to the usefulness of the input-output technique as an indispensable tool of economic analysis. George Balanchine, the late co-founder and artistic director of the New York City Ballet, was accustomed to saying, "Come to the ballet where you can *see* the music and *hear* the dance." In a similar vein, I invite the reader of this collection of articles to exploit this rich methodological approach which, for the first time, permits economists to practically *see* and *hear* the component parts of this intricately structured, elaborately assembled, and continuously changing, wonderous machine that constitutes our economy.

**NOTES**

1. *Input-Output Economics*, Oxford University Press, New York, 1986.
2. To date, seven "official" international input-output conferences have convened—the first was held in Driebergen, Holland in 1950 and the most recent took place in Innsbruck, Austria in 1979. Of the *Proceedings* of these symposia, only the papers presented in the Fourth, Sixth, and Seventh Conferences are easily accessible; the other *Proceedings* are regrettably out of print.
3. The paper, originally entitled "Über ein Ökonomisches Gleichungssystem und eine Verall-gemeinerung des Broüwerschen Fixpunktsatzes," was first published in *Ergebnisse eines Mathematischen Kolloquiums*, 8, 1937, pp. 73–83, and appeared in translation in *The Review of Economic Studies*, 13, 1945–1946.

# I

# An Overview
# of the Methodology

# 1

# Where Are We Now? A Short Account of the Development of Input-Output Studies and Their Present Trends

RICHARD STONE

When the Programme Committee for this Conference first met about two years ago they decided to invite papers on nine topics one of which was: Where are we now? At their second meeting, last summer, when the responses had come in, it was apparent that relatively few authors proposed to offer a paper for this session. It was also clear that the papers that were offered were so diverse in subject matter that they could hardly be grouped into a session and would be better accommodated elsewhere in the programme. Accordingly, the Committee invited me to open a discussion on where we now are and to seek contributions from other members of our group who would speak from the platform before the meeting was thrown open for general discussion. Happily, Professors Baranov, Ghosh, Leontief, and Tsukui have responded to my appeal. I hope that in this way different points of view will be represented.

In what follows I shall set the present stage of input-output analysis in its historical context and then try to identify what seem to me the growing points for the future. Although economists and statisticians engaged in input-output analysis will continue to be largely concerned with the problems that have preoccupied them in the past, the cumulative effects of the new developments are likely to bring about considerable changes in the subject as we know it today.

## The Background

Input-output analysis can be dated from Wassily Leontief's paper of 1936 and his famous book which followed it five years later (Leontief, 1936, 1941). As usual, it is possible to find precursors: François Quesnay, whose *Tableau économique* dating from the mid-eighteenth century has been set out in input-output form by Phillips (1955) and Barna (1975), is often referred to; the construction of equations relating

This paper was presented at the Seventh International Conference on Input-Output Techniques held in Innsbruck, Austria, April 9–13, 1979, pp. 1–39. Reprinted by permission of the author.

input and output was suggested by Walras (1874) and Dmitriev (1904); and a "chessboard" table for the USSR was published in USSRCSB (1926). But all this does not quite add up to input-output analysis. In his paper on the use of mathematics in economics, Nemchinov (1959) lists what he considers as Leontief's own initial contributions to the subject, including the all-important formulation of the first model connecting inputs and outputs, which made it possible to calculate indirect as well as direct inputs and thus to carry out the many, now familiar analyses which depend on being able to do this.

So forty years ago Leontief provided us with an input-output model and a database in the form of small tables for the United States in 1919 and 1929. The new ideas took on and were rapidly developed in a number of areas which can be outlined as follows.

### The Construction of Input-Output Tables

In the second edition of Leontief (1941), published in 1951, a table was added for the United States in 1939; and in the same year the documentation for the first official table for that country, constructed by the Bureau of Labor Statistics and relating to 1947, began to appear in USBLS (1951). Other early examples were: in Britain, Barna (1951, 1953), UKCSO (1952) and Stewart (1958); in Denmark, Denmark DS (1948, 1951); in Holland, Netherlands CBS (1946, 1952); and in Italy, Chenery, Clark, and Cao Pinna (1953). More and more countries followed suit and, at a later date, lists of tables and their main characteristics were set out in UNSO (1966, 1973).

### Input-Output Taxonomy

Work on the construction of tables naturally gave rise to discussions on questions of classification, definition, and treatment. After publishing its standardised system of national accounts, OEEC (1952), the Economics and Statistics Division of the OEEC, under Milton Gilbert, arranged meetings on specialised aspects of a complete system. In the case of input-output it seemed that the time was not ripe for agreement on taxonomy: there was not yet much experience in many countries, and what there was had not been the subject of wide discussion. Much later, the United Nations tried to deal with the main issues in UNSO (1966, 1973).

### Input-Output and the National Accounts

In the beginning input-output was often regarded as something quite separate from the national accounts and, in some countries, the group constructing input-ouput tables was distinct from the group constructing the national accounts. Initially, this may not have been a bad idea since it ensured that each subject was developed on its merits before an attempt was made to integrate the two. However, with the development of disaggregated models covering many aspects of the economy it became a nuisance because the different bodies of data tended to develop their own taxonomies and more work had to be done to make them compatible. John Utting and I contributed a paper on input-output in a system of national accounts (Stone and Utting, 1953) to the first conference on input-output techniques (it was not called that in those days) which was held at Driebergen in Holland in 1950. It was a cosy affair: the promoter was Jan Tinbergen and I think we numbered twenty-seven. A comparison with the numbers attending this conference, the seventh in the series, gives some idea of how the subject has grown in three decades.

The setting of input-output in the framework of the national accounts played an important part in the report (Stone, 1961), published by the OEEC and arising from the discussions of the 1950s. The problems were worked through in detail in terms of British experience and data by the group engaged on the Cambridge Growth Project; and the social accounting matrix (SAM), which formed the main database for the group's model of the British economy, was published in Cambridge DAE (1962b). This experience provided some ideas for the revision of the UN's system of national accounts, UNSO (1968).

### Prices and Quantities

It was known from the outset that along with the quantity model, which enables total outputs to be derived from a knowledge of final outputs, there is also a price model which enables the total cost per unit, or price, of each product to be derived from a knowledge of the primary input cost per unit. This makes it possible to study price repercussions and the transmission of inflationary pressures. Interesting early work in these areas can be found in Goodwin (1952, 1953) and in Rasmussen (1956). From a different point of view the model system provides a basis for the construction of consistent systems of index-numbers of prices and quantities as described in Stone (1956, Ch. III) and UNSO (1968, Ch. IV). This approach to index-numbers, which involves the measurement of inputs as well as outputs, leads naturally to the method of double deflation. It might be supposed that this method would always be superior to single deflation, in which a single indicator is attached to a net output weight. However, as is shown in Hill (1971, Ch. II and annex), this presupposes that indicators are not subject to error; if they are, it may be preferable to use an indicator of output or input by itself.

### Dynamics

Leontief formulated a dynamic version of the model (Leontief et al., 1953, Ch. 3) by adding a term in the change in the output vector premultiplied by a matrix of capital coefficients. This led to two developments: theoretical work on the mathematical properties of the dynamic model; and empirical work on the construction of matrices of capital coefficients. Dynamic analysis in an input-output framework was set out in Dorfman, Samuelson, and Solow (1958), and various aspects arising in connection with the Cambridge Growth Project were discussed in Cambridge DAE (1962a) and in Stone and Brown (1962). The dynamic inverse appeared in Leontief (1970a).

## The Stability of Coefficients, Updating and Projection

Input-output coefficients change and the reasons for this and the problems to which it gives rise have been studied from the beginning. The matter is discussed in Leontief et al. (1953) and, among many other studies, I will mention Arrow and Hoffenberg (1959), Tilanus (1966), and Carter (1970). In Sevaldson (1963, 1970, 1976) we find a series of studies of the stability of coefficients based on Norwegian experience. In Barker (1975a,b) we find some results of updating and projecting intermediate demand based on British experience. In Stone (1978, sect. 3.f) there is a short account of various methods that have been used.

### Regional Studies

From an early stage input-output has been used as a method of regional analysis. As far as I know, the original model was proposed by Isard (1951). In it the United States was divided into three regions each with twenty industries. Since it was a fixed-coefficient model, an industry in any region which required a given input would obtain it in fixed proportions from each region in which it was made. A different kind of model, making use of the distinction between locally and nationally balanced commodities, was described in Leontief et al. (1953, Ch. 4). As far as nationally balanced commodities are concerned, this model shows each region's vectors of imports and exports but not the source of the imports or the destination of the exports. In Moses (1955) a model complementary to the preceding one is set out which provides for each commodity a trading pattern connecting the regions pairwise.

Shortly after these theoretical developments, local and regional input-output tables began to appear. Examples are: the work of Amsterdam MBS (1953–1954) for Amsterdam; Artle (1959) for Stockholm; Bauchet (1955) for the region of Lorraine; Chenery, Clark, and Cao Pinna (1953) for Italy; Derwa (1957, 1958) for the region of Liège; Japan KEF (1958) for the Kinki region and the rest of Japan; and Moore and Peterson (1955) for the State of Utah. This kind of work has continued and by now a great many regional tables have been constructed.

From a wider point of view, nations can be regarded as the regions of a larger area and their industrial structure can be compared in terms of standardised input-output tables. A method for doing this was proposed in Chenery and Watanabe (1958). The matter was discussed at our fifth conference in 1971; however, the voluminous material provided by Ambica Ghosh (ECAFE countries), Vera Nyitrai (COMECON countries), Vittorio Paretti (EEC countries), and Jiri Skolka and others (ECE countries) was not published in the proceedings. Some information for the ECAFE region is, however, contained in Ghosh (1974).

The standardisation of tables paves the way for the construction of supra-national tables. This was first done, I believe, for the OEEC countries in 1953 by Kirschen and associates (1959).

It would not be difficult to think of other topics that have engaged the attention of input-output analysts more or less regularly over the past thirty years, although I think that the seven topics I have mentioned cover the main themes that have been with us all along. Yet there are always new themes arising out of the development of the subject. First, there is the development of the input-output model regarded as a disaggregated model mainly concerned with product flows and the primary inputs needed to sustain them. Second, there is the extension of this model to handle pollution, whether or not it is treated, and to take account of such matters as the distribution of income and questions of money and finance which were not treated in the past on a disaggregated basis in general models of the economy. Third, there is the application of input-output accounting techniques to the organisation of demographic and social data and to the building of sociodemographic models. And, finally, there are the applications of computer programming to data processing and to model construction and solution. Let us now take a look at each of these.

## The Development of the Input-Output Model

In the early stages the input-output model consisted of a matrix of intermediate product flows bordered to the right by one or more vectors of final demand and below by one or more vectors of primary inputs and other production costs, such as provisions for depreciation and indirect taxes. Input-output coefficients were calculated, as a matter of simple arithmetic, by dividing the elements in the intermediate product vectors by the corresponding output total. Apart from a certain number of arithmetical and accounting identities, these coefficients constituted the relationships of the model. They could be arranged in a matrix, usually denoted by $A$, from which the Leontief inverse, or matrix multiplier, $(I - A)^{-1}$ could be calculated, although with a good deal more time and trouble than is involved nowadays. In the quantity version of the model this inverse transforms final demands into total outputs; in the price version, its transpose transforms primary input and similar costs per unit of output into total costs per unit of output, or prices. Many other interesting calculations can be made with this inverse; the main ones are illustrated in Stone (1961, Ch. XIV).

Forty years ago Tinbergen had started to build models of the economy. The first related to Holland and various versions of it can be found in Tinbergen (1935, 1937, 1959). Shortly afterwards came his well-known studies for the League of Nations in Tinbergen (1939a,b). These early models, and the many that followed them in the same spirit, covered many aspects of the economy but in terms of a comparatively small number of macrovariables. They were dynamic and intended for purposes of policy analysis and prediction. They were the forerunners of the macroeconomic short-term forecasting models of today.

By contrast, input-output models have developed into disaggregated medium-term planning models. This transition involves a number of steps of which the following are among the more important.

### Endogenising the Exogenous Variables

Instead of taking final demands as given, we could introduce explanations, at least of some of them, into the model. Consumers' expenditures are an obvious candidate for this treatment and the linear expenditure system provides a convenient formulation since, for a fixed set of prices, expenditure on each commodity is a linear function of total expenditure. My first paper on the subject (Stone, 1954) was presented to our second conference. We now know from Deaton (1974) that this simple system has a serious defect, which it shares with all systems based on additive preferences; namely, that it imposes severe restrictions on total expenditure elasticities in relation to own-price elasticities. Thus, we have the choice of putting up with the defects of the linear expenditure system because of its convenience, adopting other demand equations despite the fact that they may not fit neatly into a general model or searching for a revised formulation which will combine both advantages. The almost ideal demand system (AIDS) proposed in Deaton and Muellbauer (1977) is a move in this direction.

Corresponding efforts can be made with respect to the other components of final demand [see, for example, the paper on foreign trade in Barker (1972) which was presented at our fifth conference]. An exception, perhaps, is government current expenditure on goods and services which seems very difficult to model and can best be based on statements of the government's intentions or some form of trend projection.

### Interactions and Feedback

Once the components of final demand cease to be fixed exogenously they are free to vary in response to what is happening in the model: just as outputs depend on final demands, so final demands depend on incomes, which in turn depend largely on outputs but also on taxes and other transfers. As a consequence it is necessary to build consumption functions into the model which will close the output-income-expenditure loop.

### Generalising Input-Output Production Functions

I mentioned in a previous section (The Stability of Coefficients, Updating and Projection) the early recognition of the fact that input-output coefficients change and the efforts that were made to update and project them. The methods used represent one way of doing something that has to be done, but for the most part they are wholly empirical and do not throw much light on the causal factors at work; for instance, it seems fairly clear that input-output coefficients must be influenced by relative prices. This suggests that the simple relationships derivable from an input-output table should be generalised and an attempt should be made to test the generalisation.

At the same time there is another issue to be considered. In the initial model, primary inputs per unit of output are either exogenous or residual. In an attempt to explain them we might set up production functions connecting value added and primary inputs. But we should then have, for no very good reason, different functions for primary and intermediate inputs. There is much to be said, therefore, for starting not from value added but from gross output and relating this to all inputs. This was done in Wigley (1970) using a vintage model of production.

This approach was carried a good deal further by Peterson (1974) in a paper presented to our sixth conference. Following Shephard (1970) he switched from production functions to cost functions and applied the generalised Leontief function suggested by Diewert (1971), which under certain conditions reduces to the fixed-coefficient hypothesis of input-output analysis. For computational reasons Peterson began by aggregating inputs into five groups (new investment goods, fuels, materials, services, and labour), and then adopted a two-stage procedure. At the first stage, producers decide on their aggregate inputs of each of the five groups, depending on the expected level of output and a set of group-price indices; and at the second stage, they decide on their demands for individual inputs, taking account of their decision at the first stage and the prices of these inputs.

The method can be modified and the stages applied in the opposite order. For instance, in the paper presented to this conference by Peterson, the first stage was to estimate the input-mix of seven fuels into fifteen industry groups and two final expenditure groups (public administration and households) on the basis of British data for the years 1955–1975. The results show the importance of technical change (or changes in tastes) and price substitution in determining this mix and the unreality of assuming that the fuels are demanded in fixed proportions. The second stage was to estimate for each of the using groups a demand equation for energy (i.e., the aggregation of the individual fuels); these equations fit well and show that in many cases relative prices, temperature, and residual time-trends, as well as output levels, are important. By combining the two stages the input of each fuel into each using group can be estimated.

### Factor Prices and Commodity Prices

The input-output price model enables commodity prices to be calculated as an accumulation of input costs per unit of output. In the simplest case, certain prices, say import prices, are assumed to rise, and the repercussions on domestic commodity prices are worked out on the assumption that domestic factor prices do not rise. Clearly this is not very likely to happen and the opposite extreme assumption (namely, that domestic factor prices rise so as to maintain the purchasing power of domestic factors) can be worked out by treating the household sector as endogenous. When this is done, domestic factor prices will rise in line with the rise in import prices and the international transmission of inflation will be immediate. But this, too, is unlikely to happen and the question is to find out by how much factor prices are likely to be affected in different parts of the economy.

This is a difficult question and the model builder may wish to keep open the option of treating factor prices as exogenous. At the same time there is an immense literature on price and wage equations which should eventually provide a basis for completing this aspect of the model.

### Dynamics

The early disaggregated planning models concentrated on a future target year without going into the question of the path from the present to that year. These models may have contained time lags in a number of their equations, and so dynamic elements were present; however, these were not sufficient to calculate a path except, perhaps, under rather restrictive assumptions about the development of the economy and the normality of the target year. For instance, in a static model it seems acceptable to determine stockbuilding by means of fixed coefficients, whereas in fact stockbuilding is likely to prove a volatile element exerting a dynamic influence, if only a transient one, on the economy.

With the introduction of dynamics, disaggregated models come to share an important feature with macroeconomic models of the Tinbergen type, which have been dynamic all along though their time horizon is usually not much more than eighteen months to two years.

### Testing the Model as a Whole

If one tries to explain an econometric model to natural scientists their first reaction is usually to enquire what has been done to test the model's performance. It can usually be said that bits of the model have been tested: that the consumption, production, or foreign trade functions give a good fit over the period of observation and even over later years if they are supplied with the subsequently observed values of the relevant determining variables. But if the model is limited so that, implicitly or explicitly, it always refers to normal years, or if important variations are fixed by assumption, it is difficult to carry out and interpret tests of the model as a whole. A complete dynamic model allows us to get over this difficulty since it enables us to make predictions, always given the expected values of the exogenous variables, rather than highly qualified projections. The model can be started off at the outset of the observation period and run through it into the future. We shall then be able to see, for example, how the consumption functions work when they are presented with disposable income and other endogenous determining variables as worked out by the model rather than with the observed values of these variables, as in the partial tests.

### Policy Simulation, Control, and Optimisation

One of the purposes of building econometric models is to contribute to policy making. This can be attempted in various ways.

The first is policy simulation, an essentially exploratory technique involving neither control nor optimisation, which consists of running the model on different assumptions and, in particular, assumptions relating to instruments considered relevant to the policy issues under discussion. The method is useful but has the disadvantage that, unless the issues and the acceptable means of tackling them are narrowly defined, the number of simulations to be performed may be very large.

The second method, which provides the model builder with a clearer picture of what he is supposed to aim at, is the targets and instruments approach introduced by Tinbergen (1952, 1956). This consists of setting up a number of targets (such as the level of the balance of payments, the number of unemployed, and so on), and then working out how to set policy instruments so as to hit these targets. An account of how this can be done and a number of illustrations of doing it drawn from the Cambridge Growth Project's static model of the British economy can be found in Livesey (1976). The application of the method to a dynamic model is a larger problem because it is necessary to think in terms of the time paths of targets and instruments. We have not yet attempted such an application but it does not seem to involve difficulties of principle.

The main problem with this method is that while it is possible to work out trade-offs between instruments, it does not provide a basis for working out trade-offs between targets.

The third method, which does allow these calculations to be made, is programming. However, this may be thought to require too much information since we must be able to set up a utility function to guide our choice of the degree to which different targets should be approximated and, furthermore, it must be possible to give the problem a particular form if it is to be solved by known computational procedures. Thus, the standard quadratic programming problem requires the variables to be nonnegative, the constraints to be linear, and the utility function to be the sum of a linear and a quadratic form which must be negative definite if the global optimum is to be unique. In the present context an application of this method would seem both difficult and uncertain: difficult because it is hard to see how the utility function could be constructed; and uncertain because of the restrictive conditions required to reach a solution.

## Extensions of the Input-Output Model

In the preceding section my intention was to trace the changes that have been taking place in the original input-output model. In this section I shall deal with what can fairly be called extensions—in the sense that they were not regarded as part of input-output analysis in the early days.

### *Accounting for Pollution*

A method of introducing pollutants and the corresponding treatment services into an input-output table was proposed by Leontief (1970b). It consists, essentially, of setting up a number of additional columns containing the cost structures of the treatment services and an equal number of rows containing the emissions of pollutants which could be handled by the corresponding service. Emissions appear among the cost elements of the various industries because they provide a measure of the output of a treatment service needed to remove them. This arrangement can be applied to final users as well as to productive activities.

This neat solution leaves for further discussion the question of how much of each pollutant should be treated. If total emissions appear in the cost structures, it is implied that they should be treated in full—but the community might prefer only to treat them partially and as a consequence have more resources to devote to the production of "regular" goods and services. Furthermore, the community is likely to be more interested in the state of the air, land, or water after treatment than in the amount of treatment carried out. These issues were discussed in Stone (1972) and Meade (1972).

From a practical point of view there is the additional point that a certain amount of pollution has always been treated either by the polluter himself or by some public service. Interesting accounts of the conceptual and statistical issues that arise in environmental measurement and of the amount of national expenditure on pollution abatement and control, which have been prepared by the U.S. Department of Commerce since 1972, are given in Cremeans (1977). Clearly, the full implementation of Leontief's original proposal would be an extremely difficult undertaking.

Papers on pollution have featured the following in our last two conferences: an application of Leontief's scheme to the problem of air pollution in Leontief and Ford (1972) at the fifth conference; and contributions by Cumberland and Stram (1976), Hartog and Houweling (1976), and Thoss (1976) at the sixth conference. Further results from the last writer's model are given by Thoss and Wiik (1978).

### The Distribution of Income

Input-output started off as a means of analysing the productive system and disaggregation was confined to branches of production and their products. For this purpose there is no need to disaggregate the income and outlay accounts of sectors; and this was not done in standard systems of national accounts either. Even the revised SNA, UNSO (1968), contains only a small number of sectors and no division of the household sector. At the time, the UN Statistical Office was working on a complementary system of distribution statistics for households, and provisional guidelines have been published in UNSO (1977).

Given the necessary information, it is possible to set up an input-output system for the income and outlay accounts for sectors which receive income, make transfers among themselves, spend for consumption, and save. This is demonstrated in Pyatt et al. (1977) and is further illustrated in the papers presented at the conference on social accounting methods in development planning held in Cambridge in the spring of 1978 under the sponsorship of the World Bank Research Program. The proceedings of this conference will eventually appear in a volume edited by Pyatt.

Another development in this field (yet one which is more akin to the sociodemographic models of the section below entitled, "New Worlds to Conquer") is to treat the generation of income as a Markov process. This was proposed by Champernowne in the 1930s in a thesis which was not published at the time but eventually appeared in revised form in Champernowne (1973) [for other publications in this field by the same author see Champernowne (1953; and 1969, Vol. 2, Ch. 18)]. Many applications of this method have been made, for instance in Vandome (1958), Esberger and Malmquist (1972), and Shorrocks (1975).

### Wealth and the Flow of Funds

As was pointed out by Stone (1966a), capital account and balance sheet data can be set up in an input-output format and used as a basis for simple linear models. This type of question was discussed at our fifth conference by Isayev (1972) and Roe (1972). The second of these authors drew some conclusions from an application of the simple model which he had worked out on the basis of British data in Cambridge DAE (1971). As was to be expected, this model provides some insight into structural relationships but is unsatisfactory as a forecasting device because of changing coefficients. No doubt something could be done, as it is in applications to the productive system, but it is doubtful if in this instance it would work very well since financial coefficients are likely to be extremely volatile, so that a more sophisticated model is required.

### International Trade and World Models

Although this subject has perhaps always been recognised as a part of input-output analysis, until about ten years ago world trade models were all quite small. Taplin (1967) recommended that a multiregional, multicommodity model be built in terms of ten or twelve regions and six categories of goods. A step in this direction is described by Thorbecke and Field (1974). Their model is a short-term, demand-oriented model based on annual data over the period 1953–1967 and appears to perform fairly well over the observation period. It contains no commodity division but analyses trade flows among ten endogenous regions, plus the Sino-Soviet bloc, which is treated as exogenous.

In the meantime a world income and trade model, MEGISTOS, was constructed, which is described in detail in Duprez and Kirschen (1970). Its purpose was to provide a coherent system of projections for the main economic aggregates at the world level in 1975. The countries of the world were grouped into twelve regional zones and linked together in terms of trade, aid, and borrowing. Another type of model, Project LINK, is discussed in Ball (1973) and Waelbroeck (1976). The first of these volumes describes the conceptual framework within which the integration of individual country models is taking place and presents some of the underlying empirical work. The second sets out the equations used in the models of the thirteen collaborating countries, as they were in late 1973, together with those of the regional models for developing areas prepared by UNCTAD.

The subject was represented at our sixth conference in papers by Tokoyama et al. (1976), Petri (1976), and Panchamukhi (1976). Another recent study is given in Nyhus (1978).

The latest contribution in this field, described in Leontief et al. (1977), is the United Nations study, *The Future of the World Economy*. This is a highly disaggregated model with fifteen regions and forty-five branches of economic activity. It is static and does not contain explicit feedback mechanisms or optimisation. It embodies exogenous population estimates and handles international trade in terms of a single international trading pool to which all exports are delivered and from which all imports are drawn. It was constructed in the first instance for the study of development in relation to environmental questions, and provision is made for calculating the cost of pollution abatement. It is, however, a general-purpose economic model which can be used to analyse the evolution of the world economy from many points of view.

**New Worlds to Conquer**

About fifteen years ago I began to work on the flow of pupils and students through the British educational system and showed in Stone (1966b) that an input-output scheme could be very useful in organising information of this kind. Later I contributed a paper, Stone (1970), on what I called demographic input-output to our fourth conference. Shortly afterwards, at the suggestion of my friend Philip Redfern, who had treated similar problems in Redfern (1967), I altered slightly the format I had been using, but the scheme remained essentially dynamic. In the later version the numbers in the opening and closing stock-vectors of the population are connected by flows from the beginning of the year to the end of it. Just as in the economic quantity model the matrix multiplier transforms final demands into total outputs, so in the demographic model it transforms the new entrants (i.e., the year's births and immigrations) into the closing population stock-vector. As in the economic case, there is also a price model which combines with the quantity model to generate the equivalent of the identity national income equals national expenditure.

As can be seen from Stone (1973) and UNSO (1975) the possibilities of applying this model in social demography are almost unlimited. In such applications the place of input-output coefficients is taken by transition proportions; and if these can be interpreted as probabilities, then the process studied is an example of an absorbing Markov chain. An analysis of the Norwegian educational system from this point of view is given in Thonstad (1969).

Among the many writers who have contributed to this kind of analysis I shall mention Andorka and Illés (1972), Coleman (1972), whose paper was presented at our fifth conference, Cooper and Schinnar (1973), Fox (1974), Rees and Wilson (1977), Schinnar (1976, 1977), and Shishido et al. (1978).

**Automating Model Building and Analysis**

Over the last generation the amount of data available for economic and social model building has greatly increased and the models have tended to become larger and larger. As can be seen from the first paper given at this conference by Barker, Peterson, and Winters, the present version of the Cambridge multisectoral dynamic model, MDM3, contains 2759 equations, of which 759 are stochastic, 7484 variables, and 12,884 coefficients. It is solved year by year by the Gauss-Seidel iterative method, each year requiring 10 to 25 iterations to obtain adequate convergence and taking 4 to 6 seconds of computing time.

The increase in the size of models has had two consequences. First, the amount of data processing, estimation, solving, and testing has grown enormously; and, second, large models are inconvenient for many experimental purposes in which the amount of detail they contain is irrelevant. Fortunately, at the same time, the speed and power of computers has also been growing and new algorithms and programming methods have been devised. These developments have made a number of procedures possible which greatly ease the task of the model builder.

## The Adjustment of Social Accounting Matrices

Social accounting matrices are compiled from a variety of sources, and this is also true of parts of them such as input-output tables. Inevitably, the data used are in some degree incomplete, inconsistent, and of varying reliability. When the pieces are assembled a number of discrepancies are revealed and there are also some gaps: not so very long ago a number of countries derived their estimates of consumers' expenditure as a residual. The discrepancies could be removed by applying the classic technique of the adjustment of conditioned observations by the method of least squares, provided we were willing to make a subjective estimate of the variance matrix of the errors. This application was suggested in Stone, Champernowne, and Meade (1942) and on a number of subsequent occasions I worked out small examples to illustrate the method. For various reasons it did not catch on— the most important, I imagine, was the amount of computing involved. This difficulty is overcome in Byron (1978) in which the problem is reformulated in terms of a quadratic loss function, and use is made of the conjugate gradient algorithm. The method is applied to a matrix of order 46 and the author remarks that it is feasible to adjust a monster matrix (e.g., order 1000).

## A Generator for Input-Output Tables and Models

Since 1962, research has proceeded at the IFO Institute in Munich on the possibilities of automating the construction of tables and the estimation of the parameters of the corresponding models; and in 1975 these ideas began to be applied to the construction of interregional tables, as reported in the paper by Gehrig et al. of the University of Frankfurt to this conference. It is recognised that tables are based on statistics taken from sources of varying reliability and on estimates and guesses made with varying degrees of expert knowledge. Detailed instructions, therefore, must be given to the computer not only on the definitions and classifications to be used and the type of tables to be constructed, but also on the means of combining the basic data. This brings us back to the above-mentioned problem concerning the consistency and reliability of data.

These developments seem likely to prove of great value because most of the large models that are built these days are intended to continue; indeed, it is only if they do continue that they, and econometric model building in general, are likely to improve. In maintaining such models there is the continuing need to introduce data for the latest year, to revise data from the past, to rebase index-numbers, and so on. Even with good worksheets and data banks this is a laborious process and one which cannot easily be undertaken at any moment as a new batch of data becomes available; and the same is true of reestimating parameters. By encouraging model builders to reduce everything to rules, the problems created by inevitable changes in staff would be diminished.

## A Programme for Model Solution

In the second paper presented to this conference by Barker, Peterson, and Winters, the experience gained in solving the Cambridge Growth Project's MDM series of models has been used to write a general programme which can be used to solve any model of the general type of MDM. This has been useful in forcing us to think through every aspect of the solution of large econometric models of the economy; and we hope it will be useful to others who would prefer to devote their energies to the basic statistics and underlying economics of their model, rather than to devising an efficient method of solving it.

### The Condensed Form

As econometric models grow in size, it becomes more and more difficult to see what is going on in them and they become less and less convenient for experimental purposes. It would be interesting to know, therefore, if the model could be boiled down to a small number of equations connecting its macrovariables without destroying its essential features and losing much more than detail.

This model of the model was termed "the condensed form" in Barker (1976, pp. 35–50) and was applied to our static model. In the first paper presented to this conference by Barker, Peterson, and Winters it is applied to MDM3, the latest version of our dynamic model. As can be seen, the results are satisfactory, doubtless because MDM3 has a comparatively simple structure although it is very large. We think this development will be useful for pedagogic and experimental purposes.

### Conclusions

I hope I have managed to convey in this short survey the immense vitality of input-output analysis: its ability to transform itself from simple beginnings into more and more complicated forms; its capacity to throw light on branches of the social sciences other than the one in which it originated; its contribution to new ways of using computers which its very success has stimulated.

It looks to me as if there is still a good deal of life in the traditional topics, although I hope that before too long we shall have settled questions of how to construct input-output tables and of input-output taxonomy. The development of the input-output model seems to be leading us in directions in which its input-output core is becoming less and less discernible. This is as it should be, because it shows the possibility of improving the very simple relationships which were used initially. However, it must be remembered that it is precisely simple relationships like these which got disaggregated econometric model-building off the ground, and that without them the first step forward might have been long delayed. The extension of input-output ideas to other aspects of economics shows the power of those ideas, but probably they too will eventually give way to more sophisticated formulations. Applications to social demography may be considered more directly connected with Markov chains than with input-output; however, there have been sessions on these applications at our last three conferences and the subject is represented at this one. Moreover, the recognition that for each quantity equation there is a corresponding cost equation seems a distinctive contribution from input-output. As regards developments involving computing, these are only to be expected in subjects which involve data processing and parameter estimation on a large scale and the solution of large systems of nonlinear equations.

It seems clear enough that input-output has developed enormously in its forty-odd years of life. In making my survey I regret that I have given a most unbalanced account of the literature in terms of sources and languages. I have tended to cite the contributions with which I am most familiar. I hope that others will correct this onesidedness.

I conclude that input-output is in a flourishing and exciting condition at the present time. With that encouraging thought I will give the floor to the next speaker.

**References**

Amsterdam Municipal Bureau of Statistics. 1953–1954. Stedelijke jaarekeningen van Amsterdam (Regional accounts for Amsterdam). *Kwartaalbericht van het Bureau van Statistick der Gemeente Amsterdam* supplement, no. 4.

Andorka, Rudolf, and János Illés. 1972. Attempts at systematization in social statistics. In *Input-Output Techniques*, Proceedings of Second Hungarian Conference on Input-Output Techniques. Budapest: Publishing House of the Hungarian Academy of Sciences.

Arrow, Kenneth J., and Marvin Hoffenberg. 1959. *A Time Series Analysis of Interindustry Demands.* Amsterdam: North-Holland.

Artle, Roland. 1959. *Studies in the Structure of the Stockholm Economy.* Stockholm: Business Research Institute, Stockholm School of Economics.

Ball, R. J., ed. 1973. *The International Linkage of National Economic Models.* Amsterdam: North-Holland.

Barker, Terence S. 1972. Foreign trade in multisectoral models. In *Input-Output Techniques*, ed. A. Bródy and A. P. Carter. Amsterdam: North-Holland.

Barker, Terence S. 1975a. Some experiments in projecting intermediate demand. In *Estimating and Projecting Input-Output Coefficients*, ed. R. I. G. Allen and W. F. Gossling. London: Input-Output Publishing Co.

Barker, Terence S. 1975b. An analysis of the updated 1963 input-output transactions table. In *Estimating and Projecting Input-Output Coefficients*, ed. R. I. G. Allen and W. F. Gossling. London: Input-Output Publishing Co.

Barker, Terence S., ed. 1976. Economic Structure and Policy. No. 2 in *Cambridge Studies in Applied Econometrics.* London: Chapman and Hall.

Barna, Tibor. 1952. The interdependence of the British economy. *Journal of the Royal Statistical Society, Series A* 115(1): 29–81.

Barna, Tibor. 1953. Experience with input-output analysis in the United Kingdom. In *Input-Output Relations*, ed. Netherlands Economic Institute. Leiden: Stenfert Kroese.

Barna, Tibor. 1975. Quesnay's *Tableau* in modern guise. *The Economic Journal* 85(339): 485–496.

Bauchet, Pierre. 1955. *Les tableaux économiques: Analyse de la region Lorraine.* Paris: Genin.

Byron, Ray P. 1978. The estimation of large social account matrices. *Journal of the Royal Statistical Society Series A*, 141(3): 359–367.

Cambridge, Department of Applied Economics. 1962a. *A Computable Model of Economic Growth.* No. 1 in *A Programme for Growth.* London: Chapman and Hall.

Cambridge, Department of Applied Economics. 1962b. *A Social Accounting Matrix for 1960.* No. 2 in *A Programme for Growth.* London: Chapman and Hall.

Cambridge, Department of Applied Economics. 1971. *The Financial Interdependence of the Economy, 1957–1966.* No. 11 in *A Programme for Growth.* London: Chapman and Hall.

Carter, Anne P. 1970. *Structural Change in the American Economy.* Boston: Harvard University Press.

Champernowne, D. G. 1953. A model of income distribution. *The Economic Journal* 63: 318–351.

Champernowne, D. G. 1969. *Estimation and Uncertainty in Economics.* 3 vols. Edinburgh and London: Oliver and Boyd.

Champernowne, D. G. 1973. *The Distribution of Income between Persons.* Cambridge: Cambridge University Press.

Chenery, Hollis B., Paul G. Clark, and Vera Cao Pinna. 1953. *The Structure and Growth of the Italian Economy.* Rome: U.S. Mutual Security Agency.

Chenery, Hollis B., and Tsunehiko Watanabe. 1958. International comparisons of the structure of production. *Econometrica* 26(4): 487–521.

Coleman, James S. 1972. Flow models for occupational structure. In *Input-Output Techniques*, ed. A. Bródy and A. P. Carter. Amsterdam: North-Holland.

Cooper, W. W., and A. P. Schinnar. 1973. A model for demographic mobility analysis under patterns of efficient employment. *Economics of Planning* 13(3): 139–173.

Cremeans, John E. 1977. Conceptual and statistical issues in developing environmental measures— Recent U.S. experience. *The Review of Income and Wealth* series 23, no. 2: 97–115.

Cumberland, John H., and Bruce N. Stram. 1976. Empirical application of input-output models to environmental problems. In *Advances in Input-Output Analysis*, ed. K. R. Polenske and J. V. Skolka. Cambridge: Ballinger.

Deaton, Angus. 1974. A reconsideration of the empirical implications of additive preferences. *The Economic Journal* 84(334): 338–348.

Deaton, Angus, and John Muellbauer. 1977. An Almost Ideal Demand System. Discussion paper no. 62/77. University of Bristol, Department of Economics.

Denmark, Department of Statistics. 1948. *Nationalproduktet og Nationalindkomsten 1930–1946. Statistiske Meddelelser*, series 4, Vol. 129, pt. 5.

Denmark, Department of Statistics. 1951. *Nationalproduktet og Nationalindkomsten 1946–1949. Statistiske Meddelelser*, series 4, Vol. 140, pt. 2.

Derwa, Léon. 1957. Une nouvelle méthode d'analyse de la structure économique. *Revue de Conseil Économique Wallon* no. 28: 16–42.

Derwa, Léon. 1958. Technique d'input-output et programmation linéaire. *Revue du Conseil Économique Wallon* no. 34: 35–58.

Diewert, W. E. 1971. An application of the Shephard duality theorem: A generalised Leontief production function. *Journal of Political Economy* 79(3): 481–507.

Dmitrief, V. K. 1974. *Economic Essays on Value, Competition and Utility.* English transl. by D. Fry. New York: Cambridge University Press.

Dorfman, Robert, Paul A. Samuelson, and Robert M. Solow. 1958. *Linear Programming and Economic Analysis.* New York: McGraw-Hill.

Duprez, C., and E. S. Kirschen, eds. 1970. *Megistos: A World Income and Trade Model for 1975.* Amsterdam: North-Holland.

Esberger, Sven Erik, and Sten Malmquist. 1972. *A Statistical Study of the Development of Incomes* (in Swedish). Lund: Berlingska Boktryckeriet.

Fox, Karl A. 1974. *Social Indicators and Social Theory.* New York: Wiley.

Ghosh, A. 1974. *Development Planning in South-East Asia.* Rotterdam: Rotterdam University Press.

Goodwin, R. M. 1952. A note on the theory of the inflationary process. *Economia Internazionale* 5(1): 3–21.

Goodwin, R. M. 1953. Static and dynamic linear general equilibrium models. In *Input-Output Relations*, ed. Netherlands Economic Institute. Leiden: Stenfert Kroese.

Hartog, H. den, and A. Houweling. 1976. Pollution, pollution abatement, and the economic structure of the Netherlands. In *Advances in Input-Output Analysis*, ed. K. R. Polenske and J. V. Skolka. Cambridge: Ballinger.

Hill, T. P. 1971. *The Measurement of Real Product.* Paris: O.E.C.D.

Isard, Walter. 1951. Interregional and regional input-output analysis: A model of a space-economy. *The Review of Economics and Statistics* 33(4): 318–328.

Isayev, B. L. 1972. Material-financial balance of a union republic. In *Input-Output Techniques*, ed. A. Bródy and A. P. Carter. Amsterdam: North-Holland.

Japan, Kansai Economic Federation. 1958. *Interregional Input-Output Table for the Kinki Area and the Rest of Japan* (in Japanese). Osaka: Kansai Economic Federation.

Kirschen, E. S. et al. 1959. *The Structure of European Economy in 1953*. Paris: O.E.E.C.

Leontief, Wassily W. 1936. Quantitative input and output relations in the economic system of the United States. *The Review of Economic Statistics* 18(3): 105–125.

Leontief, Wassily W. 1941. *The Structure of American Economy*. 1st ed. (*1919–1929*), Cambridge: Harvard University Press, 1941; 2nd ed. (*1919–1939*), New York: Oxford University Press, 1951.

Leontief, Wassily W. 1970a. The dynamic inverse. In *Contributions to Input-Output Analysis*, ed. A. P. Carter and A. Bródy. Amsterdam: North-Holland.

Leontief, Wassily W. 1970b. Environmental repercussions and the economic structure: An input-output approach. In *A Challenge to Social Scientists*, ed. Shigeto Tsuru, Asahi, Tokyo. Reprinted in *The Review of Economics and Statistics* 52: 262–271.

Leontief, Wassily W., and Daniel Ford. 1972. Air pollution and the economic structure: empirical results of input-output computations. In *Input-Output Techniques*, ed. A. Bródy and A. P. Carter, Amsterdam: North-Holland.

Leontief, Wassily W., et al. 1953. *Studies in the Structure of the American Economy*. New York: Oxford University Press.

Leontief, Wassily W., and others. 1977. *The Future of the World Economy*. New York: Oxford University Press.

Livesey, D. A. 1976. Solving the model. In *Economic Structure and Policy*, ed. T. S. Barker, no. 2 in *Cambridge Studies in Applied Econometrics*. London: Chapman and Hall.

Meade, J. E. 1972. Citizens' demands for a clean environment. *L'industria* no. 3/4: 145–152.

Moore, Frederick T., and James W. Petersen. 1955. Regional analysis: An Interindustry model of Utah. *The Review of Economics and Statistics* 37(4): 368–383.

Moses, Leon N. 1955. The stability of interregional trading patterns and input-output analysis. *The American Economic Review* 45(5): 803–832.

Nemchinov, V. S. 1959. The use of mathematical methods in economics. In *The Use of Mathematics in Economics*, ed. V. S. Nemchinov. Moscow: Publishing House of Socio-Economic Literature; English trans., ed. A. Nove, Edinburgh and London: Oliver and Boyd. 1964.

Netherlands, Central Bureau of Statistics (1946). *Statistische en econometrische onderzoekingen* (new series). Netherlands C.B.S., quarterly since 1946. See in particular: 1950, nos. 1 and 2; 1951, all numbers; 1952, no. 1.

Netherlands, Central Bureau of Statistics. 1952. *National Accounts of the Netherlands 1948–1949*. Netherlands: C.B.S., The Hague.

Nyhus, Douglas. 1978. A detailed model of bilateral commodity trade and the effects of exchange rate changes. In *Econometric Contributions to Public Policy*, ed. R. Stone and A. W. A. Peterson. London: Macmillan.

O.E.E.C. 1952. *A Standardised System of National Accounts*. Paris: O.E.E.C. *1958 Edition*, 1959.

Panchamukhi, V. R. 1976. A multisectoral and multicountry model for planning ECAFE production and trade. In *Advances in Input-Output Analysis*, ed. K. R. Polenske and J. V. Skolka. Cambridge: Ballinger.

Peterson, A. W. A. 1974. Factor demand equations and input-output analysis. Paper presented at the Sixth International Conference on Input-Output Techniques, Vienna.

Petri, Peter A. 1976. A multilateral model of Japanese-American trade. In *Advances in Input-Output Analysis*, ed. K. R. Polenske and J. V. Skolka. Cambridge: Ballinger.

Phillips, Almarin. 1955. The *Tableau Économique* as a simple Leontief model. *The Quarterly Journal of Economics* 69(1): 137–144.

Pyatt, Graham, Alan R. Roe, and associates. 1977. *Social Accounting for Development Planning with special reference to Sri Lanka*. Cambridge: Cambridge University Press.

Rasmussen, P. Nørregaard. 1956. *Studies in Inter-sectoral Relations*. Amsterdam: North-Holland.

Redfern, Philip. 1967. *Input-Output Analysis and Its Application to Education and Manpower*. CAS Occasional Paper no. 5. London: H.M.S.O.; revised ed., CSC Occasional Paper no. 5. London: H.M.S.O., 1976.

Rees, P. H., and A. G. Wilson. 1977. *Spatial Population Analysis*. London: Arnold.

Roe, Alan R. 1972. Enforcement of the balance-sheet identity in financial analysis. In *Input-Output Techniques*, ed. A. Bródy and A. P. Carter. Amsterdam: North-Holland.

Schinnar, A. P. 1976. A multidimensional accounting model for demographic and economic planning interactions. *Environment and Planning A* 8(4): 455–475.

Schinnar, A. P. 1977. An eco-demographic accounting-type multiplier analysis of Hungary. *Environment and Planning A* 9(4): 373–384.

Sevaldson, Per. 1963. Changes in input-output coefficients. In *Structural Interdependence and Economic Development*, ed. Tibor Barna. London: Macmillan.

Sevaldson, Per. 1970. The stability of input-output coefficients. In *Applications of Input-Output Analysis*, ed. A. P. Carter and A. Bródy. Amsterdam: North-Holland.

Sevaldson, Per. 1976. Price changes as causes of variations in input-output coefficients. In *Advances in Input-Output Analysis*, ed. K. R. Polenske and J. V. Skolka. Cambridge: Ballinger.

Shephard, Ronald W. 1970. *Theory of Cost and Production Functions*. Princeton: Princeton University Press.

Shishido, Shuntaro, et al. 1978. Changes in the regional distribution of population in Japan and their implications for social policy. In *Econometric Contributions to Public Policy*, ed. R. Stone and A. W. A. Peterson. London: Macmillan.

Shorrocks, A. F. 1976. Income mobility and the Markov assumption. *The Economic Journal* 86(343): 566–578.

Stewart, I. G. 1958. Input-output table for the United Kingdom, 1948. *The Times Review of Industry*, London and Cambridge Economic Bulletin, new series, no. 28, pp. vii–ix.

Stone, Richard. 1954. Linear expenditure systems and demand analysis: An application to the pattern of British demand. *The Economic Journal* 64(255): 511–527.

Stone, Richard. 1956. *Quantity and Price Indexes in National Accounts*. Paris: O.E.E.C.

Stone, Richard. 1961. *Input-Output and National Accounts*. Paris: O.E.E.C.

Stone, Richard. 1966a. The social accounts from a consumer's point of view. *The Review of Income and Wealth*, ser. 12, no. 1, pp. 1–33. Partially reprinted as Ch. XV in *Mathematical Models of the Economy and Other Essays*. London: Chapman and Hall.

Stone, Richard. 1966b. Input-output and demographic accounting: A tool for educational planning. *Minerva* 4(3): 365–380. Reprinted as Ch. XVIII in *Mathematical Models of the Economy and Other Essays*. London: Chapman and Hall.

Stone, Richard. 1970. Demographic input-output: An extension of social accounting. In *Contributions to Input-Output Analysis*, ed. A. P. Carter and A. Bródy. Vol. 1. Amsterdam: North-Holland.

Stone, Richard. 1972. The evaluation of pollution: Balancing gains and losses. *Minerva* 10(3): 412–425.

Stone, Richard. 1973. Transition and admission models in social demography. *Social Science Research* 2(2): 185–230; also in *Social Indicator Models*, ed. K. C. Land and S. Spilerman. New York: Russell Sage Foundation.

Stone, Richard. 1978. *Input-Output Analysis and Economic Planning: A Survey*. Paper presented at the International Symposium on Mathematical Programming and Its Applications in Economics (Venice, June 1978).

Stone, Richard, and J. A. C. Brown. 1962. Output and investment for exponential growth in consumption. *The Review of Economic Studies* 29(80): 241–245.

Stone, Richard, D. G. Champernowne, and J. E. Meade. 1942. The precision of national income estimates. *The Review of Economic Studies* 9(2): 111–125.

Stone, Richard, and J. E. G. Utting. 1953. The relation between input-output analysis and national accounting. In *Input-Output Relations*. Leiden: Stenfert Kroese.

Taplin, G. B. 1967. Models of world trade. *International Monetary Fund Staff Papers* 14: 433–455.

Thonstad, Tore. 1969. *Education and Manpower*. Edinburgh and London: Oliver and Boyd.

Thorbecke, Erik, and Alfred J. Field, Jr. 1974. A ten-region model of world trade. In *International Trade and Finance*, ed. W. Sellekaerts. London: Macmillan.

Thoss, Rainer. 1976. A generalized input-output model for residuals management. In *Advances in Input-Output Analysis*, ed. K. R. Polenske and J. V. Skolka. Cambridge: Ballinger.

Thoss, Rainer, and Kjell Wiik. 1978. Optimal allocation of economic activities under environmental constraints in the Frankfurt metropolitan area. In *Econometric Contributions to Public Policy*, ed. R. Stone and A. W. A. Peterson. London: Macmillan.

Tilanus, C. B. 1966. *Input-Output Experiments: The Netherlands 1948–1961*. Rotterdam: Rotterdam University Press.

Tinbergen, J. 1935. Quantitative Fragen der Konjunkturpolitik. *Weltwirtschaftliches Archiv* 42(1): 316–399.

Tinbergen, J. 1937. *An Econometric Approach to Business Cycle Problems.* Paris: Hermann et cie.

Tinbergen, J. 1939a. *A Method and its Application to Investment Activity.* Geneva: League of Nations.

Tinbergen, J. 1939b. *Business Cycles in the United States of America, 1919–1932.* Geneva: League of Nations.

Tinbergen, J. 1952. *On the Theory of Economic Policy.* Amsterdam: North-Holland.

Tinbergen, J. 1956. *Economic Policy: Principles and Design.* Amsterdam: North-Holland.

Tinbergen, J. 1959. *Selected Papers*, ed. L. H. Klaassen, L. M. Koyck, and H. J. Witteveen. Amsterdam: North-Holland.

Tokoyama, K. et al. 1976. Structures of trade, production and development. In *Advances in Input-Output Analysis*, ed. K. R. Polenske and J. V. Skolka. Cambridge: Ballinger.

U.K., Central Statistical Office. 1952. *National Income and Expenditure.* London: H.M.S.O.

U.N., Statistical Office. 1966. *Problems of Input-Output Tables and Analysis.* Studies in Methods, series F, no. 14. New York: U.N.

U.N., Statistical Office. 1968. *A System of National Accounts.* Studies in Methods, series F, no. 2, rev. 3. New York: U.N.

U.N., Statistical Office. 1973. *Input-Output Tables and Analysis.* Studies in Methods, series F, no. 14, rev. 1. New York: U.N.

U.N., Statistical Office. 1975. *Towards a System of Social and Demographic Statistics.* Studies in Methods, series F, no. 18. New York: U.N.

U.N., Statistical Office. 1977. *Provisional Guidelines on Statistics of the Distribution of Income, Consumption and Accumulation of Households.* Studies in Methods, series M, no. 61. New York: U.N.

U.S., Bureau of Labor Statistics. 1951. *The 1947 Interindustry Relations Study.* A: Summary tables. B: Sector reports. U.S. Department of Labor, Bureau of Labor Statistics; A: 50-sector tables (3), Dec. 1951; 200-sector tables (3), Oct. 1952; general explanation of the 200-sector table, B.L.S. report no. 33, June 1953. B: B.L.S. reports issued from Feb. 1953 onwards.

U.S.S.R., Central Statistics Board. 1926. Balance sheet of the national economy of the U.S.S.R. for 1923–1924. *Transactions of the U.S.S.R. Central Statistics Board*, Vol. 29, Moscow.

Vandome, Peter. 1958. Aspects of the dynamics of consumer behaviour. *Bulletin of the Oxford University Institute of Statistics* 20(1): 65–105.

Waelbroeck, Jean L., ed. 1976. *The Models of Project LINK.* Amsterdam: North-Holland.

Walras, Léon. 1874. *Éléments d'Économie Politique Pure.* 1st ed., Lausanne: Corbaz, 1874; definitive ed., Paris: Pichon et Durand-Auzias, 1926; English transl. by William Jaffé. London: Allen and Unwin, 1954.

Wigley, K. J. 1970. Production models and time trends of input-output coefficients. In *Input-Output in the United Kingdom*, ed. W. F. Gossling. London: Cass.

# II

# National and
# Regional Studies

# 2

# Experiences in Using Input-Output Techniques for Price Calculations

OLAV BJERKHOLT

The use of macroeconomic models in Norway has some distinctive features. One is the use of a detailed input-output model, called MODIS, for short-term macroeconomic analysis. In other countries input-output models are typically used for structural and long run studies while the short-term is covered by aggregate econometric models. Another feature is the role played by the model within the institutional environment. Economic policy is formulated, analyzed, and implemented within the framework of the model. The model has itself become part of the institutional environment and serves as a system for gathering, evaluating, and presenting information as well as a representation of the functioning of the economic system. I will speak here first about the origin and background of input-output models in Norwegian planning, then outline the development of Norwegian input-output models and give some details of the input-output model in current use, in particular the price part of it, and finally discuss the more important uses of price calculations.

## The Background of Input-Output Models in Norwegian Planning

The use of input-output models for economic policy purposes is a long tradition in Norway. The first use of input-output calculations were made in the 1950s. The first input-output model explicitly designed for planning needs became available in 1960. Input-output tables have been compiled as an integrated part of the national accounts for every year since 1949.

Macroeconomic planning started in Norway in 1946 with annual economic plans called national budgets. The national budget is a budget in national accounting terms for the total economy, not only for the government sector, and thus implied an extension of the political responsibility for the economic development compared to the prewar period. The pressing economic problems of this time were those of reconstruction: scarcity of materials, import constraints, rationing of consumer goods, etc. As the economy recovered from the abnormal

This article was published in *International Use of Input-Output Analysis*, edited by Reiner Stäglin. © Vandenhoeck and Ruprecht, 1982, pp. 113–129. Reprinted by permission.

postwar situation and the external environment changed, the set of policy instruments changed with less direct controls and rationing and more emphasis on indirect policy measures for managing the economy. The overall framework of the national budget as the annual economic plan has been maintained, however, for every year since 1946.

Until around 1960 the national budgets were worked out through a decentralized, administrative procedure. The various ministries worked out plan proposals for their respective sectors of the economy and submitted these to the Ministry of Finance which combined the sector plans into an overall national budget. The Ministry of Finance did more than just gather the plans together, of course. The national budget had to be checked for internal consistency and economic realism and it was a political document that needed to represent the government's targets with regard to economic development.

As already indicated the national budgeting in the early postwar period was a very detailed process because of the many constraints in the economy at this time. The task of checking internal consistency in such a plan is a problem which begs for an input-output model. As this was not available, the consistency checking was limited to looking after definitional equations and subjective plausibility considerations. When input-output tables became available a lot more could be done to corroborate the national budgets by comparison with historical observations of various relationships. From the middle of the 1950s experimental input-output calculations were performed on a 30-sector Leontief model in a cooperation between the Institute of Economics at the University of Oslo and the Central Bureau of Statistics. These calculations were derived from inverse input-output coefficients that were of great value in estimating the direct and indirect import content of various demand components.

In 1959 the first MODIS model, MODIS I, was completed and as it was designed specifically for national budgeting it could be put to work immediately, within the administrative environment. It took, of course, several years before the national budgeting process was completely reorganized with the model at the very center. For some years one may talk of a coexistence of the old administrative method of preparing the national budget and the new model-based way. But soon the model became an irreplaceable tool for the national budgeting and even more so when the further development of computer techniques increased the utilization of the model within the administrative environment.

## The Development of Norwegian Input-Output Models

The first MODIS model, MODIS I, was completed in 1959 and used from 1960 to 1965. It was a simple Leontief model with an aggregate consumption function. It had three sets of equations: input-output equations for 125 industries and imports; income equations determining wages, profits, and indirect taxes from production; and consumption equations determining the size and composition of household consumption from after-tax wages and profits. The MODIS I model had its limitations with regard to problems that could be dealt with by means of the model, above all that it did not include price relations and only very limited income relations. The model also had operational limitations, it was programmed for a first-generation computer and was cumbersome and time-consuming in use. Nevertheless, the model served an important purpose in proving the superiority of input-output techniques over administrative methods, gaining trust within the planning administration as a reliable tool in forecasting the real flows of the economy, and thus preparing the ground for more ambitious models.

The next version, MODIS II, which arrived in 1965, included a complete set of input-output relations in prices as well as quantity. The number of industries was increased to about 140, and final demand and other variables were dealt with in a considerably more detailed way.

By combining prices and quantities the results from MODIS II also displayed incomes. Furthermore, the model included relations for direct taxes and for indirect taxes and subsidies. The final results included a set of hierarchic accounts of disposable income. Disposable income for Norway was subdivided into government and private disposable income. The latter was subdivided further into disposable income for enterprises and disposable income for households which again was subdivided into disposable income for wage and salary earners and disposable income for self-employed.

The effort behind MODIS II was very ambitious with regard to completeness, in trying to build a model framework to cover the main areas of economic policy. Up to that time the whole national budget exercise was conducted in constant prices only. Prices were dealt with as a separate area of economic policy. The inclusion of price relations in MODIS II paved the way for securing the consistency of the national budget in a general equilibrium sense just as MODIS I had been a tool to secure the consistency between final demand and the composition of production.

MODIS II had, however, weaknesses both with regard to content and in terms of operationality and reliability. The model was improved and rebuilt as MODIS III in 1967. Throughout the period of MODIS III from 1967 until 1973 the use of the model by the Ministry of Finance increased tremendously. The model acquired in this period its central plan in the national budgeting process and was also used for other purposes such as medium-term planning, calculation of impact coefficients, and for ad hoc analysis of macroeconomic problems such as tax reform, consequences of the devaluation of the dollar, etc. The model was updated from new input-output tables every year. A valuable feature of the model was also its ability to calculate up to twenty solutions simultaneously.

MODIS IV was completed in 1973 and is the current version of the MODIS model. The basic input-output structure of the model differs from that of its predecessors by being based on commodity-by-industry tables rather than industry-by-industry tables. This made the description of the economy more realistic and had definite analytical advantages, in particular for price calculations. The experience from the use of MODIS III played a major role designing the user properties of the new model such as the system of communications between the model and its user.

Some numbers may indicate the amount of details in these models. The number of commodities in MODIS IV is about 200, the number of industries is about 125 with an additional 17 government production sectors. Household consumption is subdivided in 48 categories and government consumption in 65 categories. Another characteristic feature of these models is the amount of institutional details that have been embedded in them. MODIS IV distinguishes for instance between 28 direct taxes, 85 indirect taxes and subsidies, and 21 government income transfers.

One could get the impression from these numbers that the Norwegian economy is one of extensive government control executed through an all-inclusive model. That would be a misleading impression. The model has limited explanatory power. It is a very open model in the sense that much is left unexplained and represented in the model through exogenous variables. This is the case for instance for private investment, exports, wage rates, and productivity growth. The model serves mainly as a tool to secure consistency in the policy-making process. An important part of the consistency checking is provided by the input-output relations. The main behavioral element of the model is the consumption relations determining the amount as well as the composition of consumption from disposable incomes and prices.

Much emphasis has been put on the detailed representation of government instruments like taxes. The idea is that the tax rules should be represented as exact as possible within the model. This requires that the overall amount of detail in the model is sufficient to represent for instance indirect taxes in an adequate way.

An overall account shows that MODIS IV has about 2000 exogenous variables and the number of output variables are about 5000. The model is capable of handling several hundred policy alternatives simultaneously. From these numbers it is easy to see that the data handling properties of the model are quite important for its use as an efficient tool. One important property of the model is that the user may communicate with the model at different levels of aggregation. The model itself is not aggregated, only the input and output data. A user may choose aggregation levels to fit the needs of his problem by combining a high level of aggregation for some variables with a more disaggregate treatment of variables of particular interest. In this way one has tried to overcome the disadvantages of always having to deal with a very detailed model. For the presentation of model results there are similar flexibilities in giving the user a choice of edited tables at various levels of aggregation.

### The Price Model of MODIS IV: An Embodiment of the Scandinavian Model of Inflation

So far little has been said about the theoretical content of the MODIS model apart from it being an input-output model. As we are here mainly concerned with applications of the price part of the model I shall leave the rest of the model with what I have already said about it but go somewhat deeper into the price relations.

The price model of MODIS IV is the combination of two strands of thought: the input-output price model of Leontief determining supply prices by adding up cost components within a simultaneous system of equations; and, on the other hand, the Scandinavian model of inflation formulated in the 1960s, although it had its forerunners, and normally presented within a two-sector representation of the economy.

The basic assumption of the input-output model is often rendered as an assumption of constant input-output coefficients. But this is the assumption relevant for the quantity model, not for the price model. Constant input-output coefficients imply that quantity components in the same column of the input-output table change proportionately. The corresponding dual assumption for the price model is that the price components along a row of the input-output table change proportionately. This assumption is far from trivially fulfilled. It depends on the choice of price concept in the accounting and on the specification of the rows of the input-output table. In MODIS IV much emphasis has been placed on the requirements for the validity of this basic assumption of the price model.

The input-output table underlying the MODIS IV model is a commodity-by-industry table with about 200 commodities and 140 industries (including government productions sectors). It has been found quite crucial for the price model that the rows represent commodities and not industry outputs. Of the 200 commodities more than half are tradeables (i.e., they are exported or imported and often both). The price model distinguishes for each tradeable commodity between import price, domestic price, and export price. This has also been found necessary to account adequately for short-term price changes. The price concept itself is that of basic price, which is defined as producers' price less indirect taxes. The reason for this choice is that indirect taxes on a commodity are often—at least in Norway—dependent upon its destination. Typically commodities destined for exports are exempt from taxation. Some commodities may be taxed differently when used for consumption than when used for investment or intermediate input.

All these details of the specifications of the price relations have been found important enough for the validity of the results to be included in the price model of MODIS IV although they complicate the solution of the price model considerably when compared to the original Leontief version. The distinction between domestic price and export price also implies that prices cannot be solved independently from the solution of quantities.

The Scandinavian model of inflation is a model of the interrelation of prices and incomes in small, open economies. A fundamental distinction is drawn in this model between sheltered and exposed industries. The latter group consists of industries which are exposed to strong competition from abroad, either because they export most of their products or because they sell their products on the domestic market under strong foreign competition while sheltered industries, on the other hand, are those whose products are marketed at home under conditions that leave them relatively free of foreign competition.

There is a long-run version of the Scandinavian model of inflation that aims at explaining the long-term movement of wages and prices in an economy where, because of foreign trade, national wage and price trends are subject to strong price impulses from abroad. Only the short-term implications of the model are included in MODIS IV; among these is the implication that import and export prices are exogenous. Domestic prices of commodities with competition from imports are also exogenous; in the underlying theory they are assumed to follow the prices of similar imported goods, but this is not formally built into the model. Industries which produce commodities with exogenously determined prices have residually determined profits. In sheltered industries, on the other hand, prices are determined by cost-plus pricing in such a way that the profit share (profits as a share of factor income) will assume a predetermined value. For a number of commodities in the sheltered category the government influence over prices is so strong, in particular in the short run, that the prices in the model are set exogenously. These regulated and negotiated prices comprise for instance agricultural prices stipulated through income settlements and prices of commodities from government owned or dominated sectors such as electricity, transport, post and telecommunications.

Another important feature of the price model of MODIS IV, in particular from a policy point of view, is the detailed treatment of indirect taxes and subsidies. Altogether there are 85 indirect taxes and subsidies specified within the model, and they fall into different categories with regard to how they are treated in the model. Some are taxes on values, others on quantities. One is a general value added tax, others may be taxes or subsidies on one particular use of a commodity.

The price model of MODIS IV is a fully integrated part of the complete model. It can logically be considered as a separate part of the model, but it is normally not used independently from the rest of the model. In the following we shall look at some uses of the model in which the price part is of particular interest.

## The Need for Price Calculations

What is the need for price calculations for instance for a central government and what kind of price calculations are needed? Most governments would appreciate to have price forecasts from a reliable source. A number of research institutes in the major industrialized countries provide such forecasts. But how can the government use price forecasts when the price development depends upon the decisions of the government yet to be taken. Obviously, the forecasters have made, at least implicitly, some assumptions about what the government's decisions will be. This is a puzzle which was posed and answered by the late professor Ragnar Frisch. He said in an article many years ago and well known to Norwegian economists:

> How can it be possible to make a projection without knowing the decisions that will *basically influence* the course of affairs? It is as if the policy maker would say to the economic expert: "Now you ... try to guess what I am going to do, and make your estimate accordingly. On the basis of the *factual* information thus received I will then decide what to do". The shift from the on-looker view-point to the decision view-point must be founded on a much more coherent form of logic. It must be based on a decision model, i.e., a model where the *possible* decisions are built in *explicitly* as essential variables.

Frisch here rejects the idea that a government in the use of models should adopt what he calls an on-looker point of view. It should instead adopt a decision point of view (i.e., use models which can analyze the effects of government decisions). Frisch has been influential in almost every aspect in model building in Norway. No wonder then that Norwegian models adhere to his views, as they do, in particular in this case.

What motivated Frisch here are the possibilities he saw for using economic theory and mathematical techniques as scientific tools for analyzing the interrelationships between targets and instruments. He also had strong convictions on the obligation of a government to apply scientifically based economic planning to promote the welfare of the society.

The MODIS model is thus a decision model in the sense of Frisch, an instrument to be used in and analyzing economic policy, a purpose which is completely different from that of providing the best possible price forecasts for the general public. The experiences of using price calculations in Norway are therefore mainly related to the use of formulating and deciding an economic policy from an evaluation of the effects.

Since price calculations by means of input-output techniques were incorporated in the MODIS model from 1965 they have become of ever increasing importance. There are several reasons for this.

The Norwegian economy has gradually become more open with exports and imports amounting to 40–50 percent of gross domestic product. This makes the economy more exposed to inflationary impulses from abroad. The ability to assess in a fairly detailed way the impact of increasing import prices has been of particular importance in the 1970s. The breakdown of the system of fixed exchange rates in the OECD area caused a similar need to assess the impact of fluctuating exchange rates on domestic prices as well as on trade volumes, incomes, etc. Revaluation as well as devaluation has been used as a policy instrument by the Norwegian government in the 1970s.

Increasing government expenditure and changes in the tax system including the introduction of a value added tax put more emphasis on indirect taxes and subsidies and thus implied a greater impact on domestic prices, in particular food prices, from fiscal policy. Increased concern with relative incomes between and within socioeconomic groups has increased the inflationary pressure and focused attention on incomes policy. Government participation in income settlements may take different forms, but almost any kind of active government incomes policy requires a good basis of information on the price consequences of wage and salary increases.

In the 1970s the international trade within the OECD area has grown much slower than in the preceding decade. This has led to sharper competition between exporting countries and more concern with competitiveness. In Norway the cost level in manufacturing increased considerably above that of the competitors in the mid 1970s and the government was faced with the task of recovering competitiveness while maintaining a lower level of unemployment than most competitors. Price controls were among the instruments considered and used to achieve this aim. Naturally, price calculations on the effect, at least the temporary effect, of such strong measures were a necessity.

## The Use of Price Calculations by Means of Impact Tables

One way of presenting the content of an economic model is by means of impact tables. An impact table shows the impact on one or more of the endogenous variables of the model of changes in selected exogenous variables. An example of an

Table 2.1.   Some Impact Coefficients for MODIS IV (1980)

| 10% change in | % change in the price index of | | | |
| --- | --- | --- | --- | --- |
| | Private consumption | Government consumption | Gross investment | Consumer price index |
| Import prices | 2.85 | 1.19 | 3.83 | 2.91 |
| Wage rates | 2.22 | 7.88 | 3.22 | 2.23 |
| Productivity rates | −2.05 | −1.10 | −2.93 | −2.03 |
| Value added tax | 1.19 | 0.42 | 0.67 | 1.29 |
| Gasoline tax | 0.14 | 0.05 | 0.02 | 0.17 |
| Tax on liquor, wine, etc. | 0.17 | 0.00 | 0.00 | 0.09 |
| Subsidies on milk and milk-products | −0.12 | −0.01 | 0.00 | −0.07 |

impact table is given in Table 2.1. The table shows the effect on the price indices of private consumption, government consumption, gross investment, and gross domestic product of changes in import prices, wage rates, and selected indirect taxes and subsidies. The table is calculated by means of MODIS IV for 1980.

From the table one can see for instance that a 10 percent increase in import prices will increase the price index of private consumption by 2.85 percent while the price index of government consumption increases by less than half, only 1.19 percent. This reflects, of course, the difference in import content between private consumption and government consumption. The increase in the price index of gross investment on the other hand is 3.83 percent revealing the much higher import content of investment goods.

The other rows of the table show in a similar way the differential impact on the same price indices of 10 percent increases in wage rates, productivity rates, three indirect taxes, and milk subsidies. A fourth column shows the impact on the official consumer price index, a Laspeyres index with weights based on a 1973 consumer survey. The impact on this index is quite similar to but not identical with the price index of private consumption. Quite notable is the impact of the tax on liquor, wine, etc. which affects the price index of private consumption almost twice as much as the consumer price index. This reflects the well-known tendency to underestimate expenditures on liquor, wine, etc. in consumer surveys.

The results presented in impact tables are calculated by full simulation runs on the model. In practice it is both more convenient and efficient to calculate the impact coefficients by simulation than by analytic derivations from the equation system. Impact tables can be useful as an introduction to a model. They offer a shortcut to the reduced form of the equation system with numerical results that can be interpreted and applied. It should be stressed, however, that proper use of impact tables requires a deeper understanding of the theoretical content of the model. To understand fully the meaning of the impact of a specific exogenous variable on a given endogenous variable, it is necessary to have an overall knowledge of the theoretical content of the model and, in particular, which variables are exogenous and which are endogenous.

Impact tables cannot give the full content of a large-scale model. In MODIS IV there are about 2000 exogenous variables and about 5000 endogenous variables.

The impact tables can be constructed at different levels of aggregation. For some purposes more aggregate tables are convenient while for others more detailed tables are needed. In Norway impact tables for MODIS IV have been found very useful. They are published for every year in a publication of 250 pages.

The impact tables are used first of all by the Ministry of Finance before and between model runs to calibrate the use of policy instruments, to assess margins of uncertainty in external influences, etc. The impact tables can also be used by political parties to assess the effect of policy proposals. Requests from companies, research institutes, and others to the Central Bureau of Statistics for use of the MODIS model to analyze a variety of problems are often dealt with by means of impact tables which in most cases can give a fully adequate answer and thus save the user the trouble and expense of a full use of the model.

## The Use of Price Calculations in Income Settlement

Input-output price calculations have played an important role in income settlements in Norway since 1965. This is no doubt the most important use of price calculations. Such calculations have been used at two stages in the income settlement process: first at a preparatory stage prior to negotiations and later when the government has intervened in the final stage of the negotiations.

Income settlements in Norway comprise settlements between trade unions and employers' organizations—normally every second year—and settlements between the government and agricultural organizations. About two thirds of all wage and salary earners belong to a union.

The preparatory use of price calculations originated in 1965 when the government appointed a small expert committee to prepare background material for the upcoming income settlements. The expert committee constructed a small input-output model called PRIM, which could be characterized as a pedagogical simplification of the price part of the MODIS model. Calculations by means of this model illustrated the consequences of alternative income settlements with regard to prices and real incomes. The experts made assessments of all relevant background variables such as world market prices and productivity changes and then represented the outcome of the income settlements by two key parameters: the average wage and salary increase and the increase in agricultural prices. The response to the report of the expert committee from the organizations taking part in income settlements was quite favorable.

The expert committee continued its work also for the next round of income settlements. The reports of the committee did not include any recommendations, at least not explicitly, but presented the background material in a way which allowed the participants in the ensuing income negotiations to use other assessments of background variables than those adopted by the committee. This was done by means of impact tables presenting the effect of background variables as well as the key parameters on prices and real incomes.

The importance which the PRIM model acquired as a medium for discussing alternative income settlements was due in great extent to its formal simplicity. The model was transparent enough to be understood and used by the parties of the income settlements. On the other hand the model was too aggregate and simplified to represent the consequences of income settlements in more detail.

The expert committee was later replaced by a committee with representatives from the trade unions, employers' organizations, agricultural organizations, and the government. The committee has no mandate to negotiate or reach agreements on behalf of their organizations. The use of the PRIM model was later replaced by the full MODIS model. In recent years the committee has put much less emphasis on alternative outcomes of the income settlements and concentrated more on the general outlook and assessment of important background variables. The use of input-output price calculations is still, however, the central analytic framework of the committee. The use of this framework has doubtlessly done much to provide a common ground in understanding the basic interrelationships of prices and incomes.

The Norwegian system of determining wage and salary increases is based on free negotiations between trade unions and employers' organizations. Agreements are normally made for one or two years. When agreement cannot be reached by negotiations the government can by Parliamentary decision in each individual case have the agreement settled by a national arbitration committee.

The normal outcome of wage negotiations is then agreement by negotiation. The agreements are, of course, of great importance for the government's short-term economic policy, with considerable effects on price and cost development as well as on income distribution. It is of considerable concern to the government that the negotiating parties reach an agreement that the government finds consistent with the situation of the economy. The government will normally not intervene in the negotiations, but, of course, present relevant background information, etc. The establishment of the above-mentioned committee for preparing material and calculations for the income settlements has been of particular importance.

In the 1970s the central government has intervened on a number of occasions to influence the outcome of wage negotiations. The interventions have resulted in agreements that have been called "combined income settlements" or "package deals". The interventions have taken different forms from time to time, but basically they have been of two different forms. One is that the government has presented just prior to negotiations a package of policy measures that implied increased real incomes to wage earners through tax reductions, increases in food subsidies or children's allowances, etc. In return the government has wanted lower nominal wage increases. The other way is by entering the negotiations as a third party committing itself to a similar package in return for an agreed wage settlement. In 1973 and 1974 the first way was chosen, in 1975, 1976, and 1977 the second was used.

A considerable part of the wage increases are the result of wage drift not accounted for in the negotiated agreements. This creates problems for the government whose concern is with the overall wage increase rather than the negotiated part of it. The agreements normally have index clauses that provide automatic compensation for price increases. This is also a worry for the government in its effort to keep nominal increases within bounds. For some years the government stated as part of combined income settlements an intention with regard to the increase in average real disposable wage incomes (e.g., 3.5 percent from 1975 to 1976 and 2.5 percent from 1976 to 1977). This amounted to a virtual guarantee of the stated development in real incomes through tax and price policies.

The government participation in income settlements as described very briefly above would have been impossible without model tools for price and income calculations as detailed and comprehensive as the MODIS IV. Throughout the income settlements in the middle of the 1970s price and incomes calculations were used intensively for preparing and putting into effect the policies that the government had committed itself to.

## The Use of Price Calculations in Maintaining a Full Employment Economy

After the oil price shock in 1973–1974 most industrialized countries have experienced an economic development that has differed in many ways from the stability and steady growth of the preceding period. Unemployment has in many industrialized countries risen to levels comparable to the depression period of the 1930s, while at the same time inflation rates have been high and a number of countries have run into severe balance of payments difficulties.

The Norwegian government has in this period pursued a policy different from that of most other OECD countries. Employment has been maintained at a high level by counteracting slow growth in foreign demand by boosting domestic demand for private and government consumption. The choice of policy reflects a higher priority given to the target of full employment, while other countries have stressed control over inflation as the prime target in economic policy. Both economic policy and economic development in Norway in the 1970s have been influenced very much by the discovery of vast oil and gas reserves in the Norwegian sector of the North Sea. The oil and gas riches eased the balance of payments constraint. The projections of future export earnings from oil and gas allowed a policy that incurred balance of payments deficits in the middle 1970s. On the other hand the booming sector of oil related activities caused unbalances in the economy and difficulties in economic policy management.

To cope with these problems the government applied a number of selective measures and ran into well-known problems of accurately calibrating and timing the policy measures. Domestic demand tended to increase more than expected causing problems of overfull employment, severe inflationary pressure, and loss of competitiveness. The use of price calculations played a great role in policy considerations in this period.

The government has as a rule not published price forecasts except for one or two years. In the National Budget publication there are published figures since 1975, however, of gross national product and its main components in constant and current prices. With a pocket calculator the implicit price assumptions could easily be calculated. For private consumption the implicit forecasts—usually circumscribed as assumptions used in calculations—are as shown in Table 2.2.

Table 2.2.    Private Consumption Price Index (*Current Growth Rate*)

| Year | National budget (previous autumn) | National accounts |
|------|-----------------------------------|-------------------|
| 1975 | 11.6 | 11.7 |
| 1976 | 8.7 | 8.6 |
| 1977 | 8.3 | 8.6 |
| 1978 | 9.0 | 8.3 |
| 1979 | 4.4 | 5.2 |
| 1980 | 6.1 | 9.8 |
| 1981 | 10.8 | 13.6 |

As can be seen from the table there is a remarkable record of correct price assumptions in the middle 1970s. These are the years of the income policies described earlier. The results are to a large extent due to policy measures to fulfill the governments commitment as part of the income policies rather than good forecasting. From 1978 strict price and income controls were introduced for a period of 16 months in an effort to get control over inflation once and for all as it was expressed. As the table indicates the underlying inflationary pressure was stronger than expected and not only did the rate of inflation increase but also the difference between the actual rate and the assumption in the national budget. The inflation rate in 1981 was not only higher than in the preceding years, it was the highest since 1951. The government fulfilled its aim of full or nearly full employment, however, and could hardly have done so without the use of a policy oriented model as detailed as MODIS IV. When price controls have been used, as they have in some form or other for a considerable part of the period since 1974, a detailed model for price calculations is particularly useful.

A model like the price model of MODIS IV gives a good overview of the situation but it has its limitations. The most important is undoubtedly that the dynamics of the inflation process is not represented within the model.

**Some Final Comments**

The input-output price model has been applied in Norway and elsewhere for analyzing a variety of problems. The Norwegian experience is an example of intensive use of the input-output price model in short-term economic policy making, a use which is not very common in other countries. It may be true that the input-output price model has got much less attention than its quantity counterpart. The input-output relationships in prices and quantities describe dual aspects of the structure and functioning of modern economies. It is hardly meaningful to ask which aspect is more important. The main reason why the input-output price relationships are used less has got to do, I believe, with the data requirements. The price data in input-output tables are often of poor quality, in particular for nonmanufacturing sectors.

There is a lot to say about weaknesses and insufficiencies of price data in the Norwegian input-output tables too. There are two advantages worth mentioning for input-output modeling in Norway which have done a lot to overcome data deficiencies and to a large extent are prerequisites for the current use of input-output models.

The first advantage is the updatedness of the input-output tables. The first preliminary and somewhat incomplete input-output table appears already in January after the completion of the year. A second preliminary—and usually more reliable—table appears in April, the third preliminary table in November and then the final table about 18 months after the completion of the year. In some countries the most recent input-output table is several years old. In particular for economic policy use of price calculations this is a disadvantage.

The other advantage in Norway is the close cooperation between those working with national accounts, input-output tables, model building, and model use. Input-output tables are constructed as an integrated part of the national accounts and the national accounting work takes place literally on the same floor as the model-building work of the Central Bureau of Statistics. This has meant that the compilation of input-output tables has been influenced to a considerable degree by model needs. Again, this has been of particular importance for the price calculations which are quite dependent upon the value concepts used in the input-output tables, the treatment of indirect taxes, the commodity specification, etc.

In the preceding pages I have only dealt with the use of input-output price calculations in a short-term perspective. I would just like to mention at the very end that I think there may be an important future use of input-output price calculations to provide information on relative prices in a long-term perspective. Such information on future prices may be valuable for households, firms, and local government as well as for the central government for decisions taken today.

**References**

**General**

Aukrust, Odd. 1978. Econometric methods in short-term planning. In *Econometric Contributions to Public Policy*, ed. Richard Stone and William Peterson, pp. 64–83. London: Macmillan Press Ltd.

Bjerve, Petter Jakob. 1976. Trends in Norwegian planning 1945–1975. Artikler No. 84. Oslo: Central Bureau of Statistics.

Frisch, Ragnar. 1961. A survey of types of economic forecasting and programming and a brief description of the Oslo channel model. Memorandum from the Institute of Economics (May 13).

**PRIM**

Aukrust, Odd. 1970. PRIM—A model of the price and income distribution mechanism of an open economy. *Review of Income and Wealth* series 16: pp. 51–78.

Aukrust, Odd. 1977. Inflation in the open economy. In *Worldwide Inflation: Theory and Recent Experience*, ed. Lawrence B. Krause and Walter Sâlant, pp. 109–166. Washington, D.C.: Brookings.

## MODIS

Bjerkholt, Olav. 1968. A precise description of the system of equations of the economic model MODIS III. *Economics of Planning* 8(1–2): pp. 26–56.

Bjerkholt, Olav and Svein Longva. 1980. MODIS IV—A model for economic analysis and national planning. Oslo: Central Bureau of Statistics.

Sevaldson, Per. 1964. An interindustry model of production and consumption in Norway. *Income and Wealth* series x: pp. 23–50.

Sevaldson, Per. 1968. MODIS II: A macro-economic model for short-term analysis and planning. In *Macro-Economic Models for Planning and Policy-Making*, United Nations, Geneva, pp. 161–171.

# 3

# The INFORUM Model

MARGARET B. BUCKLER, DAVID GILMARTIN, AND
THOMAS C. REIMBOLD

The INFORUM, *In*terindustry *For*ecasting at the *U*niversity of *M*aryland, project uses an input-ouput model to make long-term forecasts for the American economy. The project, directed by Clopper Almon, has been in existence for nine years. The maintenance of the model and the continuing research to improve it are supported by about fifteen private companies. Recently, several agencies of the United States government and groups from two foreign countries have also become sponsors of the model.

The model divides the economy into 185 product sectors and forecasts the output of each sector and its sales to each intermediate and final demand purchaser. The structure of the model, its performance, and recent extensions of it to include prices and wages will be briefly reviewed here. Several applications will be described, including a forecast with assumptions to reflect the "energy crisis." Finally, future directions of our research will be discussed.

## The INFORUM Model

The workings of the model—the theory behind it, several applications using it, and a recent forecast with it—have been published in Almon et al. (1985), so only a brief description of the forecasting model need be given here. Annual forecasts of sales for the next fifteen years are made in real terms. Overall exogenous controls include projections of population, labor force, households, interest rates, and government purchases. The model is not a crystal ball—it is not able to determine whether 1985 will be a boom or a recession year. Rather, the user of the model assumes some level of economic activity, as reflected in the employment rate, and then runs the model to determine the effect.

Final demand equations, which utilize this exogenous information, as well as information generated by the model, have been estimated. Personal consumption expenditures per capita for each product depend upon income, relative prices, and a time trend. Investment spending by 90 sectors (aggregates of the 185 sectors) to maintain a desired capital stock is determined from the levels of output in the five

This article was published in *Advances in Input-Output Analysis*, edited by Karen R. Polenske and Jiri V. Skolka. © Ballinger Publishing Company, 1976, pp. 297–327. Reprinted by permission.

preceding years and the current year and from the cost of capital. This investment spending is then converted into purchases of producer durable equipment from the 185 product sectors by means of a capital matrix. Stock adjustment equations and other methods are used to forecast the 28 types of construction expenditures, which are translated into purchases of construction materials by assembling a construction matrix. Imports for each product depend upon domestic demand and the relative foreign-to-domestic price for that product.[1] Exports also depend upon this relative price and lagged domestic output. Inventory change for each product is a function of the inventory stock and the domestic supply of the product such that the inventory-domestic supply relation is maintained.[2] Once final demands for each product have been determined, the input-output matrix is used to solve for output.[3] Interindustry coefficients are not fixed but change over time to follow a logistic curve.

The labor force provides the limit for the size of the forecast economy. Employment is determined from five types of productivity equations estimated by industry.

### Simulation Testing

The performance of many parts of the model has been tested by simulating the period 1966–1971. Consumption, equipment investment, productivity, and across-the-row coefficient changes were intensively examined.[4] By requiring them to have satisfactory simulations as well as satisfactory regression fits, significant improvements were made in many of the equations.

To test the INFORUM model, a number of initial calculations were required. In particular, all equations to be used in the forecast simulation of the entire model required individual testing. Complete final demand and cost-of-input data are required to balance the input-output matrices for each year of the simulation. With these matrices, we are able to separate the individual components of the total error of the simulated forecasts. The matrix balancing was performed by the RAS method.[5]

The simulation period is from 1967 through 1971. All major stochastic equations of the model are, therefore, estimated through 1966 and then used to forecast the remaining years. The simulation study is designed to answer the following questions:

1. How accurate and reliable are forecasts generated by input-output models?
2. Since input-output forecasts depend as much on the prediction of input-output coefficients as on the forecasts of final demand, what portion of the total errors stems from incorrectly specified coefficients and how much is contributed by the final demand estimation?
3. How much feedback error is generated within the model and how does it affect the convergence process? (Feedback here refers to the fact that output affects investment, which in turn affects output.)

To answer these questions, four simulation tests were carried out, the results of which are shown in Table 3.1, (a) with constant 1965 base-year coefficients; (b) with logistic curve, equiproportional changes across each row; (c) with completely balanced matrices; and (d) with completely balanced matrices and final demand calculated from actual output. Test (b) represents the actual forecasting model, and its results are compared with the behavior of the other three simulations.

For each of the four simulations, the errors are recorded in Table 3.1 for four items: personal consumption expenditures (PCE), producer durable equipment (PDE), inventory change, and output. The errors are measured as a percent of actual output and represent cumulative averages over the five-year simulation period, 1967–1971. A positive error implies under-prediction, and vice versa:

$$e_i = \sum_{t=1967}^{1971} \frac{a_{it} - p_{it}}{q_{it}} \cdot 100 \qquad i = 1, 2, \ldots, 185 \qquad (3\text{--}1)$$

where $a$ refers to actual value, $p$ refers to predicted value, and $q$ stands for output.

The analysis of Table 3.1 is divided into three parts: overall behavior of the model; sector analysis of five-year average errors; and sector analysis of year-by-year results.

### Overall Behavior of the Model

In the two bottom lines of Table 3–1, the overall errors of the simulations are given. The "weighted cumulative error" ($WCE$) measures the average aggregate prediction error, and the "weighted absolute error" ($WAE$) reflects the average size of the errors.

$$WCE = \frac{\sum_{i=1}^{185} (e_i q_i)}{\sum_{i=1}^{185} q_i} \qquad (3\text{--}2)$$

$$WAE = \frac{\sum_{i=1}^{185} (|e_i| q_i)}{\sum_{i=1}^{185} q_i} \qquad (3\text{--}3)$$

where $e$ is defined in Equation (3–1), and $q$ stands for output in 1971. A quick comparison of the four different runs shows that the final demand errors remain approximately the same. But before we can conclude that output errors have little or no feedback effect on the final demand prediction, the result must be analyzed in further detail.

The general improvement as we go from test (a) to test (d) is clear; however, a large portion of the total prediction error can be attributed to coefficient misspecifications, as may be seen clearly when the overall output errors are graphed, as in Figure 3.1. In this figure, a, b, c, and d, refer to the four different simulations that were made.

Since PCE is not a function of output, its error column is the same for all four runs. The overall weighted absolute error (1.5 percent) is the largest of the three final demand components. But the fact that the average PCE error is larger than

Table 3.1.  Average Cumulative Simulation Errors as a Percent of Actual Output for 1967–1971: Energy Crisis Forecast

| Sector No. | Title | (a) Constant coefficient and predicted output | | | | (b) Predicted coefficient and predicted output | | | | (c) Balanced coefficient and predicted output | | | | (d) Balanced coefficient and actual output | | | |
|---|---|---|---|---|---|---|---|---|---|---|---|---|---|---|---|---|---|
| | | PCE | PDE | Inventory | Output | PCE | PDE | Inventory | Output | PCE | PDE | Inventory | Output | PCE | PDE | Inventory | Output |
| 1 | Dairy farm products | 1.0 | 0.0 | 2.3 | 1.5 | 1.0 | 0.0 | 2.5 | 4.0 | 1.0 | 0.0 | 2.0 | -0.4 | 1.0 | 0.0 | 2.0 | -0.4 |
| 2 | Poultry and eggs | 2.2 | 0.0 | 0.1 | -0.6 | 2.2 | 0.0 | 0.0 | -2.6 | 2.2 | 0.0 | 0.4 | 6.0 | 2.2 | 0.0 | 0.4 | 6.0 |
| 3 | Meat animals, other livestock | 0.1 | 0.0 | 0.6 | 18.0 | 0.1 | 0.0 | 0.5 | 7.2 | 0.1 | 0.0 | 0.5 | 6.2 | 0.1 | 0.0 | 0.5 | 6.2 |
| 4 | Cotton | 0.0 | 0.0 | -1.8 | -104.3 | 0.0 | 0.0 | 2.7 | -30.2 | 0.0 | 0.0 | 4.5 | 2.4 | 0.0 | 0.0 | 4.5 | 2.4 |
| 5 | Grains | 0.0 | 0.0 | 0.4 | -0.4 | 0.0 | 0.0 | 0.4 | -5.6 | 0.0 | 0.0 | 0.4 | 3.9 | 0.0 | 0.0 | 0.4 | 3.9 |
| 6 | Tobacco | 0.0 | 0.0 | -5.7 | -5.5 | 0.0 | 0.0 | -5.7 | -2.2 | 0.0 | 0.0 | -5.4 | -2.6 | 0.0 | 0.0 | -5.4 | -2.6 |
| 7 | Fruit, vegetables, other crops | 1.0 | 0.0 | -0.1 | -2.2 | 1.0 | 0.0 | -0.1 | -1.9 | 1.0 | 0.0 | -0.1 | 2.7 | 1.0 | 0.0 | -0.1 | 2.7 |
| 8 | Forestry and fishery produce | -3.2 | 0.0 | 0.1 | 6.7 | -3.2 | 0.0 | -0.2 | 17.0 | -3.2 | 0.0 | 0.3 | -2.0 | -3.2 | 0.0 | 0.3 | -2.1 |
| 9 | ND^b | 0.0 | 0.0 | 0.0 | 0.0 | 0.0 | 0.0 | 0.0 | 0.0 | 0.0 | 0.0 | 0.0 | 0.0 | 0.0 | 0.0 | 0.0 | 0.0 |
| 10 | Agr., forestry and fish services | 0.1 | 0.0 | -0.2 | 15.8 | 0.1 | 0.0 | -0.2 | 22.2 | 0.1 | 0.0 | 0.1 | 3.3 | 0.1 | 0.0 | 0.1 | 3.2 |
| 11 | Iron ores | 0.0 | 0.0 | 0.3 | -14.0 | 0.0 | 0.0 | 0.7 | -5.9 | 0.0 | 0.0 | 1.3 | 7.4 | 0.0 | 0.0 | 1.3 | 6.4 |
| 12 | Copper ore | 0.0 | 0.0 | 0.1 | 3.6 | 0.0 | 0.0 | 0.1 | 2.1 | 0.0 | 0.0 | 0.1 | 4.1 | 0.0 | 0.0 | 0.1 | 3.4 |
| 13 | Other nonferrous ores | 0.0 | 0.0 | 0.2 | -2.6 | 0.0 | 0.0 | 0.2 | -4.6 | 0.0 | 0.0 | 0.2 | 8.0 | 0.0 | 0.0 | 0.2 | 7.1 |
| 14 | Coal mining | 0.6 | 0.0 | 0.2 | 4.3 | 0.6 | 0.0 | 0.1 | 2.6 | 0.6 | 0.0 | 0.1 | 3.1 | 0.6 | 0.0 | 0.1 | 2.8 |
| 15 | Crude petroleum, natural gas | 0.0 | 0.0 | -0.1 | 1.8 | 0.0 | 0.0 | -0.1 | -4.8 | 0.0 | 0.0 | -0.2 | 0.2 | 0.0 | 0.0 | -0.2 | 0.0 |
| 16 | Stone and clay mining | -0.1 | 0.0 | -0.3 | -15.5 | -0.1 | 0.0 | -0.2 | -13.2 | -0.1 | 0.0 | 0.1 | 1.2 | -0.1 | 0.0 | 0.1 | 1.1 |
| 17 | Chemical fertilizer mining | 0.0 | 0.0 | 0.3 | 3.8 | 0.0 | 0.0 | 0.1 | -0.4 | 0.0 | 0.0 | 0.5 | 3.8 | 0.0 | 0.0 | 0.5 | 3.6 |
| 18 | New construction | 0.0 | 0.0 | 0.0 | 0.5 | 0.0 | 0.0 | 0.0 | 0.5 | 0.0 | 0.0 | 0.0 | 0.0 | 0.0 | 0.0 | 0.0 | 0.0 |
| 19 | Maintenance construction | 0.0 | 0.0 | 0.0 | 0.0 | 0.0 | 0.0 | 0.0 | 0.0 | 0.0 | 0.0 | 0.0 | 0.0 | 0.0 | 0.0 | 0.0 | 0.0 |
| 20 | Complete guided missiles | 0.0 | 0.0 | 0.0 | -2.5 | 0.0 | 0.0 | 0.0 | -1.5 | 0.0 | 0.0 | -0.0 | 0.1 | 0.0 | 0.0 | -0.0 | 0.1 |
| 21 | Ammunition | 0.9 | 0.0 | 0.9 | 6.0 | 0.9 | 0.0 | 0.9 | 6.0 | 0.9 | 0.0 | 0.9 | 2.3 | 0.9 | 0.0 | 0.9 | 2.3 |
| 22 | Other ordnance | -0.7 | 0.0 | -0.0 | 1.5 | -0.7 | 0.0 | -0.0 | 1.2 | -0.7 | 0.0 | 0.0 | -0.7 | -0.7 | 0.0 | 0.0 | -0.7 |
| 23 | Meat products | 5.4 | 0.0 | -0.0 | 5.2 | 5.4 | 0.0 | -0.1 | 6.5 | 5.4 | 0.0 | -0.1 | 6.4 | 5.4 | 0.0 | -0.1 | 6.4 |
| 24 | Dairy products | -3.7 | 0.0 | -0.2 | -7.7 | -3.7 | 0.0 | -0.2 | -7.8 | -3.7 | 0.0 | -0.2 | -4.4 | -3.7 | 0.0 | -0.2 | -4.4 |
| 25 | Canned and frozen foods | -0.4 | 0.0 | 0.7 | -5.1 | -0.4 | 0.0 | 0.7 | -1.2 | -0.4 | 0.0 | 0.7 | 0.5 | -0.4 | 0.0 | 0.7 | 0.5 |
| 26 | Grain mill products | 2.8 | 0.0 | -0.0 | 2.6 | 2.8 | 0.0 | -0.0 | 0.8 | 2.8 | 0.0 | 0.0 | 4.8 | 2.8 | 0.0 | 0.0 | 4.8 |
| 27 | Bakery products | -5.0 | 0.0 | -0.0 | -6.0 | -5.0 | 0.0 | -0.0 | -5.1 | -5.0 | 0.0 | -0.0 | -4.9 | -5.0 | 0.0 | -0.0 | -4.9 |

| | 1 | 2 | 3 | 4 | 5 | 6 | 7 | 8 | 9 | 10 | 11 | 12 | 13 | 14 |
|---|---|---|---|---|---|---|---|---|---|---|---|---|---|---|
| 28 Sugar | 10.8 | 0.7 | 0.0 | 5.4 | 10.8 | 0.7 | 0.0 | 5.4 | 11.0 | 0.7 | 0.0 | 5.4 | 13.4 | 0.7 |
| 29 Confectionery products | 1.5 | -0.0 | 0.0 | 1.3 | 1.5 | -0.0 | 0.0 | 1.3 | 2.8 | -0.0 | 0.0 | 1.3 | -0.4 | -0.0 |
| 30 Alcoholic beverages | 5.1 | 0.0 | 0.0 | 4.5 | 5.2 | 0.0 | 0.0 | 4.5 | 6.2 | 0.1 | 0.0 | 4.5 | 3.4 | 0.0 |
| 31 Soft drinks and flavorings | 6.2 | 0.1 | 0.0 | 5.3 | 6.3 | 0.1 | 0.0 | 5.3 | 5.1 | 0.1 | 0.0 | 5.3 | 4.3 | 0.1 |
| 32 Fats and oils | 3.1 | 0.2 | 0.0 | 0.5 | 3.1 | 0.2 | 0.0 | 0.5 | -5.0 | 0.2 | 0.0 | 0.5 | -4.4 | 0.2 |
| 33 Misc. food products | 3.1 | 0.2 | 0.0 | 2.9 | 3.1 | 0.2 | 0.0 | 2.9 | 2.5 | 0.2 | 0.0 | 2.9 | 0.3 | 0.1 |
| 34 Tobacco products | 3.3 | 1.5 | 0.0 | 1.1 | 3.3 | 1.5 | 0.0 | 1.1 | 2.7 | 1.3 | 0.0 | 1.1 | -0.4 | 1.5 |
| 35 Broad and narrow fabrics | -4.0 | 0.0 | 0.0 | -0.2 | -3.9 | 0.0 | 0.0 | -0.2 | -8.9 | 0.0 | -1.8 | -0.2 | -9.0 | 0.0 |
| 36 Floor coverings | 14.1 | 1.6 | 1.2 | 10.2 | 14.9 | 1.7 | 1.9 | 10.2 | 14.8 | 1.7 | 0.0 | 10.2 | 19.3 | 2.0 |
| 37 Miscellaneous textiles | 3.3 | 0.3 | 0.0 | 0.3 | 3.5 | 0.3 | 0.0 | 0.3 | -6.7 | 0.0 | 0.0 | 0.3 | -13.0 | -0.3 |
| 38 Knitting | -0.6 | -0.1 | 0.0 | 2.5 | -0.6 | -0.1 | 0.0 | 2.5 | -2.0 | -0.0 | 0.0 | 2.5 | 20.1 | -0.1 |
| 39 Apparel | -10.7 | 0.4 | 0.0 | -10.1 | -10.7 | 0.4 | 0.0 | -10.1 | -11.7 | 0.4 | 0.0 | -10.1 | -11.6 | 0.4 |
| 40 Household textiles | 3.0 | 1.5 | 0.0 | 0.9 | 3.1 | 1.5 | 0.0 | 0.9 | 1.2 | 1.3 | 0.0 | 0.9 | -0.5 | 1.1 |
| 41 Lumber and wood products | 2.2 | 0.4 | 0.0 | -0.1 | 2.3 | 0.4 | 0.0 | -0.1 | 8.3 | 0.6 | 0.0 | -0.1 | -4.3 | 0.1 |
| 42 Veneer and plywood | 5.6 | 1.0 | 0.0 | 0.0 | 5.9 | 1.0 | 0.0 | 0.0 | 0.9 | 0.3 | 0.0 | 0.0 | 2.2 | 0.4 |
| 43 Millwork and wood products | 2.6 | 0.2 | 0.0 | 0.8 | 2.7 | 0.2 | 0.0 | 0.8 | 0.9 | 0.0 | 0.0 | 0.8 | 12.7 | 0.4 |
| 44 Wooden containers | 13.5 | 10.7 | 0.0 | 0.0 | 13.8 | 10.7 | 0.0 | 0.0 | 13.0 | 9.8 | 0.0 | 0.0 | -41.2 | 3.5 |
| 45 Household furniture | -6.3 | -0.2 | -1.2 | -5.8 | -6.3 | -0.2 | -1.2 | -5.8 | -5.8 | -0.2 | -1.4 | -5.8 | -7.9 | -0.2 |
| 46 Other furniture | 7.0 | 0.1 | 5.7 | 0.8 | 8.9 | 0.1 | 7.5 | 0.8 | 9.4 | 0.1 | 9.0 | 0.8 | 10.6 | 0.2 |
| 47 Pulp mills | 2.3 | 0.6 | 0.0 | 0.0 | 2.5 | 0.6 | 0.0 | 0.0 | -10.9 | 0.1 | 0.0 | 0.0 | -15.7 | -0.0 |
| 48 Paper and paperboard mills | 1.4 | 0.1 | 0.0 | -0.0 | 1.6 | 0.1 | 0.0 | -0.0 | -6.9 | -0.3 | 0.0 | -0.0 | -5.7 | -0.2 |
| 49 Paper products, n.e.c.[a] | 0.8 | 0.2 | 0.0 | -0.4 | 0.9 | 0.2 | 0.0 | -0.4 | -5.1 | -0.0 | 0.0 | -0.4 | -3.6 | 0.0 |
| 50 Wall and building paper | 2.7 | -0.3 | 0.0 | 0.0 | 2.8 | -0.3 | 0.0 | 0.0 | -4.1 | -0.3 | 0.0 | 0.0 | -8.0 | -0.4 |
| 51 Paperboard containers | 1.5 | 0.0 | 0.0 | 0.2 | 1.6 | 0.0 | 0.0 | 0.2 | -3.0 | 0.0 | 0.0 | 0.2 | -1.5 | 0.0 |
| 52 Newspapers | 2.0 | -0.0 | 0.0 | 0.8 | 2.1 | -0.0 | 0.0 | 0.8 | 0.6 | -0.0 | 0.0 | 0.8 | 0.9 | -0.0 |
| 53 Periodicals | 2.6 | -0.1 | 0.0 | 1.6 | 2.7 | -0.1 | 0.0 | 1.6 | -0.9 | -0.4 | 0.0 | 1.6 | -6.0 | -0.6 |
| 54 Books | -1.1 | 1.1 | 0.0 | -2.0 | -1.1 | 1.1 | 0.0 | -2.0 | -6.7 | 1.1 | 0.0 | -2.0 | -10.9 | 0.9 |
| 55 Industrial chemicals | 2.1 | 0.3 | 0.0 | 0.0 | 2.4 | 0.3 | 0.0 | 0.0 | -8.1 | -0.2 | 0.0 | 0.0 | -1.8 | 0.2 |
| 56 Business forms, blank book | 1.0 | -0.1 | 0.0 | 0.6 | 1.2 | -0.1 | 0.0 | 0.6 | -7.5 | -0.2 | 0.0 | 0.6 | 7.8 | -0.0 |
| 57 Commercial printing | 1.7 | 0.0 | 0.0 | 0.3 | 1.8 | 0.0 | 0.0 | 0.3 | -2.8 | 0.0 | 0.0 | 0.3 | 1.3 | 0.0 |
| 58 Other printing, publishing | 3.1 | 0.4 | 0.0 | 2.4 | 3.2 | 0.4 | 0.0 | 2.4 | 0.2 | 0.3 | 0.0 | 2.4 | -10.5 | -0.1 |
| 59 Fertilizers | 2.9 | 0.6 | 0.0 | -0.2 | 2.9 | 0.6 | 0.0 | -0.2 | -13.5 | -1.5 | 0.0 | -0.2 | -17.5 | -1.8 |
| 60 Pesticides and agricultural chemicals | 2.6 | 0.5 | 0.0 | 0.1 | 2.6 | 0.5 | 0.0 | 0.1 | 7.0 | 0.4 | 0.0 | 0.1 | 18.5 | 0.9 |
| 61 Miscellaneous chemical products | 2.2 | 0.2 | 0.0 | 0.2 | 2.4 | 0.2 | 0.0 | 0.2 | -3.4 | 0.1 | 0.0 | 0.2 | -8.6 | -0.1 |
| 62 Plastic materials and resins | 1.9 | -0.0 | 0.0 | 0.0 | 2.2 | -0.0 | 0.0 | 0.0 | -17.4 | 0.0 | 0.0 | 0.0 | 0.3 | -0.0 |

Table 3.1 (continued).

| Sector | (a) Constant coefficient and predicted output | | | | (b) Predicted coefficient and predicted output | | | | (c) Balanced coefficient and predicted output | | | | (d) Balanced coefficient and actual output | | | |
|---|---|---|---|---|---|---|---|---|---|---|---|---|---|---|---|---|
| No. Title | PCE | PDE | Inven-tory | Output | PCE | PDE | Inven-tory | Output | PCE | PDE | Inven-tory | Output | PCE | PDE | Inven-tory | Output |
| 63 Synthetic rubber | 0.0 | 0.0 | 0.1 | 1.5 | 0.0 | 0.0 | -0.0 | -0.7 | 0.0 | 0.0 | 0.3 | 5.9 | 0.0 | 0.0 | 0.3 | 5.6 |
| 64 Cellulosic fibers | 0.0 | 0.0 | -0.7 | -31.0 | 0.0 | 0.0 | 0.2 | -6.4 | 0.0 | 0.0 | 0.3 | -1.3 | 0.0 | 0.0 | 0.3 | -1.5 |
| 65 Noncellulosic fibers | 0.0 | 0.0 | 1.0 | 10.2 | 0.0 | 0.0 | 0.7 | 5.0 | 0.0 | 0.0 | 0.4 | 1.5 | 0.0 | 0.0 | 0.4 | 1.3 |
| 66 Drugs | 1.7 | 0.0 | 0.2 | -2.0 | 1.7 | 0.0 | 0.2 | -3.3 | 1.7 | 0.0 | 0.2 | 3.2 | 1.7 | 0.0 | 0.2 | 3.2 |
| 67 Cleaning and toilet products | 2.0 | 0.0 | 0.3 | 2.1 | 2.0 | 0.0 | 0.2 | 0.6 | 2.0 | 0.0 | 0.2 | 2.4 | 2.0 | 0.0 | 0.2 | 2.4 |
| 68 Paints | -0.0 | 0.0 | -0.2 | -10.7 | -0.0 | 0.0 | -0.1 | -6.0 | -0.0 | 0.0 | -0.0 | 2.2 | -0.0 | 0.0 | -0.0 | 1.9 |
| 69 Petroleum refining | -0.7 | 0.0 | 0.3 | 0.2 | -0.7 | 0.0 | 0.1 | -5.2 | -0.7 | 0.0 | 0.2 | 0.4 | -0.7 | 0.0 | 0.2 | 0.2 |
| 70 Fuel oil | -0.7 | 0.0 | 0.0 | 9.3 | -0.7 | 0.0 | -0.0 | -9.0 | -0.7 | 0.0 | -0.0 | 0.4 | -0.7 | 0.0 | -0.0 | 0.3 |
| 71 Paving and asphalt | 0.0 | 0.0 | 0.1 | 3.1 | 0.0 | 0.0 | 0.1 | 2.6 | 0.0 | 0.0 | -0.1 | 0.9 | 0.0 | 0.0 | -0.1 | 0.9 |
| 72 Tires and inner tubes | 3.7 | 0.0 | 1.1 | 12.7 | 3.7 | 0.0 | 0.8 | 6.2 | 3.7 | 0.0 | 0.8 | 7.4 | 3.7 | 0.0 | 0.8 | 7.1 |
| 73 Rubber products | 0.9 | 0.1 | 0.2 | -0.1 | 0.9 | 0.1 | 0.4 | 3.7 | 0.9 | 0.1 | 0.5 | 4.2 | 0.9 | 0.0 | 0.5 | 3.7 |
| 74 Miscellaneous plastic products | 0.4 | 0.0 | 0.2 | 21.7 | 0.4 | 0.0 | -0.1 | -13.6 | 0.4 | 0.0 | 0.0 | 2.9 | 0.4 | 0.0 | 0.0 | 2.6 |
| 75 Leather and ind. leather products | 0.0 | 0.0 | -0.3 | -18.6 | 0.0 | 0.0 | -0.1 | -6.7 | 0.0 | 0.0 | -0.1 | -0.8 | 0.0 | 0.0 | -0.1 | -0.9 |
| 76 Footwear (except rubber) | -0.2 | 0.0 | 0.2 | 1.2 | -0.2 | 0.0 | 0.2 | 1.2 | -0.2 | 0.0 | 0.2 | -0.1 | -0.2 | 0.0 | 0.2 | -0.1 |
| 77 Other leather products | -3.4 | 0.0 | 0.4 | -9.3 | -3.4 | 0.0 | 0.9 | 0.0 | -3.4 | 0.0 | 0.4 | -2.8 | -3.4 | 0.0 | 0.4 | -2.8 |
| 78 Glass | -0.1 | 0.0 | -0.4 | 5.7 | -0.1 | 0.0 | -0.4 | 0.9 | -0.1 | 0.0 | -0.3 | 3.2 | -0.1 | 0.0 | -0.3 | 3.0 |
| 79 Structural clay products | 0.0 | 0.0 | 0.3 | -13.1 | 0.0 | 0.0 | 0.3 | -10.0 | 0.0 | 0.0 | 0.3 | 0.5 | 0.0 | 0.0 | 0.3 | 0.5 |
| 80 Pottery | 0.8 | 0.0 | 0.1 | 1.9 | 0.8 | 0.0 | 0.1 | 1.4 | 0.8 | 0.0 | 0.0 | 4.2 | 0.8 | 0.0 | 0.0 | 3.8 |
| 81 Cement, concrete, gypsum | -0.0 | 0.0 | -0.1 | -7.6 | -0.0 | 0.0 | -0.2 | -10.6 | -0.0 | 0.0 | 0.1 | 0.3 | -0.0 | 0.0 | 0.1 | 0.2 |
| 82 Other stone and clay products | -0.1 | 0.0 | -0.2 | -3.0 | -0.1 | 0.0 | -0.2 | -4.4 | -0.1 | 0.0 | -0.2 | 2.2 | -0.1 | 0.0 | -0.2 | 1.9 |
| 83 Steel | 0.0 | 0.0 | 0.0 | -6.7 | 0.0 | 0.0 | 0.0 | 0.0 | 0.0 | 0.0 | 0.0 | 4.6 | 0.0 | 0.0 | 0.0 | 3.8 |
| 84 Copper | 0.0 | 0.0 | -0.0 | 1.9 | 0.0 | 0.0 | -0.0 | 1.8 | 0.0 | 0.0 | 0.1 | 4.0 | 0.0 | 0.0 | 0.1 | 3.3 |
| 85 Lead | 0.0 | 0.0 | 0.8 | -42.2 | 0.0 | 0.0 | 1.0 | -20.3 | 0.0 | 0.0 | 1.3 | 5.8 | 0.0 | 0.0 | 1.3 | 5.1 |
| 86 Zinc | 0.0 | 0.0 | 1.4 | -27.0 | 0.0 | 0.0 | 1.4 | -20.5 | 0.0 | 0.0 | 1.6 | 7.3 | 0.0 | 0.0 | 1.6 | 6.4 |
| 87 Aluminum | -0.0 | 0.0 | 1.6 | 9.1 | -0.0 | 0.0 | 1.3 | 3.6 | -0.0 | 0.0 | 1.6 | 7.7 | -0.0 | 0.0 | 1.5 | 7.0 |
| 88 Other primary nonferrous metals | 0.0 | 0.0 | -0.1 | 24.4 | 0.0 | 0.0 | -0.1 | 18.3 | 0.0 | 0.0 | -0.0 | 6.7 | 0.0 | 0.0 | -0.0 | 5.7 |
| 89 Other nonferrous roll and draw | 0.0 | 0.0 | -0.4 | 6.7 | 0.0 | 0.0 | -0.4 | 4.1 | 0.0 | 0.0 | -0.4 | 4.2 | 0.0 | 0.0 | -0.4 | 3.2 |
| 90 Nonferrous wire drawing | -0.0 | 0.2 | -0.0 | -7.8 | -0.1 | 0.1 | -0.1 | -2.4 | -0.0 | 0.2 | 0.1 | 2.2 | -0.0 | 0.1 | 0.1 | 1.8 |

| # | Industry | | | | | | | | | | | | | | | | |
|---|----------|----|----|----|----|----|----|----|----|----|----|----|----|----|----|----|----|
| 91 | Nonferrous casting and forging | 3.4 | -0.4 | 0.0 | 0.0 | 4.4 | -0.4 | 0.0 | 0.0 | 2.9 | -0.6 | 0.0 | 0.0 | 12.6 | -0.5 | 0.0 | 0.0 |
| 92 | Metal cans | 2.8 | 0.1 | 0.0 | 0.0 | 2.8 | 0.1 | 0.0 | 0.0 | 6.0 | 0.3 | 0.0 | 0.0 | 6.9 | 0.2 | 0.0 | 0.0 |
| 93 | Metal barrels and drums | 1.9 | 0.3 | 0.0 | 0.0 | 2.3 | 0.3 | 0.3 | 0.0 | 0.2 | 0.3 | 0.1 | 0.0 | -14.1 | -0.3 | 0.1 | 0.0 |
| 94 | Plumbing and heating equipment | 2.3 | -1.1 | 0.0 | 0.5 | 2.3 | -1.1 | 0.0 | 0.5 | -5.7 | -1.0 | 0.0 | 0.5 | -12.9 | -0.9 | 0.0 | 0.5 |
| 95 | Structural metal products | 2.6 | 0.3 | 1.8 | -0.0 | 3.1 | 0.3 | 2.1 | -0.0 | -6.0 | -0.1 | 1.6 | 0.0 | 7.2 | 0.4 | 1.9 | 0.0 |
| 96 | Screw machine products | 4.9 | 0.7 | 0.0 | -0.0 | 5.9 | 0.7 | 0.0 | -0.0 | 6.7 | 0.6 | 0.0 | -0.0 | 9.4 | 0.7 | 0.0 | -0.0 |
| 97 | Metal stampings | 4.8 | 0.3 | 0.0 | -0.3 | 5.3 | 0.3 | 0.0 | -0.3 | 6.1 | 0.3 | 0.0 | -0.3 | 7.2 | 0.4 | 0.0 | -0.3 |
| 98 | Cutlery, hand tools, hardware | 3.0 | -0.7 | 0.1 | 0.9 | 3.4 | -0.7 | 0.1 | 0.9 | 3.0 | -0.8 | 0.1 | 0.9 | 0.7 | -0.8 | 0.1 | 0.9 |
| 99 | Misc. fabricated wire products | 1.7 | -0.3 | 0.0 | -0.2 | 2.2 | -0.3 | 0.0 | -0.2 | -5.7 | -0.6 | 0.0 | -0.2 | -11.3 | -0.7 | 0.0 | -0.2 |
| 100 | Pipes, valves, fittings | 0.7 | 0.3 | -0.8 | 0.0 | 1.7 | 0.3 | -0.3 | 0.0 | 1.9 | 0.2 | -3.3 | 0.0 | 2.8 | 0.2 | -3.8 | 0.0 |
| 101 | Other fabricated metal products | 3.3 | 0.4 | -0.1 | -0.1 | 3.9 | 0.4 | -0.0 | -0.1 | 3.2 | 0.3 | 0.1 | -0.1 | 8.6 | 0.6 | 0.1 | -0.0 |
| 102 | Engines and turbines | 7.0 | -0.3 | 5.5 | 0.1 | 8.3 | -0.2 | 6.2 | 0.1 | 8.8 | 0.0 | 5.3 | 0.1 | 26.0 | 0.2 | 5.8 | 0.1 |
| 103 | Farm machinery | 5.7 | 3.5 | -1.9 | 3.4 | 8.7 | 3.8 | 0.6 | 3.4 | 5.4 | 3.4 | -1.6 | 3.4 | 2.4 | 3.0 | -4.1 | 3.4 |
| 104 | Constr., mine, oilfield machinery | 1.1 | -0.1 | 0.9 | 0.0 | 1.7 | -0.1 | 1.4 | 0.0 | 0.2 | -0.1 | 0.8 | 0.0 | -3.3 | -0.1 | 1.1 | 0.0 |
| 105 | Materials handling machinery | 3.3 | 0.6 | 2.2 | 0.0 | 6.6 | 0.7 | 5.0 | 0.0 | 3.1 | 0.4 | 3.6 | 0.0 | 11.5 | 0.8 | 5.8 | 0.0 |
| 106 | Machine tools, metal cutting | 2.4 | 1.1 | 0.4 | 0.1 | 8.4 | 1.4 | 5.6 | 0.1 | 9.3 | 1.2 | 7.0 | 0.1 | 10.6 | 1.4 | 8.4 | 0.1 |
| 107 | Machine tools, metal forming | 7.2 | 2.0 | 4.0 | 0.0 | 12.3 | 2.4 | 8.3 | 0.0 | 13.3 | 2.4 | 8.2 | 0.0 | 16.0 | 2.6 | 9.0 | 0.0 |
| 108 | Other metal working mach. | 4.3 | 0.6 | 0.4 | -0.2 | 6.3 | 0.6 | 1.2 | -0.2 | 5.7 | 0.5 | 0.7 | -0.2 | 6.4 | 0.6 | 1.3 | -0.2 |
| 109 | Special industrial machinery | 2.2 | -1.3 | 3.2 | 0.0 | 3.9 | -1.3 | 4.6 | 0.0 | 0.6 | -1.1 | 2.2 | 0.0 | 2.7 | -1.2 | 4.0 | 0.0 |
| 110 | Pumps, compressors, blowers | 3.2 | -0.5 | 3.1 | 0.0 | 6.0 | -0.5 | 5.5 | 0.0 | 3.2 | -0.5 | 3.1 | 0.0 | 7.5 | -0.5 | 4.2 | 0.0 |
| 111 | Ball and roller bearings | 3.9 | -0.3 | 0.0 | 0.0 | 5.5 | -0.3 | 0.0 | 0.0 | -3.0 | -0.4 | 0.0 | 0.0 | -0.9 | -0.4 | 0.0 | 0.0 |
| 112 | Power transmission equipment | 3.2 | -0.4 | 0.0 | 0.0 | 4.7 | -0.4 | 0.0 | 0.0 | 3.3 | -0.4 | 0.0 | 0.0 | 1.0 | -0.4 | 0.0 | 0.0 |
| 113 | Industrial patterns | 5.3 | 1.1 | 3.1 | 0.0 | 9.4 | 1.3 | 6.6 | 0.0 | 8.4 | 0.9 | 6.1 | 0.0 | 4.3 | 0.8 | 6.0 | 0.0 |
| 114 | Computers and related machines | -0.5 | -0.2 | -0.1 | 0.0 | 0.1 | -0.2 | 0.3 | 0.0 | -0.1 | -0.3 | -0.5 | 0.0 | 1.5 | -0.2 | 0.6 | 0.0 |
| 115 | Other office machinery | 9.1 | 3.8 | 4.5 | 0.1 | 11.6 | 4.0 | 6.7 | 0.1 | 9.6 | 3.6 | 6.1 | 0.1 | 12.2 | 3.8 | 8.0 | 0.1 |
| 116 | Service industry machinery | 5.7 | -0.1 | 2.3 | 2.3 | 7.2 | -0.1 | 3.5 | 2.3 | 3.1 | -0.2 | 3.1 | 2.3 | 18.4 | 0.1 | 4.4 | 2.3 |
| 117 | Machine shop products | 4.2 | 0.3 | -0.0 | -0.0 | 5.1 | 0.3 | -0.0 | -0.0 | 13.4 | 0.4 | -0.0 | -0.0 | 31.8 | 0.9 | -0.0 | -0.0 |
| 118 | Electrical measuring instruments | 6.0 | -0.3 | 4.5 | 0.0 | 9.0 | -0.2 | 6.9 | 0.0 | 11.7 | -0.2 | 6.7 | 0.0 | 14.7 | -0.0 | 8.2 | 0.0 |
| 119 | Transformers and switchgear | 16.4 | 0.9 | 15.0 | 0.0 | 18.4 | 1.0 | 16.6 | 0.0 | 13.1 | 0.9 | 16.1 | 0.0 | 21.5 | 1.2 | 17.1 | 0.0 |
| 120 | Motors and generators | 3.0 | 0.3 | 0.8 | -0.0 | 4.8 | 0.4 | 1.6 | -0.0 | 6.4 | 0.5 | 0.7 | -0.0 | 2.5 | 0.3 | 1.0 | -0.0 |
| 121 | Industrial controls | 6.1 | -0.7 | 1.0 | 0.0 | 8.8 | -0.7 | 1.8 | 0.0 | 3.4 | -0.7 | 1.0 | 0.0 | 18.1 | -0.6 | 1.2 | 0.0 |
| 122 | Welding appl., graphite products | 4.5 | 0.9 | 1.5 | -0.0 | 7.3 | 1.0 | 3.7 | -0.0 | 0.7 | 0.7 | 2.5 | -0.0 | 0.6 | 0.7 | 3.1 | -0.0 |
| 123 | Household appliances | 1.8 | -0.7 | 0.1 | 1.1 | 1.9 | -0.7 | 0.2 | 1.1 | 0.8 | -0.7 | 0.3 | 1.1 | 2.0 | -0.6 | 0.3 | 1.1 |
| 124 | Electrical lighting and wiring equip. | 1.4 | 0.8 | 0.2 | -0.8 | 1.7 | 0.8 | 0.3 | -0.8 | -3.3 | 0.4 | 0.2 | -0.8 | 5.4 | 1.0 | 0.3 | -0.8 |
| 125 | Radio and TV receiving | 3.9 | 1.0 | 0.0 | 2.2 | 4.1 | 1.0 | 0.1 | 2.2 | 3.0 | 0.9 | 0.1 | 2.2 | -0.6 | 0.7 | 0.2 | 2.2 |

Table 3.1 (continued).

| No. | Title | (a) Constant coefficient and predicted output PCE | PDE | Inventory | Output | (b) Predicted coefficient and predicted output PCE | PDE | Inventory | Output | (c) Balanced coefficient and predicted output PCE | PDE | Inventory | Output | (d) Balanced coefficient and actual output PCE | PDE | Inventory | Output |
|---|---|---|---|---|---|---|---|---|---|---|---|---|---|---|---|---|---|
| 126 | Phonograph records | 14.3 | 0.0 | -2.1 | 17.1 | 14.3 | 0.0 | -2.1 | 17.1 | 14.3 | 0.0 | -2.2 | 12.9 | 14.3 | 0.0 | -2.2 | 12.8 |
| 127 | Communication equipment | 0.1 | 3.5 | 0.2 | 14.3 | 0.1 | 1.8 | 0.1 | 3.1 | 0.1 | 2.8 | 0.1 | 4.1 | 0.1 | 1.8 | 0.1 | 2.6 |
| 128 | Electronic components | -0.7 | 0.2 | 0.7 | -0.7 | -0.7 | 0.1 | 0.1 | -6.9 | -0.7 | 0.2 | 0.6 | 2.4 | -0.7 | 0.1 | 0.6 | 1.7 |
| 129 | Batteries | -0.3 | 2.0 | 1.4 | 9.1 | -0.3 | 1.5 | 0.9 | 2.5 | -0.3 | 1.7 | 0.9 | 4.2 | -0.3 | 1.0 | 0.9 | 3.2 |
| 130 | Engine electrical equipment | 0.1 | 0.0 | 5.9 | 22.8 | 0.1 | 0.0 | 4.8 | 13.0 | 0.1 | 0.0 | 5.4 | 10.4 | 0.1 | 0.0 | 5.3 | 9.9 |
| 131 | X-ray, electrical equipment, n.e.c. | -2.9 | -0.7 | -0.5 | -2.4 | -2.9 | -1.2 | -0.5 | -5.9 | -2.9 | -1.8 | -0.5 | -4.6 | -2.9 | -2.7 | -0.5 | -5.6 |
| 132 | Truck, bus, trailer bodies | 0.0 | 7.2 | 5.0 | 15.0 | 0.0 | 5.0 | 4.8 | 12.6 | 0.0 | 7.2 | 4.1 | 11.5 | 0.0 | 4.7 | 3.9 | 8.8 |
| 133 | Motor vehicles | 3.4 | 1.0 | 0.5 | 8.6 | 3.4 | 0.5 | 0.4 | 7.4 | 3.4 | 0.8 | 0.5 | 7.0 | 3.4 | 0.5 | 0.4 | 6.5 |
| 134 | Aircraft | 0.3 | 7.3 | 3.0 | 10.7 | 0.3 | 7.0 | 3.0 | 10.3 | 0.3 | 4.0 | 2.9 | 7.2 | 0.3 | 1.0 | 2.5 | 3.9 |
| 135 | Aircraft engines | 0.0 | 0.0 | 0.1 | 5.9 | 0.0 | 0.0 | 0.1 | 5.9 | 0.0 | 0.0 | 0.3 | 2.3 | 0.0 | 0.0 | 0.3 | 1.8 |
| 136 | Aircraft equipment, n.e.c. | 0.0 | 0.0 | -0.2 | 18.4 | 0.0 | 0.0 | -0.2 | 8.7 | 0.0 | 0.0 | -0.3 | 2.6 | 0.0 | 0.0 | -0.3 | 1.5 |
| 137 | Ship and boat building | 0.8 | -19.7 | 0.4 | -12.5 | 0.8 | -18.2 | 0.4 | -14.4 | 0.8 | -9.8 | 0.2 | -9.0 | 0.8 | -9.8 | 0.2 | -9.0 |
| 138 | Railroad equipment | 0.0 | -9.9 | -0.1 | -15.4 | 0.0 | -8.8 | -0.0 | -11.3 | 0.0 | -3.2 | -0.1 | -3.9 | 0.0 | -5.3 | -0.0 | -6.5 |
| 139 | Cycles, trans. equipment, n.e.c. | 33.6 | -1.5 | 1.0 | 37.6 | 33.6 | -5.3 | 1.0 | 32.9 | 33.6 | -2.2 | 1.0 | 39.9 | 33.6 | -2.7 | 1.0 | 39.3 |
| 140 | Trailer coaches | 34.6 | 1.0 | 0.5 | 35.2 | 34.6 | -0.0 | 0.5 | 34.2 | 34.6 | 0.5 | 0.5 | 36.1 | 34.6 | 0.4 | 0.5 | 35.9 |
| 141 | Engineering and scientific instr. | 0.0 | 2.0 | 0.5 | 2.6 | 0.0 | 1.8 | 0.7 | 7.3 | 0.0 | 2.1 | 0.4 | 4.0 | 0.0 | 1.4 | 0.4 | 2.6 |
| 142 | Mechanical measuring devices | -0.1 | 3.1 | 0.5 | 0.8 | -0.1 | 3.2 | 1.0 | 9.8 | -0.1 | 3.7 | 0.9 | 7.4 | -0.1 | 3.0 | 0.9 | 6.1 |
| 143 | Optical and ophthalmic goods | 4.0 | 3.6 | 10.2 | 29.8 | 4.0 | 1.1 | 9.9 | 26.9 | 4.0 | 4.1 | 9.9 | 19.1 | 4.0 | 2.7 | 9.8 | 17.5 |
| 144 | Medical and surgical instruments | 1.0 | -1.1 | 0.7 | 1.8 | 1.0 | -2.3 | 0.6 | 0.4 | 1.0 | -2.9 | 0.5 | -0.0 | 1.0 | -3.0 | 0.6 | -0.2 |
| 145 | Photographic equipment | -0.3 | 0.6 | 0.4 | 10.6 | -0.3 | -1.0 | -0.2 | -5.0 | -0.3 | 0.4 | 0.0 | 1.0 | -0.3 | -0.1 | 0.0 | 0.4 |
| 146 | Watches and clocks | 0.5 | -0.0 | -2.0 | 8.9 | 0.5 | -0.0 | -2.0 | 1.3 | 0.5 | -0.0 | -2.0 | -1.9 | 0.5 | -0.0 | -2.0 | -2.0 |
| 147 | Jewelry and silverware | 8.5 | 0.0 | 0.0 | 7.1 | 8.5 | 0.0 | 0.0 | 7.0 | 8.5 | 0.0 | 0.0 | 9.1 | 8.5 | 0.0 | 0.0 | 9.1 |
| 148 | Toys, sport, musical instruments | -4.7 | -0.3 | 0.1 | -6.6 | -4.7 | -0.1 | 0.1 | -5.3 | -4.7 | -1.1 | 0.1 | -5.9 | -4.7 | -1.1 | 0.1 | -5.9 |
| 149 | Office supplies | 0.6 | 0.0 | -0.1 | -9.2 | 0.6 | 0.0 | -0.1 | -9.3 | 0.6 | 0.0 | -0.0 | 0.9 | 0.6 | 0.0 | -0.0 | 0.8 |
| 150 | Misc. manufacturing, n.e.c. | 0.3 | 1.2 | 0.6 | -0.3 | 0.3 | 1.2 | 0.4 | -3.5 | 0.3 | 0.7 | 0.6 | 0.3 | 0.3 | 0.5 | 0.6 | -0.0 |
| 151 | Railroads | -0.0 | 0.0 | 0.0 | -10.0 | -0.0 | 0.0 | 0.0 | -5.7 | -0.0 | 0.0 | 0.0 | 2.5 | -0.0 | 0.0 | 0.0 | 2.3 |
| 152 | Busses and local transit | 3.6 | 0.0 | 0.0 | -9.8 | 3.6 | 0.0 | 0.0 | -9.8 | 3.6 | 0.0 | 0.0 | 4.5 | 3.6 | 0.0 | 0.0 | 4.5 |
| 153 | Trucking | 0.3 | 0.0 | 0.0 | 8.4 | 0.3 | 0.0 | 0.0 | -2.0 | 0.3 | 0.0 | 0.0 | 2.9 | 0.3 | 0.0 | 0.0 | 2.7 |
| 154 | Water transportation | -1.9 | 0.0 | 0.0 | -21.2 | -1.9 | 0.0 | 0.0 | -0.1 | -1.9 | 0.0 | 0.0 | -1.1 | -1.9 | 0.0 | 0.0 | -1.3 |

| | | | | | | | | | | | | | | | |
|---|---|---|---|---|---|---|---|---|---|---|---|---|---|---|---|
| 155 Airlines | 5.7 | 0.0 | 0.0 | 16.5 | 5.7 | 0.0 | 0.0 | 12.9 | 5.7 | 0.0 | 0.0 | 7.9 | 5.7 | 0.0 | 7.8 |
| 156 Pipelines | -5.7 | 0.0 | 0.0 | -9.5 | -5.7 | 0.0 | 0.0 | -14.9 | -5.7 | 0.0 | 0.0 | -5.4 | -5.7 | 0.0 | -5.5 |
| 157 Freight forwarding | -0.0 | 0.0 | 0.0 | -2.1 | -0.0 | 0.0 | 0.0 | 7.1 | -0.0 | 0.0 | 0.0 | 2.4 | -0.0 | 0.0 | 2.2 |
| 158 Telephone and telegraph | 2.4 | 0.7 | 0.0 | 6.1 | 2.4 | 0.3 | 0.0 | -5.8 | 2.4 | 0.5 | 0.0 | 3.5 | 2.4 | 0.3 | 3.2 |
| 159 Radio and TV broadcasting | 0.0 | 0.0 | 0.0 | 3.4 | 0.0 | 0.0 | 0.0 | 1.6 | 0.0 | 0.0 | 0.0 | 1.8 | 0.0 | 0.0 | 1.6 |
| 160 Electric utilities | 1.0 | 0.0 | 0.0 | 3.6 | 1.0 | 0.0 | 0.0 | -3.5 | 1.0 | 0.0 | 0.0 | 1.8 | 1.0 | 0.0 | 1.7 |
| 161 Natural gas | -0.4 | 0.0 | 0.0 | 2.0 | -0.4 | 0.0 | 0.0 | -3.5 | -0.4 | 0.0 | 0.0 | 0.3 | -0.4 | 0.0 | 0.1 |
| 162 Water and sewer services | -0.8 | 0.0 | 0.0 | -4.2 | -0.8 | 0.0 | 0.0 | 0.4 | -0.8 | 0.6 | 0.0 | -0.1 | -0.8 | 0.4 | -0.2 |
| 163 Wholesale trade | 0.0 | 0.6 | 0.0 | 6.2 | 0.0 | 0.4 | 0.0 | -0.3 | 0.0 | 0.2 | 0.0 | 0.3 | 0.0 | 0.1 | -0.1 |
| 164 Retail trade | 0.2 | 0.1 | 0.0 | 8.4 | 0.2 | 0.1 | 0.0 | -0.0 | 0.2 | 0.0 | 0.0 | 0.8 | 0.2 | 0.0 | 0.6 |
| 165 Banks, credit agencies, brokers | 0.8 | 0.0 | 0.0 | 12.1 | 0.8 | 0.0 | 0.0 | 6.0 | 0.8 | 0.0 | 0.0 | 1.4 | 0.8 | 0.0 | 1.3 |
| 166 Insurance | -2.7 | 0.0 | 0.0 | 0.4 | -2.7 | 0.0 | 0.0 | -0.5 | -2.7 | 0.0 | 0.0 | -3.1 | -2.7 | 0.0 | -3.1 |
| 167 Owner-occupied dwellings | 1.4 | 0.0 | 0.0 | 1.4 | 1.4 | 0.0 | 0.0 | 1.4 | 1.4 | 0.0 | 0.0 | 1.4 | 1.4 | 0.0 | 1.4 |
| 168 Real estate | 1.3 | 0.0 | 0.0 | -0.0 | 1.3 | 0.0 | 0.0 | -1.7 | 1.3 | 0.0 | 0.0 | 2.0 | 1.3 | 0.0 | 1.9 |
| 169 Hotel and lodging places | -3.1 | 0.0 | 0.0 | -5.8 | -3.1 | 0.0 | 0.0 | -3.2 | -3.1 | 0.0 | 0.0 | -2.6 | -3.1 | 0.0 | -2.7 |
| 170 Personal and repair services | -1.9 | 0.0 | 0.0 | -1.7 | -1.9 | 0.0 | 0.0 | -1.8 | -1.9 | 0.0 | 0.0 | -1.7 | -1.9 | 0.0 | -1.7 |
| 171 Business services | 0.6 | 0.0 | 0.0 | 11.4 | 0.6 | 0.0 | 0.0 | 9.3 | 0.6 | 0.0 | 0.0 | 1.7 | 0.6 | 0.0 | 1.5 |
| 172 Advertising | 0.0 | 0.0 | 0.0 | -4.4 | 0.0 | 0.0 | 0.0 | -1.0 | 0.0 | 0.0 | 0.0 | 1.9 | 0.0 | 0.0 | 1.7 |
| 173 Auto repair | 1.7 | 0.0 | 0.0 | 3.1 | 1.7 | 0.0 | 0.0 | 3.4 | 1.7 | 0.0 | 0.0 | 2.4 | 1.7 | 0.0 | 2.3 |
| 174 Movies and amusements | 2.7 | 0.0 | 0.0 | 4.0 | 2.7 | 0.0 | 0.0 | 5.1 | 2.7 | 0.0 | 0.0 | 3.7 | 2.7 | 0.0 | 3.7 |
| 175 Medical services | 4.2 | 0.0 | 0.0 | 4.1 | 4.2 | 0.0 | 0.0 | 4.3 | 4.2 | 0.0 | 0.0 | 4.2 | 4.2 | 0.0 | 4.2 |
| 176 Private schools and NPO | -4.3 | 0.0 | 0.0 | -3.4 | -4.3 | 0.0 | 0.0 | -3.6 | -4.3 | 0.0 | 0.0 | -4.3 | -4.3 | 0.0 | -4.3 |
| 177 Post office | 0.3 | 0.0 | 0.0 | -0.3 | 0.0 | 0.0 | 0.0 | 1.2 | 0.3 | 0.0 | 0.0 | 1.2 | 0.3 | 0.0 | 1.0 |
| 178 Federal government enterprises | -0.0 | 0.0 | 0.0 | 3.1 | -0.0 | 0.0 | 0.0 | -0.8 | -0.0 | 0.0 | 0.0 | 1.5 | -0.0 | 0.0 | 1.4 |
| 179 NA[b] | 0.0 | 0.0 | 0.0 | 0.0 | 0.0 | 0.0 | 0.0 | 0.0 | 0.0 | 0.0 | 0.0 | 0.0 | 0.0 | 0.0 | 0.0 |
| 180 ST and LOC electric utilities | 2.7 | 0.0 | 0.0 | -14.2 | 2.7 | 0.0 | 0.0 | -10.5 | 2.7 | 0.0 | 0.0 | 3.6 | 2.7 | 0.0 | 3.5 |
| 181 Noncompetitive imports | 3.6 | 0.0 | 0.0 | -10.5 | 3.6 | 0.0 | 0.0 | -10.0 | 3.6 | 0.0 | 0.0 | 3.7 | 3.6 | 0.0 | 3.7 |
| 182 Business travel (dummy) | 0.0 | 0.0 | 0.0 | 8.1 | 0.0 | 0.0 | 0.0 | 5.6 | 0.0 | 0.0 | 0.0 | 1.9 | 0.0 | 0.0 | 1.6 |
| 183 Office supplies (dummy) | 0.0 | 0.0 | 0.0 | -17.2 | 0.0 | 0.0 | 0.0 | -17.4 | 0.0 | 0.0 | 0.0 | 0.3 | 0.0 | 0.0 | 0.1 |
| 184 Unimportant ind. (dummy) | 0.0 | 0.0 | 0.0 | -11.0 | 0.0 | 0.0 | 0.0 | -13.0 | 0.0 | 0.0 | 0.0 | 4.2 | 0.0 | 0.0 | 3.5 |
| 185 Computer rental (dummy) | 0.0 | 0.0 | 0.0 | 38.3 | 0.0 | 0.0 | 0.0 | 35.8 | 0.0 | 0.0 | 0.0 | 1.9 | 0.0 | 0.0 | 1.5 |
| Wtd. cum. error | 0.6 | 0.3 | 0.1 | 3.3 | 0.6 | 0.2 | 0.1 | 0.4 | 0.6 | 0.3 | 0.2 | 2.1 | 0.6 | 0.2 | 1.8 |
| Wtd. abs. error | 1.5 | 0.4 | 0.2 | 6.5 | 1.5 | 0.3 | 0.2 | 4.3 | 1.5 | 0.4 | 0.2 | 3.1 | 1.5 | 0.3 | 2.8 |

[a] n.e.c. = not elsewhere classified.
[b] No definition.

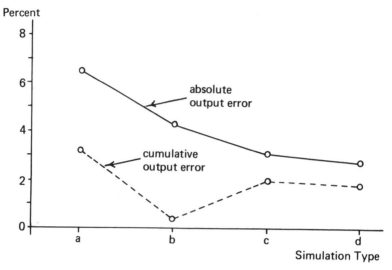

Fig. 3.1. Overall prediction errors of complete model simulations.

the one of PDE (0.4 percent) or inventory changes (0.2 percent) does not indicate inferior PCE equations but rather reflects the fact that PCE is a much larger component of total final demand than PDE or inventory change. Indeed, the errors for PDE are zero in many sectors simply because some of those particular sectors do not produce capital goods at all.

Because more and more of the forecasting error is removed as we introduce more precise coefficients, the test results in Table 3–1 help to isolate individual error components. The next section, therefore, contains a sector analysis of the simulated forecasts. The behavior of test (b), the model with the across-the-row changed coefficients (which reflects the performance of the standard INFORUM model), will be evaluated by comparing its results with both the constant and balanced coefficient simulations. Finally, the fourth simulation, which uses exogenous outputs in the final demand equations, will be used to evaluate the effect of feedback errors.

## Sector Analysis of Five-Year Average Errors

Only six PCE sectors show relatively large forecasting errors. But the inadequate forecasting behavior of each equation can be understood, for these sectors are affected by data problems, style-determined shifts, introduction of new products, and changes in taste. All other PCE equations are well behaved and only a few show any consistent underpredicting or overpredicting. Also, the across-sector offsetting effect is quite strong, as is proven by the small weighted cumulative error. If a given industry is underpredicted, a related industry will be overpredicted. In short, if measured as a percent of actual output, the overall PCE error is only 0.6

percent (cumulative) and 1.5 percent (absolute). These rather small aggregate errors occur despite the relatively large errors of the six sectors.

There are about 45 sectors that sell more than 10 percent of their output to PDE. Nine of these are equipment-supplying sectors and show unsatisfactory simulation results. Comparisons of simulations (a) and (b) show, however, that the across-the-row changes remove a substantial portion of the errors. Still, a large portion of these errors stems from a few inadequate investment functions, notably from those for Sector 160, Electric utilities (which affect PDE for Sector 119, Transformers and switchgears), and Sectors 163, Wholesale trade, and 164, Retail trade (which affect PDE for Sector 46, Other furniture). Eight sectors (103, 104, 105, 109, 114, 116, 127, and 133) show less than 10 percent error. When comparing the PDE column with the PCE column, however, we must bear in mind that the PCE share of many sectors is larger than the PDE portion of output and that all errors are measured as a percent of actual output. This definition makes all errors additive, so that it is possible to analyze each final demand component's contribution to the total output error. In short, the analysis of the PDE errors has shown that only a few investment equations are responsible for a large portion of the PDE errors, and that at the same time a significant portion of the errors is caused by the constant-coefficient assumption.

The individual errors of the inventory equations are small when measured as a percent of actual output. But again, we must realize that inventory change is a small portion of total output. The percentage error of the inventory equations is quite high. It is, however, well accepted that the change in inventory is the most difficult final demand component to forecast; but because it is a relatively small portion of final demand, the large percentage prediction errors can be tolerated. When measured as a percent of actual output, very few sectors show larger than 3 percent errors, and most of the significant errors belong to small sectors. For example, Sector 133, Motor vehicles, one of the large-inventory sectors, only misses by 0.5 percent, while the largest error—10.2 percent—belongs to Sector 143, Optical and ophthalmic products, a rather small sector accounting for only 0.06 percent of total output. Although the overall behavior of the inventory change errors is satisfactory (0.1 percent average output error), we must note that the equations are estimated through 1971, and therefore this test is not an altogether reliable indication of how well they forecast.

An examination of the output error columns shows a large number of sizeable underpredictions and overpredictions. Fortunately, the largest errors belong to the small, volatile output sectors, such as Sector 4, Cotton (−104.3 percent—probably a data problem) and Sector 44, Wooden containers (−41.2 percent). These sizeable final demand errors naturally result in output errors. But sectors that sell chiefly to intermediate use will have output errors arising from both the final demand errors of their customer sectors and the errors in the input-output coefficients. It is therefore these sectors from which we can best evaluate the across-the-row change technique. With this technique, all coefficients in a row are changed in the same proportion, and that proportion is determined from a logistic curve. The curve is fitted to the ratio of actual intermediate use to what intermediate use would have

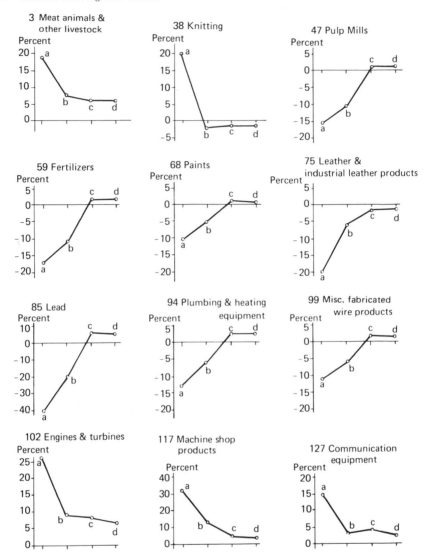

Note: The horizontal axis refers to the four simulations.

Fig. 3.2. Selected output errors of complete model simulations.

been had coefficients remained constant. The curves were fitted over the period for which data were available—1958–1971. They are used in the INFORUM model on rows where annual time series on individual coefficients are not available. Their effect is shown graphically in Figure 3.2, where the vertical axis shows the percent output error from Table 3–1, while the horizontal axis represents the four tests (a) through (d). Sectors with large output errors and large intermediate use were selected for these plots.

All of the industries shown respond favorably to the coefficient-change technique, test (b). The strongest improvement is shown by Sector 3, Meat animals and other livestock products; Sector 38, Knitting; Sector 75, Leather and industrial leather products; Sector 102, Engines and turbines; Sector 117, Machine shop products; and Sector 127, Communication equipment. For the remaining sectors, improvement from across-the-row changes is about one-half of the potential error reduction. This conclusion follows from the fact that the prediction error of test (b)—across-the-row changes—is about halfway between the constant-coefficient and the balanced-coefficient error. Finally, the feedback error effect seems to be rather insignificant: tests (c) and (d) give very similar results. Using the actual output instead of the output predicted by the model in the final demand equations does not appreciably change the forecasts.

### Developing an Income Side

The model forecasts sales in real terms. Substantial progress has been made toward completing the wage, price, and income side of the model. Industry wages are estimated as a function of the consumer price index, the unemployment rate, productivity, and the level of employment. Prices are estimated as a function of unit labor and materials costs and a distributed lag on output.[6]

### Prices

The price computations rely on the quantity forecasts and, once made, enter into the forecasts of consumer demands and of input-output coefficients. Prices are calculated monthly, so that they can respond quickly and dramatically to a change in supply or demand; annual simulations would sleep through the interesting action. In every industry, the specification of the price equation is the same:

$$P_{jt} = c_j + \sum_{k=1}^{18} u_k(UMC_{j,t-k} + ULC_{j,t-k}) + \sum_{k=1}^{18} v_i Q_{j,t-k} \tag{3-4}$$

where

$c_j$ = constant

$P_{jt}$ = price index for industry $j$ in month $t$ (1967 = 1.0)

$Q_{jt}$ = constant-dollar output of industry $j$ in month $t$

$UMC_{jt}$ = unit material cost of industry $j$ in month $t$

$ULC_{jt}$ = unit labor cost of industry $j$ in month $t$

$u_k, v_k$ = distributed lag weights

$k$ = time lag in months

A basic requirement for each industry is that it set a price to cover its costs. The prices of materials and labor are determined by forces beyond the control of the industry. The industry adjusts by choosing the mix of inputs and by setting the price of output. But the adjustment is not made immediately or all at once. In the simulations, it occurs over an 18-month period. When the adjustment is fully registered, the price may have risen proportionately less than cost or proportionately more. This characteristic—the elasticity of price with respect to cost—varies from industry to industry, as does the pattern of the response over time. The shape and sum of these distributed lags are estimated by making a regression analysis of each industry's experience. The analysis will be described later.

The forecasting of prices requires forecasting of the unit materials and labor costs for each industry. But the unit materials cost of an industry is nothing else than the ratio of the average price of the inputs to the price of the output. These input prices are the very same product prices that are the subject of the forecast. The industry's input column of the input-output matrix gives the weights for forming the appropriate average. An important feature to note is that this calculation is not a simultaneous determination of all materials costs and prices. The simultaneity is broken by making one month's prices depend on past months' costs, not on the current month's cost. An industry price is set at the beginning of each month on the basis of prior experience. Once this month's prices are in hand, 185 unit cost figures can be formed with the updated input-output table and set aside for the next month's price computations.

To each industry's unit materials cost must be added its unit labor cost, calculated as the ratio of average wage to labor productivity. Labor productivity is generated in the quantity model to arrive at employment by industry. The wages are the dependent variables of another set of equations and are discussed below.

Cost push is half the story. Demand conditions are the other half. If demand for an industry's product is strong, producers will pass along cost increases. Slack demand will eat away at the cost-induced increases in price, making the observed price rise less than would otherwise be expected. To incorporate this effect into price forecasts, changes in outputs are allowed to have a transitory effect on prices; the long-run effects are near zero, in accord with the idea that the production occurs on a relatively flat portion of the long-run average-cost curves. The distributed lag specification allows both the short-run and long-run characteristics; the weights are positive on recent increments in output to reflect the bidding-up of price and negative on increments of more distant months as long-run adjustments are made. The outputs that enter the industry price equations for forecasting are derived from the quantity model.

**Wages**

The industry wages are the subject of regression analysis. The explanatory variables are the consumer price index, the national unemployment rate, aggregate labor productivity, and employment in the particular industry. In each of the four variables, a change is allowed 12 months to have its full effect on the wage. Three of these variables are forecast in the quantity model; the fourth, the consumer price index, is an average of all the prices and is calculated from the forecasts. Again, the simultaneity is broken by excluding the current month's consumer price index from the wage computation and making wages depend wholly on lagged values.

The consumer price index and aggregate productivity are included in the wage equations to set the overall wage level; workers in all industries participate in increases coming from these two forces. The reciprocal of the unemployment rate enters to catch the short-term pressures of the labor market.

For all four variables, polynomial distributed lags were estimated. In the forecast, the wages are thereby made to reflect recent price experience as well as technological progress, productivity, and the varying levels of industry demand. The unit labor cost can then be added to the unit materials cost to complete the measure of unit production cost used in the price equations.[7]

## The Simulations

The computations simulate several interactions that are important in any effort to anticipate price changes. Money wages rise roughly in proportion to aggregate productivity increase plus inflation. For industries whose prices have risen more than the average, household demand declines relative to what would otherwise have occurred. Producers substitute relatively cheaper inputs for relatively more expensive ones where possible. Thus, both intermediate and final demands respond to relative prices, and output levels are affected. These results are carried into the investment decisions. Because the capital stock needed for production will decline with output in industries experiencing rapid price increases, productivity will grow more slowly and employment more quickly. Meanwhile, all the prices will be felt by the producing industries as costs of production. Thus, inflated prices cause further inflation. As they work through demand adjustments, they create self-correcting forces. The complete simulations show the net effect as it appears over a period of several years.

## Regression Analysis

Before discussing some of the simulations, it would be appropriate to present the results of the regression analysis of the historical behavior of industry prices. To estimate the lags between unit cost and price and between output and price, monthly times series were needed for three variables for each input-output sector: the price, the total payroll, and the quantity of output. Prices came from the U.S. Department of Labor. Wholesale prices were used where possible; consumer prices were used for all services. Industry payrolls were taken from the U.S. Department of Labor data on employment, earnings, and hours. Payrolls were made as comprehensive as possible, including overtime pay. Since the salaries of supervisors and executives are not covered, these employees were counted in at the same average earnings as production workers. This method gave the derived unit labor cost at the more nearly correct magnitude, although the variable certainly fails to reflect shifts between these two classes of employees. The third set of industry data,

Table 3.2. Characteristics of Price Equations

| | Sector | | | Unit cost elasticity | Average lag | Output elasticity |
|---|---|---|---|---|---|---|
| Number | Title | $\bar{R}^2$ | $\rho$ | | | |
| 7 | Fruits, vegetables, and other crops | 0.548 | 0.55 | 0.75 | 15.4 | 0.22 |
| 15 | Crude petroleum, natural gas | 0.547 | 0.91 | 0.75 | 11.4 | 0.08 |
| 16 | Stone and clay mining | 0.985 | 0.94 | 1.01 | 7.3 | 0.25 |
| 17 | Chemical fertilizer mining | 0.646 | 0.96 | 1.14 | 13.6 | 0.25 |
| 21 | Ammunition | 0.935 | 0.84 | 0.98 | 11.5 | −0.11 |
| 25 | Canned and frozen foods | 0.950 | 0.92 | 1.08 | 6.9 | 0.17 |
| 26 | Grain mill products | 0.725 | 0.86 | 0.75 | 3.9 | 0.25 |
| 27 | Bakery products | 0.966 | 0.94 | 1.25 | 3.8 | 0.13 |
| 28 | Sugar | 0.602 | 0.83 | 1.01 | 5.4 | 0.25 |
| 29 | Confectionery products | 0.808 | 0.87 | 0.89 | 4.4 | 0.23 |
| 30 | Alcoholic beverages | 0.963 | 0.91 | 0.75 | 9.9 | −0.00 |
| 31 | Soft drinks and flavorings | 0.982 | 0.88 | 1.25 | 10.4 | 0.23 |
| 32 | Fats and oils | 0.562 | 0.80 | 1.25 | 3.6 | −0.12 |
| 35 | Broad and narrow fabrics | 0.581 | 0.95 | 0.94 | 10.3 | −0.02 |
| 36 | Floor coverings | 0.818 | 0.95 | 0.75 | 4.8 | −0.10 |
| 38 | Knitting products | 0.849 | 0.95 | 0.75 | 3.6 | −0.16 |
| 39 | Apparel | 0.972 | 0.92 | 1.24 | 7.5 | 0.18 |
| 40 | Household textiles | 0.689 | 0.94 | 0.75 | 7.1 | 0.03 |
| 43 | Millwork and wood products | 0.901 | 0.98 | 0.80 | 5.3 | 0.13 |
| 44 | Wooden containers | 0.973 | 0.76 | 0.75 | 9.2 | −0.01 |
| 45 | Household furniture | 0.885 | 0.94 | 1.04 | 8.3 | 0.19 |
| 46 | Other furniture | 0.986 | 0.87 | 0.83 | 9.7 | 0.20 |
| 47 | Pulp mill products | 0.538 | 0.95 | 0.75 | 13.1 | 0.0 |
| 48 | Paper and paperboard mills | 0.962 | 0.94 | 0.75 | 9.9 | 0.19 |
| 59 | Fertilizers | 0.968 | 0.85 | 0.91 | 9.2 | 0.23 |
| 61 | Misc. chemical products | 0.704 | 0.98 | 0.75 | 15.3 | 0.25 |
| 65 | Noncellulosic fibers | 0.484 | 0.94 | 0.75 | 8.4 | −0.07 |
| 66 | Drugs | 0.832 | 0.95 | 0.75 | 2.3 | −0.06 |
| 67 | Cleaning and toilet products | 0.986 | 0.91 | 0.75 | 7.1 | 0.13 |
| 69 | Petroleum refining | 0.621 | 0.86 | 1.08 | 5.6 | −0.12 |
| 74 | Misc. plastic products | 0.757 | 0.93 | 0.75 | 13.3 | 0.17 |
| 75 | Leather and industrial leather products | 0.653 | 0.95 | 1.25 | 12.0 | 0.24 |
| 76 | Footwear (except rubber) | 0.990 | 0.74 | 1.25 | 6.0 | 0.22 |
| 77 | Other leather products | 0.974 | 0.74 | 0.90 | 8.0 | −0.02 |
| 78 | Glass | 0.916 | 0.96 | 1.25 | 5.4 | 0.14 |
| 82 | Other stone and clay products | 0.853 | 0.95 | 1.25 | 6.0 | −0.01 |
| 83 | Steel | 0.917 | 0.95 | 1.25 | 10.1 | 0.25 |
| 88 | Primary nonferrous metals | 0.490 | 0.99 | 0.75 | 6.2 | 0.25 |
| 93 | Metal barrels and drums | 0.952 | 0.85 | 1.05 | 7.4 | 0.12 |
| 94 | Plumbing and heating equipment | 0.898 | 0.92 | 0.97 | 2.6 | −0.00 |
| 95 | Screw machine products | 0.923 | 0.98 | 1.25 | 4.2 | 0.25 |
| 98 | Cutlery, hand tools, hardware | 0.953 | 0.86 | 1.07 | 7.1 | 0.21 |
| 100 | Pipes, valves, fittings | 0.936 | 0.94 | 1.25 | 13.0 | 0.07 |

| | Sector | | | | | |
|---|---|---|---|---|---|---|
| Number | Title | $\bar{R}^2$ | $\rho$ | Unit cost elasticity | Average lag | Output elasticity |
| 102 | Engines and turbines | 0.961 | 0.95 | 0.94 | 3.4 | 0.22 |
| 103 | Farm machinery | 0.979 | 0.95 | 1.03 | 5.9 | 0.24 |
| 104 | Construction, mine, oilfield machinery | 0.991 | 0.92 | 1.18 | 12.3 | 0.25 |
| 106 | Machine tools, metal cutting | 0.993 | 0.71 | 1.16 | 9.8 | −0.25 |
| 107 | Machine tools, metal forming | 0.978 | 0.60 | 0.75 | 8.8 | −0.15 |
| 108 | Other metal working machines | 0.949 | 0.97 | 1.07 | 9.3 | 0.16 |
| 110 | Pumps, compressors, blowers | 0.946 | 0.99 | 1.25 | 9.0 | 0.25 |
| 112 | Power transmission equipment | 0.768 | 1.00 | 1.25 | 5.2 | 0.25 |
| 113 | Industrial patterns | 0.961 | 0.97 | 1.25 | 6.9 | 0.25 |
| 114 | Computers and related machines | 0.468 | 0.87 | 0.75 | 11.1 | −0.14 |
| 117 | Machine shop products | 0.893 | 0.95 | 1.25 | 5.8 | 0.25 |
| 118 | Electrical measuring instruments | 0.940 | 0.70 | 0.86 | 11.8 | 0.25 |
| 119 | Transformers and switchgears | 0.472 | 0.98 | 0.88 | 2.3 | −0.04 |
| 121 | Industrial controls | 0.782 | 0.95 | 1.22 | 10.0 | −0.10 |
| 123 | Household appliances | 0.926 | 0.90 | 0.84 | 4.4 | 0.03 |
| 124 | Electrical lighting and wiring equipment | 0.889 | 0.92 | 1.25 | 15.9 | 0.20 |
| 126 | Phonograph records | 0.429 | 0.98 | 0.75 | 4.9 | 0.25 |
| 127 | Communication equipment | 0.905 | 0.86 | 0.75 | 3.9 | 0.22 |
| 129 | Batteries | 0.704 | 0.97 | 0.75 | 14.9 | 0.13 |
| 131 | X-ray and other electrical equipment, n.e.c.[a] | 0.947 | 0.82 | 1.08 | 8.0 | 0.25 |
| 133 | Motor vehicles | 0.965 | 0.74 | 0.82 | 12.0 | −0.05 |
| 138 | Railroad equipment | 0.817 | 0.97 | 1.25 | 8.3 | 0.25 |
| 139 | Cycles and transportation equipment, n.e.c. | 0.529 | 0.98 | 0.75 | 6.5 | 0.03 |
| 145 | Photographic equipment | 0.886 | 0.97 | 0.75 | 15.4 | 0.17 |
| 148 | Toys, sport, musical instruments | 0.985 | 0.91 | 0.94 | 11.8 | 0.13 |
| 149 | Office supplies | 0.420 | 0.85 | 0.75 | 10.9 | 0.13 |
| 150 | Misc. manufacturing, n.e.c.[a] | 0.932 | 0.91 | 0.75 | 16.0 | 0.24 |
| 155 | Airlines | 0.912 | 0.87 | 0.88 | 13.8 | 0.13 |
| 158 | Telephone and telegraph | 0.549 | 0.81 | 0.75 | 14.5 | 0.03 |
| 160 | Electric utilities | 0.918 | 0.88 | 0.75 | 15.8 | 0.10 |
| 161 | Natural gas | 0.743 | 0.95 | 0.75 | 12.7 | 0.25 |
| 162 | Water and sewer services | 0.976 | 0.93 | 1.11 | 10.8 | 0.25 |
| 165 | Banks, credit agencies and brokers | 0.863 | 0.98 | 0.88 | 11.3 | 0.25 |
| 166 | Insurance | 0.978 | 0.97 | 1.25 | 2.9 | 0.25 |
| 167 | Owner-occupied dwellings | 0.994 | 0.85 | 0.85 | 3.7 | 0.09 |
| 168 | Real estate | 0.950 | 0.98 | 0.75 | 15.1 | 0.07 |
| 170 | Personal and repair services | 0.984 | 0.98 | 1.03 | 9.0 | 0.25 |

[a] n.e.c. = not elsewhere classified.

monthly output, comes from the Board of Governors of the Federal Reserve System's Industrial Production Index program. Both seasonalized and unadjusted series were used, the latter for the denominator of the unit labor cost and the former for output, the independent variable. The other explanatory variable, unit cost, was formed for each month by weighting all prices for the month according to the industry's input coefficients and adding payroll per unit of output, that is, by adding unit materials costs to unit labor costs. Where the price variable was missing, there could not, of course, be an equation. The period covered by the data was generally 1958 to 1971. The results of the regression are shown in Table 3.2 for typical sectors. The first column shows the $\bar{R}^2$; the second column, the $\rho$ or autocorrelation coefficient of the errors. Since the data are monthly, and no lagged values of the dependent variable are used, it is not surprising that the $\rho$ values are high. They are used to move the forecasts gradually from the last actual value back to the value forecast in the equation.

The unit cost, output, and price were all represented as index numbers with the average value during 1967 set to unity. The sum of the individual lags could therefore be interpreted as the elasticities. If the sum of the 18 weights on unit cost is 0.9, then 90 percent of cost increases will be permanently recorded in a higher price for the output. Such a result would imply that between 1958 and 1971 the industry had been unable to pass along all increases in cost; that inability is carried forward into the forecasts. The third column of Table 3.2 shows these elasticities of price with respect to cost. The estimation method constrained them to lie between 0.75 and 1.25. The weighted average length of lag in months, $\sum k u_k / \sum u_k$ is shown in the fourth column.

When it comes to the output variable, the weights have the analogous interpretation, and so does the sum of the weights. However, the elasticities of price with respect to output, shown in column 5 of Table 3.2, are much less than unity. They should be near zero, meaning that if enough time is allowed for installing new capacity, a much greater quantity could be provided at nearly as low a price. In the months immediately following an increase in output, the price may rise, but the effect is transitory.

### Oil and Food in the U.S. Inflation

The price-wage model will be illustrated by using it to answer the question of how much of the 1973 inflation can really be traced to uncontrollable factors in the agricultural and crude oil sectors. To answer this question, it is necessary to simulate the economy under the hypothetical condition that price increases in these sectors did not occur. In the government price control program, the fourth quarter of 1972 was adopted as a base period for measuring cost changes. The simulations use November 1972 to match this base. The hypothetical case, then, consisted of making a complete simulation forecast for the years 1971 to 1976, but for seven sectors—Dairy farms (raw milk); Poultry and eggs; Meat animals; Cotton; Grains; Fruits and vegetables and other crops; and Crude petroleum and natural gas (crude oil)—the results from the price equations were replaced after November 1972 by that month's price. Under this assumption, it appears that wholesale prices would have risen only 5 percent in 1973. The actual increase was over 18 percent. The comparison is misleading, however, for the model cannot quite reproduce the actual rate. If the second simulation forecast is made using actual raw materials price movements during 1973, wholesale prices appear to rise 15 percent instead of 18 percent. This difference is the error of the model. The difference between 5 percent and 15 percent is attributable to inflation in the raw materials sectors whose prices were allowed to change.

A third simulation (in which agricultural prices remain at their November 1972 levels, but inflation is allowed to occur in crude oil prices) traces out a path between the first two. Under these conditions, wholesale prices would have risen 6 percent during 1973. Compared with 5 percent, the additional aggregate effect is much less dramatic than the headlines and price quotations of November and December 1973. During 1974, however, a strong effect can be expected. The first simulation led to a prediction of 9 percent, compared with the original 4 percent. Since the second simulation differs from the third only by the agricultural price changes, the further observation can be made that those price changes will increase this 9 percent to 14 percent. The best prediction for 1974 is this 14 percent inflation, a rate hardly less than the rate for 1973. In each year, two-thirds of the increase is traceable to raw goods; but in 1973, oil played a negligible part, while in 1974, its part will be as large as that of the agricultural sectors. The full course of the simulation is shown in Figure 3.3.

## Applications of INFORUM

Forecasts with the INFORUM model have been used to aid business and government planning. A model such as this one provides a consistent framework for assessing the performance of various sectors of the economy. Altering various assumptions under which the model is run and comparing the effects on demand and employment is a simple application of the model. Although this approach suggests answers to some of the questions that concern users of the model, the standard format is often an incomplete answer to many specific problems. In many cases, the model can be expanded and tailored to the problem at hand. Several questions to which the model has been adapted include forecasting company profits, selecting a merger partner, increasing the product detail, and assessing environmental impacts. These areas have been fully described by Almon et al. (1985), and will be only briefly outlined here. A topic of more current interest—the energy crisis—will then be discussed in the context of the INFORUM model.

One sponsor has used the model with a profit model of its own to evaluate the effect of alternative assumptions and government policies on the company's profits. Its specific products are related to one of the 185 sectors in the INFORUM model, or to particular cells in the row of that sector. This relationship is used to determine the growth sales of the company's products. The information is then fed into the company's model to forecast profits.

A similar approach was used to aid another company in the selection of a merger partner. Here, the sales of various product lines of prospective partners were related to the sectors in the INFORUM model. Then a composite sales forecast

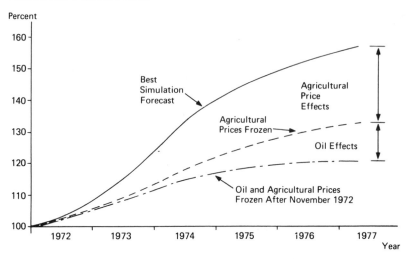

Fig. 3.3. Wholesale price index simulations, 1972–1977.

was made for each company. This type of approach was also used to study diversification and the sensitivity of the products under consideration to cyclical effects. The use of quadratic programming enabled us to specify a trade-off between rapid growth and risk aversion. We hope to apply this technique to the analysis of portfolio management.

Often many business decisions require knowledge of the demand for a product at a much more detailed level than it is possible to incorporate into a standard model. For example, investment decisions by a paper company may depend not merely on how much paper will be required by book printing, but on what kind of paper— whether it is a grade used to make paperbacks, or a much higher quality used to make textbooks. In such a case, when detailed product information on end use is available, coefficients can be derived to show how much of that product is sold to each of the 185 sectors in the model. In this way, the number of rows in the model can be expanded in areas of special interest. When information on such coefficients is available over time, trends in the coefficients can be estimated. Future demand for the products is obtained from the forecast coefficients and the forecasts of outputs of the 185 sectors of the model. Studies of this nature have been done for paper and plastics.

The Environmental Protection Agency has developed a model known as SEAS—Strategic Environmental Assessment System—to study the impact of different growth policies and compositions on the environment. The INFORUM model is the "front end," which provides forecasts of outputs to this system. The SEAS model then forecasts the generation of pollutants and their regional impacts.

### The Energy Crisis

The Arab oil embargo, shortages of domestic refinery capacity, and rapid growth in the demand for petroleum products combined in late 1973 to create an "energy crisis." Predictions of a severe recession were commonplace. Quite naturally, many users of the INFORUM model wished to know what the model had to say about the consequences of this energy crisis. Accordingly, imports of crude and refined petroleum were revised to reflect the embargo. Crude imports were reduced by about 20 percent; refined petroleum imports were reduced by 18 percent. The 1974 price of crude petroleum was exogenously introduced at double its 1973 level and fed into the price model to determine the effect on other prices. This new price forecast was introduced into the real model. Hence, the consumption forecasts were altered to reflect the revised relative prices according to their respective price elasticities.

As it was then structured, the model's results were not encouraging. It required an unreasonably large increase in domestic crude petroleum production to meet the needs of the economy for 1974. Alternatively, if all demands were reduced to a level of possible crude petroleum production, a large amount of unemployment was generated. However, the model had not taken a significant factor into account. It had assumed a zero elasticity of intermediate demand with respect to the relative price of petroleum. Certainly there are opportunities for the substitution of coal for oil, and of less petroleum-intensive products and services for those relatively more petroleum-intensive. There are other potentials for conserving, such as reducing fuel oil used for space heating. Because we lacked specific information about the magnitudes of the elasticities, we introduced some assumptions. These assumed elasticities are shown in Table 3.3.

Rerunning the model with the assumed elasticities permitted intermediate demand to react to relative price changes. Now, with a 112 percent increase in the price of crude petroleum, domestic production is kept to 2.7 percent of its 1973 level. Tables 3.4 and 3.5 and Figure 3.4 show the format of the INFORUM forecasts. Table 3.4, which shows the forecast made in April 1974, lists final demands, employment, and prices. It also lists the important assumptions that were exogenous to this forecast. Table 3.5 lists the sales of energy-related sectors to the important purchasers. Figure 3.4 shows historical and forecast energy outputs.

Some problems with the data should be noted here. The base year of the model is 1971, the most recent year for which we have complete information on outputs, in this case from the *Annual Survey of Manufactures: 1971* (1973). However, we have drawn on other statistics published in the *Survey of Current Business (SCB)* (1974)

Table 3.3.  Assumed Relative Price Elasticities for Intermediate Demand

| Energy-related sector | Substitute | Elasticity |
| --- | --- | --- |
| Fuel oil used for electricity | Coal | $-2.0$ |
| Fuel oil used for process heat | Coal | $-0.3$ |
| Fuel oil used for space heat[a] | None | $-0.3$ |
| Gasoline and other petroleum products | None | $-0.1$ |
| Trucking | Railroads | $-0.2$ |
| Electric utilities | None | $-0.2$ |
| Noncellulosic fibers | Cellulosic fibers | $-0.2$ |

[a] Strout, Alan M. 1961. Weather and the Demand for Space Heat. *The Review of Economics and Statistics* 43(2) [May]: 185–192.

# 14 Coal Mining

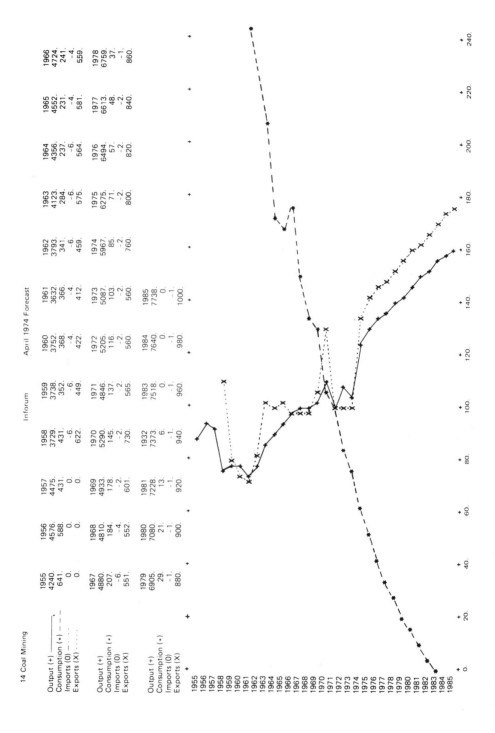

Inforum                    April 1974 Forecast

| | 1955 | 1956 | 1957 | 1958 | 1959 | 1960 | 1961 | 1962 | 1963 | 1964 | 1965 | 1966 |
|---|---|---|---|---|---|---|---|---|---|---|---|---|
| Output (+) ——— | 4240. | 4576. | 4475. | 3729. | 3738. | 3752. | 3632. | 3793. | 4123. | 4356. | 4552. | 4724. |
| Consumption (+) - - - | 641. | 588. | 431. | 431. | 352. | 368. | 366. | 341. | 284. | 237. | 231. | 241. |
| Imports (0) —— | 0. | 0. | 0. | -6. | -6. | -4. | -4. | -6. | -6. | -6. | -4. | -4. |
| Exports (X) ······ | 0. | 0. | 0. | 622. | 449. | 422. | 412. | 459. | 575. | 564. | 581. | 559. |

| | 1967 | 1968 | 1969 | 1970 | 1971 | 1972 | 1973 | 1974 | 1975 | 1976 | 1977 | 1978 |
|---|---|---|---|---|---|---|---|---|---|---|---|---|
| Output (+) | 4880. | 4810. | 4933. | 5290. | 4846. | 5205. | 5087. | 5967. | 6275. | 6494. | 6613. | 6759. |
| Consumption (+) | 207. | 184. | 178. | 145. | 137. | 116. | 103. | 85. | 71. | 57. | 48. | 37. |
| Imports (0) | -6. | -4. | -2. | -2. | -2. | -2. | -2. | -2. | -2. | -2. | -2. | -1. |
| Exports (X) | 551. | 552. | 601. | 730. | 565. | 560. | 560. | 760. | 800. | 820. | 840. | 860. |

| | 1979 | 1980 | 1981 | 1982 | 1983 | 1984 | 1985 |
|---|---|---|---|---|---|---|---|
| Output (+) | 6905. | 7080. | 7228. | 7373. | 7518. | 7640. | 7738. |
| Consumption (+) | 29. | 21. | 13. | 6. | 0. | 0. | 0. |
| Imports (0) | -1. | -1. | -1. | -1. | -1. | -1. | -1. |
| Exports (X) | 880. | 900. | 920. | 940. | 960. | 980. | 1000. |

## 15 Crude Petroleum, Nat. Gas

Output (+) ———
Imports (0) — · —
Exports (X) ·····

| | 1955 | 1956 | 1957 | 1958 | 1959 | 1960 | 1961 | 1962 | 1963 | 1964 | 1965 | 1966 |
|---|---|---|---|---|---|---|---|---|---|---|---|---|
| Output (+) | 10061. | 10726. | 10679. | 10046. | 10842. | 11164. | 11495. | 11897. | 12357. | 12609. | 13001. | 13948. |
| Imports (0) | 0. | 0. | 0. | -1302. | -1302. | -1354. | -1431 | -1604. | -1646. | -1735. | -1802. | -1786. |
| Exports (X) | 0. | 0. | 0. | 32. | 15. | 13. | 14. | 11. | 11. | 10. | 12. | 26. |

| | 1967 | 1968 | 1969 | 1970 | 1971 | 1972 | 1973 | 1974 | 1975 | 1976 | 1977 | 1978 |
|---|---|---|---|---|---|---|---|---|---|---|---|---|
| Output (+) | 14844. | 15414. | 15699. | 16812. | 16705. | 17055. | 17490. | 17957. | 18904. | 19681. | 20479. | 21183. |
| Imports (0) | -1731. | -1897. | -2048. | -2066. | -2555. | -3075. | -4106. | -3200. | -3350. | -3416. | -3482. | -3548. |
| Exports (X) | 134. | 49. | 32. | 51. | 40. | 68. | 72. | 50. | 50. | 53. | 56. | 59. |

| | 1979 | 1980 | 1981 | 1982 | 1983 | 1984 | 1985 |
|---|---|---|---|---|---|---|---|
| Output (+) | 21888. | 22659. | 23430. | 24168. | 24977. | 25795. | 26597. |
| Imports (0) | -3614. | -3680. | -3746. | -3812. | -3878. | -3944. | -4010. |
| Exports (X) | 62. | 65. | 68. | 71. | 74. | 77. | 80. |

April 1974 Forecast

Inforum

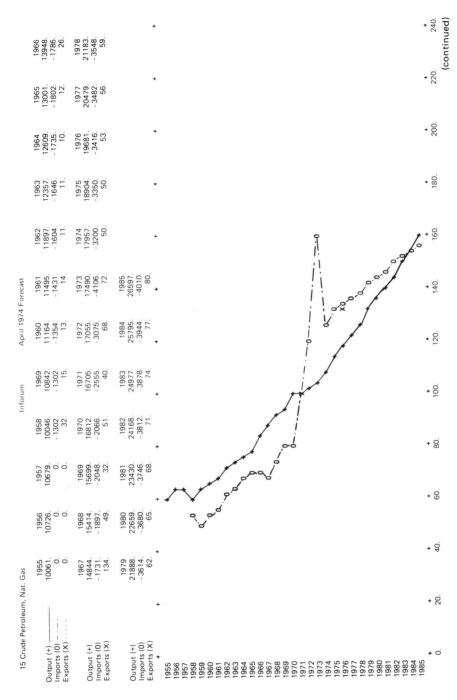

Fig. 3.4. Historical and forecast energy outputs.

(continued)

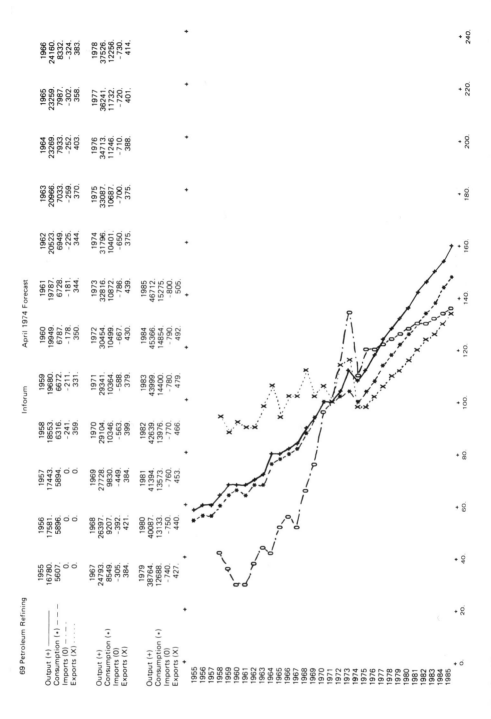

69 Petroleum Refining

Inforum      April 1974 Forecast

| | 1955 | 1956 | 1957 | 1958 | 1959 | 1960 | 1961 | 1962 | 1963 | 1964 | 1965 | 1966 |
|---|---|---|---|---|---|---|---|---|---|---|---|---|
| Output (+) | 16780. | 17581. | 17443. | 18553. | 19680. | 19949. | 19787. | 20523. | 20966. | 23269. | 23259. | 24160. |
| Consumption (•) | 5607. | 5896. | 5894. | 6316. | 6672. | 6787. | 6728. | 6949. | 7033. | 7933. | 7987. | 8332. |
| Imports (0) | 0. | 0. | 0. | -241. | -211. | -178. | -181. | -225. | -259. | -252. | -302. | -324. |
| Exports (X) | 0. | 0. | 0. | 359. | 331. | 350. | 344. | 344. | 370. | 403. | 358. | 383. |

| | 1967 | 1968 | 1969 | 1970 | 1971 | 1972 | 1973 | 1974 | 1975 | 1976 | 1977 | 1978 |
|---|---|---|---|---|---|---|---|---|---|---|---|---|
| Output (+) | 24793. | 26397. | 27728. | 29104. | 29341. | 30454. | 32816. | 31796. | 33087. | 34713. | 36241. | 37526. |
| Consumption (•) | 8549. | 9207. | 9830. | 10346. | 10364. | 10499. | 10872. | 10401. | 10687. | 11246. | 11732. | 12256. |
| Imports (0) | -305. | -392. | -449. | -563. | -588. | -667. | -786. | -650. | -700. | -710. | -720. | -730. |
| Exports (X) | 384. | 421. | 384. | 399. | 379. | 430. | 439. | 375. | 375. | 388. | 401. | 414. |

| | 1979 | 1980 | 1981 | 1982 | 1983 | 1984 | 1985 |
|---|---|---|---|---|---|---|---|
| Output (+) | 38764. | 40087. | 41394. | 42639. | 43999. | 45366. | 46712. |
| Consumption (•) | 12688. | 13133. | 13573. | 13976. | 14400. | 14854. | 15275. |
| Imports (0) | -740. | -750. | -760. | -770. | -780. | -790. | -800. |
| Exports (X) | 427. | 440. | 453. | 466. | 479. | 492. | 505. |

Output (+) ————
Consumption (•) — — —
Imports (0) — · —
Exports (X) ········

72

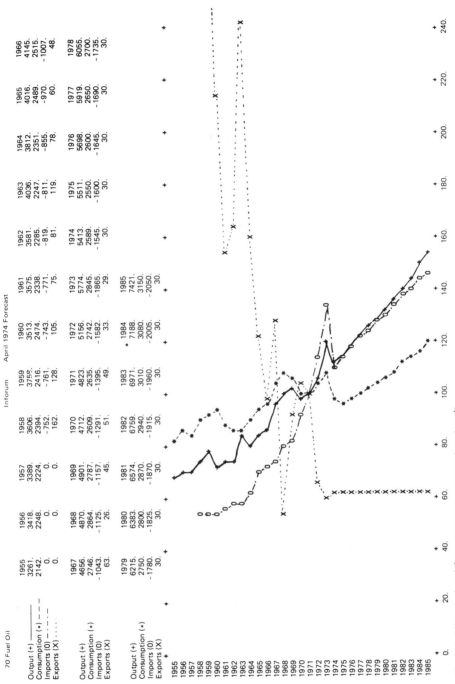

Fig. 3.4. Historical and forecast energy outputs (*continued*).

Table 3.4.  Energy Crisis Forecast. GNP Summary[a] (Billions of 1971$)

| | 1971 | 1972 | 1973 | 1974 | 1975 | 1976 | 1978 | 1980 | 1983 | 1985 |
|---|---|---|---|---|---|---|---|---|---|---|
| Gross national product | 1050.36 | 1099.12 | 1159.96 | 1182.80 | 1237.67 | 1285.28 | 1362.86 | 1440.53 | 1548.44 | 1612.89 |
| Personal consumption expenditures | 664.90 | 692.02 | 730.84 | 742.48 | 768.26 | 794.73 | 854.37 | 911.81 | 992.99 | 1041.79 |
| Durable goods | 103.48 | 112.09 | 120.89 | 114.90 | 119.00 | 122.95 | 132.38 | 141.17 | 153.91 | 161.18 |
| Nondurable goods | 278.12 | 284.83 | 299.87 | 304.98 | 313.52 | 324.27 | 344.78 | 364.57 | 393.75 | 412.36 |
| Services | 283.30 | 296.07 | 312.47 | 322.09 | 335.12 | 346.66 | 375.40 | 403.24 | 441.89 | 464.33 |
| Gross private domestic investment | 151.97 | 170.28 | 188.16 | 186.67 | 207.52 | 218.08 | 220.01 | 231.77 | 246.44 | 254.92 |
| Structures | 80.94 | 89.32 | 91.67 | 87.97 | 94.59 | 99.74 | 98.75 | 102.64 | 107.66 | 110.29 |
| Residential | 42.58 | 50.98 | 49.04 | 38.37 | 43.68 | 47.77 | 46.66 | 49.75 | 52.80 | 54.30 |
| Producers' durable equipment | 67.41 | 75.22 | 84.39 | 89.74 | 100.55 | 106.06 | 112.13 | 120.10 | 130.18 | 136.73 |
| Agriculture | 5.37 | 5.42 | 5.57 | 5.72 | 6.61 | 6.79 | 6.83 | 6.76 | 7.06 | 7.34 |
| Mining | 2.40 | 2.51 | 2.66 | 2.84 | 3.22 | 3.47 | 3.69 | 3.91 | 4.22 | 4.46 |
| Construction | 2.53 | 2.96 | 3.19 | 3.31 | 3.79 | 4.03 | 4.15 | 4.40 | 4.58 | 4.78 |
| Nondurable goods | 7.80 | 8.51 | 9.84 | 11.17 | 11.82 | 12.25 | 12.87 | 13.57 | 14.34 | 15.10 |
| Durable goods | 7.81 | 9.30 | 10.97 | 12.71 | 14.20 | 15.41 | 15.10 | 15.38 | 16.18 | 17.04 |
| Transportation | 4.29 | 5.61 | 5.70 | 5.83 | 7.38 | 7.67 | 8.25 | 9.01 | 10.30 | 10.96 |
| Communication | 7.76 | 8.91 | 9.87 | 10.89 | 11.72 | 12.53 | 13.88 | 15.34 | 17.26 | 18.23 |
| Utilities | 7.24 | 7.49 | 8.30 | 9.53 | 9.92 | 10.37 | 11.39 | 12.35 | 13.71 | 14.41 |
| Trade | 8.53 | 9.31 | 10.07 | 10.69 | 12.24 | 13.23 | 14.31 | 16.06 | 17.67 | 18.91 |
| Finance and services | 6.64 | 7.37 | 9.85 | 8.14 | 10.48 | 10.77 | 11.48 | 12.25 | 12.81 | 12.97 |
| Inventory change | 3.60 | 5.69 | 11.95 | 8.81 | 12.18 | 12.08 | 8.95 | 8.84 | 8.39 | 7.69 |
| Exports of goods and services | 66.13 | 66.55 | 71.75 | 78.43 | 82.95 | 90.32 | 100.47 | 106.78 | 117.19 | 124.16 |
| Merchandise | 36.88 | 36.17 | 39.03 | 43.15 | 45.40 | 49.90 | 55.43 | 57.96 | 62.31 | 65.11 |
| Margins and scrap | 5.90 | 5.50 | 5.93 | 6.55 | 6.89 | 7.56 | 8.39 | 8.77 | 9.42 | 9.84 |
| Travel | 2.46 | 2.85 | 3.11 | 3.39 | 3.77 | 4.06 | 4.44 | 4.69 | 5.05 | 5.30 |
| Passenger fares, other transport | 3.71 | 4.05 | 4.36 | 4.49 | 4.67 | 4.92 | 5.28 | 5.56 | 6.00 | 6.28 |
| Other private services | 1.97 | 2.03 | 2.21 | 2.47 | 2.62 | 2.90 | 3.26 | 3.42 | 3.70 | 3.88 |
| Investment income | 12.01 | 12.90 | 14.07 | 15.29 | 16.58 | 17.91 | 20.67 | 23.58 | 28.22 | 31.49 |
| Government | 3.21 | 3.24 | 3.24 | 3.24 | 3.24 | 3.24 | 3.24 | 3.24 | 3.24 | 3.24 |

| | | | | | | | | | | |
|---|---|---|---|---|---|---|---|---|---|---|
| Imports of goods and services | −65.41 | −66.27 | −71.85 | −70.33 | −72.18 | −74.77 | −80.90 | −89.02 | −101.68 | −110.25 |
| Merchandise | −47.77 | −47.09 | −52.14 | −49.56 | −50.36 | −51.39 | −54.32 | −59.42 | −67.33 | −72.70 |
| Noncompetitive merchandise | −3.10 | −3.21 | −3.33 | −3.38 | −3.52 | −3.64 | −3.88 | −4.10 | −4.42 | −4.61 |
| Margins and scrap | 5.32 | 5.89 | 6.52 | 6.19 | 6.30 | 6.43 | 6.81 | 7.45 | 8.45 | 9.13 |
| Travel | −4.29 | −4.44 | −4.79 | −4.97 | −5.25 | −5.50 | −6.11 | −6.66 | −7.36 | −7.72 |
| Passenger fares, other transport | −4.30 | −4.32 | −4.69 | −4.81 | −4.98 | −5.22 | −5.71 | −6.24 | −7.11 | −7.68 |
| Other private services | −0.87 | −0.90 | −0.96 | −0.93 | −0.94 | −0.95 | −0.99 | −1.06 | −1.16 | −1.23 |
| Investment income | −3.06 | −3.86 | −4.53 | −5.25 | −6.03 | −6.83 | −8.53 | −10.36 | −13.33 | −15.46 |
| Government | −7.42 | −7.42 | −6.82 | −6.22 | −5.62 | −5.62 | −5.62 | −5.62 | −5.62 | −5.62 |
| Foreign long-term capital to U.S. | 2.27 | 2.31 | 2.50 | 2.69 | 2.88 | 3.00 | 3.23 | 3.46 | 3.81 | 4.04 |
| U.S. long-term capital abroad | −6.35 | −6.59 | −6.96 | −7.32 | −7.69 | −7.91 | −8.35 | −8.79 | −9.45 | −9.89 |
| U.S. govt. long-term capital flows | −2.19 | −3.84 | −3.82 | −3.80 | −3.79 | −3.79 | −3.79 | −3.79 | −3.79 | −3.79 |
| Government purchases | 232.76 | 235.33 | 239.90 | 244.28 | 249.76 | 255.57 | 267.51 | 277.86 | 292.32 | 301.24 |
| Public construction | 30.10 | 28.19 | 28.41 | 30.05 | 30.97 | 31.78 | 33.08 | 34.20 | 35.73 | 36.71 |
| National defense | 69.71 | 66.54 | 65.40 | 64.30 | 63.19 | 63.82 | 65.07 | 66.32 | 68.19 | 69.44 |
| Nondefense federal | 24.10 | 25.98 | 26.46 | 27.83 | 29.20 | 29.94 | 31.42 | 32.90 | 35.29 | 36.88 |
| State and local | 108.85 | 114.53 | 119.48 | 121.89 | 126.13 | 129.75 | 137.60 | 144.06 | 152.67 | 157.74 |
| Education | 52.54 | 54.93 | 57.82 | 58.68 | 60.48 | 61.94 | 64.56 | 66.18 | 68.49 | 70.23 |
| Productivity ($1000 per employee) | 12.64 | 12.88 | 13.25 | 13.31 | 13.68 | 13.89 | 14.18 | 14.53 | 15.04 | 15.37 |
| Private sector | 13.88 | 14.15 | 14.59 | 14.66 | 15.11 | 15.36 | 15.73 | 16.15 | 16.80 | 17.22 |
| Employed persons (millions) | 81.96 | 84.14 | 86.32 | 87.63 | 89.23 | 91.27 | 94.76 | 97.79 | 101.52 | 103.50 |
| Private industry | 65.61 | 67.70 | 69.70 | 70.89 | 72.23 | 73.96 | 76.82 | 79.32 | 82.32 | 83.82 |
| Civilian government | 13.53 | 13.96 | 14.18 | 14.36 | 14.66 | 14.98 | 15.62 | 16.16 | 16.91 | 17.40 |
| Defense | 1.14 | 1.09 | 1.03 | 0.97 | 0.91 | 0.91 | 0.91 | 0.91 | 0.91 | 0.91 |
| Military | 2.81 | 2.48 | 2.43 | 2.39 | 2.34 | 2.33 | 2.32 | 2.31 | 2.29 | 2.28 |
| Aggregate price forecasts | | | | | | | | | | |
| Wholesale price index (1967 = 100) | 114.14 | 121.02 | 136.69 | 153.98 | 163.18 | 170.36 | 178.47 | 181.86 | 187.37 | 197.75 |
| Gross national product | 143.02 | 151.53 | 164.38 | 183.02 | 197.23 | 208.86 | 221.34 | 227.44 | 231.64 | 231.08 |
| Personal consumption expenditures | 121.67 | 129.17 | 139.80 | 156.55 | 168.41 | 178.49 | 189.53 | 195.26 | 205.85 | 224.89 |
| Durable goods | 113.98 | 118.54 | 124.57 | 133.49 | 145.59 | 155.83 | 168.23 | 173.21 | 175.29 | 174.06 |
| Nondurable goods | 133.38 | 142.60 | 162.13 | 185.77 | 196.04 | 201.95 | 209.66 | 213.34 | 215.62 | 215.06 |
| Services | 148.25 | 157.47 | 166.91 | 186.08 | 201.86 | 216.92 | 232.38 | 240.98 | 248.21 | 248.75 |

Table 3.4. (continued).

| | 1971 | 1972 | 1973 | 1974 | 1975 | 1976 | 1978 | 1980 | 1983 | 1985 |
|---|---|---|---|---|---|---|---|---|---|---|
| Gross private fixed investment | 141.47 | 149.18 | 160.07 | 174.52 | 190.18 | 201.53 | 212.87 | 217.07 | 218.40 | 216.04 |
| Business structures | 164.05 | 173.05 | 186.43 | 207.00 | 225.60 | 237.86 | 247.52 | 250.75 | 251.19 | 247.67 |
| Producers' durable equipment | 127.98 | 133.88 | 141.81 | 153.54 | 168.38 | 180.06 | 193.12 | 198.12 | 200.02 | 198.24 |
| Residential construction | 148.35 | 159.34 | 175.29 | 190.09 | 204.20 | 213.16 | 220.97 | 223.77 | 224.39 | 221.77 |
| Exports | 126.47 | 132.46 | 152.93 | 171.28 | 180.42 | 189.66 | 199.25 | 203.25 | 204.99 | 203.73 |
| Government | 174.19 | 183.91 | 199.24 | 223.43 | 240.94 | 253.30 | 266.19 | 271.66 | 274.57 | 272.75 |
| Federal | 172.06 | 182.25 | 198.41 | 220.52 | 237.06 | 249.31 | 262.02 | 267.18 | 269.87 | 268.24 |
| State and local | 176.98 | 186.41 | 201.90 | 226.71 | 244.62 | 257.08 | 270.80 | 276.99 | 280.57 | 278.95 |
| Public construction | 194.51 | 205.37 | 219.46 | 251.39 | 273.02 | 287.01 | 298.85 | 303.03 | 304.24 | 300.59 |
| Exogenous assumptions | | | | | | | | | | |
| Consumption | | | | | | | | | | |
| Disposable per capita income 58$ | 2679.00 | 2770.00 | 2889.00 | 2940.00 | 3030.00 | 3104.00 | 3286.67 | 3440.00 | 3610.00 | 3680.00 |
| Population—series E (millions) | 207.05 | 208.80 | 210.50 | 212.16 | 213.92 | 215.79 | 219.79 | 224.13 | 231.04 | 235.70 |
| School age population | 59.91 | 59.61 | 59.18 | 58.75 | 58.32 | 57.97 | 56.32 | 54.62 | 53.35 | 53.46 |
| Plant and equipment investment | | | | | | | | | | |
| Long-term interest rate | 0.02 | 0.03 | 0.02 | 0.02 | 0.03 | 0.03 | 0.02 | 0.02 | 0.02 | 0.02 |
| Short-term interest rate | −0.00 | −0.00 | 0.02 | 0.02 | 0.02 | 0.02 | 0.02 | 0.01 | 0.01 | 0.01 |
| Rent-/construction cost index | 0.79 | 0.76 | 0.75 | 0.77 | 0.80 | 0.80 | 0.80 | 0.80 | 0.80 | 0.80 |
| Households (millions) | 64.37 | 66.67 | 67.64 | 68.87 | 70.08 | 71.54 | 74.44 | 77.30 | 81.45 | 84.21 |
| Investment tax credit | 0.00 | 0.07 | 0.07 | 0.07 | 0.07 | 0.07 | 0.07 | 0.07 | 0.07 | 0.07 |
| Foreign trade | | | | | | | | | | |
| Average foreign currency price | 1.00 | 1.01 | 1.03 | 1.05 | 1.09 | 1.12 | 1.13 | 1.13 | 1.12 | 1.11 |
| Employment | | | | | | | | | | |
| Labor force (millions) | 86.93 | 88.99 | 90.58 | 92.17 | 93.78 | 95.39 | 98.62 | 101.81 | 105.69 | 107.72 |
| Civilian unemployment rate | 5.90 | 5.60 | 4.83 | 5.05 | 4.97 | 4.43 | 4.00 | 4.03 | 4.02 | 3.99 |

[a] Numbers have been scaled to 1971 published levels and may not add to totals.

Table 3.5.  Selected Interindustry Sales

| | | Inforum seller 14 (energy crisis forecast, coal mining) | | | | |
|---|---|---|---|---|---|---|
| Buyer | | 1971 | 1973 | 1974 | 1980 | 1985 |
| 14 | Coal mining | 897.8 | 942.5 | 1105.6 | 1311.7 | 1433.7 |
| 35 | Broad and narrow fabrics | 22.8 | 25.3 | 26.8 | 32.0 | 34.1 |
| 48 | Paper and paperboard mills | 146.0 | 158.1 | 172.0 | 210.2 | 228.6 |
| 55 | Industrial chemicals | 136.1 | 155.1 | 160.1 | 214.4 | 251.8 |
| 81 | Cement, concrete, gypsum | 128.2 | 135.7 | 140.3 | 172.1 | 190.8 |
| 83 | Steel | 772.4 | 859.6 | 866.3 | 918.0 | 910.4 |
| 133 | Motor vehicles | 24.2 | 27.7 | 25.4 | 31.2 | 34.5 |
| 160 | Electric utilities | 1618.9 | 1714.1 | 2162.3 | 2644.1 | 2960.3 |
| | Sum of intermediate flows | 4106.9 | 4407.4 | 5063.1 | 6036.2 | 6607.1 |
| | | Sales to final demand | | | | |
| 186 | Personal consumption | 137.2 | 103.1 | 85.2 | 21.1 | 0.0 |
| 187 | Defense expenditures | 41.2 | 41.2 | 41.2 | 45.3 | 49.4 |
| 189 | Education | 22.6 | 26.8 | 27.4 | 32.2 | 34.7 |
| 193 | Change in inventories | −44.8 | −72.0 | −31.8 | 14.2 | 10.0 |
| 194 | Exports | 565.3 | 560.0 | 760.0 | 900.0 | 1000.0 |
| | Sum of sales to final demand | 738.8 | 679.9 | 904.3 | 1043.8 | 1131.4 |
| | Total | 4845.7 | 5087.3 | 5967.4 | 7079.9 | 7738.4 |

| | | Seller 15 (crude petroleum, nat. gas) | | | | |
|---|---|---|---|---|---|---|
| Buyer | | 1971 | 1973 | 1974 | 1980 | 1985 |
| 15 | Crude petroleum, nat. gas | 118.6 | 124.2 | 127.5 | 160.9 | 188.8 |
| 69 | Petroleum refining | 15830.0 | 17705.1 | 17154.7 | 21627.9 | 25201.9 |
| 160 | Electric utilities | 104.4 | 116.4 | 120.1 | 150.0 | 173.5 |
| 161 | Natural gas | 3308.1 | 3523.5 | 3598.2 | 4211.9 | 4821.6 |
| | Sum of intermediate flows | 19412.9 | 21524.9 | 21056.6 | 26227.4 | 30481.2 |
| | | Sales to final demand | | | | |
| 193 | Change in inventories | −193.0 | 0.0 | 50.0 | 46.6 | 45.8 |
| 195 | Imports | −2555.1 | −4106.1 | −3200.0 | −3680.0 | −4010.0 |
| | Sum of sales to final demand | −2708.1 | −4034.6 | −3100.0 | −3568.4 | −3884.2 |
| | Total | 16704.7 | 17490.3 | 17956.6 | 22658.9 | 26597.1 |

| | | Seller 69 (petroleum refining) | | | | |
|---|---|---|---|---|---|---|
| Buyer | | 1971 | 1973 | 1974 | 1980 | 1985 |
| 69 | Petroleum refining | 3070.5 | 3434.2 | 3327.5 | 4195.1 | 4888.4 |
| 3 | Meat animals, other livestock | 141.2 | 148.0 | 155.4 | 178.2 | 198.7 |
| 5 | Grains | 691.1 | 697.9 | 716.2 | 871.9 | 995.2 |
| 7 | Fruit, vegetables, other crops | 270.8 | 307.3 | 320.9 | 392.2 | 447.4 |
| 19 | Maintenance construction | 211.2 | 243.7 | 242.7 | 306.5 | 356.0 |
| 68 | Paints | 145.5 | 166.2 | 169.3 | 195.8 | 210.0 |
| 70 | Fuel oil | 4823.1 | 5774.0 | 5412.6 | 6383.3 | 7420.6 |
| 71 | Paving and asphalt | 342.8 | 398.7 | 411.8 | 551.5 | 666.2 |
| 83 | Steel | 138.6 | 167.4 | 160.0 | 171.2 | 172.2 |
| 152 | Busses and local transit | 215.5 | 287.2 | 286.0 | 346.7 | 386.0 |
| 155 | Airlines | 1258.0 | 1635.1 | 1742.9 | 2594.0 | 3272.8 |
| 158 | Telephone and telegraph | 140.0 | 161.5 | 156.9 | 218.1 | 272.7 |
| 163 | Wholesale trade | 1369.1 | 1594.8 | 1524.9 | 2022.6 | 2452.0 |
| 164 | Retail trade | 909.2 | 1047.1 | 998.3 | 1291.2 | 1536.0 |
| 170 | Personal and repair services | 156.4 | 175.7 | 168.4 | 208.7 | 236.0 |
| 171 | Business services | 176.1 | 213.0 | 209.1 | 276.8 | 333.1 |

Table 3.5 (*continued*).

| 173 | Auto repair | 177.5 | 193.8 | 183.2 | 221.2 | 255.1 |
|---|---|---|---|---|---|---|
| | Sum of intermediate flows | 16870.2 | 19723.3 | 19220.4 | 24169.8 | 28332.8 |

| | | Sales of construction activities | | | | |
|---|---|---|---|---|---|---|
| 18 | Highways | 455.9 | 407.2 | 424.7 | 419.2 | 414.9 |
| 20 | Conservation and development | 254.0 | 246.3 | 277.5 | 363.8 | 436.5 |
| | Sum of sales to construction | 1357.5 | 1390.6 | 1456.6 | 1649.8 | 1793.3 |

| | | Sales to final demand | | | | |
|---|---|---|---|---|---|---|
| 186 | Personal consumption | 10364.0 | 10872.3 | 10400.7 | 13133.0 | 15275.1 |
| 187 | Defense expenditures | 229.3 | 229.3 | 229.3 | 252.2 | 275.2 |
| 189 | Education | 270.2 | 320.2 | 326.9 | 385.0 | 414.9 |
| 194 | Exports | 378.8 | 439.4 | 375.0 | 440.0 | 505.0 |
| 195 | Imports | −588.2 | −786.0 | −650.0 | −750.0 | −800.0 |
| | Sum of sales to final demand | 11113.3 | 11702.3 | 11119.2 | 14267.7 | 16585.5 |
| | Total | 29340.9 | 32816.2 | 31796.3 | 40087.3 | 46711.5 |

| | | Seller 70 (fuel oil) | | | | |
|---|---|---|---|---|---|---|
| Buyer | | 1971 | 1973 | 1974 | 1980 | 1985 |
| 5 | Grains | 56.3 | 66.6 | 71.6 | 86.6 | 97.7 |
| 48 | Paper and paperboard mills | 75.4 | 103.7 | 95.7 | 120.1 | 135.4 |
| 55 | Industrial chemicals | 25.9 | 37.5 | 39.1 | 52.4 | 61.5 |
| 81 | Cement, concrete, gypsum | 26.3 | 35.4 | 32.3 | 40.4 | 46.1 |
| 83 | Steel | 94.9 | 134.2 | 120.6 | 130.2 | 132.6 |
| 151 | Railroads | 360.8 | 496.4 | 551.7 | 669.0 | 748.1 |
| 152 | Busses and local transit | 85.3 | 133.2 | 134.7 | 163.1 | 181.6 |
| 153 | Trucking | 746.6 | 1022.9 | 1049.9 | 1466.1 | 1822.2 |
| 154 | Water transportation | 478.7 | 526.8 | 472.9 | 473.2 | 473.1 |
| 160 | Electric utilities | 476.9 | 736.1 | 370.6 | 517.2 | 696.7 |
| 163 | Wholesale trade | 46.8 | 63.9 | 60.1 | 80.1 | 98.2 |
| 164 | Retail trade | 196.0 | 264.5 | 248.0 | 322.4 | 387.7 |
| 165 | Banks, credit agencies, brokers | 34.5 | 48.7 | 45.6 | 60.9 | 72.0 |
| 166 | Insurance | 38.4 | 46.8 | 42.6 | 53.4 | 62.6 |
| 171 | Business services | 49.2 | 69.7 | 67.3 | 89.6 | 109.0 |
| 175 | Medical services | 52.8 | 69.2 | 65.3 | 82.8 | 98.1 |
| 176 | Private schools and NPO | 69.3 | 97.7 | 93.6 | 120.3 | 141.2 |
| | Sum of intermediate flows | 3372.8 | 4576.4 | 4170.1 | 5289.2 | 6241.5 |

| | | Sales to final demand | | | | |
|---|---|---|---|---|---|---|
| 186 | Personal consumption | 2635.0 | 2845.0 | 2589.5 | 2800.0 | 3150.0 |
| 189 | Education | 85.9 | 101.8 | 90.3 | 45.6 | 24.3 |
| 192 | General state and local gov. | 50.7 | 58.0 | 51.9 | 28.5 | 16.1 |
| 194 | Exports | 49.0 | 29.4 | 30.0 | 30.0 | 30.0 |
| 195 | Imports | −1395.1 | −1865.0 | −1545.0 | −1825.0 | −2050.0 |
| | Sum of sales to final demand | 1450.3 | 1197.6 | 1242.5 | 1094.2 | 1179.1 |
| | Total | 4823.1 | 5774.0 | 5412.6 | 6383.3 | 7420.6 |

to update the energy sectors. We introduced data on personal consumption expenditures, inventory change, imports, and exports. In mid April of 1974, complete data for 1973 had not been published in the *SCB*; therefore, many of these 1973 numbers are estimates based on data through October. While domestic production of refined products increased almost 13 percent between 1971 and 1973, domestic supply (production plus imports) of crude petroleum increased only 5.7 percent. Domestic production of crude fell by more than 2 percent.

With output of refined petroleum and imports and inventory change of crude petroleum in agreement with published statistics, the model generates domestic crude petroleum demand that is 7 percent higher than published crude petroleum production in 1973. Statistics on crude imports are likely to be accurate, because this information reported in the *SCB* is collected by the Tariff Commission. There was essentially no change in reported inventory stocks. A negative change of more than a billion dollars would have been necessary for the output generated by the model to conform with published data. We feel it would be even more unrealistic to assume that the sales of crude per unit of refined petroleum declined drastically. Indeed, preliminary reports of the *Census of Manufactures, 1972* [1974, Table 4] indicate a slight increase in the historical trend of this coefficient. For these reasons, we have not forced the 1973 output generated by the model to conform with published data. Rather, it seems more realistic to assume that the published data are inaccurate.

There are still limitations to this forecast. The model has not addressed itself to the problem of complementary goods—how does the price of gasoline affect the demand for automobiles? The consumption equation for automobiles had an upward taste-change factor of 0.6 percent per year. We have removed this factor. There are problems of time lags in the shift to a substitute energy source.

All in all, we feel that the elasticities are realistic and that, with a little allowance for adjustment problems, the forecasts show that the economy could operate at a high level of employment within the limits of available petroleum. (Layoffs in the automotive industry have come, not from a lack of fuel to *produce* the cars, but from a shift in demand to cars that the industry was not equipped to make.)

## Future Directions—A System of World Input-Output Models

Any model, to remain useful, must be adapted to address new questions, such as the energy crisis, and be improved where it is weak. During the last several years, we have devoted some effort to a more reasonable specification of the foreign trade equations, particularly to take fluctuating currencies into account. Although we are satisfied with the import equations, problems remain with exports. Currently, exports are related to lagged domestic output and the relative foreign to domestic price. Obviously, they should instead be related to foreign demand. Although this task is simple enough in the historical period, in order to forecast exports we would need to forecast world demand by product.

We are planning, therefore, to construct a system of input-output forecasting models for the major countries, which are interconnected through their trade sectors. Previous experience with an input-output model of Austria convinced us that our approach is feasible, provided we have the help and cooperation of economists in those countries.

**NOTES**

1. Domestic demand = output + imports − exports.
2. Domestic supply = output + imports + exports.
3. The Seidel iterative solution method has been adapted to solve the simultaneous equations for outputs, imports, and inventory change.
4. Some of the results are given in the book by Almon et al. (1985).
5. The balancing method is explained by Almon et al. (1985).
6. Brian O'Connor, a former research student at the University of Maryland, has estimated the other components of income. When his work has been coordinated with the estimates of the present INFORUM model, disposable income, now exogenous, can be generated within the model. Alternatively, we will be able to test whether the present tax structure generates a disposable income that is consistent with a desired level of employment.
7. The wage equations were estimated by David Belzer, a research student at the University of Maryland. He is completing the integration of the income side into the real model.

### References

Almon, Clopper, Margaret Buckler, Lawrence C. Horwicz, and Thomas C. Reimbold. 1974. *1985: Interindustry Forecasts of the American Economy*. Lexington: Lexington Books, D.C. Heath and Company.

Strout, Alan M. 1961. Weather and the demand for space heat. *The Review of Economics and Statistics* 43(2) [May]: 185–192.

U.S. Bureau of the Census. 1973. *Annual Survey of Manufactures, 1971: General Statistics for Industry Groups and Industries*. M71(AS)-1. Washington, D.C.: U.S. Government Printing Office.

U.S. Bureau of the Census. 1974. *Census of Manufactures, 1972: Preliminary Reports—Industry Series*. Washington, D.C.: U.S. Government Printing Office.

U.S. Department of Commerce. 1974. Current business statistics. *Survey of Current Business* 54(2) [February]: S1–S40.

# 4

# Long-Range Forecasting
# with a Regional Input-Output Model

WILLIAM H. MIERNYK

For some planning purposes national forecasts are adequate, and the recent interindustry projections by Almon (1966) and the Interagency Growth Study (1966) project fill an important need. But one of the outstanding characteristics of the American economy is its geographic diversity. Certain types of economic activities—generally those serving local or regional markets—are almost ubiquitous. But other sectors serve national markets, and within these sectors there is considerable regional specialization. This specialization provides the rationale for regional input-output studies.

One method of estimating the regional impacts of changes in national final demand is to partition a national model into a multiregional model with national and local sectors (Leontief, 1965; Leontief, 1966, pp. 223–257; Miller, 1966, pp. 105–125). Conceptually, models of this type—which Isard (1951) has described as "balanced" models—will show the differential impacts on all regions of exogenous changes. But the data problems involved in implementing multiregional and interregional models are formidable, and given the variety of ways in which regions can be defined there is at present an important place for specialized regional input-output tables that are small-scale versions of the national model. This paper will be concerned with the latter.

Regional research has become a "growth industry" since the end of World War II. And as Tiebout (1957, p. 140) pointed out a decade ago: "It is not too much of an overstatement to say that post-World War II regional research has been almost completely dominated by regional applications of input-output models". At the time Tiebout (1957, p. 142) wrote most regional studies were concerned with analyzing local impacts, regional balances of payments, or interregional flows. Even now input-output models are not used extensively for making regional forecasts. There have been some regional interindustry forecasts, and these will be discussed, but it is necessary first to describe two broad types of regional input-output studies, and the potential of each for regional forecasting.

This article was published in *The Western Economic Journal*, Volume 6, Number 3, June 1968, pp. 165–176. © Western Economic Association International. Reprinted by permission.

**I**

One method of constructing a regional input-output table makes use of "adjusted" national coefficients. Although several adjustment techniques have been used—including some which are described in rather vague terms—only one will be discussed here. This is the method that makes use of location quotients. The location quotient for industry $i$ on industry $j$, $(LQ_{ij})$, is defined as the ratio of regional employment in industry $i$ to regional employment in industry $j$ divided by the ratio of national employment in industry $i$ to national employment in industry $j$. If $LQ_{ij} \geqslant 1$, the national input coefficient is considered to be representative of the region and is transferred directly from the national to the regional table. If $LQ_{ij} < 1$, the national coefficient is multiplied by the location quotient to obtain the regional coefficient, and the difference between the computed coefficient and the national coefficient is transferred to the import row of that sector. National coefficients can thus be adjusted downward but not upward. This procedure is followed until a complete table of direct input coefficients has been constructed for the region. A Leontief matrix is derived from the direct coefficients and inverted. The inverse matrix is then postmultiplied by a specified "bill of goods" to obtain a transactions table.

The second approach relies on direct surveys to obtain data for the measurement of interindustry flows. Regional sectors are defined and made to correspond as closely as possible to national sectors. Next, a random sample of establishments in each sector is selected and data on interindustry transactions are obtained by means of interviews. Final demand and total sales by sector are obtained from secondary sources or estimated from state tax and employment records. With the exogenous portion of the table specified, the sample data of interindustry transactions can be blown up to complete the transactions table.[1]

The only advantage that can be claimed for the "adjusted" coefficient method is that a regional input-output table can be constructed in a relatively short time and at relatively low cost. This method does not adequately allow for differences in industry and product-mixes among regions. Even if it did, the table derived could not be used for consistent regional forecasting since it provides no information about changes in direct coefficients over time. The problem of estimating regional from national coefficients is further compounded by different aggregations of establishments into sectors in national and regional models. Also, employment is reported on a different basis than that used in defining interindustry sectors so that only rough estimates can be used in computing location quotients. In brief, while a regional input-output table based on "adjusted" national coefficients can give a rough *description* of a regional economy it is virtually useless for projection purposes.

The major disadvantage of a regional input-output table based on direct survey data is that its construction is expensive and time-consuming. But if time and money are available—and if the selected sample of businessmen in the region is willing to cooperate in providing detailed data on sales and purchases—the model can be used for making long-range, consistent regional forecasts. Before discussing such forecasts, it will be useful to emphasize the differences between a regional and the national model by a direct comparison.

## II

The most significant difference between a regional economy in the United States and the national economy, from a trade point of view, is that the former is quite open while the latter can be considered as virtually closed. This can be shown by comparing interindustry transactions and imports in the United States with those in the Colorado River Basin (CRB), a region covering about one-twelfth of the land area of the United States (see Table 4.1). The structural differences between the region and the nation are only partly revealed by Table 4.1. A considerable amount of two-way aggregation was required before the two tables could be compared. In some cases, notably agriculture and mining, the regional table is more disaggregated than the national table. In most cases, however, the 82-sector national table shows more industry detail than the regional table. The highest degree of disaggregation possible was used to construct Table 4.1, but this still required compression of some of the detailed transactions given in the two tables.[2]

On the average, interindustry transactions in the United States accounted for 49 percent of total inputs in 1958. The comparable figure for the Colorado River Basin in 1960 was 32 percent. But in the national table imports accounted for only 1.7 percent of total inputs while the comparable figure for the Colorado River Basin was 22 percent. Table 4.1 shows that there is wide variation around these averages, and the ranges for both interindustry transactions and imports are greater in the region than in the nation.

The highest degree of interdependence on the input side in the nation is in the food and kindred products sector where interindustry purchases accounted for more than 70 percent of total transactions, compared with 50 percent in the CRB. But in primary metals, interindustry transactions in the Colorado River Basin accounted for more than 87 percent of total inputs or 27 percentage points more than the U.S. average. At the other extreme, interindustry purchases accounted for more than 66 percent of the total in U.S. textiles and apparel, but in the Colorado River Basin—where there are very few textile or apparel producers—such purchases accounted for only 7 percent of the total.

On the trade side there are equally significant differences, even in the service sectors which are among the most rapidly growing parts of the economy. As a nation we import few services, but regions regularly import services. And in some cases these account for a significant part of total purchases. In the Colorado River Basin, for example, almost 66 percent of total inputs to the oil field services sector were imported, a substantial part of which consisted of labor, management, and technical services. Similarly, in wholesale and retail trade—where only the margin is counted as a transaction—the nation imported nothing in 1958 while imports of the trade sector amounted to more than 21 percent of total input transactions in the regional table.

Table 4.1. Interindustry Transactions, Imports, and Value Added as Percent of Total Gross Output (or Outlays), U.S., 1958, and Colorado River Basin, 1960

| Sector | Interindustry transactions | | | Imports | | | Value added | | |
|---|---|---|---|---|---|---|---|---|---|
| | U.S. | CRB | Difference[a] | U.S. | CRB | Difference[a] | U.S. | CRB | Difference[a] |
| 1. Food and kindred products | 70.3 | 49.5 | −20.8 | 4.1 | 23.7 | +19.6 | 25.6 | 26.7 | + 1.1 |
| 2. Textiles and apparel | 66.4 | 7.0 | −59.4 | 2.0 | 37.8 | +35.8 | 31.6 | 55.2 | +23.6 |
| 3. Chemicals and petroleum products | 66.2 | 33.6 | −32.6 | 2.3 | 36.6 | +34.3 | 31.6 | 29.8 | − 1.8 |
| 4. Livestock and products | 64.9 | 49.4 | −15.5 | 0.8 | 12.7 | +11.9 | 34.3 | 37.9 | + 3.6 |
| 5. Lumbar and wood products | 62.4 | 49.9 | −12.5 | 5.1 | 11.4 | + 6.3 | 32.5 | 38.6 | + 6.1 |
| 6. Primary metals | 60.5 | 87.5 | +27.0 | 3.7 | 3.0 | − 0.7 | 35.8 | 9.5 | −26.3 |
| 7. Fabricated metals | 59.0 | 14.2 | −44.8 | 0.6 | 47.8 | +47.2 | 40.4 | 37.9 | − 2.5 |
| 8. Contract construction | 58.2 | 45.9 | −12.3 | 0 | 22.4 | +22.4 | 41.8 | 31.7 | −10.1 |
| 9. Leather and products | 57.6 | 30.1 | −27.5 | 1.1 | 14.3 | +13.2 | 41.3 | 55.6 | +14.3 |
| 10. Pulp and paper | 57.6 | 60.4 | + 2.8 | 6.8 | 10.3 | + 3.5 | 35.6 | 29.2 | − 6.4 |
| 11. Furniture and fixtures | 57.3 | 27.4 | −29.9 | 0 | 35.0 | +35.0 | 42.7 | 37.7 | − 5.0 |
| 12. All other manufacturing | 57.0 | 25.8 | −31.2 | 1.6 | 18.1 | +16.5 | 41.4 | 56.1 | +14.7 |
| 13. Agricultural services | 55.3 | 31.8 | −23.5 | 0 | 28.4 | +28.4 | 44.7 | 39.8 | − 4.9 |
| 14. Printing and publishing | 51.7 | 15.3 | −36.4 | 0.3 | 23.2 | +22.9 | 47.9 | 61.5 | +13.6 |
| 15. Oil field services | 51.4 | 12.8 | −38.6 | 0 | 65.9 | +65.9 | 48.6 | 21.3 | −27.3 |
| TOTAL | *49.0* | *32.3* | *− 16.7* | *1.7* | *22.0* | *+ 20.3* | *49.3* | *45.7* | *− 3.6* |
| 16. Stone, clay, and glass products | 48.4 | 37.1 | −11.3 | 1.4 | 17.9 | +16.5 | 50.2 | 44.9 | − 5.3 |
| 17. All other agriculture | 47.2 | 35.2 | −12.0 | 2.3 | 16.6 | +14.3 | 50.5 | 48.2 | − 2.3 |
| 18. All other services | 45.1 | 21.6 | −23.5 | 0.2 | 26.5 | +26.3 | 54.7 | 51.9 | − 2.8 |
| 19. Forestry | 41.6 | 10.0 | −31.6 | 19.3 | 6.4 | −12.9 | 39.1 | 83.6 | +44.5 |
| 20. Coal | 41.5 | 11.1 | −30.4 | 0.1 | 24.0 | +23.9 | 58.4 | 64.8 | + 6.4 |
| 21. Transportation | 33.3 | 26.7 | − 6.6 | 4.0 | 32.8 | +28.8 | 62.7 | 40.5 | −22.2 |
| 22. All other mining | 32.3 | 30.2 | − 2.1 | 11.0 | 23.4 | +12.4 | 56.7 | 46.4 | −10.3 |
| 23. Rentals and finance | 31.2 | 10.5 | −20.7 | 0.1 | 4.2 | + 4.1 | 68.7 | 85.2 | +16.5 |
| 24. Utilities | 30.0 | 24.2 | − 5.8 | 0.3 | 27.2 | +26.9 | 69.7 | 48.5 | −21.2 |
| 25. Oil and gas | 29.7 | 8.1 | −21.6 | 8.8 | 49.5 | +40.7 | 61.5 | 42.4 | −19.1 |
| 26. Trade | 27.2 | 26.3 | − 0.9 | 0 | 21.4 | +21.4 | 72.8 | 52.3 | −20.5 |

[a] CRB minus U.S.

*Source*: U.S. data (*Survey of Current Business*, 1965, pp. 33–39). CRB data are based on six subbasin tables prepared for the Federal Water Pollution Control Administration by the Universities of Colorado. Denver and New Mexico, and the Economic Research Service of the U.S. Department of Agriculture.

The point need not be labored further. Some of the differences in structure between a region and the national economy could be accounted for by applying a location quotient to national coefficients. Still the aggregation problem alone would insure that not all of these differences would be taken into account by this method. Thus even for descriptive purposes a regional input-output table based on adjusted national coefficients can provide only a rough approximation.[3] But more important, for present purposes, even if such tables accurately *described* the structure of the regional economy they could not be used effectively for forecasting. The objective of this discussion of the shortcomings of regional tables based on "adjusted" national coefficients is to drive home the point that if one wishes to make interindustry regional forecasts it is necessary to follow the expensive and time-consuming path of constructing a regional transactions table from survey data.[4]

## III

The method of interindustry forecasting described in this section is a simple one, but it requires data that can only be obtained from a representative sample of establishments in each of the sectors of the transactions table. Strictly speaking, it is not a dynamic model, such as the one used by Almon in making his national projections, but there are no fixed coefficients in the system. The necessary data for using a fully dynamic model were not available in the CRB study.[5]

Data deficiencies are not the only problem one faces in regional input-output forecasting. There is the added problem imposed by the open nature of the regional economy. Thus in projecting regional input coefficients two kinds of changes must be considered. These are: (1) technological change, as in the national model, and (2) changes in trade patterns, especially in a growing region. The effects of the latter changes on the structure of the regional economy can easily outweigh those of the former.

The forecasts are made by projecting coefficients, by the procedure described below, and relating these to independent projections of final demand to obtain intermediate transactions and total gross output for each sector in the system. A theoretical case can be made for the use of *marginal* coefficients, but this presupposes the availability of a set of transactions tables for different time periods. Moreover, a recent empirical study by Tilanus shows that because of sampling variation marginal coefficients are too unstable to be used for forecasting.[6] Thus the object becomes one of projecting *average* flow coefficients to the target year.

The first step is to identify and select a subsample of "best practice" establishments in each sector. This can be done by computing output per man-hour or some other measure of efficiency for each sample establishment.[7] After the best practice establishments have been identified new input coefficients are computed for each sector from the subsamples. The assumption is made that the average input pattern of the most efficient establishments at time $t$ will approximate the average input patterns of all establishments at $t + k$, where $k$ is a time period to be determined. At this point a certain amount of judgment must be exercised.

If it could be assumed that the best practice establishment were about ten years younger than the average establishment in each sector, one might be justified in concluding that the best practice input coefficients represent those that would be found in the average establishment ten years later. But the matter is not so simple as this. There is no direct relationship between age of establishment and efficiency. Thus there is no way for the analyst to avoid making a judgment about the length of time it will take for the best practice coefficients to become the averages for each sector. What is ordinarily done is to assume that the best practice input coefficients lie somewhere between the average coefficients for the base year and those that might be expected in the target year. The two points established for each sector can thus form the basis for "judicious extrapolation." This might sound like a euphemism for simple guesswork. But the serious analyst will do more than extrapolate on the basis of hunches. Industry experts can be consulted and if engineering and industry productivity data are available they can be used in the extrapolation of regional coefficients. Also, in some cases historical price data can be used to estimate the substitution of some inputs for others. This can be done, for example, in extrapolating the substitutes of capital for labor inputs. There is no simple formula for the extrapolation of best practice input coefficients, however, and the honest analyst will be the first to admit that the final outcome represents his best judgment.[8]

The projection of input coefficients to reflect anticipated changes in technology, and to a limited extent the effects of changes in relative prices, clearly poses a set of difficult problems. But as noted earlier changing trade patterns can have an even greater influence on the structure of a regional economy. Over time, output and employment in a region are affected by changes in national and local final demand, industrial mix, and regional shares.[9] The effects of such shifts must be taken into account when making long-range regional projections. This involves the use of techniques, based upon location theory, that are employed in making impact studies. One approach to the trade pattern problem was illustrated by Isard and Kuenne in their analysis of the direct and indirect effects of the location of a new steel mill on the greater New York-Philadelphia region.[10]

Isard and Kuenne based their ten-year projection on a known change in the economic base of the region. By analyzing agglomeration patterns in other regions they were able to identify the sectors that would be expected to grow as the result of the location of a new steel mill in the area. Other regional analysts will not always be in the fortunate position of working from a known change to anticipated changes. But effects similar to those discussed by Isard and Kuenne must be taken into account in long-range regional projection. In the general regional case it might be necessary to make a series of preliminary projections before the final projection is completed. After the input coefficients have been adjusted for technical change, and final demand has been projected independently, total gross output can be computed for the target year. Analysis of the total projected sales, sector by sector, can suggest some of the changes in trade patterns that might be anticipated. Because of scale problems, for example, some inputs might be imported by a sector at a given level of that sector's total gross output. But if there is a sufficient increase in demand within the region there can be a shift from imports to interindustry transactions between the base and target years. This will be illustrated by a specific example.

The demand for paper products—newsprint, wrapping paper, boxes, etc.—is ubiquitous. But paper and most paper products are manufactured in fairly large establishments which tend to be localized. In a sparsely populated region paper products will be imported, and sometimes these have to be shipped long distances. As the population grows, and as industry develops in that region, it might become more economical to produce some paper products locally than to continue to import them. This can be done, of course, only if the necessary resources are available within the region. And since paper products compete in national markets, transport savings must be sufficient to offset any initial diseconomies due to relatively small-scale production.[11] The projection of trade patterns involves an attempt to determine the feasibility of the location of new industries in the region, and the level of regional demand needed to induce such locations.

Shifts from imports to interindustry transactions will not be limited to goods-producing industries. As Table 4.1 shows, the Colorado River Basin is a substantial importer of services. Much construction activity in the region is also conducted by firms located elsewhere. As a region develops, an increasing share of services will be provided by establishments located in the region. All changes of this kind will affect regional input patterns and must be taken into account in making long-range projections.

There can also be changes in output patterns which will affect the structure of the regional economy. At an early stage of its development a region might be a heavy exporter of raw materials. As the region develops, manufacturing or other intermediate processing establishments might locate closer to the raw materials. Under these circumstances export sales of the sector producing raw materials will decline, interindustry transactions will rise, and the new intermediate processing sector will export products with more value added. Interregional shifts on the output side are probably more difficult to anticipate than those on the input side. But since they can significantly affect interindustry transactions the analyst must do everything he can to anticipate them.

Regrettably, there is no analytical model that will readily identify shifts in regional inputs and outputs. It is necessary to fall back on analysis of long-run trends, location theory, and the advice of industry and marketing experts. Those who have a preference for rigorous models that completely avoid judgmental influences will do well to steer clear of regional forecasting.

## IV

In spite of the qualifications mentioned above it can be argued that long-range regional forecasts based on an input-output model are superior to more highly aggregated projections. If the assumptions underlying regional forecasts are clearly spelled out, and if both changing technology and changing trade patterns are taken into account, the resulting projections can be useful to planners and policymakers. The major advantages of regional interindustry projections, as is true of their national counterparts, is the vast amount of detail which they provide and their internal consistency. As is also true of the national projections, if one disagrees with the assumptions upon which the regional projections are based he can substitute his own assumptions and derive new projections with a relatively small expenditure of time and money.

The results of projecting one set of regional input coefficients are given in Table 4.2. The calculations shows the effects on total gross output by sector when the projected input coefficients were applied to base-year final demands. Data in the table relate to the Lower Main Stem Sub-Basin of the Colorado River Basin for 1960.[12]

The changes in Table 4.2 are not directly comparable with those analyzed by Anne Carter (1966, 1967) for the national economy, since the national changes do not reflect shifts in trade patterns. But where technological influences can be isolated from trade pattern changes the regional changes are in the same direction—although not necessarily of the same magnitude—as those found in the national study.

Of particular significance in the regional table are the shifts within agriculture since this is an important activity in the region. Some sectors would have produced less in 1960 with projected coefficients but others would have produced more. Outside agriculture, all sectors would have produced more in 1960 with projected coefficients than they actually did.

Perhaps the outstanding characteristic of these projections is that, with a few exceptions, the changes are small. Where the changes are relatively large—in sectors 1, 3 and 15 through 17—they are due primarily to anticipated changes in trade patterns. Changes in the remaining sectors reflect, in the main, differences between best practice and average input coefficients in the base year without extrapolation to the final target year. The actual projections, which are not given here, show much larger changes, but these obviously are primarily attributable to anticipated increases in final demand.

An important point is that no one needs to settle for the projections given in Table 4.2. There might be disagreement, for example, about the length of time it will take for the projected coefficients to become "new" average coefficients. But this period can easily be varied before extrapolations are made. Once the basic tables for each of the subbasins are published, input coefficients can be revised on the basis of alternative sets of assumptions and new projections derived in a short time. The only part of the process that requires a great deal of time is the construction of the transactions tables and the computation of best practice coefficients. Once these are available, and an initial set of projections has been made to provide bench marks, the projections can be varied—or completely revised—in a matter of hours. And the computation of new projections, once the data are fed to a computer, involves only a matter of minutes.

One final comment is in order. Even the relatively modest shifts in interdependence shown in Table 4.2 suggest a net gain in investment in this region between the base year and that represented by the projected coefficients. It is simply

Table 4.2. Total Gross Output, Lower Main Stem SubBasin, 1960 Actual and "Projected" Based on Best Practice Coefficients (*Thousands of Dollars*)

| Sector | Total Gross Output[a] (1) 1960 coefficients | (2) Projected coefficients | (3) Differences (2) − (1) | (4) Percent change |
|---|---|---|---|---|
| 1. Range livestock | $    8,069 | $    5,885 | $ − 2,184 | − 27.1 |
| 2. Feeder livestock | 18,574 | 18,672 | 98 | 0.5 |
| 3. Dairy | 1.965 | 1,434 | −    531 | − 27.0 |
| 4. Forage, feed and food crops | 11,324 | 11,159 | −    165 | −  1.5 |
| 5. Cotton | 13,943 | 14,001 | 58 | 0.4 |
| 6. Vegetable and melon products | 28,563 | 28,591 | 28 | 0.1 |
| 7. Citrus crops | 3,370 | 3,373 | 3 | 0.1 |
| 8. Forestry | 3,496 | 3,386 | −    110 | −  3.2 |
| 9. All other agriculture | 6,560 | 6,678 | 118 | 1.8 |
| 10. Uranium | 7,283 | 7,382 | 99 | 1.4 |
| 11. All other mining | 20,137 | 21,501 | 1,364 | 6.8 |
| 12. Food and kindred products | 15,889 | 16,343 | 454 | 2.8 |
| 13. Lumber and wood products | 14,197 | 14,934 | 737 | 5.2 |
| 14. Chemicals | 19,617 | 20,637 | 1,020 | 5.2 |
| 15. Printing and publishing | 7,979 | 9,321 | 1,342 | 16.8 |
| 16. Fabricated metals | 1,706 | 2,533 | 827 | 48.5 |
| 17. Stone, clay and glass | 15,182 | 16,822 | 1,640 | 10.8 |
| 18. All other manufacturing | 33,836 | 34,041 | 205 | 0.6 |
| 19. Wholesale trade | 39,971 | 41,740 | 1,769 | 4.4 |
| 20. Service stations | 13,818 | 14,718 | 900 | 6.5 |
| 21. Eating and drinking places | 35,411 | 35,672 | 261 | 0.7 |
| 22. All other retail | 93,625 | 94,515 | 890 | 1.0 |
| 23. Agricultural services | 19,961 | 20,454 | 493 | 2.5 |
| 24. Lodging | 122,312 | 122,481 | 169 | 0.1 |
| 25. All other services | 152,330 | 155,005 | 2,675 | 1.8 |
| 26. Transportation | 48,275 | 50,900 | 2,625 | 5.4 |
| 27. Electric energy | 30,488 | 33,542 | 3,054 | 10.0 |
| 28. All other utilities | 31,513 | 33,223 | 1,710 | 5.4 |
| 29. Contract construction | 159.430 | 168,232 | 8,802 | 5.5 |
| 30. Rentals and finance | 96,750 | 101,005 | 4,255 | 4.4 |
| TOTAL | $1,075,574 | $1,108,180 | $32,606 | 3.0 |

[a] Assuming constant (1960) final demand.

*Source*: See Table 4.1.

assumed that this investment will be forthcoming. There is no explicit investment term in the interindustry equations of this model. This is an important deficiency which, it is hoped, can be corrected in future regional studies.

No attempt will be made to defend the projections used to illustrate the method discussed in this paper. They undoubtedly are on the conservative side due to understatement of technological influences. But we are just beginning to learn something about the effects of technological change on national input coefficients.

There is a pressing need for investigation of these effects at the regional level. The only advantages claimed for the method discussed here are that: (a) it takes into account significant differences in economic structure among regions thus avoiding the criticism of earlier regional input-output studies that assumed constancy of regional production functions and interarea coefficients, and (b) it is a flexible method that can easily accommodate a variety of assumptions to produce alternative sets of projections.

A major drawback of the approach is that it is relatively costly to implement. Furthermore, it is far from certain that the collection of input-output data by small, independent research organizations represents an efficient use of resources. An ideal setup would be to have all state data reported in the censuses of manufactures, agriculture, business, mining, and transportation on a standardized input-output basis. This would permit the construction of state tables showing detailed shipments by sector to all other states. From a data point of view this would be a modular system that would permit states to be combined into any desired set of regions. Since all interstate transactions would balance, the result would be a functional and flexible multiregional model of the national economy.[13]

After tables of this kind have been constructed for a number of census years it would be possible to measure trend changes in average regional input coefficients. Since the trend changes would reflect both technological and trade pattern influences these coefficients could be used for forecasting. For a given set of regions the forecasts would be consistent with national projections. But independent projections could be made for any region consisting of a collection of states.

This arrangement would not solve all regional input-output problems. In some cases, such as New England, economic regions are coterminous with state boundaries. But there are other regions, such as the Colorado River Basin or Appalachia, whose boundaries cross state lines and which are defined by county lines. It would not be feasible to collect input-output data on a county basis, however, and if an interindustry analysis were desired for a region of this type, state data would have to be disaggregated. It is possible that limited surveys would be needed to determine shipments into and out of the specialized region.

A highly disaggregated classification of sectors would be required for the state tables to avoid some of the problems discussed in this paper. Such a proposal would require business establishments to record the source of purchases and the destination of sales, and, in some situations, purchases and sales would have to be traced through several layers of transactions. But the problems which such situations suggest are not insuperable or even of a high order of difficulty. The availability of data disaggregated along spatial as well as sectoral lines would add greatly to the power of analytical tools currently available. And the gains to users of statistics would appear to far outweigh the additional costs and inconveniences entailed by a new form of record-keeping.

NOTES

1. The standard procedure is to expand output (row) and input (column) data separately followed by reconciliation of discrepancies. The latter requires the exercise of judgment which is generally based on evaluations of the reliability of data obtained from sales versus purchasing departments in the sample establishments.
2. The regional data in Table 4.1 were aggregated from six subregional tables each covering a subbasin of the CRB. Kenneth L. Shellhammer supervised the tedious clerical task of aggregating the regional table, and the subsequent task of making the national and regional tables comparable. John H. Chapman, Jr., who was involved in the original CRB study, provided invaluable assistance during the aggregation of the subbasin tables.
3. For a detailed discussion of the difficulty of estimating regional from national coefficients see (Isard, 1960, pp. 327–371; Tiebout, 1957, pp. 141–146). The limitations of such a model for projection purposes are discussed by Isard (1960, pp. 338–343).
4. It should be emphasized that the time and expense are involved in constructing the basic transactions table, not in the use of this table after it has been completed.
5. What was missing in the CRB study is a table of capital coefficients to supplement the table of transactions on current account. Sample data on capital transactions and capacity utilization in West Virginia have been collected by the Regional Research Institute of West Virginia University and are to be used to implement a regional version of Almon's dynamic model.
6. Tilanus (1967) analyzed marginal input coefficients for the Netherlands derived from 13 consecutive input-output tables constructed between 1948 and 1960.
7. Since man-hour data are not available at the regional level a substitute measure must be employed. In the Colorado River Basin study, for example, two measures were used. One was output per employed worker, and the second was output divided by labor plus capital inputs, using depreciation data to estimate the latter. While both of these measures are quite rough they give some indication of variations in efficiency among establishments, and this is the only purpose for which they were used.
8. For further details on the method of projection, and a hypothetical example (Miernyk, 1965, pp. 117–125).
9. For illustrations of these effects on a county-by-county basis from 1940 through 1960 (Miernyk, 1965; Huston, 1967).
10. In their analysis, Isard and Kuenne (1953) used national coefficients since regional coefficients were not available. They thought this would not entail "serious distortions" because in their view the resource characteristics and industrial composition of the region were not significantly different from national averages. They also included households in the endogenous portion of their model as an approximate offset to import leakages which were not specifically taken into account.
11. The paper illustration was chosen because it actually occurred during the Colorado River Basin study. Other trade pattern changes were projected on the basis of a series of industry location studies and upon crop pattern studies by agricultural experts.
12. It should be emphasized that the actual projections of total gross output by sector differ significantly from the changes shown in Column (2) of Table 4.2. Because of anticipated increases in total final demand every sector in the subbasin is expected to register an increase in total sales between the base and target years. The declines shown in Column (3) and (4) reflect only anticipated changes in trade and crop patterns.
13. The CRB study is a small-scale prototype of such a model since all transactions among subbasins are balanced.

References

Almon, C., Jr. 1966. *The American Economy to 1975*. New York: Harper & Row.
Ashby, L. D. 1965. *Growth Patterns in Employment by County*. 8 vols. U.S. Department of Commerce, Office of Business Economics. Washington.
Carter, A. P. The Economics of Technological Change. *Scientific American 214* (April): 25–31.

Carter, A. P. 1966. Changes in the Structure of the American Economy, 1947 to 1958 and 1962. *Rev. Econ. Stat.*, 49 (May): 209–224.

Huston, D. B. 1967. The Shift and Share Analysis of Regional Growth: A Critique. *Southern Econ. Jour.*, 34 (April): 577–581.

Isard, W. 1951. Interregional and Regional Input-Output Analysis: A Model of a Space Eonomy. *Rev. Econ. Stat.*, 33 (November): 318–328.

Isard, W. 1960. *Methods of Regional Analysis: An Introduction to Regional Science.* New York: John Wiley.

Isard, W., and R. E. Kuenne. 1953. The Impact of Steel Upon the Greater New York-Philadelphia Industrial Region. *Rev. Econ. Stat.*, 35 (November): 289–301.

Leontief, W. *et al.* 1965. The Economic Impact—Industrial and Regional—of an Arms Cut. *Rev. Econ. Stat.*, 47 (August): 217–241.

Leontief, W. *et al.* 1966 *Input-Output Economics.* New York: Oxford University Press. (Second edition 1986.)

Miernyk, W. H. 1965. *The Elements of Input-Output Analysis.* Boulder: University of Colorado.

Miller, R. E. 1966. Interregional Feedback Effects in Input-Output Models: Some Preliminary Results. *Regional Science Assn. Papers*, 17: 105–125.

The Transactions Table of the 1958 Input-Output Study and Revised Direct and Total Requirements Data. 1965. *Survey of Current Business*, September: 33–49.

Tiebout, C. M. 1957. Regional and Interregional Input-Output Models: An Appraisal. *Southern Econ. Jour.*, 24 (October): 140–147.

Tilanus, C. B. 1967. Marginal Versus Average Input Coefficients in Input-Output Forecasting. *Quart. Jour. Econ.*, 81 (February): 140–145.

U.S. Department of Labor. Bureau of Labor Statistics. 1966. Interagency Growth Study Project. *Projections 1970.* Bulletin No. 1536. Washington.

# 5

## The Implementation of a Multiregional
## Input-Output Model for the United States

KAREN R. POLENSKE

Within the last decade, an increasing number of policy decisions at all levels of government have required an extensive knowledge of regional economic conditions. This requirement will undoubtedly expand in the future. Evidence of the need for economic analysis applied at a regional level is readily apparent in many of the proposals presently being discussed in relation to energy, water, fuel requirements, transportation, environmental control, employment, arms de-escalation, inflation, and so on. A multiregional input-output model can be used to assist in a rigorous analysis of these and other economic issues.

A multiregional input-output model that includes 44 regions and 78 industries has been formulated and implemented at the Harvard Economic Research Project. To implement the model, a comprehensive set of regional data, which uses a common industrial and regional classification scheme and covers all industries and regions in the economy, has been assembled for the United States. Five major sets of multiregional input-output data have been compiled for each state: base-year outputs, employment, and payrolls; 1963 interindustry flows; 1963 interregional trade flows; base-year final demands; and 1970 and 1980 projected final demands. A consistent set of multiregional input-output tables has been assembled for each state for 1963, and supplemental historical state final demand estimates have been made for 1947 and 1958. For each of the three years, the state estimates are consistent at the aggregate level with the respective Office of Business Economics (OBE) national input-output tables. The six components of final demand have been projected to 1970 and 1980 and have been made consistent with the national 1970 and 1980 projections published by the Bureau of Labor Statistics (BLS). Given the base-year technology and interregional trade data and the projected sets of final demands, 1970 and 1980 regional outputs and shipments of commodities among regions have been generated simultaneously using a multiregional input-output model described later in this paper. The sets of data that were assembled and calculated during the course of this study are listed in Table 5.1.

For all except the interregional trade flows, the data were assembled for 86 industries in 51 regions (50 states plus the District of Columbia)[1] using the OBE

This article was published in *Input-Output Techniques*, edited by Andrew Bródy and Anne P. Carter. © North-Holland Publishing Company, 1972, pp. 171–189. Reprinted by permission.

Table 5.1. Multiregional Input-Output Data for the United States

| Name of matrix | Matrix dimension | | Years | |
|---|---|---|---|---|
| 1. Payrolls, employment, and output[a] | | | | |
| Payrolls | 51 × 86 | 1947 | 1958 | 1963 |
| Employment | 51 × 86 | 1947 | 1958 | 1963 |
| Output | 51 × 86 | 1947 | 1958 | 1963 |
| 2. Final demands (6 matrices for each year) | 51 × 86 | 1947 | 1958 | 1963 |
| 3. Projected final demands[a] (6 matrices for each year) | 51 × 86 | | 1970 | 1980 |
| 4. Regional input-output tables (51 matrices) | 83 × 83 | 1963 | | |
| 5. Interregional trade flows (61 matrices) | 44 × 44 | 1963 | | |
| 6. Projected outputs and interregional trade 1970 and 1980 projected outputs (2 matrices) | 44 × 78 | | 1970 | 1980 |
| 1970 and 1980 projected interregional trade flows (122 matrices) | 44 × 44 | | 1970 | 1980 |

[a] These data were assembled for the project by Jack Faucett Associates, Inc.

input-output industrial classification scheme and the regional classification given in Table 5.A.2 in the appendix. Since there are 4386 figures for each particular component, a complete multiregional input-output set of data for a single year contains more than 300,000 numbers. A considerable amount of research effort was of course required just to assure that as the data were assembled, an internal consistency was maintained between the state figures and the national aggregates.

National income accounts were first officially published for the United States in the early 1930s. The first input-output table published by the government was one for 1947, prepared by the Bureau of Labor Statistics (U.S. Department of Labor, 1947; U.S. Bureau of Economic Analysis, 1970). Since 1950, two additional national tables have been published—the 1958 table in 1965 (U.S. Department of Labor, 1965) and the 1963 table in 1969 (U.S. Department of Labor, 1969), these by the Office of Business Economics. As far as is known, the only sets of official interregional tables available for any country are the 1960 and 1965 Japanese tables prepared by the Ministry of International Trade and Industry (MITI, 1970; MITI, n.d.). With the expansion of interest in regional economic research around the world, however, it is obvious that regional economic accounts will be assembled in more countries in the future. The United States government has never prepared either complete and consistent sets of regional income accounts or multiregional input-output tables, although consideration is presently being given to the preparation of both. The multiregional input-output (MRIO) accounting system used for the present study is described here in detail.

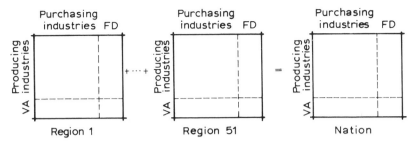

Fig. 5.1. Regional input-output tables.

## A Multiregional Accounting System

For the implementation of the United States multiregional input-output model, three basic sets of data are required: interindustry flows, final demands, and interregional flows.

Figure 1 shows how the first two sets of data are organized. All of the intermediate and final current account purchases made by industries and by public and private consumers located within a particular region are represented in the figure by the large rectangular blocks, which will be referred to hereafter as regional input-output tables. Each row of a specific regional table shows the total distribution of a commodity to the intermediate and final users within that region. Each column of the table shows the total purchases of goods, services, and value added by the particular intermediate or final purchaser located in that region. It should be noted that the multiregional accounting system described here provides a detailed specification of goods, services, and value added by the region in which they are consumed, regardless of their region of origin. To complete the multiregional system, the accounts must be expanded to include trade among regions.

The large square within each regional table represents the interindustry sales and purchases. The rows specify the producing industry but do not designate the region in which the good was produced. The columns specify the purchasing industry which is actually located within the region. For the present model, the number of producing industries exactly equals the number of purchasing industries.

The rectangle at the right of each regional table represents the purchases by final consumers (public and private) in the region. These include personal consumption expenditures, gross private capital formation, net inventory change, net foreign exports, and federal, state, and local government purchases of goods and services.

The rectangle along the bottom of each regional table represents the payments to factors of production: wages and salaries, rent, depreciation, taxes, etc. All of these are lumped in the 1963 national input-output table and are referred to as value added.

The small square in the lower right-hand corner of each regional table represents essentially the wage and salary payments to domestic household employees and

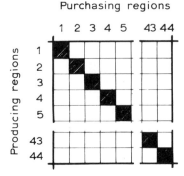

Purchasing regions

Fig. 5.2. Interindustry coefficient matrices.

government employees. In addition, some balancing items for the foreign sector are included here.

A complete set of these regional account data was prepared for 51 regions for 1963. As mentioned earlier, final demand data, consistent with this accounting framework, were assembled for 1947, 1958, 1963, 1970, and 1980. The technology and interregional trade flows were compiled for 1963 only.

A word of caution is required to those users accustomed to working with a balanced national input-output table. For a particular regional input-output table, the sum of all the elements in each column gives the total input requirements of each industry in the region (i.e., total value of production), while the sum of all elements in each row of the table gives the total consumption by users within a region. The sums of corresponding rows and columns of a regional table will not necessarily be equal, with the differences being attributable to interregional trade. In the national table the sums of corresponding rows and columns must be equal, since the total consumption of output must equal the total production for each industry.

Figure 5.2 portrays the final organization of the regional input-output tables. For the implementation of the MIRO model, the tables had to be aggregated from the 51 to the 44 regions for which transportation data were available. Each interindustry flow was divided by the total regional output of the respective industry; therefore, the regional input coefficients give the total input of a specific good or service required by an industry within a particular region per unit of output of that industry. In Figure 5.2 the industrial input coefficient matrices for each region have been rearranged into a large block diagonal matrix with all blocks on the diagonal having entries and all blocks off the diagonal being zero. Each block has dimensions $78 \times 78$ to include the 78 producing industries only.

The industrial purchases by intermediate and final users within a region form one part of the multiregional accounting system. To complete the accounts, a set of interregional trade flows is also required. Trade flows were assembled for 61 of the 86 industries and for 44 regions, which are direct aggregates of the 51 regions. For the MRIO model, shipments were not estimated for any of the construction, service, dummy, or value added industries. Interregional trade flows were assembled only for products that use the normal modes of freight transportation and that are included in the regular transportation statistics.

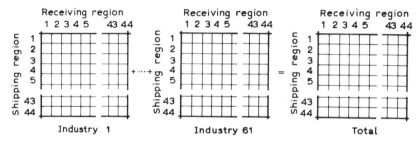

Fig. 5.3. Interregional commodity trade flows.

Figure 5.3 shows the organization of the interregional trade flows.

Each of the 61 commodity trade matrices, representing shipments of the products of a single industry, is square and shows for the products of one industry the total shipments among all the regions in the economy. The same regions are listed along the rows and columns of each matrix. For example, the first row of the first commodity trade matrix shows the shipments to each region of the products of Industry 1 produced by the first region. In the MRIO accounting system, this production coincides with the total of the first column in the first regional input-output table shown in Figure 5.1. The first column of the first commodity matrix in Figure 5.3 shows the shipments of the products of Industry 1 into the first region from all the regions in the economy. The sum of the column equals total consumption of the products of Industry 1 in Region 1, and in the MRIO accounting system, this equals the row sum of the first regional input-output table shown in Figure 5.1. Consistency is thus maintained between the data in the regional input-output table and the flows in the interregional trade tables.

The 61 commodity matrices for the United States give a complete specification of the interregional shipments among the 44 regions. For the implementation of the MRIO model, each interregional trade flow, represented in Figure 5.3 was divided by the sum of the column and these trade coefficients were rearranged into a large matrix composed of $44 \times 44$ diagonal block matrices, as shown in Figure 5.4. Each of the diagonal block matrices is square, with 78 rows and columns—one row and one column for each industry. Although this is not represented in the figure, rows and columns 4, 11, 12, and 65 through 78 are actually zero in all except the blocks of intraregional matrices because no interregional shipments could be estimated for these service industries. In all the blocks of matrices, the elements off the diagonal are zero.

As the preceding discussion has shown, the introduction of regions into the input-output framework creates a fairly complicated set of accounts. In the appendix, the economic relationships between the interregional trade flows and the regional input-output tables are given in mathematical terms.

## The United States Economy

The multiregional input-output model for the United States was implemented at the Harvard Economic Research Project in the fall of 1970 for 78 industries and 44

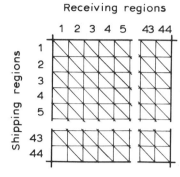

Fig. 5.4. Multiregional input-output trade matrix.

regions. As each of the three major data components (final demands, interindustry purchases, and interregional trade flows) was assembled, consistency checks were made between the state or regional estimates and the national input-output data. Consistent 1970 and 1980 projections were then made both of industrial outputs and of interregional commodity flows. Although there has not yet been time to evaluate the data in detail, industry by industry, region by region, year by year, a few summary characteristics are immediately apparent.

**Final Demands**

For the present study, final demands were assembled for 51 regions and 86 industries for each of the six components of the gross national product (GNP): personal consumption expenditures, gross private capital formation, net inventory change, gross exports, federal government purchases, and state and local government net purchases of goods and services. The 1947, 1958, and 1970 distributions of GNP among what are generally considered to be the four major components are presented in Figure 5.5 for the nine census regions.[2] The increased public expenditures that were made during 1958, a recession year, are readily apparent from the figure. For the United States as a whole, personal consumption expenditure have remained a fairly constant proportion of the gross national product. Its proportion was the highest in 1947, when the increased private consumption expenditures that occurred at the end of World War II represented 68 percent of the total GNP.

The state estimates prepared for the study indicate that there are wide variations in the distribution of GNP from state to state. For example, in 1958, the personal consumption expenditures percentage varies from a low of 43 for Vermont to a high of 72 for Rhode Island and Wisconsin. For the nine census regions, some of these wide disparities cancel each other out, but even at this more aggregate level, significant regional differences in the distribution of GNP among the components can be seen. Some of these differences seem to be lessening over time. In 1947, there was a variation of 17 percentage points between the region with the highest personal consumption percentage (Region 2) and the one with the lowest (Region 8). By 1958, this variation had been reduced to 10 percentage points.

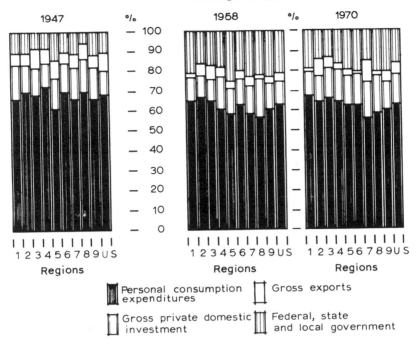

Fig. 5.5. Distribution of major components of the GNP for the nine census regions: 1947, 1958, and 1970.

## Technology

For each of the 51 regions, a complete set of 1963 direct input coefficients is now available. After the data had been assembled, a brief analysis of variations in regional technology was conducted. The calculations showed that large differences occur from region to region in the input structure in the agricultural industries, food products, wood and paper products, printing, petroleum products, primary nonferrous manufacturing, other transportation equipment, broadcasting, and state and local government enterprises. Regional variation in technology seems to be concentrated in primary industries and industries producing primary commodities. Future research, however, should be directed toward a comprehensive examination of the technology matrices.

## Outputs, Employment, and Payrolls

Output, employment, and payroll data were assembled for 1947, 1958, and 1963. These data had to be used frequently to estimate other components of the multiregional input-output accounts. The three sets of data together provide considerably more information than is evident at first glance. Payroll divided by employment is a measure of wages, output divided by employment is a measure of labor productivity, and output divided by payroll is a measure of unit labor costs. For regional economic studies, of course, aggregate payrolls and employment should be stratified by type of employee.

The time profile for 1947 to 1958 to 1963 does provide some indication of expanding and contracting industries or regions. During this period, there has been a movement of employment away from primary production, light industry, machine products, transportation and trade, and personal services toward the new industries of electronics, aircraft, ordnance, packaging, and the financial, business, utility, and mass media services. Employment in the basic industries, such as iron, oil, construction, vehicles, and consumer nondurables, seems to have changed little in relative importance.

State shares of employment, payrolls, and output have also changed from 1947 to 1963. The percentage of total United States employment has increased significantly for states in the West and South, and a corresponding decrease has occurred for states in the Midwest, Middle Atlantic, and New England regions. Although the total relative employment in a state may have decreased, this will not have occurred, of course, for all industries within that state. Complete information for all industries within each state is contained in the matrices listed under number 1 in Table 5.1.

### Projections

For the implementation of the multiregional input-output model for the United States, the 1970 and 1980 projected final demands were used together with the 1963 technical and trade coefficients to obtain 1970 and 1980 regional outputs and interregional trade flows. The outputs for 1947, 1958, and 1963 were aggregated to 44 regions for comparison with the 1970 and 1980 outputs calculated with the column coefficient model. The percentage distribution of total output among the 44 regions is shown in Table 5.2 for each of the five years.

The percentages for states in the West North Central region, such as North Dakota (12), South Dakota (13), Nebraska (14), and Kansas (15), decrease steadily between 1947 and 1980. It is apparent, however, from the table that the 1947 to 1963 trends are not always continued on to 1970 and 1980. In California, for example, the percentage of total United States output increased from 7.5 in 1947 to 10.4 in 1963, but this percentage is shown to decrease in 1970 to 9.8 and to increase again to 10.4 in 1980. Although only summary information can be presented in this paper, it should be emphasised that outputs for the five years are available for each of 78 industries in the 44 regions.

### Interregional Trade

Interregional trade data are now available for 1963, 1970, and 1980 for each of the 61 industries that produce goods that are shipped by the normal modes of

Table 5.2. Regional Percentage Distribution of Outputs 1947, 1958, 1963, 1970, 1980

| Region | Year | | | | |
|---|---|---|---|---|---|
| | 1947 | 1958 | 1963 | 1970 | 1980 |
| 1 Maine | 0.48 | 0.45 | 0.40 | 0.38 | 0.37 |
| 2 Vermont | 0.21 | 0.17 | 0.17 | 0.17 | 0.17 |
| 3 Massachusetts, Rhode Island, New Hampshire, Connecticut | 5.64 | 5.12 | 5.08 | 5.11 | 5.05 |
| 4 New York | 11.67 | 11.25 | 10.47 | 11.00 | 10.49 |
| 5 Pennsylvania, New Jersey | 10.78 | 10.11 | 9.57 | 9.95 | 9.90 |
| 6 Michigan, Ohio | 11.03 | 10.69 | 11.09 | 11.60 | 11.61 |
| 7 Indiana, Illinois | 10.85 | 10.07 | 9.54 | 9.93 | 9.69 |
| 8 Wisconsin | 2.50 | 2.35 | 2.31 | 2.44 | 2.39 |
| 9 Minnesota | 2.21 | 1.94 | 1.88 | 1.84 | 1.83 |
| 10 Iowa | 2.23 | 1.73 | 1.64 | 1.55 | 1.47 |
| 11 Missouri | 2.57 | 2.48 | 2.40 | 2.38 | 2.27 |
| 12 North Dakota | 0.46 | 0.33 | 0.27 | 0.25 | 0.23 |
| 13 South Dakota | 0.51 | 0.36 | 0.32 | 0.27 | 0.25 |
| 14 Nebraska | 1.11 | 0.92 | 0.86 | 0.75 | 0.72 |
| 15 Kansas | 1.48 | 1.28 | 1.18 | 1.03 | 1.00 |
| 16 Delaware, Maryland | 1.61 | 1.91 | 1.96 | 1.91 | 2.00 |
| 17 District of Columbia | 0.52 | 0.49 | 0.52 | 0.52 | 0.50 |
| 18 Virginia | 1.70 | 1.78 | 1.84 | 1.66 | 1.70 |
| 19 West Virginia | 0.95 | 0.80 | 0.78 | 0.70 | 0.73 |
| 20 North Carolina | 2.17 | 2.17 | 2.31 | 2.37 | 2.33 |
| 21 South Carolina | 0.92 | 0.91 | 0.96 | 1.01 | 1.05 |
| 22 Georgia | 1.55 | 1.75 | 1.85 | 1.91 | 1.95 |
| 23 Florida | 1.19 | 1.99 | 2.19 | 2.11 | 2.31 |
| 24 Kentucky | 1.30 | 1.34 | 1.30 | 1.22 | 1.24 |
| 25 Tennessee | 1.43 | 1.47 | 1.54 | 1.56 | 1.58 |
| 26 Alabama | 1.23 | 1.26 | 1.30 | 1.27 | 1.31 |
| 27 Mississippi | 0.65 | 0.63 | 0.68 | 0.65 | 0.67 |
| 28 Arkansas | 0.68 | 0.63 | 0.70 | 0.71 | 0.74 |
| 29 Louisiana | 1.48 | 1.64 | 1.66 | 1.72 | 1.75 |
| 30 Oklahoma | 1.17 | 1.07 | 1.08 | 1.00 | 0.99 |
| 31 Texas | 4.84 | 5.44 | 5.24 | 5.51 | 5.53 |
| 32 Montana | 0.40 | 0.38 | 0.38 | 0.31 | 0.30 |
| 33 Idaho | 0.35 | 0.33 | 0.31 | 0.27 | 0.27 |
| 34 Wyoming | 0.22 | 0.22 | 0.23 | 0.21 | 0.20 |
| 35 Colorado | 0.75 | 0.92 | 0.93 | 0.84 | 0.87 |
| 36 New Mexico | 0.28 | 0.43 | 0.39 | 0.35 | 0.35 |
| 37 Arizona | 0.36 | 0.56 | 0.67 | 0.61 | 0.65 |
| 38 Utah | 0.38 | 0.48 | 0.48 | 0.41 | 0.40 |
| 39 Nevada | 0.11 | 0.17 | 0.24 | 0.22 | 0.22 |
| 40 Washington | 1.49 | 1.56 | 1.45 | 1.35 | 1.37 |
| 41 Oregon | 0.96 | 0.96 | 0.97 | 0.85 | 0.81 |
| 42 California | 7.55 | 9.50 | 10.44 | 9.78 | 10.35 |
| 43 Alaska | — | — | 0.11 | 0.09 | 0.08 |
| 44 Hawaii | — | — | 0.34 | 0.26 | 0.27 |

transportation. Calculations were made for inflows as percent of consumption and outflows as percent of production. For the 1970 and 1980 projected interregional trade flows, the percentages for the inflows will be exactly the same as the 1963 figures because this was the assumption adopted for the MRIO model. The outflow as percent of production figures changes only slightly between 1970 and 1980. When the 1963 and the 1970 or 1980 percentages are compared, however, significant variations can be found.

Since the calculations for the MRIO model have only recently been completed, there has not been time for an extensive review of the data. But the entire set of multiregional input-output data assembled for this study can be used with the MRIO model to conduct thorough regional economic analyses. A comprehensive, systematic analysis of regional economic problems is certainly required for the successful implementation of many government programs and for the investigation of some of the problems confronting American industries.

## Appendix 1. The Multiregional Input-Output Model

A number of spatially differentiated, general equilibrium models could be specified to estimate interregional trade and regional outputs for an economy. Three fixed coefficient models were considered for use in this study: the row coefficient, column coefficient, and point estimate gravity trade models. Previously, only a limited amount of multiregional input-output data has been available, and consistent sets of regional input-output data for two or more years were nonexistent; consequently, empirical testing of the models has been restricted. Now, however, the sets of Japanese tables for 1960 and 1965, published in Japan by the Ministry of International Trade and Industry (MITI, 1970; MITI, n.d.), provide some data for testing of the models. Because the U.S. data assembled for the present study cover more industries and regions than previous sets of multiregional input-output data, new computation techniques had to be developed for handling such large-scale calculations (Cohen et al., 1970). Improvements in the techniques will undoubtedly be made as more experience in working with MRIO models is accumulated.

The column coefficient model was the first fixed coefficient model to be used in multiregional input-output studies. It was originally implemented by Chenery (1953) in a two-region study of the Italian economy and later by Moses (1955) for a three-region analysis of the American economy. The present form of the gravity model was first tested for individual commodity shipments and is discussed in a paper given by Leontief and Strout at the 1961 conference (Leontief and Strout, 1966) and by this writer in a paper given at the 1968 conference (Polenske, 1970a).

At present the row coefficient model has been tested only by this writer in two separate tests (Polenske, 1966; 1970b). Because in the last test the estimates using the row coefficient model were extremely inaccurate, the model was not considered for the final implementation of the United States model. The results for both the column and gravity trade models were reasonable, and the conclusion was that ". . . the tests reveal no discernible difference in the overall predictive ability of the column coefficient and the point estimate gravity trade models" (Polenske, 1970b, p. 82).

Both models were operational when the 9-region, 10-industry Japanese data were tested. In the last stages of computations with the 44-region, 78-industry United States data, however, problems were encountered in trying to obtain a convergent solution for the gravity model. The time required to determine the causes of the problems and find solutions to them was not available, and it was therefore determined that since the Japanese tests had shown that both models provided estimates with similar degrees of accuracy, the column coefficient model would be employed for the final round of the United States computations.

Because all of the models were tested during this study, notations are given for all three, although the complete set of equations will be presented only for the column coefficient model, the one which was finally selected to implement the U.S. model. The notations for the models follow:[3]

$\hat{\phantom{x}}$ indicates a block diagonal matrix;

$\Delta$ indicates the change between a base-year and the given year;

$n$ designates the number of regions;

$m$ designates the number of commodities.

Although the column coefficient model is a linear model, the $\Delta$s are used because they are required later when this model is compared with the linearized form of the gravity trade model.

## Appendix 2. Matrix Notation

$\Delta X$    column vector $(mn \cdot 1)$ giving the change in production. Each element describes the change in output of commodity $i$ produced in region $g$.

$\Delta Y$    column vector $(mn \cdot 1)$ giving the change in total final demand. Each element describes the change in the total amount of commodity $i$ consumed by final users in region $g$ regardless of the place where the good was produced.

$\hat{A}$    block diagonal matrix $(mn \cdot mn)$ with $n$ square matrices $(m \cdot n)$ of input coefficients along the diagonal describing the structure of production in each region. For the United States MRIO model, separate regional technical coefficients were assembled for each region.

$S, T$    each is a square matrix $(nm \cdot nm)$ filled with diagonal matrices $(m \cdot m)$. The $t_i^{gh}$ elements relate outflows from the region $g$ to the production in the region while the $s_i^{gh}$ elements relate inflows into region $h$ to the total consumption in the region.

$R$    square matrix $(nm \cdot nm)$ filled with diagonal matrices $(m \cdot m)$. Each $r_i^{gh}$ element describes the fraction of total production of commodity $i$ in region $g$ that is exported to region $h$. The sum of each row of this matrix must equal 1, since the coefficients are proportions of total production.

$C$    square matrix $(nm \cdot nm)$ filled with diagonal matrices $(m \cdot m)$. Each element $c_i^{gh}$ describes the fraction of total consumption of commodity $i$ in region $h$ that is imported from region $g$. The sum of each column of this matrix must equal 1, since the coefficients are proportions of total consumption. It is assumed that each industry within region $h$ will consume the same fraction as imports:

$$c_i^{gh} = c_{i1}^{gh} = c_{i2}^{gh} = \cdots = c_{im}^{gh}$$

## Appendix 3. Column Coefficient Model

The column coefficient model in matrix form is written as:

$$\Delta X^{0r} = A^r \Delta X^{r0} + \Delta Y^r \tag{5-1}$$

$$\Delta X^{g0} = \sum_{r=1}^{n} \hat{C}^{gr} \Delta X^{0r} \tag{5-2}$$

$$\sum_{r=1}^{n} \Delta X^{0r} = \sum_{r=1}^{n} \Delta X^{r0} \tag{5-3}$$

where $r = 1, \ldots, n;\ g = 1, \ldots, n - 1$.

Equations (5–1) and (5–2) can be combined to obtain:

$$\Delta X^{g0} = \sum_{r=1}^{n} \hat{C}^{gr} A^r \Delta X^{r0} + \sum_{r=1}^{n} \hat{C}^{gr} \Delta Y^r \tag{5-4}$$

where $g = 1, \ldots, n - 1$.

When the technical coefficients, trade coefficients, and final demands are given, the set of $mn$ unknown outputs can be solved using the $mn$ equations. The complete set of equations is written as:

$$(I - C\hat{A})\Delta X = C\Delta Y \tag{5-5}$$

which when solved for $\Delta X$ becomes

$$\Delta X = (I - C\hat{A})^{-1} C\Delta Y \tag{5-6}$$

The column coefficient model can be easily compared with the row coefficient and the point estimate gravity trade models. Some of the differences and similarities in the three fixed coefficient models are shown in the summary in Table 5.A.1. As shown in that table, the three systems of equations can be rearranged until the three trade matrices are in the form, $R'$, $C^{-1}$, and $S^{-1}T'$. Both the row and column coefficient models are linear. Although the interregional trade equation for the gravity model is nonlinear, it is linearized for the actual implementation of the model. The three fixed coefficient models were tested on the 10-industry, 9-region data for Japan (Polenske, 1970b) before the United States MRIO model was implemented.

## NOTES

1. In the case of the interregional trade flows, detailed data could be obtained for only 44 regions and 61 industries, because the service and value added industries do not trade commodities.

Table 5.A.1. Summary of the Multiregional Input-Output Models

| | Row coefficient model | Column coefficient model | Gravity model[a] |
|---|---|---|---|
| Trade coefficient equation | $x_i^{gh} = r_i^{gh} x_i^{g0}$ | $x_i^{gh} = c_i^{gh} x_i^{0h}$ | $x_i^{gh} = \dfrac{x_i^{g0} x_i^{0h}}{x_i^{00}} \cdot q_i^{gh}$ |
| Equation system in matrix form | $R'\Delta X = \hat{A}\Delta X + \Delta Y$ <br> $(R' - A)\Delta X = \Delta Y$ <br> $\Delta X = (R' - \hat{A})^{-1}\Delta Y$ | $\Delta X = C(\hat{A}\Delta X + \Delta Y)$ <br> $(I - C\hat{A})\Delta X = C\Delta Y$ <br> $\Delta X = (I - C\hat{A})^{-1}C\Delta Y$ <br> or $\Delta X = (C^{-1} - \hat{A})^{-1}\Delta Y$ | $T'\Delta X = S(\hat{A}\Delta X + \Delta Y)$ <br> $(T' - S\hat{A})\Delta X = S\Delta Y$ <br> $\Delta X = (T' - S\hat{A})^{-1}S\Delta Y$ <br> or $\Delta X = (S^{-1}T' - \hat{A})^{-1}\Delta Y$ |

[a] The system of equations for the gravity model incorporates a simplified version of the basic gravity trade coefficient equation. The elements of $S$ and $T'$ in the gravity model are defined as:

$$s_i^{gh} = x_i^{g0}\left[1 - \frac{x_i^{00} x_i^{gh}}{x_i^{g0} x_i^{0h}}\right] \quad \text{for} \quad g \neq k, \quad s_i^{gh} = 1 \quad \text{for} \quad g = k$$

$$t_i^{gh} = x_i^{0g}\left[1 - \frac{x_i^{00} x_i^{hg}}{x_i^{h0} x_i^{0g}}\right] \quad \text{for} \quad h \neq k, \quad t_i^{gh} = \quad \text{for} \quad h = k$$

where $k$ can be assigned arbitrarily.

Table 5.A.2. Regional Classification

| Regions | | |
|---|---|---|
| Census | MRIO | States |
| 1 | 1 | Maine |
| | 2 | Vermont |
| | 3 | Massachusetts, Rhode Island, New Hampshire, Connecticut |
| 2 | 4 | New York |
| | 5 | Pennsylvania, New Jersey |
| 3 | 6 | Michigan, Ohio |
| | 7 | Indiana, Illinois |
| | 8 | Wisconsin |
| 4 | 9 | Minnesota |
| | 10 | Iowa |
| | 11 | Missouri |
| | 12 | North Dakota |
| | 13 | South Dakota |
| | 14 | Nebraska |
| | 15 | Kansas |
| 5 | 16 | Delaware, Maryland |
| | 17 | District of Columbia |
| | 18 | Virginia |
| | 19 | West Virginia |
| | 20 | North Carolina |
| | 21 | South Carolina |
| | 22 | Georgia |
| | 23 | Florida |
| 6 | 24 | Kentucky |
| | 25 | Tennessee |
| | 26 | Alabama |
| | 27 | Mississippi |
| 7 | 28 | Arkansas |
| | 29 | Louisiana |
| | 30 | Oklahoma |
| | 31 | Texas |
| 8 | 32 | Montana |
| | 33 | Idaho |
| | 34 | Wyoming |
| | 35 | Colorado |
| | 36 | New Mexico |
| | 37 | Arizona |
| | 38 | Utah |
| | 39 | Nevada |
| 9 | 40 | Washington |
| | 41 | Oregon |
| | 42 | California |
| | 43 | Alaska |
| | 44 | Hawaii |

2. Gross private capital formation and net inventory change equal the gross private domestic investment component; and federal, state, and local government purchases equal the government component. The states within each census region are presented in Table 5.A.2 in the appendix.
3. For a more complete presentation of the models, the reader should refer to the two papers mentioned earlier (Polenske, 1970a; 1970b).

## References

Chenery, H. B. 1953. Regional analysis. In *The Structure and Growth of the Italian Economy*, ed. H. B. Chenery and P. Clark, pp. 97–129. Rome: United States Mutual Security.

Cohen, C. P., P. W. Solenberger, and G. Tucker. 1970. Iterative and inversion techniques for solving large-scale mutiregional input-output models. No. 17 (June), EDA Report. Harvard Economic Research Project.

Leontief, W. and A. Strout. 1986. Multiregional input-output analysis. In *Input-Output Economics*, 2nd ed., ed. W. Leontief, pp. 129–161. New York: Oxford University Press.

Ministry of International Trade and Industry (MITI). 1970. The interregional input-output table of Japan. (June 22). Tokyo.

Ministry of International Trade and Industry (MITI). n.d. Interregional input-output table for Japan. In *Trade and Industry of Japan*. Tokyo.

Moses, L. N. 1955. The stability of interregional trading patterns and input-output analysis. *The American Economic Review* 45(5): 803–832.

Polenske, K. R. 1966. A case study of transportation models used in multiregional analysis. Unpublished Ph.D. dissertation, Harvard University.

Polenske, K. R. 1970a. Empirical implementation of a multiregional input-output gravity trade model. In *Contributions to Input-Output Analysis*, ed. A. P. Carter and A. Bródy. Amsterdam: North Holland.

Polenske, K. R. 1970b. An empirical test of interregional input-output models: estimation of 1963 Japanese production. *The American Economic Review* 60(2): 76–82.

Polenske, K. R. 1970c. A multiregional input-output model for the United States. No. 21 (December) revised, EDA Report. Harvard Economic Research Project.

U.S. Department of Commerce. Office of Business Economics. 1969. Input-output structure of the U.S. economy: 1963. *Survey of Current Business* 49(11): 16–47.

U.S. Department of Commerce. Office of Business Economics. 1965. The transactions table of the 1958 input-output study and revised direct and total requirements data. *Survey of Current Business* 45(9): 33–49.

U.S. Department of Labor. 1947. Bureau of Labor Statistics. n.d. Listing of the 1947 input-output table, 450-order. Washington, D.C. (Revised by U.S. Bureau of Economic Analysis, 1970.)

# 6

# Application of Input-Output Analysis for Less Developed Countries (LDCs)

VICTOR BULMER-THOMAS

## Input-Output Analysis in LDCs

The spread of input-output techniques some thirty years ago was greeted with enthusiasm in the developing world, where the decolonisation movement coincided with the belief that policymakers and planners could influence economic performance over the medium- and even long-run. Input-output analysis was seen as a powerful tool in economic transformation and a means by which the national economy could become less dependent on the international trade cycle.

A large number of less developed countries (LDCs) constructed tables in the 1950s and 1960s, but the initial enthusiasm did not survive into the 1970s. The input-output data base, for example, used by the University of Bradford (Bradford, 1981) lists many countries for which tables were constructed in the earlier period and for which no repeat exercise was carried out.

The problem has not been the construction of the table itself, since a surprisingly large number of LDCs have been able to prepare at least one table. All Latin American countries, for example, with the exception of Honduras and Paraguay, have prepared input-output tables at some time, although in many cases the exercise has been limited to a single table which is now seriously out of date. It is fair to say, however, that in only a handful of cases (e.g., South Korea, Taiwan) has input-output analysis become an integral part of official planning and policymaking with adequate resources devoted to the preparation of up-to-date tables.

Some years ago, a table was prepared for what was then Tanganyika by two economists, who concluded that the exercise was not really worthwhile in view of the undeveloped nature of intersectoral transactions (Peacock, 1957). I do not, however, believe that this accounts for the change of heart and lack of enthusiasm felt by many LDCs with respect to input-output analysis.[1]

The reason for the change is accounted for primarily by three considerations. First, LDCs find the costs of table construction very high, while the benefits seem much more modest; secondly, there has been an unacceptably long lag before tables have been published (the Brazilian table for 1970, for example, was not published until 1979); and, thirdly, the uses to which the tables have been put have frequently

This article was published in *International Use of Input-Output Analysis*, edited by Reiner Stäglin. © Vandenhoeck and Ruprecht, 1982, pp. 199–230. Reprinted by permission.

been inappropriate. The result is that input-output analysis has become something to which many LDCs pay lip service, but to which they do not feel strongly committed.

In my recent book on input-output analysis (Bulmer-Thomas, 1982), I have argued that input-output techniques do have an important role to play in LDCs and have offered suggestions to deal with the three considerations listed above. First, however, it is necessary to understand in more detail why LDCs feel less than enthusiastic about the input-output method.

Although the cost of table construction is considerable, this would not matter if a sufficiently high level of benefits could be foreseen. Similarly, a major reason for long gestation lags is lack of resources devoted to the projects. It follows that the crucial consideration is the apparent lack of uses or applications of input-output analysis suitable for LDCs.

The inappropriate nature of many applications in turn had several causes. First, there are difficulties posed by the assumptions of the input-output model; with a single table and each cell presented as a point estimate, it is necessary to assume fixed coefficients of production and constant returns to scale; this is clearly inappropriate over the medium-run in LDCs, where industrial firms typically work well below capacity and enjoy economies of scale.[2] Industrial activity, furthermore, is typically bimodal in nature with a small number of firms in the "formal" sector producing a high proportion of gross output under capital-intensive techniques, while a large number of firms in the "informal" sector account for the remainder of production using labour-intensive techniques. Aggregation theory (e.g., Morimoto, 1971) suggests that putting both kinds of establishments in one sector will cause aggregation bias, but this is in practice what has almost invariably happened in LDCs.

Secondly, the main thrust of advance in input-output research has been towards the needs of advanced countries. The dynamic model, the pollution model, and the interregional model are all examples of this. Even the most traditional use of input-output analysis (as a medium-term forecasting model) is not particularly appropriate for LDCs, since methods for updating and projecting input-output coefficients are still far from satisfactory.[3]

Thirdly, the emphasis in LDCs since the 1973 oil crisis has shifted from the medium- and long-run toward short-run stabilisation questions. Problems of internal and external disequilibrium have become intense over the short-run and there is a widespread belief in official circles that input-output analysis does not have much to say about short-run issues. The lack of official interest in input-output methods is not, of course, confined to LDCs (it is, for example, found in the United Kingdom), but LDCs lack the independent research institutions which have been responsible in advanced countries for so much of the progress in recent years.

As a result of these difficulties, cases have occurred where input-output tables in LDCs have been stillborn—gathering dust on the shelves of government offices. This is an unhappy state of affairs, which can only be resolved through the development of suitable applications. Critics will argue that such applications exist and will point to the many papers presented at international input-output conferences using LDCs as case studies. Even here, however, one notes an excessive stress on linkage analysis which is of questionable relevance for LDCs (e.g., McGilvray, 1977), while the very recent work on income distribution has also come under attack (Taylor, 1979a).

## Policy Problems Facing LDCs

If input-output analysis is to become firmly rooted in the LDC policy-making process, it must become adapted to LDC requirements. In this section, therefore, I shall sketch out briefly the main policy problems facing LDCs. No attempt will be made at this stage to see whether or not input-output models can be adapted to solve these problems; this more ambitious task will be postponed until the latter section entitled "Appropriate Uses of Input-Output Analysis for LDCs" and onward.

For much of the first development decade (1960–70), the accent continued to be on the aggregate real rate of growth of the economy; measured by this narrow criterion, the performance of the developing world has not been unsatisfactory. According to the most recent Development Report, for example, (World Bank, 1981), the average annual rate of growth of GNP per person was positive for all LDC regions in the 1960s and 1970s (other than Low-Income Sub-Saharan Africa in the 1970s), while the average for all LDCs in the 1970s (2.7 percent) surpassed that of industrial market countries (2.5 percent).

Paradoxically, however, the more successful has been the growth performance of LDCs, the more the emphasis has shifted away from growth to other criteria. The reason for this will be examined shortly, but it would be foolish to dispense altogether with the growth criterion; improving the growth performance remains an important objective of policymaking in LDCs and one to which input-output analysis should address itself.

The disenchantment with growth stemmed from the growing perception that the distribution of additional benefits was extremely unequal. The dissatisfaction referred not only to the size distribution of income, but also to the functional distribution, the interregional distribution, and the rural-urban distribution. Efforts to rationalise an increased inequality in income distribution by reference to the Kuznets hypothesis[4] have been seen to rest on flimsy theoretical foundations (Anand, 1981) and are in any case politically unacceptable.

In cross-section studies, it is always possible to find LDCs with the *same* levels of income per head, but *different* indexes of inequality (however defined). One implication of this is that policy does have a role to play in determining the distribution of benefits and so the latter remains another important objective for LDCs. The means by which input-output analysis can serve this objective will be discussed below.

In the previous section, the current emphasis in LDCs on questions of short-run stabilisation policy was stressed. The successive wave of oil price increases has brought to the front the problem of external equilibrium for the LDC oil importers. Unlike the advanced countries, the developing countries have sought to cure external disequilibrium, not by a retreat into recession, but by heavy borrowing on international capital markets.[5] At the same time, the refusal to go into recession has exposed them to inflationary pressures, which in some cases have been exacerbated by the high levels of public (internal) borrowing.

The crisis in Poland in 1981 brought home to international capital lenders the dangers of an overexposed position. It seems safe to predict that requests for additional borrowing by heavily indebted countries (e.g., Brazil) will come under closer scrutiny and several countries (e.g., Costa Rica) have been forced to default on their external public debt and to request renegotiation.

The problem of internal[6] and external disequilibrium, therefore, which in the case of LDCs are closely related, have become critical and many LDCs have been and will be compelled to sacrifice growth in pursuit of short-run equilibrium. Stabilisation programmes, however, continue to be advocated by the International Monetary Fund (IMF) (Killick, 1981), are notoriously unpopular, and by no means always successful. It is of some importance, therefore, to know whether input-output analysis can make a contribution in this field as well.

## Use of Macromodels for Resolution of Policy Problems

The three policy objectives listed above (growth, distribution, stabilisation) have not typically been approached through input-output analysis in LDCs. Instead, the macromodels used have been characterised by aggregation of the product market into one, or sometimes two, sectors.[7] The product market has been linked to the factor market in some simple way and sometimes the two have been integrated with the bond and money markets (Behrman, 1979).

The rate of growth objective has been met traditionally through the Harrod-Domar model or the Two-Gap model (for a good review, see Taylor, 1979b). Recently, the work of Shaw (1973) and McKinnon (1973) on financial repression has been formalised into a macromodel which determines the real rate of growth from interaction of the product with the financial market (for example, Fry, 1980).

Policy issues related to the distribution of benefits have posed problems for macromodellers, because they require greater detail in the factor market than has normally been provided. The functional distribution of income has usually been tackled through a classical-inspired macromodel which assumes fixed real wages (for example, Chapter 10 of Taylor, 1979b), but more sophisticated models have also been attempted (Taylor, 1980). Other questions of distribution (e.g., the size distribution of income) have usually been ignored in macromodels for LDCs.

The stabilisation question has been tackled by two rather different, although related, models.[8] The first, which I shall refer to as the old orthodoxy, stresses the need for equilibrium in the money market. Both internal and external disequilibrium are traced to excessive expansion of the money supply and a good example of a macromodel embodying this approach is the monetarist model of inflation for Latin America (e.g., Vogel, 1974). This type of model has been used widely to provide the necessary theoretical underpinning for IMF-inspired stabilisation programmes.

The other model, which I shall call the new orthodoxy, also concentrates on disequilibrium in the money market, but recognizes that the money supply may not be an exogenous variable in small, open economies. This approach, stemming from the work of Mundell and Johnson on the monetary approach to the balance of payments (Frenkel, 1976), uses a macromodel to trace out that combination of the exchange rate, domestic credit creation and the money supply which can be expected to achieve internal and external equlibrium.[9]

In addition to the macromodels listed above, there is a nonorthodox point of view associated particularly with Lance Taylor and economists working on Brazil (Taylor, 1981). This approach works with a macromodel that looks at policy questions in terms of an IS-LM diagram; the diagram, however, is set in the price-output rather than the interest-output plane and the IS curve is sometimes seen as upward sloping (cf., Leff, 1980), while the LM curve may be backward-bending. Despite the interest generated by these models, however, they have not yet found official favour and remain in a somewhat separate category from the previous models.

It would appear from the above description that macromodelling for LDCs is a rich field which is continually adapting to the changing needs of the developing world. Is a "macro" framework appropriate, however, for the policy problems faced by these countries?

The most serious weakness of this "macro" approach is the aggregated treatment of the product market. The use of a one- or two-sector model involves assumptions about resource mobility and relative price effects which are inappropriate in the case of LDCs; time and time again, studies have demonstrated that the results of a one-sector model can be reversed when two or more sectors are used.[10] This is so, because the use of a multisector approach allows one to recognise some of the more obvious market imperfections which operate in LDCs.

The assumption of product homogeneity has implications for other aspects of macromodelling. Price determination, for example, must be resolved either by generalised excess demand or by a "mark-up" model over prime costs. In neither case is scope given to relative price effects and the propagation mechanism found in, for example, the structuralist theory of inflation.

The aggregated treatment of the product market forces macromodels to work with parameters which may not be stable even in the short-run. A fairly typical macromodel, for example, will work with a capital-output ratio, a wage rate, an import coefficient, and so on. Each of these[11] is a weighted average of very unequal parameters [12] so that stability depends on the assumption that the weights do not change. This may not be so, as has frequently been demonstrated in the case of the capital-output ratio (e.g., Myrdal, 1968).

There is one further limitation placed on macromodels by the assumption of product homogeneity. Any model must have some *adjustment* mechanism, by means of which equilibrium is restored following an initial disturbance[13]; macromodels almost invariably focus on either the investment-saving relationship or the supply-demand balance in the money market. Any divergence ex ante will be corrected ex post and the manner of correction will represent the adjustment mechanism.

In advanced countries, there are good reasons for focussing on either of these relationships. The investment decision, for example, is largely divorced from the act of saving so that an ex ante divergence can easily arise. In LDCs, however, the separation of the two is not so obvious and much saving is automatically translated into investment so that a divergence ex ante is less likely. Meanwhile, in the money market lack of public control over the money supply makes this adjustment mechanism less interesting from a policy point of view.

There *are* relationships in LDCs which would make a good basis for studying the adjustment mechanism. An example is provided by the food supply-demand balance, in which correction of an initial disturbance sets in motion a chain of events that reveals sharply the problems faced by LDCs. This sort of adjustment mechanism, however, is ruled out by the assumption of product homogeneity, since such a model requires at the very least two sectors.

Despite these weaknesses, which in principle could be corrected through the use of a multisector framework, the macromodel does have one advantage for LDCs over input-output analysis which makes it deceptively attractive.[14] It integrates the product with the financial market in a way which input-output analysis has so far failed to do. This means that a macromodel can study the interaction of real and financial variables, an interaction which is clearly of great importance in LDCs. The advance of input-output analysis in LDCs may well depend on achieving a similar integration within a multisector framework. I shall return to this important point in the section entitled, "The Integration of Real and Financial Variables in Input-Output Analysis."

## Obstacles to Use of Input-Output Analysis in LDCs

If the use of macromodels is inappropriate for tackling many of the policy issues faced by LDCs, then much will need to be done to improve the various stages involved in input-output analysis before the developing countries can regard it as an adequate substitute. In this section, I shall point to the major obstacles faced by LDCs in using input-output techniques and suggest, when possible, ways in which the obstacles can be overcome.

### Construction of Input-Output Tables

The most serious problems which LDCs face are the costs of construction and the inappropriate nature of the international classification of activities. There are also difficulties related to the availability of appropriate data (e.g., on factor inputs by sector) and to the absence of data (e.g., for nonmarketed production).

The costs of construction are high not only because of the need for a census of production in many fields, but also because data processing work can be extremely time-consuming. Given the long gaps between construction of tables in LDCs, it is usually the case that the construction team has had little or no experience of this work. As a result, over-ambitious plans are made which greatly extend the costs of data processing.

An example will make this clear. All input-output textbooks agree that basic prices are the superior method for valuing transactions in input-output tables (United Nations, 1973). They do not, however, always make it clear that this superiority is based on the assumption that one knows the different margins, tax rates, etc., applied to each cell in any given row of the table. If, as often happens in LDCs, only a single margin is used for all cells in a row, one does not improve the uniformity of valution of transactions, but one does add considerably to the work of data processing (Bulmer-Thomas, 1982, Ch. 6).[15]

The United Nations classification scheme for activities, which is widely used in LDCs, requires that establishments be grouped together on the basis of their principal product. This means that informal and formal sector activities will be aggregated together since their product descriptions match; thus, a small establishment using unpaid family labour may end up in the same sector as a huge corporation using capital-intensive techniques.

The difficulties involved in ruthlessly separating informal from formal sector activities are not insuperable. The problems are two-fold: first, one must identify the production function for informal sector firms (e.g., Dar Al-Handasah, 1979); and, secondly, one must identify the user of the finished product. The latter is difficult, because expenditure surveys do not identify purchases by type of producing establishment; this means that the output of each informal sector activity must be allocated from the supply side, but the nature of the product will usually make this possible.[16]

### Assumptions of Input-Output Model

In the first section of this article it was pointed out that a single table presented in the form of a point estimate forces the input-output model to work with fixed coefficients of production and constant returns to scale. As a description of production relations in LDCs, this is generally thought to be inadequate.

First, it should be made clear that the assumption refers only to intermediate inputs. There is no reason why factor demand equations cannot be specified to include nonconstant returns to scale and/or relative price effects. Developing countries usually have a time-series for some gross outputs and factor use and this information can be combined to give a realistic treatment for the factor market.

Estimation of input-output coefficients from time-series work has not proved very satisfactory even in advanced countries and it is very doubtful if this method would be of service to LDCs. It is also unrealistic to think that a time-series of input-output tables will suddenly become available for LDCs.

The alternative to a time-series approach is estimation of coefficients from a cross section. Considerable progress has been made in this field (Gerking, 1976)[17] and it may offer some assistance to LDCs. The method assumes that the point estimate in each cell of an input-output table comes from a number of establishments whose absorptions are individually known and which are available on a comparable basis.

If we denote the output of the $r^{th}$ establishment in the $j^{th}$ sector by $x_j^{(r)}$ and its absorption of the $i^{th}$ input $w_{ij}^{(r)}$, we obtain the following estimating equation:

$$w_{ij}^{(r)} = b_{ij} + m_{ij}x_j^{(r)} + \varepsilon_{ij}^{(r)} \qquad (6\text{-}1)$$

where $b_{ij}$ is a constant, $m_{ij}$ is a marginal input-output coefficient and $\varepsilon_{ij}$ is a random variable with zero mean. This equation can only be estimated by Ordinary Least Squares (OLS) (see note 17), and the parameter estimates will be biased and inconsistent if $x_j^{(r)}$ and $\varepsilon_{ij}^{(r)}$ are not independent, but this may be a tolerable price to pay for avoiding the assumption of constant returns to scale.[18]

It should not be difficult for input-output teams to obtain data from establishments in a comparable form. The normal conditions of confidentiality surrounding census or survey returns will have to be observed and adjustments will need to be carried out on absorption at the establishment level. In the age of high-speed computers, this is in any case the most appropriate level at which to adjust input-output data (Burdekin, 1978).

### Secondary Production

The lack of a one-to-one correspondence between activities and commodities means that secondary production cannot be ignored in input-output work and the position of LDCs is no exception. For many developing countries, there will be no solution other than the tedious one of constructing a make matrix separately from an absorption matrix and combining the two using the commodity and/or technology assumption to form symmetric input-output tables (Armstrong, 1975).

Symmetric tables are either commodity × commodity or industry × industry. If only the latter are required, it may be possible for LDCs to short-cut the conventional treatment of secondary production; the latter in developing countries consists almost entirely of commodities sold to final demand.[19] It follows that the quadrant of the absorption matrix referring to intermediate transactions can be thought of as either commodity × industry or industry × industry. In that case, it is possible to prepare industry × industry tables by a simple adjustment of the final demand columns from a commodity to an industry reference.

### Determination of Imports

Most LDCs are in fact small, open economies with foreign trade representing a substantial proportion of GDP. This means that imports cannot be determined exogenously, but must instead enter into the input-output model endogenously. The most usual treatment of this problem for advanced countries involves writing the balance equation for the output vector ($x$) and the import vector ($m$) as follows:

$$x = \hat{d}Ax + \hat{d}h + e \tag{6-2}$$
$$m = [I - \hat{d}]Ax + [I - \hat{d}]h$$

where $A$ is the technology coefficient matrix, $h$ is a vector of home final demend, $e$ is a vector of exports and $d$ is a diagonal matrix whose $i^{\text{th}}$ element shows the ratio of production to total supply (domestic and imported) in the $i^{\text{th}}$ sector.

Equation (6-2) can be used to determine imports endogenously and it gives good results when imports are competitive and when the division between home and imported goods is the same for all entries in any given row of the table. This, however, is precisely what cannot be assumed in LDCs; import substitution occurs first in consumer goods so that the division for final demand between home and imported goods may not be the same as for intermediate demand; even for intermediate demand, the replacement of imports by local production does not affect all firms at the same time so that each cell in any given row of the table may be subject to a different division between home production and imports.

It has long been recognised (United Nations, 1973) that the most effective solution to this problem involves construction of an import matrix with imports separately identified by source *and* by use. The problem is that expenditure surveys (whether of firms or households) typically do not provide a reliable breakdown of purchases by origin. Without a reliable import matrix, however, LDCs are forced to revert to Eq. (6–2).

A possible solution (Bulmer-Thomas, 1982, Ch. 7) is to estimate an import matrix from the supply side (i.e., to use trade data at the most detailed level in order to identify the user). This task is laborious and for advanced countries is virtually impossible. For developing countries, however, the task is not so formidable. because the recurrence of balance-of-payments problems has led very often to straight controls on imports and a body of information on their final destination. Many LDCs, for example, code imports by destination and this information is extremely useful in preparing import matrices from the supply side.

### Updating and Projecting Technical Coefficients

The long gestation lags experienced in producing input-output tables in LDCs, coupled with a process of rapid structural change, mean that technical coefficients derived even from a current table can be seriously out of date. This problem becomes even more acute when the input-output model is used in a projection or forecasting exercise.

It was stated above that mechanical methods of adjusting input-output coefficients are being viewed with increasing suspicion. The success of these methods, however, improves enormously when exogenous information is included (Lynch, 1979). Thus, the updating exercise can be carried out with more confidence, the greater the proportion of transactions (by value) estimated exogenously.

Here, paradoxically, LDCs may be at an advantage over advanced countries. The key coefficients in the table often refer to large establishments over which the government exercises considerable direct or indirect control. In any case, it should not be difficult to organise a supplementary survey for the purpose of identifying changes in the key coefficients; this survey can take advantage of the concentrated nature of industrial production in LDCs.

The task of projecting input-output coefficients is more formidable. I shall argue in the next section that medium-term planning with an input-output model is not the most pressing requirement for LDCs, so that the difficulty of projecting coefficients may not prove so serious. However, that is not much comfort for those countries committed to projections using input-output analysis, so the problems cannot be side-stepped.

The method of gathering information exogeneously is unlikely to work for a projection, unless firms are willing to commit themselves to revealing their future plans (Fisher, 1975). The use of trends in row and column multipliers (based on mechanical methods) is also of dubious utility, unless the multipliers themselves can be determined endogenously (Parikh, 1979). One possible short-cut is provided by Tilanus' Statistical Correction Method (Tilanus, 1966), in which information on the coefficient matrix is lost in order to provide a forecast of output, incomes and employment, etc.[20] I am, however, acutely aware that this remains an unsatisfactory part of input-output research.

The above description of the obstacles to the advance of input-output analysis in LDCs has been of necessity brief, but it should be sufficient to demonstrate that there is much which can be done by way of improvement. The backwardness and (economic ) smallness of LDCs gives them, in some respects, an advantage over advanced countries; The degree of industrial concentration is usually greater than in advanced countries so that a survey of as little as 500 firms can give one considerable insights into intersectoral transactions. Key coefficients can usually be identified quite easily and updated using exogenous information.

There remain formidable problems, of course, which have barely been touched upon in this section. Examples are nonmarketed production and the measurement of future or current coefficients adjusted for import substitution. As pointed out in the first section, however, many LDCs *have* succeeded in constructing tables; what they have not done so successfully is to prepare them in a manner appropriate to their needs.

## Appropriate Uses of Input-Output Analysis for LDCs

In what follows, I shall assume that an input-output table has been constructed, that it is not yet out of date and that policymakers and planners in the country concerned wish to make use of it. What are the most appropriate uses to which it can be put?

Two of the most frequent answers to this question are provided by linkage analysis and economy-wide development planning. The former has been advocated as an empirical counterpart to Hirschman's industrialisation hypothesis (Hirschman, 1958), while the latter is seen as an opportunity to give depth to the sectoral "stage" in the planning process so that both feasibility and consistency can be achieved.

The algebra of both linkage analysis (Yotopoulos, 1973) and economy-wide development planning (Taylor, 1975) have been worked out in great detail. What is not so clear is their justification in the context of LDC development.

The problems of using linkage analysis as an industrialisation strategy are now familiar. The ranking of sectors in terms of output, incomes, and employment creation is not consistent (Panchamukhi, 1975), while confusion arises over interpretation of linkages as *potential* or *actual* (Bulmer-Thomas, 1978). Linkage analysis pays no attention to comparative advantage and, indeed, implies policies contradicted by international trade theory (Riedel, 1975).

Medium-term or long-term economy-wide development planning assumes a degree of sophistication which is just not present in most LDCs. The ability to project technological coefficients remains very unsatisfactory; uncertainty over earnings from primary product exports casts a shadow over the planning exercise and few governments in LDCs work with a time horizon anything like as long as that demanded by medium-term planning.[21]

I shall argue shortly that there is a role for economy-wide development planning over the short run (one to two years). This role, however, is less important than others, which have not received sufficient prominence either from policymakers or input-output specialists. Therefore, it is worth considering what those uses might be.

Economic performance is still measured primarily by macroeconomic measures obtained from the national accounts. The latter are prepared on an annual basis for all LDCs, but suffer from a number of well-known weaknesses. An input-output table, if properly constructed, serves as a check on the reliability of the national accounts and reconciliation of the two independent estimates serves as a useful device for both the national accounts and the input-output team.

It is frequently the case, however, that the input-output table is subordinate to the national accounts. This is true of advanced countries as well, although it could be argued that it does not matter so much in their case since the national accounts' estimates are fairly reliable. In LDCs the failure to make use of the independent information provided by an input-output table represents a waste of scarce resources.

It has long been a "stylised fact" of LDC life that economic development is not susceptible to Keynesian demand-management policies. This is because there are sectoral and factor bottlenecks on the supply-side, which cannot be eliminated in the short run. Identification of these bottlenecks is crucial for two reasons: first, long-run development will be harmed unless they can be identified and overcome; and, secondly, short-run development can be promoted if demand can be switched to sectors which are *not* subject to bottlenecks and for which the derived demand for primary inputs is feasible.[22]

It should by now be clear that a macroeconomic framework is unsuitable for this kind of study. The assumption of product homogeneity rules out the identification of excess demand at the sectoral level; this, however, is precisely the function fulfilled by economy-wide development planning using input-output techniques. With a Leontief inverse, the sectoral and factor implications of any given bill of final goods can be worked out and their feasibility can then be assessed from supply-side considerations.

Much of the literature on linkage analysis is written as if the industrialisation process in LDCs had only just begun. Historically, that is correct since the priority given to industry is comparatively recent (except in Latin America and parts of Asia). On the other hand, even the poorest LDCs have succeeded in creating an industrial base in the last twenty to thirty years which extends into many lines of production.[23]

The rise of industry in both absolute and proportional terms has created many problems. It has produced distortions in factor prices, which work against agriculture with damaging consequences. The high cost and inefficiency of much industrial activity in LDCs is notorious and grave difficulties are being experienced in expanding production through exports.

It is neither desirable nor possible to roll back the clock and commence industrialisation all over again. Instead, it can be argued that one of the highest priorities faced by LDCs is to rationalise their existing industrial system so as to promote those sectors with the greatest chances of long-run success and to curb those sectors which offer little hope of long-run comparative advantage. In reallocating resources in this way, LDCs could be satisfying criteria of both efficiency and growth.

Input-output techniques have a very powerful role to play in this rationalisation of industry. First, they can be used to measure the import intensity of import-substituting industrialisation (ISI), so that the *net* rather than *gross* savings of foreign exchange can be determined in each line of production.[24] This draws the attention of planners to those sectors whose contribution to a foreign exchange scarcity is minimal.

When measuring the import intensity of ISI, it is conventional to use a formula such as:

$$m = A_m[I - A_D]^{-1}[h^d + e] + h^m = M[h^d + e] + h^m \qquad (6\text{--}3)$$

where $m$, $h^d$, $h^m$, $e$ are vectors of imports, home final demand (net of imports), final demand imports, and exports, respectively. $A_m$, $A_D$ are matrices of import and domestic input-output coefficients, respectively, and the column sums of $M(= A_m[I - A_D]^{-1})$, therefore, measure the direct and indirect import requirement when final demand for the sector at the head of the column changes by unity. Unfortunately, use of $M$ as a measure of import intensity misses an important aspect of industrialisation in LDCs—the dependence on imported capital equipment. It is possible, however, to adapt the model to allow for capital goods (Bulmer-Thomas, 1982, Ch. 15).

The rapid growth of industry in LDCs has been in response to the incentives made by tariff, quota, and exchange rate changes. Producers, however, do not respond to *nominal* tariffs, for example, because these affect the prices which consumers pay, but only indirectly the profits which producers receive. Therefore, economists have invented the concept of the Effective Rate of Protection (ERP) in order to measure the incentives actually available to producers (Corden, 1966).

Because the ERP typically differs substantially from the nominal rate of protection and because governments set the latter, it is of some importance to policymakers to know the incentives offered; these may, for example, be far greater than is needed to stimulate production, and the higher the ERP for industry, the greater the burden on nonindustrial sectors.[25]

The ERP can only be measured accurately using input-output techniques. This is not often realised, because various short-cuts exist which make it possible to measure the ERP even without an input-output table.[26] In the most accurate measure (Corden, 1971), the value of nontraded inputs is broken down into its traded and factor component, which requires not only an input-output table, but also a Leontief inverse (Bulmer-Thomas, 1982, Ch. 16).

Neither of the measures discussed so far is a very good guide to inefficiency in industrial production. The measurement of import intensity, for example, focusses on only one scarce factor of production (foreign exchange), while the ERP does not measure inefficiency, because it does not work with shadow prices. Fortunately, there are alternative models available which LDCs can use.

Eastern European economists have pioneered the use of efficiency price models, which allow one to determine the price vector which would rule under certain assumptions regarding the rate of profit (Bródy, 1970). For example, it could be of interest to know what the price vector would be if the rate of profit were equalised in all sectors and was in turn equal to the social cost of capital. A comparison of this price vector with the actual price vector would reveal those sectors where above-average profits were being made.

The efficiency price model would seem eminently suitable for LDCs; it is flexible and can permit all sorts of assumptions regarding the formulation of profits (Brown, 1977). I know of no studies, however, which apply it to LDCs. Instead, the price model is used to study the inflationary process, although the resulting price vector has to be interpreted as one of long-run supply and, therefore, may differ from actual prices because of the neglect of short-run demand considerations.

An alternative measure of inefficiency is provided by the Domestic Resource Cost (DRC) criterion. I can afford to be brief here, because the criterion is well known and is in fact quite widely used in LDCs although not so much by policymakers and planners (Bruno, 1972). The criterion starts from the solution to the price model for an open economy, which can be written for the $j^{th}$ (traded) sector using shadow prices as:

$$p_j = \mu_j p_\$ + \sum_i \lambda_{ij} \pi_i \qquad (6\text{--}4)$$

where $\mu_j$ is the direct and indirect import content of unit production in the $j^{th}$ sector, $\lambda_{ij}$ is the direct and indirect $i^{th}$ factor content, $\pi_i$ is the shadow price of the $i^{th}$ factor, $p_j$ is the shadow price of the $j^{th}$ commodity and $p_\$$ is the shadow price of foreign exchange.

If the shadow price of the $j^{th}$ commodity can be approximaed by marginal import cost or marginal export revenue in dollar terms $(mr_j)$, we can manipulate Eq. (6–4) to get:

$$p_\$ = \sum_i \lambda_{ij} \pi_i / (mr_j - \mu_j) \qquad (6\text{--}5)$$

The numerator in Eq. (6–5) measures the domestic factor content involved in producing one unit of commodity $j$, while the denominator measures the net saving of foreign exchange; $p_\$$ in Eq. (6–5), therefore, measures the efficiency with which domestic resources are converted into foreign exchange.

To make use of the DRC criterion, one does not even need to know the shadow price of foreign exchange. A high $p_\$$ means that a sector is inefficient in converting domestic resources into foreign exchange, while a low one means it is efficient. If one *does* know the shadow price of foreign exchange, sectors can be ranked and those with a $p_\$$ less than the shadow price are prima facie candidates for promotion.

The implications of the DRC criterion for the study of comparative advantage should be clear. Sectors with a low $p_s$ appear to offer the best hope for industrial exports and should receive favourable treatment. Other methods have also been proposed for studying comparative advantage in LDCs; examples are Tinbergen's Semi-Input-Output Method (Kuyvenhoven, 1978) and linear programming techniques (Bruno, 1971). Both approaches make use of input-output tables, but neither has the flexibility or simplicity of the DRC criterion.

In the section "Policy Problems Facing LDCs," I talked about the three principal problems faced by policymakers in LDCs (growth, distribution, stability). The rationalisation of industry should contribute to the growth objective and the role of input-output techniques in short-run stabilisation programmes will be discussed in the next section. In the remainder of this section, I shall limit myself to the use of input-output techniques to study issues of distribution in LDCs.

The distribution of benefits refers to the distribution of income by size, function, institution, activity, and region. The latter can be thought of as rural-urban or as separate geographical units. All these categories of distribution can be of great economic, social, and political importance in LDCs, although most of them cannot be adequately handled in a macroeconomic framework.

Input-output analysis has long had a regional dimension. These models (Richardson, 1972) explicitly recognise the interregional feedback mechanism which operates when firms in one region buy from or sell to firms in another region. In LDCs, however, industry is frequently concentrated in a single region (or even city) so that the link between regions provided by intermediate goods is limited to the purchase or primary products for processing by industry.

This sort of linkage hardly justifies the use of an elaborate interregional model. There *are* important interregional feedbacks in LDCs, however; they operate not through intermediate goods, but through consumer goods. A rise in income in one region will lead to a rise in demand for consumer goods in other regions; this sort of feedback is not captured by the usual interregional model, but I have shown elsewhere (Bulmer-Thomas, 1982, Ch. 14) how the model can be modified to suit the needs of LDCs.

Good progress has been made in modelling the rural-urban distribution of income in LDCs (Pyatt, 1977). Input-output techniques, however, have still not been applied to the real resource transfer from agriculture which lies at the heart of so many dual economy models (Lee, 1971). A precondition for application of input-output techniques in this field is the separation of the production account into rural and urban activities.

The functional and size distribution of income in LDCs have been explored extensively in the last decade and the results are promising. There has been in recent years a switch from models which change the distribution of income exogenously (e.g., Paukert, 1974) and work out the implications of the change on output, incomes, employment, etc., to models which determine the distribution of income (and some overall index of inequality) endogenously (e.g., Adelman, 1977). The latter models are much more appropriate because they do not beg the question of how the distribution of income is changed and they can be used to determine income distribution from changes in policy instruments.

The models have been criticised (Taylor, 1979a) for being unnecessarily complicated in the sense that a one-sector model will give similar results. It is true, of course, that many determinants of income distribution (e.g., education) are not represented in the large-scale models and that the results of the models have been disappointing for those who believe that policy changes can shift the distribution of income in a much more equal direction. At the same time, sectors *do* vary in terms of capital-labour ratios and consumption policies *are* different by income group so that only a multisector framework can give results which may be treated with confidence. If one-sector models concur in their general conclusions, this is merely fortuitous.

### Input-Output Analysis and Short-Run Stabilisation Policy

The distinguishing features of stabilisation programmes are their short-run nature and their integration of real and financial variables. They are unlikely to be applied where the disequilibrium to be attacked has been identified as "structural" and the mechanism of adjustment to equilibrium usually involves a contraction of output, as well as shifts in the functional distribution of income.

In the section entitled, "Use of Macromodels for Resolution of Policy Problems," it was noted that the framework for handling these policy questions in LDCs was macroeconomic. Although these models have changed substantially over the years and although nonorthodox options are available, it is still worth considering whether input-output techniques have a role to play given the problems of applying macromodels to LDCs.

The great merit of input-output analysis is that its multisector framework widens the possibilities for LDCs when conducting a stabilisation programme. Output contraction may be avoidable through a shift in expenditure which implies in turn a different derived demand for scarce resources; at the same time, the causal process by which policy changes are supposed to push the economy toward equilibrium is more clearly spelled out.

There are formidable problems, however, in the way of using input-output analysis in this sphere. The solution to the price model, for example, yields long-run supply prices, which are unlikely to be of interest to policymakers concerned with an anti-inflation programme. The links between the balance-of-payments monetary account, the public sector deficit, credit creation, and the money supply are totally ignored in the input-output model and the rate of interest is nowhere to be seen.

There have been some efforts to increase the realism of the price model. Mathur (1977), for example, has introduced the Hicksian distinction between fix- and flex-price commodities into the input-output model, but the flexible prices have to be specified exogenously. Radhakrishna (1979) refers to the introduction of demand-side considerations into price formation in the model, but no details are given.

Rather than attempting to replace the "macro" framework by a multisectoral one, some economists have attempted to combine the two. In a study of Portugal, for example (Abel, 1977), the input-output price model is used to determine the "price" of each category of final demand and these prices are then used in a short-run Keynesian demand model to determine the impact of simulated changes in policy variables on the economy.

The superiority of this approach over the simple macro-model is derived from the fact that

> The traded goods content of the different categories of final demand varies widely. Therefore a devaluation will have different impacts on the relative prices of those sectors (Abel, 1977, p. 126).

The experiment, however, is not entirely successful, because the input-output model does not in fact track the course of actual price changes very well and because many of the advantages of using a multisectoral framework are not exploited.

It is doubtful if input-output analysis will ever play more than a secondary role in stabilisation policy. It was not designed with such a purpose in mind and it continues to suppress much of the information which is considered essential in designing a stabilisation programme. This does not mean, however, that LDCs need be at the mercy of draconian, neoconservative policy perceptions. If the disequilibrium is "structural," short-run stabilisation policies are irrelevant and if it is not structural, then heterodox models can be used (Taylor, 1981).

## The Integration of Real and Financial Variables in Input-Output Analysis

If money is not simply a "veil," then financial market conditions can affect the performance of the real economy and vice versa. The quantity circuit in the input-output model makes no reference to financial conditions, so that it is impossible to predict whether the quantity solution implies developments in the financial market which are consistent and feasible.

Policymakers and planners in LDCs cannot afford to avoid the question of consistency and feasibility. The relationship between financial market conditions and real growth and distribution is being increasingly stressed in the development literature and policymakers are compelled, therefore, to use a framework which recognises this interdependence. The standard, open macromodel (Dornbusch, 1980) has the great merit for LDCs of integrating the product with the factor, bond, and money markets.

Efforts have been made in the past to integrate the quantity circuit in input-output analysis with the flow of funds (Stone, 1971). This work has not been followed up seriously, because the assumption of fixed coefficients between sectoral outputs and financial instruments is too strong.

Therefore, a fresh approach is required and the following line of development suggests itself. The input-output quantity circuit embodies the essential ingredients to determine supply and demand in the money market; the principal determinants of changes in the money supply in LDCs are movements in the monetary account of the balance of payments and the public sector deficit. The latter can be determined quite successfully by the solution to an input-output model, while the former can be derived from the current account balance together with knowledge of net capital inflows. Since the current account balance can be closely approximated in the solution to an input-output model, the implication for the money supply should not be too difficult to work out.

On the money demand side, the determinants are income and the expected rate of inflation.[27] Again, the input-output model (both quantity and price circuits) can offer assistance, so that money demand implications can be worked out. Thus, the product market can at least be integrated with the money market.

It is not appropriate to carry out this sort of analysis on a full-sized input-output table—the number of tax rates, import requirements, etc., becomes overwhelming. Small tables (e.g., ten sectors), however, would be quite adequate. The input-output model could then be seen as a more detailed description of the product market within a "macro" model—rather like the Wharton model for the United States.


**Conclusions**

Input-output analysis cannot yet be said to have established itself in official circles in LDCs—with a few notable exceptions. Input-output techniques, when applied in LDCs, are in the main "academic" and are regarded by policymakers with some suspicion. This accounts for the low priority given to input-output table construction and the preference given to national accounts estimates and related macromodels.

The use of macromodels in LDCs, however, raises serious difficulties. In particular, the assumption of product homogeneity does not do justice to the realities of LDCs, while at the same time it prevents the model from exploring many avenues of interest to LDCs. These weaknesses can only be overcome within a multisector framework.

Input-output analysis is, in many ways, ideally suited to filling this role. The techniques must, however, be modified before they will gain widespread acceptance in official policymaking circles. The tables must be prepared more quickly so as to reduce the risk of a stillbirth; this means that the method of construction adopted may have to be less ambitious. The classification scheme must serve the needs of the country concerned, even if this runs counter to international recommendations. Foreign trade entries must be articulated in sufficient detail to allow for the impact of import substitution on technological, domestic, and trade coefficients.

Top priority should be given, once the table has been constructed, to revision of the national accounts, identification of supply-side bottlenecks, the rationalisation of industry and the distribution of benefits. Less priority should be given to linkage analysis and—at least for the present—medium-term development planning.

Policymakers in LDCs face three principal objectives—growth, distribution, and stability. The first two can be well served by input-output analysis, but the latter continues to cause problems. The input-output model is not well equipped to handle policy issues raised by stabilisation programmes and there are no signs of immediate improvements. Integration, however, of the input-output model into a macromodel with linked product, factor, bond, and money markets offers exciting possibilities for the future.

## NOTES

1. Even by LDC standards, Tanganyika was very undeveloped in 1957 and modern Tanzania remains one of the poorest countries of the world with agriculture still accounting for over 50 percent and manufacturing less than 10 percent of GDP. In such an economy, the scope for intersectoral transactions is indeed small, but this economic structure is no longer typical of LDCs.

2. The reason for singling out *industrial* firms is that interindustry transactions normally represent the largest and most important block of an input-output table.

3. An excellent survey on the "state of the art" can be found in Lecomber. Since Lecomber's article, only minor modifications have been made, while pessimism over mechanical methods of updating has increased (Lynch, 1979).

4. Kuznets hypothesised an inverse relationship between levels of income per head and the degree of equality up to some critical level of income. Beyond that critical level he hypothesised a positive relationship. Hence, the hypothesis suggests a U-shaped relationship between an index of equality and income per head (Kuznets, 1955).

5. That is why, as previously noted, the average annual rate of growth of LDCs, even in per capita terms, surpassed that of industrial market economies.

6. Internal disequilibrium is taken here to refer to high rates of inflation. Unemployment (and underemployment) has usually been considered long-term and structural in LDCs, so that its elimination does not enter into a short-run stabilisation programme.

7. Although the input-output model is "macro" in terms of coverage, it is "micro" in terms of approach. I shall therefore use the term "macromodel" to refer to one or two sector models in contrast with the multisectoral input-output model.

8. Some countries (e.g., Argentina under Perón in 1973), when faced with internal or external disequilibrium, have identified the root cause as "structural" and have attempted to expand their way out of difficulty (Sutton, 1981). This approach clearly does not rely on a macromodel, but at the same time it normally has not made use of an input-output framework. The analysis has been typically informal rather than formal.

9. This new orthodoxy has taken the Southern Cone countries of Latin America by storm and is applied with particular vigour in Uruguay, where Robert Mundell acts as economic adviser.

10. A good example is provided by the market for credit. With integrated capital markets and a single rate of interest, one-sector models of the economy can be manipulated to show that a rise in interest is output reducing and price increasing (Taylor, 1979b). With segmented capital markets, however, financial repression and *two*-goods-producing sectors one can show exactly the opposite (Galbis, 1977).

11. The wage rate need not, of course, be a parameter, but it remains true that wage rates vary substantially within LDCs both by sector and even by occupation.

12. It is not uncommon for the development process to be viewed as a narrowing of sectoral differentials. Net labour productivities, for example, vary enormously by sector in LDCs, but much less so in advanced countries. Development is characterised, therefore, as a process in which weighted averages are changing over time and it seems unwise to employ an approach in model-building which is forced to assume these weighted averages are constant.

13. My use of the phrase "adjustment mechanism" should not be confused with the *closure* rule, by which an economic model is made determinate (cf., Taylor, 1979b). The closure rule draws upon some behavioural or technological feature of the economy which can be regarded as "stable" (e.g., a consumption function). Sen (1963) and others have pointed out how the choice of a closure rule can affect the results of a model, but this is clearly true of all economic models—not just macromodels.

14. Another attraction is that it demands much less data. This, however, should only be an overwhelming argument in the most pressing circumstances.

15. A helpful list of suggestions for constructing input-output tables in LDCs can be found in McGilvray (1981).

16. It is a common complaint against integration of the formal with the informal sector that the former does not buy from the latter (Bromley, 1979). This information, however, is very useful in determining the allocation of informal sector output, since it must go either to itself or to final demand (Bulmer-Thomas, 1982, Ch. 11).

17. Gerking's study focuses on cross-section work, because he is anxious to use econometric techniques and avoid deterministic input-output models. His preferred method of estimation is Two Stage Least Squares (TSLS), but this (in his model) can only be applied to equations lacking a constant term (Gerking, 1976), so that TSLS assumes constant returns to scale.

18. The removal of the assumption of fixed coefficients of production is not yet realistic for LDCs. Work has been done on introducing price effects into input-output estimation for advanced counries (De Boer, 1976), but this work is still in its infancy. In any case, the assumption of fixed *technological* coefficients is probably acceptable for LDCs, since price effects have their greatest impact on the division of each coefficient between home and imported goods.

19. Examination of the make matrix of different countries reveals the different kinds of secondary products. The first kind represents a clustering in the nontraded commodity rows of the matrix and reflects the system of valuation adopted; if basic or producers' prices have been chosen, there will be a large number of entries—most of which would disappear if a delivered or purchasers' price basis of valuation were chosen. The second kind represents a clustering close to the diagonal of the industrial sector entries; when import substitution has not yet moved into intermediate goods, these entries will refer to consumer (or capital) goods. Thus, for LDCs using delivered or purchasers' prices, with industrialisation limited to consumer (or capital) goods production, industry × industry input-output tables can be prepared very easily.

20. The model predicts the future output vector ($_2x$) from current technology coefficients ($_1A$) and future final demand ($_2f$), i.e.

$$_2x = {_1}\hat{C}[I - {_1}A]^{-1}{_2}f$$

where $\hat{C}$ is a diagonal matrix whose $i^{th}$ element records the ratio of observed to predicted value of any variable (e.g., intermediate output) in the current year.

21. Some years ago I was involved in a development planning exercise in Morocco using input-output techniques. After nearly three years, the whole exercise was suspended when the price of phosphate collapsed from its previously high levels.

22. Thus, it is not true to say that demand-management policies cannot work in LDCs. They will not work, however, if a (one-sector) macromodel is used, because such a model cannot identify sectoral bottlenecks.

23. The World Development Report for 1981 estimates that industry accounted for 36 percent of GDP in both middle-income oil-importing and low-income (including China and India) countries in 1979. The figures in 1960 were 32 percent and 17 percent, respectively. The figures are less impressive if one excludes China and mining, but the picture is clear; the industrialisation process is well under way.

24. Input-output techniques can also be used to measure the contribution to growth of ISI itself (Tyler, 1976) in a way which is more accurate than the method originally proposed by Chenery (1960). This use of input-output analysis, however, while of great interest to economic historians, is not of much interest to policymakers, which is why I have not given it prominence in the main body of the paper.

25. A high ERP for industry shifts the internal terms of trade against agriculture through its impact on the exchange rate and the price of manufactured goods. At the same time, zero protection on exportable agricultural production (negative in the case of commodities subject to export taxes) combined with taxes on inputs gives much of agriculture a negative ERP.

26. In the "true Corden" method, for example (Corden, 1971), nontraded inputs are lumped together with value added, so that all that is required is knowledge of the traded inputs into each sector. A census of industrial production would serve for this purpose.

27. The expected rate of inflation is preferred to the rate of interest in money demand equations for LDCs, because the latter is often subject to government control.

## References

Abel, A. B., and L. M. Beleza. 1978. Input-output pricing in a Keynesian model as applied to Portugal. *Journal of Development Economics* 5: pp. 125–138.

Adelman, I., and S. Robinson. 1977. *Income Distribution Policies in Developing Countries, A Case Study of Korea*. Stanford: Stanford University Press.

Anand, S., and S. M. Kanbur. 1981. Inequality and Development: A Critique. St. Catherine's College, Oxford (mimeo).

Armstrong, A. G. 1975. Technology assumptions in the construction of U.K. input-output tables. In *Estimating and Projecting Input-Output Coefficients*, ed. R. I. G. Allen and W. F. Gossling. London: Input-Output Publishing Co.

Behrman, J., and J. Hanson. 1979. The use of econometric models in developing countries. In *Short-term Macroeconomic Policy in Latin America*, ed. J. Behrman and J. Hanson. Cambridge, Mass.: Ballinger Publishing Co.

de Boer, P. M. C. 1976. On the relationship between production functions and input-output analysis with fixed value shares. *Weltwirtschaftliches Archiv*, Band 112: pp. 754–759.

Bradford University, Input-Output Research Group. 1981. A Six-Sector World Input-Output Model with Data Base. Annex 1. Bradford (mimeo).

Bródy, A. 1970. *Proportions, Prices and Planning*. Amsterdam: North-Holland.

Bromley, R. 1979. *The Urban Informal Sector: Critical Perspectives*. London: Pergamon.

Brown, A. A., and J. A. Licari 1977. Price formation models and economic efficiency. In *The Socialist Price Mechanism*, ed. A. Abouchar. North Carolina: Duke University Press.

Bruno, M. 1971. Optimal patterns of trade and development. In *Studies in Development Planning*, ed. H. Chenery. Cambridge: Harvard University Press.

Bruno, M. 1972. Domestic resource costs and effective protection—Clarification and synthesis. *Journal of Political Economy* 80: pp. 16–33.

Bulmer-Thomas, V. 1978. Trade, structure and linkages in Costa Rica. *Journal of Development Economics* 5: pp. 73–86.

Bulmer-Thomas, V. 1982. *Input-Output Analysis: Sources, Methods and Applications for Developing Countries*. London: John Wiley & Co.

Burdekin, R. 1978. *The Construction of the 1973 Scottish Input-Output Tables*. Winchester: IBM (U.K.) Scientific Centre.

Chenery, H. 1960. Patterns of industrial growth. *American Economic Review*. 50: pp. 624–654.

Corden, M. 1966. The structure of a tariff system and the effective protective rate. *Journal of Political Economy* 74: pp. 221–237.

Dar Al Handasah. 1979. Étude d'identification et d'évaluation des possibilités d'investissements industriels, élaboration du tableau entrées-sorties 1975. Vols. 1–3. Beirut.

Dornbusch, R. 1980. *Open Economy Macro-economics*. New York: Basic Books.

Fisher, W. H. 1975. Ex ante as a supplement or alternative to RAS in updating intput-output coefficients. In *Estimating and Projecting Input-Output Coefficients*, ed. R. I. G. Allen and W. F. Gossling. London: Input-Output Publishing Co.

Frenkel, J., and H. G. Johnson, eds. 1976. *The Monetary Approach to the Balance of Payments*. London: Allen and Unwin.

Fry, M. J. 1980. Savings, investment, growth and the cost of financial repression. *World Development* 8(4): pp. 317–325.

Galbis, V. 1977. Financial intermediaries and economic growth in LDCs. *Journal of Development Studies* 13(2): pp. 58–72.

Gerking, S. D. 1976. Estimation of stochastic input-output models. *Studies in Applied Regional Science*. Vol. 3. Leiden: Martinus Nijhoff Social Sciences Division.

Hirschmann, A. 1958. *The Strategy of Economic Development*. New Haven: Yale University Press.

Killick, A. 1981. IMF Stabilisation Programme. London: Overseas Development Institute, Working Paper No. 6.

Kuyvenhoven, A. (1978). *Planning with the Semi-Input-Output Method*. Leiden: Martinus Nijhoff.

Kuznets, S. 1955. Economic growth and income inequality. *American Economic Review* 45: pp. 1–28.

Lecomber, J. R. C. 1975. A critique of methods of adjusting, updating and projecting matrices. In *Estimating and Projecting Input-Output Coefficients*, ed. R. I. G. Allen and W. F. Gossling. London: Input-Output Publishing Co.

Lee, T. H. 1971. *Intersectoral Capital Flows in the Economic Development of Taiwan, 1895–1960*. Ithaca: Cornell University Press.

Leff, N., and K. Sato. 1980. Macroeconomic adjustments in developing countries: instability, short-run growth and external dependency. *Review of Economics and Statistics* LXII(2): pp. 170–179.

Lynch, R. 1979. An assessment of the RAS method for updating input-output tables. Paper presented to the Seventh International Conference on Input-Output Techniques, April. Innsbruck.

McGilvray, J. 1977. Linkages, key sectors and development strategy. In *Structure, System and Economic Policy*, ed. W. Leontief. Cambridge: Cambridge University Press.

McGilvray, J. W., and W. I. Morrison. 1981. The compilation of input-output tables in developing countries. Paper presented to Seventeenth General Conference of the International Association for Research in Income and Wealth. Gouvieux, France.

McKinnon, R. 1973. *Money and Capital in Economic Development*. Washington: The Brookings Institution.

Mathur, P. N. 1977. A study of sectoral prices and their movement in the British economy in an input-output framework. In *Structure, Systems and Economic Policy*, ed. W. Leontief. Cambridge: Cambridge University Press.

Morimoto, Y. 1971. A note on weighted aggregation in input-output analysis. *International Economic Review* 12: pp. 138–143.

Myrdal, G. 1968. *Asian Drama*. New York: Pantheon.

Panchamukhi, V. R. 1975. Linkages in industrialization: a study of selected developing countries in Asia. *Journal of Development Planning* 8: pp. 121–165.

Parikh, A. 1979. Forecasts of technology matrices using the RAS method. Paper presented to the Seventh International Conference on Input-Output Techniques. April. Innsbruck.

Paukert, F., J. Skola, and J. Maton. 1974. Redistribution of income, patterns of consumption and employment: a case study for the Philippines. Paper presented to the Sixth International Conference on Input-Output Techniques. April. Vienna.

Peacock, A., and D. Dosser. 1957. Input-output analysis in an underdeveloped country: A Case Study. *Review of Economic Studies*. October.

Pyatt, G., and A. R. Roe with Lindley, R. M., Round, J. I., et al. 1977. *Social Accounting for Development Planning; with Special Reference to Sri Lanka*. Cambridge: Cambridge University Press.

Radhakrishna, R., and A. Sarma. 1979. Analysis of sectoral price movements in a developing economy: effects of movement in agricultural prices and production industrial prices, demand pattern and income distribution. Paper presented to the Seventh International Conference on Input-Output Techniques. April. Innsbruck.

Richardson, H. 1972. *Input-Output and Regional Economics*. London: Weidenfeld and Nicholson.

Riedel, J. 1975. Factor proportions, linkages and the open developing economy. *Review of Economics and Statistics* 57: pp. 487–494.

Sen, A. K. 1963. Neo-classical and neo-Keynesian theories of distribution. *Economic Record* 39: pp. 53–64.

Shaw, E. 1973. *Financial Deepening in Economic Development*. New York: Oxford University Press.

Stone, R. 1971. *The Financial Interdependence of the Economy 1957–66*. Cambridge: Chapman & Hall.

Sutton, M. 1981. The costs and benefits of stabilisation programmes: some Latin American experiences. London: Overseas Development Institute, Working Paper No. 3.

Taylor, L. 1975. Theoretical foundations and technical implications. In *Economy-Wide Models and Development Planning*, ed. C. Blitzer, P. Clark, and L. Taylor. Oxford: Oxford University Press.

Taylor, L., and F. Lysy. 1979a. Vanishing income redistributions: Keynesian clues about model surprises in the short run. *Journal of Development Economics* 6: pp. 11–30.

Taylor, L. 1979b. *Macro Models for Developing Countries*. New York: McGraw-Hill.

Taylor, L., E. Bacha, E. Cardoso, and F. Lysy. 1980. *Models of Growth and Distribution for Brazil*. New York: Oxford University Press.

Taylor, L. 1981. IS/LM in the tropics: Diagrammatics of the new structuralist macro critique. In

*Economic Stabilization in Developing Countries*, ed. W. R. Cline and S. Weintraub. Washington: The Brookings Institution.

Tilanus, C. B. 1966. *Input-Output Experiments, The Netherlands 1948–61.* Rotterdam: Rotterdam University Press.

Tyler, W. G. 1976. *Manufactured Export Expansion and Industrialization in Brazil.* Tubingen: J. C. B. Mohr (Paul Siebeck).

United Nations. 1973. Input-output tables and analysis. Series F, No. 14, Rev. 1. New York.

Vogel, R. 1974. The Dynamics of Inflation in Latin America, 1950–1969. *American Economic Review* LXIV(1): pp. 102–114.

World Bank. 1981. World Development Report 1981. Washington.

Yotopoulos, P., and J. Nugent. 1973. A balanced growth version of the linkage hypothesis: a test. *Quarterly Journal of Economics* 87: pp. 157–171.

# 7

# Input-Output Data in the United Nations System of National Accounts

ABRAHAM AIDENOFF

## Introduction

In March 1968 the Statistical Commission of the United Nations adopted a new system of national accounts (SNA). The new system represents a revision and extension of the 1952 version of the SNA.[1] The SNA gives international guidelines for purposes of economic accounting in countries with market economies, and furnishes the basis for the international reporting of comparable national accounting data by these countries.

A major extension of the national accounts in the new SNA is the integration into the system of complete data for purposes of input-output analysis. This paper indicates how the input-output data are integrated into the new system and summarizes the recommendations of the new SNA with respect to input-output analysis. To give a frame of reference for dealing with these topics, the structure of the entire system is outlined first, before we focus on the production and other accounts of the system, on the supply and disposition of goods and services. These accounts are of course of special interest in the case of input-output data and analysis.

## The Overall Structure of the System

Table 7.1 illustrates the structure of, and the connections between, each of the accounts of the new SNA. Extensive use is made of matrices in describing the new system since they furnish an effective and compact means of setting out its scope, configurations of flows and classifications, and the links between the flows and classifications.

The new system covers all flows and stocks in an economy. In addition to the production and disposition of goods and services and the receipt and outlay of incomes, which are traditionally covered in national accounting, the new SNA encompasses the flow of funds, all types of capital gains and losses, and complete

This article was published in *Applications of Input-Output Analysis*, edited by Anne P. Carter and Andrew Bródy, © North-Holland Publishing Company, 1970, pp. 349–368, Volume 2. Reprinted by permission.

balance sheets. The balance sheet accounts of the institutional sector at the beginning of a period are linked with their balance sheet accounts at the end of the period through the capital finance accounts of the institutional sectors and their revaluation accounts for the capital gains and losses. These balance sheets exhibit the depreciated stocks (i.e., the two-dimensional concept) of fixed assets. Of particular interest for capital-output coefficients are data of the system on the gross stock (i.e., one-dimensional concept) of fixed assets of industries and other categories of producers of goods and services. The transition from the beginning of a period to the end of the period in these stocks is shown in the capital formation and revaluation accounts of the producers.

As is customary, the accounts with respect to flows are separated into current and capital accounts. A new feature of the system is the division of both the capital and current accounts into accounts concerning the supply and disposition of goods and services and accounts relating to the receipt and disbursement of funds. The former accounts are the production, consumption expenditure, and capital formation accounts of the system; the latter accounts are the income and outlay and capital finance accounts of the institutional sectors. In view of the differences in subject, the two sets of accounts differ in the transactor units and classifications that are used. In the case of the producers in an economy, for example, establishments, which direct and carry on the production of differing categories of goods or services, are used as the units of observation and classification in the accounts on the production and use of goods and services. Institutional units, which control and manage the disposition of income, other sources of finance, and property, are used for this purpose in the accounts on income and outlay and capital finance. The establishments are divided into industries, government services, and private nonprofit services to households in view of their differing modes of production; and the establishments of each of these groups are classified according to the major kind of goods or services they produce. The institutional units are grouped into sectors and subsectors (e.g., nonfinancial corporate and quasicorporate enterprises, financial institutions, general government, households) in the light of their sources and uses of income and other funds and types of property. The distinctions that are made in the new SNA between accounts on the supply and disposition of goods and services and accounts on other transactions are of considerable assistance in integrating input-output data into the former accounts.

While these two sets of accounts are differentiated one from the other, they are bound together into the new SNA by common flows. For example, value added (specifically, compensation of employees and operating surplus) is first classified according to the major kind of activity of the establishments where it originates in the production accounts of the system; then it is classified according to sector and major kind of activity of the parent institutional units of the establishments, in the income and outlay accounts. The finance of the final consumption, which is classified according to object or purpose in the consumption expenditure accounts, is shown in the income and outlay accounts of households, general government, and private non-profit institutions. And gross capital formation is first classified according to the kind of economic activity of the establishments that make the outlays, in the capital formation accounts; then it is shown, financed by the parent

**Table 7.1**   An illustration of the new SNA (hypothetical values)

| Section | Category | Subcategory | Detail | # | 1 | 2 | 3 | 4 | 5 | 6 | 7 | 8 | 9 | 10 |
|---|---|---|---|---|---|---|---|---|---|---|---|---|---|---|
| Opening assets | Financial assets | | Currency and deposits | 1 | | | | | | | | | | |
| | | | Securities | 2 | | | | | | | | | | |
| | | | Other financial claims | 3 | | | | | | | | | | |
| | Net tangible assets | | All categories | 4 | | | | | | | | | | |
| Production | Commodities | Commodities, basic value | Products of agriculture and mining | 5 | | | | | | | | | | |
| | | | Products of manufacturing and construction | 6 | | | | | | | | | | |
| | | | Services of trans., comm., & distribution | 7 | | | | | | | | | | |
| | | | Other commodities | 8 | | | | | | | | | | |
| | | Commodity taxes, net | Products of agriculture and mining | 9 | | | | | | | | | | |
| | | | Products of manufacturing and construction | 10 | | | | | | | | | | |
| | | | Services of trans., comm., & distribution | 11 | | | | | | | | | | |
| | | | Other commodities | 12 | | | | | | | | | | |
| | Activities | Industries | Agriculture and mining | 13 | | | | | 26 | 0 | | 0 | −1 | |
| | | | Manufacturing and construction | 14 | | | | | 0 | 267 | | 0 | 1 | |
| | | | Transport, communications, distribution | 15 | | | | | | | 1 | 71 | | |
| | | | Other industries | 16 | | | | | | | 1 | 0 | 61 | |
| | | Government services | Public administration and defence | 17 | | | | | | | | 0 | | |
| | | | Health, education, other social services | 18 | | | | | | | | 0 | | |
| | | | Other government services | 19 | | | | | | | | 0 | | |
| | | Services of private n-p institutions | All services | 20 | | | | | | | | 0 | | |
| Consumption | Expenditure | Household goods and services | Food, beverages, tobacco | 21 | | | | | | | | | | |
| | | | Clothing and household goods and services | 22 | | | | | | | | | | |
| | | | Other goods and services | 23 | | | | | | | | | | |
| | | Government purposes | Public administration and defence | 24 | | | | | | | | | | |
| | | | Health, education, other social purposes | 25 | | | | | | | | | | |
| | | | Other government purposes | 26 | | | | | | | | | | |
| | | Purposes of private n-p institutions | All purposes | 27 | | | | | | | | | | |
| | Income and outlay | Value added | Compensation of employees | 28 | | | | | | | | | | |
| | | | Operating surplus | 29 | | | | | | | | | | |
| | | | Consumption of fixed capital | 30 | | | | | | | | | | |
| | | | Indirect taxes, net | 31 | | | | | 0 | 1 | | | 8 | |
| | | Institutional sector of origin | Enterprises, corporate and quasi-corporate | 32 | | | | | | | | | | |
| | | | General government | 33 | | | | | | | | | | |
| | | | Households and private n-p institutions | 34 | | | | | | | | | | |
| | | Form of income | Wages and salaries | 35 | | | | | | | | | | |
| | | | Employers' contributions | 36 | | | | | | | | | | |
| | | | Entrepreneurial income | 37 | | | | | | | | | | |
| | | | Operational surplus n. e. c. | 38 | | | | | | | | | | |
| | | | Property income | 39 | | | | | | | | | | |
| | | | Direct taxes on income | 40 | | | | | | | | | | |
| | | | Social security contributions | 41 | | | | | | | | | | |
| | | | Current transfers by enterprises | 42 | | | | | | | | | | |
| | | | Social security benefits | 43 | | | | | | | | | | |
| | | | Social assistance grants | 44 | | | | | | | | | | |
| | | | Other current transfers by government | 45 | | | | | | | | | | |
| | | | Current transfers by households etc. | 46 | | | | | | | | | | |
| | | | Current transfers by the rest of the world | 47 | | | | | | | | | | |
| | | Institutional sector of receipt | Enterprises, corporate and quasi-corporate | 48 | | | | | | | | | | |
| | | | General government | 49 | | | | | | | | | | |
| | | | Households and private n-p institutions | 50 | | | | | | | | | | |
| Accumulation | Increase in inventories | Industries | Agriculture and mining | 51 | | | | | | | | | | |
| | | | Manufacturing and construction | 52 | | | | | | | | | | |
| | | | Transport, communications, distribution | 53 | | | | | | | | | | |
| | | | Other industries | 54 | | | | | | | | | | |
| | | Government services | Public administration and defence | 55 | | | | | | | | | | |
| | Fixed capital formation | Industries | Agriculture and mining | 56 | | | | | | | | | | |
| | | | Manufacturing and construction | 57 | | | | | | | | | | |
| | | | Transport, communications, distribution | 58 | | | | | | | | | | |
| | | | Other industries | 59 | | | | | | | | | | |
| | | Government services | Public administration and defence | 60 | | | | | | | | | | |
| | | | Health, education, other social services | 61 | | | | | | | | | | |
| | | | Other government services | 62 | | | | | | | | | | |
| | | Services of private n-p institutions | All services | 63 | | | | | | | | | | |
| | Capital finance | Indl. cap. form., land etc. | Industrial capital formation | 64 | | | | | | | | | | |
| | | | Land, mineral rights etc. | 65 | | | | | | | | | | |
| | | Capital transfers | All categories | 66 | | | | | | | | | | |
| | | Financial assets | Currency and deposits | 67 | | | | | | | | | | |
| | | | Securities | 68 | | | | | | | | | | |
| | | | Other financial claims | 69 | | | | | | | | | | |
| | | Institutional sectors | Non-financial enterprises (corp. & quasi-corp.) | 70 | 8 | 235 | 151 | 91 | | | | | | |
| | | | Financial institutions | 71 | 184 | 62 | 102 | 11 | | | | | | |
| | | | General government | 72 | 103 | 145 | 141 | −110 | | | | | | |
| | | | Households and private n-p institutions | 73 | 1 | | 85 | 701 | | | | | | |
| Rest of the world | Current transactions | | All categories | 74 | | | | | 15 | 29 | 5 | 2 | | |
| | Capital transactions | | All categories | 75 | 31 | 51 | 115 | −32 | | | | | | |
| Revaluations | Financial assets | | Securities | 76 | | | | | | | | | | |
| | | | Other financial claims | 77 | | | | | | | | | | |
| | Net tangible assets | | All categories | 78 | | | | | | | | | | |
| Closing assets | Financial assets | | Currency and deposits | 79 | | | | | | | | | | |
| | | | Securities | 80 | | | | | | | | | | |
| | | | Other financial claims | 81 | | | | | | | | | | |
| | Net tangible assets | | All categories | 82 | | | | | | | | | | |

NOTE—In the columns: opening and closing assets are balanced by opening and closing liabilities, and net tangible assets are balanced by net worth.

| Row | 12 | 13 | 14 | 15 | 16 | 17 | 18 | 19 | 20 | 21 | 22 | 23 | 24 | 25 | 26 | 27 | 28 | 29 | 30 | 31 | 32 | 33 | 34 | 35 | 36 | 37 | 38 | 39 | 40 | 41 |
|---|---|---|---|---|---|---|---|---|---|---|---|---|---|---|---|---|---|---|---|---|---|---|---|---|---|---|---|---|---|---|
| 6 | | 1 | 19 | 0 | 3 | 0 | 0 | 0 | | 14 | 2 | 1 | | | | | | | | | | | | | | | | | | |
| 7 | | 8 | 118 | 11 | 8 | 7 | 3 | 1 | 0 | 34 | 27 | 5 | | | | | | | | | | | | | | | | | | |
| 8 | | 1 | 13 | 9 | 2 | 1 | 0 | 0 | 0 | 15 | 12 | 11 | | | | | | | | | | | | | | | | | | |
| 9 | | 1 | 14 | 3 | 2 | 1 | 2 | 1 | 0 | 2 | 12 | 18 | | | | | | | | | | | | | | | | | | |
| 10 | | 0 | 9 | | | | | | | −1 | | | | | | | | | | | | | | | | | | | | |
| 11 | | 0 | 2 | 1 | 1 | | | | | 4 | 2 | 4 | | | | | | | | | | | | | | | | | | |
| 12 | | | | 1 | | | | | | | | | | | | | | | | | | | | | | | | | | |
| 13 | | | | | 2 | 0 | 0 | 0 | | | 3 | 1 | | | | | | | | | | | | | | | | | | |
| 16 | 1 | | | | | | | | | | | | | | | | | | | | | | | | | | | | | |
| 17 | | 6 | | | | | | | | | | | | | | | | | | | | | | | | | | | | |
| 18 | | | | | | | | | | | | | 20 | | | | | | | | | | | | | | | | | |
| 19 | | | | | | | | | | 0 | | | | 16 | | | | | | | | | | | | | | | | |
| 20 | | | | | | | | | | | | | | | 6 | | | | | | | | | | | | | | | |
| 21 | | | | | | | | | | | 1 | | | | | 1 | | | | | | | | | | | | | | |
| 29 | | 9 | 64 | 29 | 24 | 10 | 10 | 4 | 2 | | | | | | | | | | | | | | | | | | | | | |
| 30 | | 5 | 23 | 13 | 14 | | | | | | | | | | | | | | | | | | | | | | | | | |
| 31 | | 1 | 6 | 5 | 6 | 0 | 1 | 0 | 0 | | | | | | | | | | | | | | | | | | | | | |
| 32 | | −1 | 9 | 1 | 6 | | | | | | | | | | | | 109 | 31 | | | | | | | | | | | | |
| 33 | | | | | | | | | | | | | | | | | 29 | 2 | | | | | | | | | | | | |
| 34 | | | | | | | | | | | | | | | | | 14 | 22 | | | | | | | | | | | | |
| 35 | | | | | | | | | | | | | | | | | | | | | 101 | 27 | 13 | | | | | | | |
| 36 | | | | | | | | | | | | | | | | | | | | | 8 | 2 | 1 | | | | | | | |
| 37 | | | | | | | | | | | | | | | | | | | | | | | 18 | | | | | | | |
| 38 | | | | | | | | | | | | | | | | | | | | | 31 | 2 | 0 | | | | | | | |
| 39 | | | | | | | | | | | | | | | | | | | | | | | 4 | | | | | | | |
| 48 | | | | | | | | | | | | | | | | | | | | | | | | | | 31 | 16 | | | |
| 49 | | | | | | | | | | | | | | | | | | | | 29 | | | | | | 2 | 4 | 28 | 9 | |
| 50 | | | | | | | | | | | | | | | | | | | | | | | | 141 | 11 | 18 | 0 | 25 | | |
| 74 | | | | | | 1 | | | | | 2 | | | | | | | | | | | | | 0 | 0 | | | 6 | 4 | |

**Table 7.1** An illustration of the new SNA (hypothetical values)—*continued*

| | | | | | 42 | 43 | 44 | 45 | 46 | 47 | 48 | 49 | 50 |
|---|---|---|---|---|---|---|---|---|---|---|---|---|---|
| Opening assets | | Financial assets | Currency and deposits | 1 | | | | | | | | | |
| | | | Securities | 2 | | | | | | | | | |
| | | | Other financial claims | 3 | | | | | | | | | |
| | | Net tangible assets | All categories | 4 | | | | | | | | | |
| Production | Commodities | Commodities, basic value | Products of agriculture and mining | 5 | | | | | | | | | |
| | | | Products of manufacturing and construction | 6 | | | | | | | | | |
| | | | Services of trans., comm., & distribution | 7 | | | | | | | | | |
| | | | Other commodities | 8 | | | | | | | | | |
| | | Commodity taxes, net | Products of agriculture and mining | 9 | | | | | | | | | |
| | | | Products of manufacturing and construction | 10 | | | | | | | | | |
| | | | Services of trans., comm., & distribution | 11 | | | | | | | | | |
| | | | Other commodities | 12 | | | | | | | | | |
| | Activities | Industries | Agriculture and mining | 13 | | | | | | | | | |
| | | | Manufacturing and construction | 14 | | | | | | | | | |
| | | | Transport, communications, distribution | 15 | | | | | | | | | |
| | | | Other industries | 16 | | | | | | | | | |
| | | Government services | Public administration and defence | 17 | | | | | | | | | |
| | | | Health, education, other social services | 18 | | | | | | | | | |
| | | | Other government services | 19 | | | | | | | | | |
| | | Services of private n-p institutions | All services | 20 | | | | | | | | | |
| Consumption | Expenditure | Household goods and services | Food, beverages, tobacco | 21 | | | | | | | | | 68 |
| | | | Clothing and household goods and services | 22 | | | | | | | | | 58 |
| | | | Other goods and services | 23 | | | | | | | | | 41 |
| | | Government purposes | Public administration and defence | 24 | | | | | | | | 20 | |
| | | | Health, education, other social purposes | 25 | | | | | | | | 16 | |
| | | | Other government purposes | 26 | | | | | | | | 6 | |
| | | Purposes of private n-p institutions | All purposes | 27 | | | | | | | | | 1 |
| | Income and outlay | Value added | Compensation of employees | 28 | | | | | | | | | |
| | | | Operating surplus | 29 | | | | | | | | | |
| | | | Consumption of fixed capital | 30 | | | | | | | | | |
| | | | Indirect taxes, net | 31 | | | | | | | | | |
| | | Institutional sector of origin | Enterprises, corporate and quasi-corporate | 32 | | | | | | | | | |
| | | | General government | 33 | | | | | | | | | |
| | | | Households and private n-p institutions | 34 | | | | | | | | | |
| | | Form of income | Wages and salaries | 35 | | | | | | | | | |
| | | | Employers' contributions | 36 | | | | | | | | | |
| | | | Entrepreneurial income | 37 | | | | | | | | | |
| | | | Operational surplus n. e. c. | 38 | | | | | | | | | |
| | | | Property income | 39 | | | | | | | 21 | 11 | 4 |
| | | | Direct taxes on income | 40 | | | | | | | 10 | | 21 |
| | | | Social security contributions | 41 | | | | | | | | | 9 |
| | | | Current transfers by enterprises | 42 | | | | | | | 0 | | |
| | | | Social security benefits | 43 | | | | | | | | 10 | |
| | | | Social assistance grants | 44 | | | | | | | | 4 | |
| | | | Other current transfers by government | 45 | | | | | | | | 11 | |
| | | | Current transfers by households etc. | 46 | | | | | | | | | |
| | | | Current transfers by the rest of the world | 47 | | | | | | | | | 1 |
| | | Institutional sector of receipt | Enterprises, corporate and quasi-corporate | 48 | | | | | | | | | |
| | | | General government | 49 | | | | 7 | | 0 | | | |
| | | | Households and private n-p institutions | 50 | 0 | 10 | 4 | 3 | | 1 | | | |
| Accumulation | Increase in inventories | Industries | Agriculture and mining | 51 | | | | | | | | | |
| | | | Manufacturing and construction | 52 | | | | | | | | | |
| | | | Transport, communications, distribution | 53 | | | | | | | | | |
| | | | Other industries | 54 | | | | | | | | | |
| | | Government services | Public administration and defence | 55 | | | | | | | | | |
| | Fixed capital formation | Industries | Agriculture and mining | 56 | | | | | | | | | |
| | | | Manufacturing and construction | 57 | | | | | | | | | |
| | | | Transport, communications, distribution | 58 | | | | | | | | | |
| | | | Other industries | 59 | | | | | | | | | |
| | | Government services | Public administration and defence | 60 | | | | | | | | | |
| | | | Health, education, other social services | 61 | | | | | | | | | |
| | | | Other government services | 62 | | | | | | | | | |
| | | Services of private n-p institutions | All services | 63 | | | | | | | | | |
| | Capital finance | Indl. cap. form., land etc. | Industrial capital formation | 64 | | | | | | | | | |
| | | | Land, mineral rights etc. | 65 | | | | | | | | | |
| | | Capital transfers | All categories | 66 | | | | | | | | | |
| | | Financial assets | Currency and deposits | 67 | | | | | | | | | |
| | | | Securities | 68 | | | | | | | | | |
| | | | Other financial claims | 69 | | | | | | | | | |
| | | Institutional sectors | Non-financial enterprises (corp. & quasi-corp.) | 70 | | | | | | | 13 | | |
| | | | Financial institutions | 71 | | | | | | | 3 | | |
| | | | General government | 72 | | | | | | | | 1 | |
| | | | Households and private n-p institutions | 73 | | | | | | | | | 10 |
| Rest of the world | | Current transactions | All categories | 74 | | 0 | | | 1 | 1 | | | |
| | | Capital transactions | All categories | 75 | | | | | | | | | |
| Revaluations | | Financial assets | Securities | 76 | | | | | | | | | |
| | | | Other financial claims | 77 | | | | | | | | | |
| | | Net tangible assets | All categories | 78 | | | | | | | | | |
| Closing assets | | Financial assets | Currency and deposits | 79 | | | | | | | | | |
| | | | Securities | 80 | | | | | | | | | |
| | | | Other financial claims | 81 | | | | | | | | | |
| | | Net tangible assets | All categories | 82 | | | | | | | | | |

42 43 44 45 46 47 48 49 50 51

NOTE—In the columns: opening and closing assets are balanced by opening and closing liabilities, and net tangible assets are balanced by net worth.

| 53 | 54 | 55 | 56 | 57 | 58 | 59 | 60 | 61 | 62 | 63 | 64 | 65 | 66 | 67 | 68 | 69 | 70 | 71 | 72 | 73 | 74 | 75 | 76 | 77 | 78 | 79 | 80 | 81 | 82 | |
|---|---|---|---|---|---|---|---|---|---|---|---|---|---|---|---|---|---|---|---|---|---|---|---|---|---|---|---|---|---|---|
|  |  |  |  |  |  |  |  |  |  |  |  |  |  |  |  |  | 23 | 68 | 12 | 172 |  | 52 |  |  |  |  |  |  |  | 1 |
|  |  |  |  |  |  |  |  |  |  |  |  |  |  |  |  |  | 32 | 153 | 9 | 263 |  | 36 |  |  |  |  |  |  |  | 2 |
|  |  |  |  |  |  |  |  |  |  |  |  |  |  |  |  |  | 126 | 120 | 125 | 146 |  | 77 |  |  |  |  |  |  |  | 3 |
|  |  |  |  |  |  |  |  |  |  |  |  |  |  |  |  |  | 304 | 18 | 133 | 206 |  |  |  |  |  |  |  |  |  | 4 |
| 0 | 0 | 0 | 0 |  |  |  |  |  |  |  |  |  |  |  |  |  |  |  |  |  | 1 |  |  |  |  |  |  |  |  | 5 |
| 1 | 0 | 0 | 2 | 10 | 8 | 14 | 0 | 2 | 1 | 0 |  |  |  |  |  |  |  |  |  |  | 34 |  |  |  |  |  |  |  |  | 6 |
| 0 | 0 | 0 | 0 | 0 | 1 | 0 | 0 | 0 | 0 | 0 |  |  |  |  |  |  |  |  |  |  | 11 |  |  |  |  |  |  |  |  | 7 |
|  |  |  | 0 | 1 | 0 | 2 | 0 | 0 | 0 | 0 |  |  |  |  |  |  |  |  |  |  | 5 |  |  |  |  |  |  |  |  | 8 |
|  |  |  |  |  |  |  |  |  |  |  |  |  |  |  |  |  |  |  |  |  |  |  |  |  |  |  |  |  |  | 9 |
|  |  |  | 0 | 0 | 0 | 0 |  |  |  |  |  |  |  |  |  |  |  |  |  |  | −1 |  |  |  |  |  |  |  |  | 10 |
|  |  |  |  |  |  |  |  |  |  |  |  |  |  |  |  |  |  |  |  |  |  |  |  |  |  |  |  |  |  | 11 |
|  |  |  |  |  |  |  |  |  |  |  |  |  |  |  |  |  |  |  |  |  |  |  |  |  |  |  |  |  |  | 12 |
|  |  |  |  |  |  |  |  |  |  |  |  |  |  |  |  |  |  |  |  |  |  |  |  |  |  |  |  |  |  | 13 |
|  |  |  |  |  |  |  |  |  |  |  |  |  |  |  |  |  |  |  |  |  |  |  |  |  |  |  |  |  |  | 14 |
|  |  |  |  |  |  |  |  |  |  |  |  |  |  |  |  |  |  |  |  |  |  |  |  |  |  |  |  |  |  | 15 |
|  |  |  |  |  |  |  |  |  |  |  |  |  |  |  |  |  |  |  |  |  |  |  |  |  |  |  |  |  |  | 16 |
|  |  |  |  |  |  |  |  |  |  |  |  |  |  |  |  |  |  |  |  |  |  |  |  |  |  |  |  |  |  | 17 |
|  |  |  |  |  |  |  |  |  |  |  |  |  |  |  |  |  |  |  |  |  |  |  |  |  |  |  |  |  |  | 18 |
|  |  |  |  |  |  |  |  |  |  |  |  |  |  |  |  |  |  |  |  |  |  |  |  |  |  |  |  |  |  | 19 |
|  |  |  |  |  |  |  |  |  |  |  |  |  |  |  |  |  |  |  |  |  |  |  |  |  |  |  |  |  |  | 20 |
|  |  |  |  |  |  |  |  |  |  |  |  |  |  |  |  |  |  |  |  |  |  |  |  |  |  |  |  |  |  | 21 |
|  |  |  |  |  |  |  |  |  |  |  |  |  |  |  |  |  |  |  |  |  |  |  |  |  |  |  |  |  |  | 22 |
|  |  |  |  |  |  |  |  |  |  |  |  |  |  |  |  |  |  |  |  |  | 2 |  |  |  |  |  |  |  |  | 23 |
|  |  |  |  |  |  |  |  |  |  |  |  |  |  |  |  |  |  |  |  |  |  |  |  |  |  |  |  |  |  | 24 |
|  |  |  |  |  |  |  |  |  |  |  |  |  |  |  |  |  |  |  |  |  |  |  |  |  |  |  |  |  |  | 25 |
|  |  |  |  |  |  |  |  |  |  |  |  |  |  |  |  |  |  |  |  |  |  |  |  |  |  |  |  |  |  | 26 |
|  |  |  |  |  |  |  |  |  |  |  |  |  |  |  |  |  |  |  |  |  |  |  |  |  |  |  |  |  |  | 27 |
|  |  |  |  |  |  |  |  |  |  |  |  |  |  |  |  |  |  |  |  |  |  |  |  |  |  |  |  |  |  | 28 |
|  |  |  |  |  |  |  |  |  |  |  |  |  |  |  |  |  |  |  |  |  |  |  |  |  |  |  |  |  |  | 29 |
|  |  |  |  |  |  |  |  |  |  |  |  |  |  |  |  |  | −13 | −1 | −1 | −4 |  |  |  |  |  |  |  |  |  | 30 |
|  |  |  |  |  |  |  |  |  |  |  |  |  |  |  |  |  |  |  |  |  |  |  |  |  |  |  |  |  |  | 31 |
|  |  |  |  |  |  |  |  |  |  |  |  |  |  |  |  |  |  |  |  |  |  |  |  |  |  |  |  |  |  | 32 |
|  |  |  |  |  |  |  |  |  |  |  |  |  |  |  |  |  |  |  |  |  |  |  |  |  |  |  |  |  |  | 33 |
|  |  |  |  |  |  |  |  |  |  |  |  |  |  |  |  |  |  |  |  |  |  |  |  |  |  |  |  |  |  | 34 |
|  |  |  |  |  |  |  |  |  |  |  |  |  |  |  |  |  |  |  |  |  | 0 |  |  |  |  |  |  |  |  | 35 |
|  |  |  |  |  |  |  |  |  |  |  |  |  |  |  |  |  |  |  |  |  | 0 |  |  |  |  |  |  |  |  | 36 |
|  |  |  |  |  |  |  |  |  |  |  |  |  |  |  |  |  |  |  |  |  |  |  |  |  |  |  |  |  |  | 37 |
|  |  |  |  |  |  |  |  |  |  |  |  |  |  |  |  |  |  |  |  |  |  |  |  |  |  |  |  |  |  | 38 |
|  |  |  |  |  |  |  |  |  |  |  |  |  |  |  |  |  |  |  |  |  | 11 |  |  |  |  |  |  |  |  | 39 |
|  |  |  |  |  |  |  |  |  |  |  |  |  |  |  |  |  |  |  |  |  | 1 |  |  |  |  |  |  |  |  | 40 |
|  |  |  |  |  |  |  |  |  |  |  |  |  |  |  |  |  |  |  |  |  |  |  |  |  |  |  |  |  |  | 41 |
|  |  |  |  |  |  |  |  |  |  |  |  |  |  |  |  |  |  |  |  |  |  |  |  |  |  |  |  |  |  | 42 |
|  |  |  |  |  |  |  |  |  |  |  |  |  |  |  |  |  |  |  |  |  |  |  |  |  |  |  |  |  |  | 43 |
|  |  |  |  |  |  |  |  |  |  |  |  |  |  |  |  |  |  |  |  |  |  |  |  |  |  |  |  |  |  | 44 |
|  |  |  |  |  |  |  |  |  |  |  |  |  |  |  |  |  |  |  |  |  |  |  |  |  |  |  |  |  |  | 45 |
|  |  |  |  |  |  |  |  |  |  |  |  |  |  |  |  |  |  |  |  |  |  |  |  |  |  |  |  |  |  | 46 |
|  |  |  |  |  |  |  |  |  |  |  |  |  |  |  |  |  |  |  |  |  | 1 |  |  |  |  |  |  |  |  | 47 |
|  |  |  |  |  |  |  |  |  |  |  |  |  |  |  |  |  |  |  |  |  |  |  |  |  |  |  |  |  |  | 48 |
|  |  |  |  |  |  |  |  |  |  |  |  |  |  |  |  |  |  |  |  |  |  |  |  |  |  |  |  |  |  | 49 |
|  |  |  |  |  |  |  |  |  |  |  |  |  |  |  |  |  |  |  |  |  |  |  |  |  |  |  |  |  |  | 50 |
|  |  |  |  |  |  |  |  |  |  |  | 0 |  |  |  |  |  |  |  |  |  |  |  |  |  |  |  |  |  |  | 51 |
|  |  |  |  |  |  |  |  |  |  |  | 5 |  |  |  |  |  |  |  |  |  |  |  |  |  |  |  |  |  |  | 52 |
|  |  |  |  |  |  |  |  |  |  |  | 1 |  |  |  |  |  |  |  |  |  |  |  |  |  |  |  |  |  |  | 53 |
|  |  |  |  |  |  |  |  |  |  |  | 0 |  |  |  |  |  |  |  |  |  |  |  |  |  |  |  |  |  |  | 54 |
|  |  |  |  |  |  |  |  |  |  |  |  |  |  |  |  |  |  |  | 0 |  |  |  |  |  |  |  |  |  |  | 55 |
|  |  |  |  |  |  |  |  |  |  |  | 2 |  |  |  |  |  |  |  |  |  |  |  |  |  |  |  |  |  |  | 56 |
|  |  |  |  |  |  |  |  |  |  |  | 11 |  |  |  |  |  |  |  |  |  |  |  |  |  |  |  |  |  |  | 57 |
|  |  |  |  |  |  |  |  |  |  |  | 9 |  |  |  |  |  |  |  |  |  |  |  |  |  |  |  |  |  |  | 58 |
|  |  |  |  |  |  |  |  |  |  |  | 16 |  |  |  |  |  |  |  |  |  |  |  |  |  |  |  |  |  |  | 59 |
|  |  |  |  |  |  |  |  |  |  |  |  |  |  |  |  |  |  |  | 0 |  |  |  |  |  |  |  |  |  |  | 60 |
|  |  |  |  |  |  |  |  |  |  |  |  |  |  |  |  |  |  |  | 2 |  |  |  |  |  |  |  |  |  |  | 61 |
|  |  |  |  |  |  |  |  |  |  |  |  |  |  |  |  |  |  |  | 1 |  |  |  |  |  |  |  |  |  |  | 62 |
|  |  |  |  |  |  |  |  |  |  |  |  |  |  |  |  |  |  |  | 0 |  |  |  |  |  |  |  |  |  |  | 63 |
|  |  |  |  |  |  |  |  |  |  |  |  |  |  |  |  |  | 28 | 1 | 5 | 10 |  |  |  |  |  |  |  |  |  | 64 |
|  |  |  |  |  |  |  |  |  |  |  |  |  |  |  |  |  | 0 | 1 | 0 | −1 |  |  |  |  |  |  |  |  |  | 65 |
|  |  |  |  |  |  |  |  |  |  |  |  |  |  |  |  |  |  |  |  |  |  |  |  |  |  |  |  |  |  | 66 |
|  |  |  |  |  |  |  |  |  |  |  |  |  |  |  |  |  | −1 | −5 | 2 | 16 |  | 8 |  |  |  |  |  |  |  | 67 |
|  |  |  |  |  |  |  |  |  |  |  |  |  |  |  |  |  | 5 | 4 | 0 | −6 |  | 5 |  |  |  |  |  |  |  | 68 |
|  |  |  |  |  |  |  |  |  |  |  |  |  |  |  |  |  | 10 | 24 | 7 | 2 |  | 5 |  |  |  |  |  |  |  | 69 |
|  |  |  |  |  |  |  |  |  |  |  |  | 0 | 0 | 2 | 14 |  |  |  |  |  |  |  | −13 | 2 | 26 | 8 | 224 | 167 | 130 | 70 |
|  |  |  |  |  |  |  |  |  |  |  |  | 0 | 11 | 1 | 9 |  |  |  |  |  |  |  | −2 | 1 | 2 | 195 | 61 | 112 | 16 | 71 |
|  |  |  |  |  |  |  |  |  |  |  |  | 2 | 3 | 5 | 5 |  |  |  |  |  |  |  | −11 | 0 | 17 | 116 | 139 | 146 | −90 | 72 |
|  |  |  |  |  |  |  |  |  |  |  |  | −2 | 0 |  | 9 |  |  |  |  |  |  |  | 0 | 0 | −1 | 1 |  | 94 | 708 | 73 |
|  |  |  |  |  |  |  |  |  |  |  |  |  |  |  |  |  |  |  |  |  |  |  |  |  |  |  |  |  |  | 74 |
|  |  |  |  |  |  |  |  |  |  |  |  | 0 | 6 | 0 | 11 |  |  |  |  | 1 |  | −1 | 1 | −2 | 37 | 50 | 127 | −33 | 75 |
|  |  |  |  |  |  |  |  |  |  |  |  |  |  |  |  | −3 | −4 | 0 | −18 |  | −2 |  |  |  |  |  |  |  | 76 |
|  |  |  |  |  |  |  |  |  |  |  |  |  |  |  |  | 1 | 1 | 1 | 1 |  | 0 |  |  |  |  |  |  |  | 77 |
|  |  |  |  |  |  |  |  |  |  |  |  |  |  |  |  | 17 | 4 | 5 | 16 |  |  |  |  |  |  |  |  |  | 78 |
|  |  |  |  |  |  |  |  |  |  |  |  |  |  |  |  | 22 | 63 | 14 | 188 |  | 60 |  |  |  |  |  |  |  | 79 |
|  |  |  |  |  |  |  |  |  |  |  |  |  |  |  |  | 34 | 153 | 9 | 239 |  | 39 |  |  |  |  |  |  |  | 80 |
|  |  |  |  |  |  |  |  |  |  |  |  |  |  |  |  | 137 | 145 | 133 | 149 |  | 82 |  |  |  |  |  |  |  | 81 |
|  |  |  |  |  |  |  |  |  |  |  |  |  |  |  |  | 336 | 23 | 145 | 227 |  |  |  |  |  |  |  |  |  | 82 |

institutional units of the establishments classified according to sector, in the capital finance accounts. The inclusion in the new system of accounts that are specific to major schemes of classification furnishes the means to show the transformations in data from one classification to another, or the interactions between these classifications.

## The Accounts on Goods and Services

This section deals with the manner in which input-output data have been made an integral part of the production, consumption expenditure, and capital formation accounts of the new SNA. The production accounts of the new system consist of accounts on commodities (rows and columns 5 through 12 of Table 7.1) and accounts on activities (rows and columns 13 through 20). The first set of accounts portrays the sources and destinations of the supply of commodities, while the second group of accounts concerns the cost structure of the domestic production of goods and services. The new SNA recommends the use of many more categories of activities and commodities than are enumerated in Table 7.1. About 70 categories are suggested at a medium level of classification of kind of activity, and about 150 at the most detailed level of this classification. The use of at least as many, and in some cases more, categories is suggested in the case of the classification of commodities.

Commodities are goods and services that are normally marketed. Included among commodities are goods and services produced by industries (i.e., producers who usually market their products at a price that is intended to yield a profit) and imported. Besides, government services and private nonprofit services to households, two additional categories of producers, may supply some commodities. The bulk of the production of these two classes of activities is noncommodities that the government and private nonprofit bodies themselves consume. While the intermediate consumption of industries consists entirely of commodities, the intermediate consumption of the government and private nonprofit services may include some noncommodities. The final consumption of households encompasses commodities and some other goods and services. The gross capital formation of the three classes of producers essentially consists of commodities.

The focus of attention in input-output analysis is, of course, on commodities and industries. The segregation, in the new SNA, of commodities from other goods and services and of industries from government and private nonprofit services, therefore, contributes to the usefulness of the data of the system for purposes of input-output studies. For this purpose it would also be advantageous to gather and compile the data on the cost structure of industries and on the disposition of the supply of commodities in terms of the same unit of observation and classification, either commodities or industries. However, in general this is neither practicable nor desirable.

Establishments often produce commodities that are not characteristic of their major kind of economic activity, and data on the intermediate and primary inputs into the various categories of commodities which these establishments produce cannot be gathered directly. Nor can the intermediate and final uses to which the establishments of various industries send their products be observed directly. Further, an industrial classification of outputs and inputs and a commodity classification of the supply and disposition of goods and services are required for certain types of economic analysis. For example, data on outputs and inputs classified by kind of economic activity are wanted for correlation with data on employment, wages and salaries, and fixed assets. On the other hand, commodity classifications of data are required when the production and disposition of commodities are to be linked with statistics of prices, external trade, and commodity markets.

The approach that is taken in the new SNA, to the problem outlined above, is to include data on the cost structures of industries and on the supply and disposition of commodities that are practical to gather and compile and that may be converted into commodity-by-commodity or industry-by-industry input-output tables without undue difficulties. For this purpose, establishments are defined as units that engage in the production of the most homogeneous groups of goods and services for which the necessary input data can be gathered, and the classification of commodities is bound to the classification by kind of economic activity. The categories of the commodity classification are defined as the characteristic products of each kind of economic activity. Most important, a submatrix (columns 5 through 12 and rows 13 through 16, in Table 7.1) is included in the system to portray the contributions of the gross outputs of the various industries to the supplies of the various commodities. Submatrices are also provided showing the supplies of commodities from imports and from government and private nonprofit services.

The requirements of input-output analysis have been taken into account in specifying the manner in which commodities are to be valued as outputs and as inputs. The gross outputs of industries are valued at producers' values (i.e., market prices at the producing establishments) at the moment of production; and the producers' values are divided into basic values and net commodity taxes. Net commodity taxes are net indirect taxes (i.e., indirect taxes minus subsidies) which vary from one use to another within the same category of commodities. The use of basic values at the moment of production thus contributes to uniformity in valuing the gross outputs of the commodity of a given category, irrespective of the specific commodity produced and the use to which it is put. Basic values are the best feasible measure of the quantities of commodities produced and distributed. In order to work with basic values in input-output analysis, the purchasers' values assigned to intermediate consumption and final uses are partitioned into basic values, distributive-trade and transport margins, and net commodity taxes.

As may be observed from Table 7.1, import duties are classified into protective duties (row 31 and columns 5 through 8) and revenue-raising duties (row 31 and columns 9 through 12). It is useful to classify the latter type of import duty as a commodity tax in order to measure imported and domestically produced goods and services on the same basis. Not shown in Table 7.1 is the distinction drawn, in the system, between competitive and complementary imports. Competitive imports are those for which a domestic industry exists; complementary imports are those that, if needed, must be imported. Unless complementary and competitive imports are distinguished, input-output analysis will work as if any commodity would be obtained in given proportions from domestic production and imports, independently of the final product mix. In general, this cannot be true. Different vectors of final demand will imply varying proportions of complementary imports.

The treatment of scraps and wastes in the new SNA is also designed to be suitable for purposes of input-output studies. Scraps and wastes are distinguished from other items in the classification of commodities; and transactions in these items are treated as interchanges (sales, purchases) among final-use accounts unless the scraps and wastes are a by-product of commodities. For example, the scrapping of machinery appears as a negative component of fixed capital formation and a positive component of the increase in stocks. However, the scrap resulting from the machining of metal parts is classed as commodity production.

Besides the gross outputs and inputs of the various industries and the supply and disposition of the various commodities, the new system delineates relationships among other categories that are of interest in input-output analysis. These categories include various functional components of final demand and various relevant types of commodities. For example, information is included concerning the kind of economic activity of the industries, government services, and private non-profit services that make fixed capital outlays on the various commodities and on the object of household consumption expenditure on the various commodities.

## The Input-Output Tables of the System

The standard tables of the new SNA that contain input-output data are illustrated by Tables 7.2 and 7.3. Table 7.2 concerns the supply and disposition of commodities. It exhibits the type of data shown in rows and columns 5 through 12 of Table 7.1. Table 7.3 deals with the gross output and cost structure of industries. It exhibits the type of data covered in rows and columns 13 through 16 of Table 7.1. Observe that Table 7.2 does not set out a number of the cross-classifications of data that appear at the intersections of rows 5 through 12 with the columns of Table 5.1. These cross-classifications of data are considered to be of interest for purposes of special input-output studies only. Tables 7.2 and 7.3 are more comprehensive and, in certain respects, more detailed than the other standard tables and accounts of the system relating to the production, consumption expenditure, and capital formation accounts. The other standard tables and accounts, which deal with the traditional data on the composition of, and expenditure on, the gross domestic product, consist of a selection from, and a consolidation or condensation of, the data of these accounts.

## Input-Output Analysis Based on the Data of the System

This section outlines the suggestions in the new SNA as to how the data illustrated in Tables 7.2 and 7.3 might be used for purposes of input-output analysis. A mathematical discussion of this topic is given in the Appendix. Since industries frequently engage in the production of subsidiary ("uncharacteristic") products, the construction of a square input-output table from the data illustrated in Tables 7.2 and 7.3 involves the transfer of the gross output of the subsidiary products and the corresponding inputs. The object of these transfers is to show either the inputs of commodities into the production of commodities in a commodity-by-commodity matrix or the inputs from industries in an industry-by-industry matrix. If outputs alone are transferred, the resulting input-output coefficient matrix cannot be said to relate either to commodities or to industries. In this case, each output can be interpreted either as (1) a commodity output plus the subsidiary production of the industry of which that commodity is the characteristic product; or (2) as an industry output plus the output of the industry's characteristic product by other industries.

The transfer of gross outputs is a relatively simple matter, since the outputs of uncharacteristic products appear as the off-diagonal elements in the industry-by-commodity submatrix of Table 7.2. The transfer of inputs is much more difficult in view of the peculiarities of, and lack of data on, the cost structure of the uncharacteristic products of industries.

The direct collection of data below the establishment level on the cost structure of subsidiary products, even where these products are not joint products with, or by-products of, the characteristic products of the establishment, is difficult and costly and involves making questionable estimates. There are also marked limits on the possibilities of compiling reliable data about these cost structures from the experience of establishments engaged in the manufacture of the subsidiary or joint products only. Even making full use of available special data, it is, therefore, usually necessary to fall back on assumptions as to the cost structure of the subsidiary products.

The two types of assumptions that are most commonly used are suggested in the revised SNA: (1) the assumption of a commodity technology, namely that the cost structure of the subsidiary products is the same as the cost structure of the establishments where these items are the characteristic products; and (2) the assumption of an industry technology, namely that the cost structure of the subsidiary products is the same as the cost structure of the establishment where they are in fact produced. While either assumption at best approximates the actual situation, the commodity technology assumption is deemed appropriate in the case of subsidiary independent products, and the industry technology assumption suitable in the case of by-products and probably subsidiary joint products.

The mathematical appendix to this article sets out the derivation of input-output coefficient matrices from the standard input-output tables of the system using the commodity technology assumption, the industry technology assumption, or a mix of the two assumptions. The use of objective across-the-board methods of transferring gross outputs and inputs is favoured over the use of more ad hoc flexible hand methods. In applying the mathematical methods of transfer, the use of a mix of the commodity and industry technology assumptions is appropriate. Employing any special information on cost structures that may be available will incorporate the valuable aspects of the hand methods. It may be seen from the

Table 7.2.  Sources and Destinations of Commodity Supplies (*Hypothetical Values*)

| Source of supply | | Agriculture, forestry, fishing 1 | Mining 2 | Food, beverages, tobacco 3 | Textiles, wearing apparel, leather 4 | Rubber, chemicals, petroleum prods. 5 | Basic metals 6 | Metal prods., machi- |
|---|---|---|---|---|---|---|---|---|
| Agriculture, forestry, fishing | 1 | 1695 | | | | | | |
| Mining | 2 | | 974 | 3 | | | | |
| Food, beverages, tobacco | 3 | | | 3637 | | 21 | | |
| Textiles, wearing apparel, leather | 4 | | | | 2907 | 5 | | |
| Rubber, chemicals, petroleum products | 5 | | 1 | 69 | 8 | 2917 | 1 | |
| Basic metal industries | 6 | | | | | 12 | 2639 | |
| Metal products, machinery, equipment | 7 | | | | 3 | 8 | 85 | 79 |
| Manufacturing n.e.c. | 8 | | 4 | | 6 | 7 | 1 | |
| Gas, electricity, water | 9 | | | | | 109 | | |
| Construction | 10 | | | | | | 1 | |
| Transport and communication | 11 | | | | | | 1 | |
| Distribution | 12 | | | 27 | 28 | | | |
| Services | 13 | | | | | | | |
| General government services | | | | | | | | |
| Services of private nonprofit institutions | | | | | | | | |
| Import duties | | 25 | 1 | 26 | 21 | 17 | 3 | |
| Imports, c.i.f. | | 741 | 129 | 592 | 200 | 423 | 151 | 3 |
| Total | | 2461 | 1109 | 4354 | 3173 | 3519 | 2882 | 84 |
| Destination of supply | | | | | | | | |
| Intermediate consumption, industries | | 942 | 867 | 819 | 1463 | 2495 | 2478 | 33 |
| Intermediate consumption, general govt. serv. | | 15 | 20 | 26 | 20 | 137 | | 5 |
| Intermediate cons., services of priv. nonprofit industries | | | | | | | | |
| Final cons. in domestic market, households | | 1434 | 192 | 3266 | 1257 | 333 | | 79 |
| Increase in inventories | | 28 | −18 | 48 | 46 | 62 | 116 | 20 |
| Gross fixed capital formation | | | 19 | | | | | 186 |
| Exports | | 42 | 29 | 195 | 387 | 492 | 288 | 17 |
| Total | | 2461 | 1109 | 4354 | 3173 | 3519 | 2882 | 845 |

| ...mmodities | | | | | | Complementary commodities | | | | | |
| Manufacturing n.e.c. | Gas, electricity, water | Construction | Transport and communication | Distribution | Services | Agriculture, forestry, fishing | Mining | Food, beverages, tobacco | Basic metals | Manufacturing n.e.c. | Commodity taxes, net |
| 8 | 9 | 10 | 11 | 12 | 13 | 14 | 15 | 16 | 17 | 18 | |
|---|---|---|---|---|---|---|---|---|---|---|---|
| 16 | | | | | | | | | | | −147 |
| 3 | 3 | 1 | | | | | | | | | 8 |
| 1 | 1 | | | | | | | | | | 353 |
| 5 | 1 | | | 7 | | | | | | | 105 |
| 20 | 61 | 1 | | | | | | | | | 63 |
| 6 | 54 | 1 | | | | | | | | | 7 |
| 21 | 3 | 38 | | | | | | | | | 329 |
| 2983 | 8 | 17 | | | | | | | | | 85 |
| | 1187 | | | | | | | | | | 37 |
| 13 | | 3151 | | | | | | | | | 3 |
| | | | 3313 | | | | | | | | 33 |
| | | | | 3818 | | | | | | | 77 |
| | | | 1 | | 4920 | | | | | | 583 |
| 20 | | | | | | | | | | | 1288 |
| 419 | | | 558 | | 291 | 457 | 423 | 130 | 204 | 116 | |
| 3507 | 1318 | 3209 | 3872 | 3825 | 5211 | 457 | 423 | 130 | 204 | 116 | 2824 |
| 2352 | 707 | 1000 | 1668 | 845 | 1340 | 316 | 423 | | 204 | 116 | 1613 |
| 112 | 66 | 197 | 81 | 41 | 385 | | | | | | 72 |
| 621 | 459 | 312 | 981 | 2800 | 2862 | 141 | | 130 | | | 1159 |
| 77 | −1 | | | 29 | | | | | | | |
| 88 | 86 | 1698 | 110 | 11 | 169 | | | | | | 59 |
| 257 | 1 | 2 | 1032 | 99 | 455 | | | | | | −79 |
| 3507 | 1318 | 3209 | 3872 | 3825 | 5211 | 457 | 423 | 130 | 204 | 116 | 2824 |

Table 7.3.   Industrial Outputs and Costs (*Hypothetical Values*)

| | | Agriculture, forestry, fishing | Mining | Food, beverages, tobacco | Textiles, wearing |
|---|---|---|---|---|---|
| | | 1 | 2 | 3 | |
| Gross output at basic values | | 1711 | 984 | 3661 | 29 |
| Commodity taxes, net | | −147 | 8 | 353 | 1( |
| Gross output at producers' prices | | 1564 | 992 | 4014 | 30 |
| | | | | | |
| Competitive intermediate inputs | | | | | |
| Agriculture, forestry, fishing | 1 | 76 | | 641 | 2 |
| Mining | 2 | 3 | 19 | 18 | |
| Food, beverages, tobacco | 3 | 281 | | 473 | |
| Textiles, wearing apparel, leather | 4 | 8 | 6 | 12 | 11 |
| Rubber, chemicals, petroleum products | 5 | 175 | 33 | 143 | 2 |
| Basic metals | 6 | | 28 | 7 | |
| Metal products, machinery, equipment | 7 | 73 | 44 | 90 | ( |
| Manufacturing n.e.c. | 8 | 12 | 30 | 126 | ( |
| Gas, electricity, water | 9 | 12 | 21 | 26 | |
| Construction | 10 | 51 | 46 | 11 | |
| Transport and communication | 11 | 69 | 36 | 81 | ( |
| Distribution | 12 | 54 | 20 | 53 | |
| Services | 13 | 56 | 19 | 100 | |
| | | | | | |
| Competitive intermediate inputs | | | | | |
| Agriculture, forestry, fishing | 14 | | | 149 | |
| Mining | 15 | | | | |
| Food, beverages, tobacco | 16 | | | | |
| Basic metals | 17 | | | | |
| Manufacturing n.e.c. | 18 | | | | |
| Commodity taxes, net | | 13 | 12 | 891 | |
| | | | | | |
| Primary inputs | | | | | |
| Compensation of employees | | 341 | 570 | 436 | 6 |
| Operating surplus | | 367 | 50 | 343 | 1 |
| Consumption of fixed capital | | 120 | 50 | 61 | |
| Indirect taxes, net | | −147 | 8 | 353 | 1( |
| Total costs | | 1564 | 992 | 4014 | 30. |

| Industries | | | | | | | | |
|---|---|---|---|---|---|---|---|---|
| Rubber, chemicals, petroleum products | Basic metal industries | Metal prods., machinery, equipment | Manufacturing n.e.c. | Gas, electricity, water | Construction | Transport and comm. | Distribution | Services |
| 5 | 6 | 7 | 8 | 9 | 10 | 11 | 12 | 13 |
| 3087 | 2770 | 8087 | 3044 | 1296 | 3171 | 3317 | 3873 | 4926 |
| 63 | 7 | 329 | 85 | 37 | 3 | 33 | 77 | 583 |
| 3150 | 2777 | 8416 | 3129 | 1333 | 3174 | 3350 | 3950 | 5509 |
| | | | | | | | | |
| 14 | | | | | | | | |
| 197 | 120 | 24 | 102 | 271 | 34 | 41 | 7 | 6 |
| 18 | | | 4 | | | 32 | 7 | |
| 64 | 4 | 59 | 69 | 1 | 8 | 16 | 84 | 16 |
| 784 | 164 | 290 | 174 | 93 | 95 | 133 | 121 | 75 |
| 24 | 812 | 1353 | 17 | 14 | 202 | 15 | 3 | 1 |
| 104 | 101 | 2091 | 111 | 57 | 165 | 217 | 78 | 143 |
| 90 | 30 | 253 | 747 | 17 | 466 | 33 | 203 | 305 |
| 88 | 119 | 98 | 59 | 85 | 7 | 7 | 84 | 71 |
| 16 | 15 | 35 | 16 | 3 | 553 | 58 | 114 | 63 |
| 117 | 117 | 159 | 98 | 47 | 79 | 670 | 98 | 36 |
| 57 | 120 | 120 | 100 | 23 | 54 | 55 | 89 | 43 |
| 117 | 109 | 303 | 155 | 27 | 122 | 76 | 121 | 57 |
| | | | | | | | | |
| 94 | | | | | | | | |
| 333 | 73 | 17 | | | | | | |
| | 204 | | | | | | | |
| 19 | | | 97 | | | | | |
| 22 | 10 | 38 | 24 | 41 | 27 | 116 | 142 | 256 |
| | | | | | | | | |
| 464 | 475 | 2493 | 895 | 295 | 1061 | 1353 | 1616 | 2073 |
| 370 | 232 | 592 | 305 | 105 | 256 | 151 | 984 | 1369 |
| 95 | 65 | 162 | 71 | 217 | 42 | 344 | 122 | 412 |
| 63 | 7 | 329 | 85 | 37 | 3 | 33 | 77 | 583 |
| 3150 | 2777 | 8416 | 3129 | 1333 | 3174 | 3350 | 3950 | 5509 |

mathematical appendix that square output and input matrices are required in applying the commodity technology assumption, but not the industry technology assumption. The use of a mix of assumptions, rather than the commodity technology approach only, in applying the mechanical techniques of transfer makes it possible to employ output and input matrices that give, at least in part, a more detailed classification for commodities than for industries.

The techniques outlined above may be employed in preparing a commodity-by-commodity or industry-by-industry input-output coefficient matrix. The new SNA emphasizes the value of compiling commodity-by-commodity matrices. A common aim of input-output analysis is to work out the direct and indirect requirements for the supply of goods and services coupled with given levels of final demand. For this purpose a commodity-by-commodity matrix multiplier is best, since demands are expressed in terms of commodities. Where the interest is in primary inputs in relation to outputs, commodity outputs might be transformed into industry outputs by means of the output matrix. With this sequence of calculations an industry-by-industry input-output coefficient matrix is unnecessary. This approach appears to be the natural one since it is likely that industry outputs adapt themselves to the demand for commodities.

### Appendix

#### The Basis of Input-Output Calculations

The information contained in Tables 7.2 and 7.3 of this paper can be rearranged schematically as in Table 7.A.1.

Table 7.A.1.   A Schematic Arrangement of Input-Output Data

|  | $U$ | $e$ | $q$ |
|---|---|---|---|
| $V$ |  |  | $g$ |
|  | $v'$ |  | $\eta$ |
| $q'$ | $g'$ | $\eta$ | |

In this table, the capital letters denote matrices, the small roman letters denote vectors and the small greek letter denotes a scalar (a single number). Vectors are written as column vectors and row vectors are written as transposed column vectors by the attachment of a prime ($'$) superscript. In what follows, the interchange of rows and columns (transposition) in a matrix is also indicated by a prime superscript.

The first row and column relate to commodities. The row shows the absorption of commodities as intermediate inputs by industries (the elements of $U$) and by final buyers, net, (the elements of $e$). The matrix $U$ has commodities in the rows and industries in the columns. Its typical element, in row $j$ and column $k$, say, represents the amount of commodity $j$ used up in production by industry $k$. The row sums of this matrix represent the total industrial intermediate use of the various commodities; the column sums represent the intermediate use of all commodities by the various industries. The vector $e$ has, again, commodities in the rows and a single column. Each element of this vector represents the net final use of a particular commodity, that is to say, the use for private and government consumption, for capital formation, and for net exports.

Since imports have been deducted in the row, the first column of the table shows the sources of commodities in the various domestic industries. The matrix $V$ has industries in the rows and commodities in the columns. Its typical element, in row $k$ and column $j$, say, represents the production of commodity $j$ by industry $k$. This matrix is strongly diagonal because the overwhelming proportion of the output of most industries consists of their own characteristic products. It is not strictly diagonal, however, because many industries have a certain amount of subsidiary production.

The column vector, $q$, at the end of the first row has as elements the domestic output of each of the commodities; this vector is repeated as a row vector at the bottom of the first column.

The second row and column relate to industries. Whereas the column sums of $V$ give the domestic outputs of the various commodities, the row sums of $V$ give the domestic outputs of the various industries. These row totals are the elements of the vector, $g$, of industry outputs. The second column shows the costs that make up the value of these outputs: the column sums of $U$ represent the costs of intermediate inputs, and the elements of the row vector $y'$ represents the costs of primary inputs (value added).

The final row and column relate to everything else in the economy. They contain, in the row, the value added in each industry and, in the column, the net final expenditure on each commodity. These sums are necessarily equal and are represented by the symbol $\eta$. This product total is less than the domestic product because it does not contain product arising in government services and in the services of private nonprofit institutions.

### Input-Output Relationships

In terms of Table 7.A.1, input-output analysis rests on six relationships, three of which are arithmetic identities and three of which are assumptions relating to the technical conditions of production.

The first relationship is

$$q = Ui + e \qquad (7\text{–}A1)$$

where $i$ denotes the unit column vector so that $Ui$ denotes the row sum of $U$. Eq. (7–A2) states that the domestic production of each commodity is absorbed either as intermediate product or as net final product.

The second relationship is

$$q = V'i \qquad (7\text{–}A2)$$

that is to say, the output of each commodity is equal to the sum of the amounts made in each of the industries.

The third relationship is

$$g = Vi \tag{7–A3}$$

that is to say each industry's gross output is equal to the sum of its outputs of each commodity.

The fourth relationship is

$$U = B\hat{g} \tag{7–A4}$$

where $B$ is a matrix of coefficients of dimensions commodity $\times$ industry and $\hat{g}$ denotes a diagonal matrix with the elements of $g$ in the diagonal. Eq. (7–A4) states that intermediate inputs of commodities are proportional to the industry outputs into which they enter.

The fifth relationship is

$$V' = C\hat{g} \tag{7–A5}$$

where $C$ is a matrix of coefficients of dimensions commodity $\times$ industry. Eq. (7–A5) states that each industry makes commodities in its own fixed proportions.

The final relationship is

$$V = D\hat{g} \tag{7–A6}$$

where $D$ is a matrix of coefficients of dimensions industry $\times$ commodity. Eq. (7–A6) states that commodities come in their own fixed proportions from the various industries.

### Technology Assumptions

The various assumptions that can be used in transferring inputs and outputs can be expressed as follows.

#### The Assumption of a Commodity Technology

From (7–A1), (7–A2), (7–A4), and (7–A5) we can write

$$
\begin{aligned}
q &= Ui + e \\
&= Bg + e \\
&= BC^{-1}q + e \\
&= (I - BC^{-1})^{-1}e
\end{aligned} \tag{7–A7}
$$

where $I$ denotes the unit matrix, and

$$
\begin{aligned}
g &= C^{-1}(I - BC^{-1})^{-1}e \\
&= (I - C^{-1}B)^{-1}C^{-1}e
\end{aligned} \tag{7–A8}
$$

Thus, if we denote the input-output coefficient matrix by $A$, we find that $A = BC^{-1}$

for a commodity-by-commodity table; and that $A = C^{-1}B$ for an industry-by-industry table. We also see from (7–A8) that if we wish to analyse industry outputs we must convert the net final demand for commodities, $e$, into the net final demand for industry outputs, represented by $C^{-1}e$.

It will be noticed that both (7–A7) and (7–A8) involve the matrix inverse, $C^{-1}$. Since a matrix can only be inverted if it is square, the assumption of a commodity technology can only be used if the number of industries is equal to the number of commodities.

*The Assumption of an Industry Technology*

From (7–A1), (7–A3), (7–A4), and (7–A6) we can write

$$q = Ui + e$$
$$= Bg + e$$
$$= BDg + e$$
$$= (I - BD)^{-1}e \qquad (7–A9)$$

and

$$g = D(I - BD)^{-1}e$$
$$= (I - DB)^{-1}De \qquad (7–A10)$$

Again if we denote the input-output coefficient matrix by $A$, we find that $A = BD$ for a commodity-by-commodity table; and that $A = DB$ for an industry-by-industry table. By comparison with (7–A7) and (7–A8) we can see that the role played in them by $C^{-1}$ is now played by $D$. Thus, with the assumption of an industry technology, it is no longer essential that the number of industries be equal to the number of commodities.

*Mixed Assumptions*

It is not necessary to rely wholly on the extreme assumptions above. They can be mixed usefully as the following example illustrates. Suppose that the elements of $V$ are divided into two parts, $V_1$ and $V_2$, so that

$$V = V_1 + V_2 \qquad (7–A11)$$

The elements of $V_1$ are outputs that it seems reasonable to treat on the assumption of a commodity technology; those of $V_2$ are ones to be treated on the assumption of an industry technology. Many, if not most, ordinary subsidiary products are likely to come into $V_1$; by-products, in particular, are likely to come into $V_2$. The formation of $V_1$ and $V_2$ may well involve splitting individual elements of $V$ since these elements may contain a mixture of products not all of which are to be treated in the same way.

With this decomposition of $V$, it would seem reasonable to write

$$g_1 = V_1 i = C_1^{-1}q_1 \qquad (7–A12)$$

and

$$g_2 = V_2 i = D_2 q \tag{7–A13}$$

From (7–A13) we can write

$$q_2 = V_2' i = \hat{q} D_2' i$$

$$= \widehat{D_2' i} q \tag{7–A14}$$

where $\widehat{D_2' i}$ denotes a diagonal matrix formed from the vector $D_2' i$. Since $q_1 = q - q_2$, a combination of (7–A12), (7–A13), and (7–A14) leads to

$$g = g_1 + g_2$$

$$= C_1^{-1}(q - \widehat{D_2' i} q) + D_2 q$$

$$= [C_1^{-1}(I - \widehat{D_2' i}) + D_2] q \tag{7–A15}$$

Eq. (7–A15), when combined with the second row of (7–A7) or (7–A9), gives

$$q = B[C_1^{-1}(I - \widehat{D_2' i}) + D_2] q + e$$

$$= \{I - B[C_1^{-1}(I - \widehat{D_2' i}) + D_2]\}^{-1} e$$

$$= (I - BR)^{-1} e \tag{7–A16}$$

Since $g = Rq$, it follows that

$$g = R(I - BR)^{-1} e = (I - RB)^{-1} Re \tag{7–A17}$$

In this case the input-output coefficient matrix, $A$, for a commodity-by-commodity table, is given by $BR$ where $R = [C_1^{-1}(I - \widehat{D_2' i}) + D_2]$. Since $R$ involves the matrix inverse $C_1^{-1}$ it follows that $C_1$ and $D_2$ must be square matrices of the same order.

Evidently, some care is needed in mixing assumptions. Thus, if $V_1$ is associated with ordinary subsidiary products, it might be thought that the assumption of fixed market shares should be used, in which case (7–A12) would be replaced by

$$g_1 = V_1 i = D_1 q_1 \tag{7–A18}$$

At the same time it might seem preferable to treat the by-products contained in $V_2$ as commodities linked to the outputs of the producing industries, in which case (7–A13) would be replaced by

$$q_2 = V_2' i = C_2 g \tag{7–A19}$$

By forming

$$g_2 = V_2 i = \hat{g} C_2' i = \widehat{C_2' i} g \tag{7–A20}$$

we can proceed as before and obtain

$$g = g_1 + g_2$$

$$= D_1 q_1 + \widehat{C_2' i} g$$

$$= D_1(q - C_2 g) + \widehat{C_2' i} g$$

$$= (I + D_1 C_2 - \widehat{C_2' i})^{-1} D_1 q \tag{7–A21}$$

whence (from (7–A7))

$$q = B(I + D_1C_2 - \widehat{C_2'i})^{-1}D_1q + e$$
$$= [I - B(I + D_1C_2 - \widehat{C_2'i})^{-1}D_1]^{-1}e$$

Let

$$S = (I + D_1C_2 - \widehat{C_2'i})^{-1}D_1$$

then

$$q = (I - BS)^{-1}e \qquad\qquad (7\text{–}A22)$$

Since $g = Sq$, it follows that

$$g = S(I - BS)^{-1}e = (I - SB)^{-1}Se \qquad\qquad (7\text{–}A23)$$

In this case the input-output coefficient matrix, $A$, for a commodity-by-commodity table is given by $BS$. Apart from possible singularity, this expression can be formed even if $D_1$ and $C_2$ are not square matrices and so, also, can the expression $(I - BS)^{-1}$.

## A Comparison of the Assumptions

Although (7–A7) and (7–A10) are superficially different, the differences can all be traced to uncharacteristic production, to off-diagonal elements in the output matrix $V$. If such production did not exist, $V$ would be a diagonal matrix and we should find that $C = D = I$. In this case the four equations reduce to

$$q = g = (I - B)^{-1}e \qquad\qquad (7\text{–}A24)$$

so that $A = B$ and there is no distinction between commodity outputs and industry outputs.

Turning now to (7–A16) and (7–A17), we can see that if $V_2 = 0$ then $C_1 = C$ and $D_2 = 0$, in which case (7–A16) and (7–A17) reduce to (7–A7) and (7–A8). On the other hand if $V_1 = 0$ then $C_1 = 0$ and $D_2 = D$ in which case (7–A12) cannot be formed, (7–A13) is equal to a combination of (7–A3) and (7–A6), and (7–A16) and (7–A17) are replaced by (7–A9) and (7–A10). Exactly the opposite happens with (7–A22) and (7–A23). If $V_2 = 0$ then $D_1 = D$ and $C_2 = 0$ in which case (7–A22) and (7–A23) reduce to (7–A9) and (7–A10). But if $V_1 = 0$ then $D_1 = 0$ and $C_2 = C$, in which case (7–A18) cannot be formed and (7–A22) and (7–A23) are replaced by (7–A7) and (7–A8). In spite of this symmetry in limiting cases, the models leading to (7–A16) and (7–A17) on the one hand and to (7–A22) and (7–A23) on the other are capable of giving substantially different results in practice.

The two models just compared represent only a very simple example of the method described on the assumption of mixed technologies in the preceding section. Much more complicated representations of the relationships of production can be elaborated by considering $g$ a certain transform of $q$, or $q$ a certain transform of $g$. In general it would seem preferable to work with the first sort of transform since, as has been said, it seems more reasonable to suppose that industry outputs adapt themselves to the demand for commodities than to suppose that commodity outputs adapt themselves to the demands placed on industries.

**NOTE**

1. The 1952 version of the SNA is set out in *A System of National Accounts and Supporting Tables*, Studies in Methods, Series F, No. 2, Rev. 2, United Nations, New York, 1964.

# 8

# Demographic Input-Output: An Extension of Social Accounting

RICHARD STONE

## Social Accounting

The term "social accounting" was introduced into economics by J. R. Hicks in the preface to Hicks (1952), which first appeared in 1942. It means, in the words of its inventor, "nothing else but the accounting of the whole community or nation, just as private accounting is the accounting of the individual firm". It has always seemed to me a very good term, preferable to its rival "national accounting", as a comprehensive expression for accounts relating to regions, to countries or to groups of countries.

Yet I am not sure that those of us who have adopted it, including myself, have fully recognised its implications. Social accounts are still thought of mainly, if not exclusively, as statements connecting economic flows and stocks expressed in money terms. Sometimes these statements include a table showing the numbers of people engaged in different branches of productive activity, but apart from this no attempt is made to account for human flows and stocks. In other words, what we have been doing so far is no more than economic accounting.

Whatever may be thought about terminology, there can be little doubt that with the progress of economic and social modelling a need has grown up for a coherent framework within which demographic, educational, manpower and other data relating to human stocks and flows can be coordinated and analysed. An accounting framework is well suited to this purpose since all who flow into a period of time, either as survivors from the preceding period or through birth and immigration during the period, must also flow out of that period, either through death and emigration during the period or as survivors into the following period. Under the name of demographic accounting, I have recently been trying to develop such a framework (Stone, 1966; Stone et al., 1968). The general purpose I have had in mind is the coordination of information on human stocks and flows; the specific purpose is the organisation of the relevant part of this information for educational planning.

In this paper I shall begin by describing the general structure of a system of

This article was published in *Contributions to Input-Output Analysis*, edited by Anne P. Carter and Andrew Bródy. © North-Holland Publishing Company, 1970, Volume 1, pp. 293–319. Reprinted by permission.

demographic accounts, starting with a very simple numerical example and gradually elaborating it with special reference to the early school ages. I shall then show how the system can be used as a basis for the development of Markov-chain models of the type described in Stone (1966). Finally, I shall discuss some of the problems involved in using such models in the context of educational planning.

## Demographic Matrices and Accounts

A demographic matrix is simply a means of presenting information on population flows. In a particular year, the inflow of population into a particular area is made up partly of those living in that area at the end of the preceding year, who flow in at the beginning of the year; partly of the births of the year, who flow in from the outside world during the year; and partly of the immigrants of the year, who also flow in from the outside world during the year. Conversely, the outflow from the area in the given year consists partly of the deaths of the year, who flow out to the outside world during the year; partly of the emigrants of the year, who also flow out to the outside world during the year; and partly of those who are still living in the area at the end of the year, who flow out into the following year. If these flows are set out in matrix form, with outflows (outputs) in the rows and inflows (inputs) in the columns, the result is a demographic matrix. An example of such a matrix, relating to the male population of England and Wales in 1963–1964 and 1964–1965, is given in Table 8.1.

In looking at this table and all the other numerical examples given in this paper, two points should be noted. The first is that the flows to the outside world consist only of deaths; emigrations are not shown, as they should be, simply because the existing statistics make it impossible to show them. The best that can be done with the data available is to treat migration in each year as a residual, that is as the difference between the population at the beginning of the year and the population at the end, making allowance for births and deaths. In Table 8.1 and in all

Table 8.1.   An Example of a Demographic Accounting Matrix: Flow Totals in England and Wales, 1963–1964 and 1964–1965 (*Thousands of Males*)

|  |  | Outside world | England and Wales | | Total outflows |
|---|---|---|---|---|---|
|  |  | Deaths | 1964 | 1965 |  |
| Outside world | Births | | 451.1 | 443.2 | — |
|  | Net immigrations | | 34.2 | 34.6 | — |
| England and Wales | 1963 | 292.4 | 22937.7 | | 23230.1 |
|  | 1964 | 274.8 | | 23148.2 | 23423.0 |
| Total inflows | | — | 23423.0 | 23626.0 | |

subsequent tables this difference is entered as net immigration. This means of course that the totals are smaller than they should be: the row totals are smaller because one outflow, emigration, is missing; and the column totals are smaller because one inflow, immigration, is reduced. Indeed, as will be seen later, in some cases emigrations exceed immigrations, so that this inflow actually becomes negative.

The second point concerns the accuracy of the figures. These are given to the nearest 100 not because it is possible to achieve this degree of precision with demographic any more than with economic data, but because when the flows are deconsolidated some of their components are very small. If one did not work throughout to a reasonable number of significant figures, these components would either disappear or not add up to the totals, as anybody used to dealing with economic accounts knows.

In Table 8.1 the outflows of each year, that is the deaths of the year and the survivors into the succeeding year, are shown in the dated rows; and the inflows into each year, that is the births and net immigrations of the year and the survivors from the preceding year, are shown in the dated columns. Thus if we take the row for 1963 we see that 292,400 males died in the course of 1963 and 22,937,700 survived into 1964; and if we take the column for 1964 we see that 451,100 were born during 1964, 34,200 were net immigrants and 22,937,700 entered the year as survivors from 1963.

As can be seen from Table 8.1, in a demographic matrix the row and column for two succeeding years do not as a rule balance, for obvious reasons: only by the oddest coincidence could the births and immigrations of one year equal the deaths and emigrations of the preceding year. If, however, we concentrate on a single year, 1964 in the example, we can see that what flows in (shown in the 1964 column) is equal to what flows out (shown in the 1964 row): the inflows, as we have seen, consist of 451,100 births *plus* 34,200 net immigrations *plus* 22,937,700 survivors from 1963, totalling 23,423,000; and the outflows consist of 274,800 deaths *plus* 23,148,200 survivors into 1965, totalling 23,423,000, as above. Thus the information relating to the flows into and out of any one year can be presented as an account. This is done for 1964 in Table 8.2.

Tables 8.1 and 8.2 show demographic accounting at its most schematic. The information they contain becomes meaningful only if it is disaggregated. This can be done in as much detail as one likes, without changing the structure of the matrix, simply by subdividing the row and column for each year into as many rows and columns as one has categories to distinguish. The primary subdivision is by age. In

Table 8.2.  Demographic Accounts for England and Wales: Flows into and out of 1964 (*Thousands of Males*)

| Inflows | | Outflows | |
|---|---|---|---|
| Births of the year | 451.1 | Deaths of the year | 274.8 |
| Survivors from 1963 | 22937.7 | Survivors into 1965 | 23148.2 |
| Net immigrants of the year | 34.2 | | |
| Total | 23423.0 | Total | 23423.0 |

turn, each age group can then be subdivided by activity, educational qualification, social unit, income group or any other characteristic that may seem interesting. In the next two sections we shall see the effect of disaggregation first by age and then by activity.

## Age Groups

When subdividing the population by age, the way one groups the ages will depend on the particular purpose one has in mind. If one is trying, as I am, to work out a tool for educational planning, one will want in the first instance to separate the school ages from the rest. In Britain the compulsory school ages go from 5 to 15, and the official statistics on school attendance (as distinct from further education) cover the ages from 2 to 19. On this basis the population could as a first step be divided into three age groups: 0–5, from birth to age of compulsory entry into school; 6–19, the remaining school ages up to and including the noncompulsory sixth form; and 20 and over, when nobody is at school. If we do this with the population flows shown in Table 8.1, we obtain Table 8.3.

In this and all subsequent numerical examples, age in any calendar year is reckoned as "age at 31 December". For example, a child who is aged 5 on 31 December 1963 must have had his fifth birthday some time in 1963 and will have his sixth birthday some time in 1964. Therefore he will be recorded among the 5-year-olds of 1963 and among the 6-year-olds of 1964; that is, he will flow from age group 5 in 1963 into age group 6 in 1964. The same criterion applies to the flows to the outside world: a child who dies in 1963 aged 4 but who if he had survived would have had his fifth birthday within the year, will be included among the 5-year-old deaths in 1963; that is, he will flow from age group 5 in 1963 to the outside world. If this two-dimensional movement, in time and in space, is kept in mind, Table 8.3 is not difficult to interpret.

The construction of Table 8.3 can be explained by concentrating on the flows into and out of 1964. Taking the column for ages 0–5: 451,100 were born during the year; −1500 were net immigrants (that is, in 1964 the emigrations of children age 0–5 exceeded the immigrations by 1500); and 2,029,100 flowed in from 1963 aged 0–4 and became 1–5 during 1964; the total inflow was 2,478,700.

Taking now the corresponding row: 11,500 died within 1964; 2,091,200 aged 0–4 survived to become 1–5 in the course of 1965; and 376,000 aged 5 survived to become 6 in 1965; the total outflow was, again, 2,478,700.

Parallel statements can be made for the two older age groups in 1964. At each age, inflows and outflows balance. This becomes more evident when the information is presented in accounting form. The set of accounts for 1964 that can be derived from Table 8.3 is shown in Table 8.4.

The entries in the first three panels of Table 8.4 are taken directly from Table 8.3. The fourth panel is simply a summary of the first three. It can also be viewed as a deconsolidation of Table 8.2.

The information contained in Tables 8.3 and 8.4 does not add much to what we had learned from Tables 8.1 and 8.2. It begins to be interesting only when the age

Table 8.3.   A First Deconsolidation of the Matrix: Three Age Groups, 1963–1964 and 1964–1965 (*Thousands of Males*)

| | Outside world | England and Wales | | | | | | Total outflows |
| | Deaths | 1964 Age 0–5 | 1964 Age 6–19 | 1964 Age 20+ | 1965 Age 0–5 | 1965 Age 6–19 | 1965 Age 20+ | |
|---|---|---|---|---|---|---|---|---|
| **Outside world** | | | | | | | | |
| Births | | 451.1 | | | 443.2 | | | — |
| Net immigrations | | −1.5 | 1.0 | 34.7 | −1.7 | 7.2 | 29.1 | — |
| **England and Wales** | | | | | | | | |
| 1963 Age 0–5 | 11.9 | 2029.1 | 370.0 | | | | | 2411.0 |
| 1963 Age 6–19 | 2.9 | | 4623.0 | 362.0 | | | | 4987.9 |
| 1963 Age 20+ | 277.6 | | | 15553.6 | | | | 15831.2 |
| 1964 Age 0–5 | 11.5 | | | | 2091.2 | 376.0 | | 2478.7 |
| 1964 Age 6–19 | 3.0 | | | | | 4657.0 | 334.0 | 4994.0 |
| 1964 Age 20+ | 260.3 | | | | | | 15690.0 | 15950.3 |
| Total inflows | — | 2478.7 | 4994.0 | 15950.3 | 2532.7 | 5040.2 | 16053.1 | |

Table 8.4.   Demographic Accounts for Three Age Groups, 1964 (*Thousands of Males*)

| | | Ages 0–5 | | | |
|---|---|---|---|---|---|
| Inflows | | | Outflows | | |
| Births: age 0 | | 451.1 | Deaths: ages 0–5 | | 11.5 |
| Survivors from 1963: | | | Survivors into 1965: | | |
| attaining ages 1–5 | | 2029.1 | aged 0–4 | 2091.2 | |
| Net immigrants: | | | | | |
| ages 0–5 | | −1.5 | aged 5 | 376.0 | |
| | | | | | 2467.2 |
| Total | | 248.7 | Total | | 2478.7 |
| | | Ages 6–19 | | | |
| Survivors from 1963: | | | Deaths: ages 6–19 | | 3.0 |
| attaining age 6 | 370.0 | | | | |
| attaining ages 7–19 | 4623.0 | | Survivors into 1965: | | |
| | | 4993.0 | aged 6–18 | 4657.0 | |
| | | | aged 19 | 334.0 | |
| Net immigrants: | | | | | 4991.0 |
| ages 6–19 | | 1.0 | | | |
| Total | | 4994.0 | Total | | 4994.0 |
| | | Ages 20+ | | | |
| Survivors from 1963: | | | Deaths: ages 20+ | | 260.3 |
| attaining age 20 | 362.0 | | | | |
| attaining ages 21+ | 15553.6 | | Survivors into | | |
| | | 15915.6 | 1965: | | |
| | | | aged 20+ | | 15690.0 |
| New immigrants: | | | | | |
| ages 20+ | | 34.7 | | | |
| Total | | 15950.3 | Total | | 15950.3 |
| | | All ages | | | |
| Births: age 0 | | 451.1 | Deaths: all ages | | 274.8 |
| Survivors from 1963: | | | Survivors into 1965: | | |
| attaining ages 1–5 | 2029.1 | | aged 0–5 | 2467.2 | |
| attaining ages 6–19 | 4993.0 | | aged 6–19 | 4991.0 | |
| attaining ages 20+ | 15915.6 | | aged 20+ | 15690.0 | |
| | | 22937.7 | | | 23148.2 |
| Net immigrants: | | | | | |
| all ages | | 34.2 | | | |
| Total | | 23423.0 | Total | | 23423.0 |

groups are taken apart by years of age. As an example, we can see what happens if we deconsolidate the youngest age group, leaving the two older ones, 6–19 and 20+, as they were. This is done in Table 8.5.

Let us again look at the flows into and out of 1964. The flow into age 0 consists only of births, 451,100, which come in from the outside world; there may well have been some immigration of 0-year-olds, but there are no estimates of this, either actual or implicit. The flow out of age 0 is made up of two parts: the deaths of infants born during the year, 8900, which go to the outside world; and the survivors from these births, 442,200, who become the 1-year-olds of 1965.

The flow into age 1 consists of 429,400 survivors from the 0-year-olds of 1963; again, nothing is known about migration at this age. The flow out of age 1 consists of 1400 deaths *plus* 428,000 survivors who become the 2-year-olds of 1965.

The flow into age 2 consists of 420,700 survivors from the 1-year-olds of 1963 *less* 200 net emigrations. The flow out of age 2 consists of 500 deaths *plus* 420,000 survivors who become the 3-year-olds of 1965.

This pattern is repeated for the later ages. Although very simple, it does give a certain amount of information of a purely demographic kind: age distribution of the population, birth rates, death rates, survival rates. A more complex picture emerges when each age is subdivided by activity.

**Activities**

Information about activities can be introduced into a demographic matrix by subdividing the row and column for each age into several rows and columns, one for each activity that is relevant at that age. Thus for young children the centre of activity is either the home or a nursery school, except for a few who go to special schools for the handicapped. When they reach the age of compulsory education the great majority go to primary school. As they grow older and reach secondary school age, the types of school they can attend become more varied and the number of school categories in the matrix increases: secondary modern, grammar, comprehensive, etc.

In theory, during the ages of compulsory education all children should be at school. In practice there is always a small minority which does not go to school: children so handicapped that they are unsuited even to special school; children of nomadic families, such as gypsies and bargees; or children living in remote districts, whose parents can satisfy the authorities that they are capable of providing at home the required formal education.

After the age of compulsory education most children leave school. Those who continue in the higher forms tend to specialise either on the arts side or on the science side, so that each school category can be subdivided into streams. At the same time other fields of activity open up. The two most obvious are further education and employment. Full-time further education takes place in universities, technical colleges, teacher training colleges, art and music schools, naval and military academies, etc.; and each of these categories can, again, be subdivided into streams, which at this stage will of course be much more numerous than the school streams. Similarly, those who enter employment will be working in different industries and, within each industry, different occupations, and can be classified accordingly.

After the age of 30 very few people will be found as students in any branch of full-time education. The great majority of men and a large minority of women will be in

Table 8.5. A Further Deconsolidation of the Matrix: Eight Age Groups, 1963–1964 and 1964–1965 (*Thousands of Males*)

| | | Outside world | England and | | | | | | | |
|---|---|---|---|---|---|---|---|---|---|---|
| | | | 1964 | | | | | | | |
| | | Deaths | Age 0 | Age 1 | Age 2 | Age 3 | Age 4 | Age 5 | Age 6–19 | Age 20+ |
| Outside world | | | | | | | | | | |
| Birth | | | 451.1 | | | | | | | |
| Net immigrations | | | 0.0 | 0.0 | −0.2 | −0.7 | −0.8 | 0.2 | 1.0 | 34.7 |
| England and Wales | | | | | | | | | | |
| | Age 0 | 0.1 | | 429.4 | | | | | | |
| | Age 1 | 1.6 | | | 420.7 | | | | | |
| | Age 2 | 0.5 | | | | 408.0 | | | | |
| 1963 | Age 3 | 0.3 | | | | | 395.0 | | | |
| | Age 4 | 0.2 | | | | | | 376.0 | | |
| | Age 5 | 0.2 | | | | | | | 370.0 | |
| | Age 6–19 | 2.9 | | | | | | | 4623.0 | 362.0 |
| | Age 20+ | 277.6 | | | | | | | | 15553.6 |
| | Age 0 | 8.9 | | | | | | | | |
| | Age 1 | 1.4 | | | | | | | | |
| | Age 2 | 0.5 | | | | | | | | |
| 1964 | Age 3 | 0.3 | | | | | | | | |
| | Age 4 | 0.2 | | | | | | | | |
| | Age 5 | 0.2 | | | | | | | | |
| | Age 6–19 | 3.0 | | | | | | | | |
| | Age 20+ | 260.3 | | | | | | | | |
| Total inflows | | — | 451.1 | 429.4 | 420.5 | 407.3 | 394.2 | 376.2 | 4994.0 | 15950.3 |

some form of employment, until their central activity returns in most cases to the home upon retirement.

By now our matrix has expanded to several hundred rows and columns. But these developments by no means exhaust the possibilities. For example, a great deal of technical and professional education is acquired through part-time courses, either within or without the system of formal full-time education. If our aim is to find out what qualifications each generation brings with it into adult life and what use it makes of them, we must try to record part-time educational activities in the matrix by separating the people who, while in employment, follow such courses from those who do not.

So far I have concentrated exclusively on the activity aspect of demographic accounting. But taking a line through economic accounting, in which one is concerned with activities when dealing with real flows, but with institutional sectors when dealing with financial flows, one should recognise that a similar dichotomy arises in demographic accounting. That is to say, one should be

Wales

| | 1965 | | | | | | | Total outflows |
|---|---|---|---|---|---|---|---|---|
| Age 0 | Age 1 | Age 2 | Age 3 | Age 4 | Age 5 | Age 6–19 | Age 20+ | |
| 443.2 | | | | | | | | — |
| 0.0 | 0.0 | −0.5 | 0.3 | −0.7 | −0.8 | 7.2 | 29.1 | — |
| | | | | | | | | 438.5 |
| | | | | | | | | 422.3 |
| | | | | | | | | 408.5 |
| | | | | | | | | 395.5 |
| | | | | | | | | 376.2 |
| | | | | | | | | 370.2 |
| | | | | | | | | 4987.9 |
| | | | | | | | | 15831.2 |
| | 442.2 | | | | | | | 451.1 |
| | | 428.0 | | | | | | 429.4 |
| | | | 420.0 | | | | | 420.5 |
| | | | | 407.0 | | | | 407.3 |
| | | | | | 394.0 | | | 394.2 |
| | | | | | | 376.0 | | 376.2 |
| | | | | | | 4657.0 | 334.0 | 4994.0 |
| | | | | | | | 15690.0 | 15950.3 |
| 443.2 | 442.2 | 427.5 | 420.3 | 406.3 | 393.2 | 5040.2 | 16032.1 | |

interested not only in the passage of human beings through a succession of activities but also in their passage through a succession of family or institutional groups of which they are members at successive ages of life. The systematic introduction of this second, purely social, aspect might prove an essential element in understanding and forecasting the demand for education beyond the age at which it is compulsory.

At this point our matrix will contain not hundreds but thousands of rows and columns. I need hardly say that I have no such monument tucked up my sleeve. A group of us in Cambridge have however made a start by constructing a number of fairly detailed matrices for the school population of England and Wales in the years 1960–1961 to 1964–1965, five for boys and five for girls. In these matrices the population up to the age of 19 is divided by years of age and subdivided by types of school and, for the older ages, types of course. A full example for the year 1963–1964 can be found in Stone et al. (1968). Here I shall limit myself to a partial example relating to the youngest age group, children from 0 to 5, in the years 1963–

Table 8.6.   A Detailed Matrix for Ages 0 to 5: Activities, 1963–1964 and 1964–1965 (*Thousands of Male*

| | | | Outside world | | | | | 1964 | | | | | England an |
|---|---|---|---|---|---|---|---|---|---|---|---|---|---|
| | | | Deaths | Age 0 | Age 1 | Age 2 | | Age 3 | | Age 4 | | Age 5 | |
| | | | | Home | Home | Home | School | Home | School | Home | School | Home | School |
| **Outside world** | | | | | | | | | | | | | |
| Births | | | | 451.1 | | | | | | | | | |
| Net immigrations | | | 0.0 | | 0.0 | −0.2 | | −0.7 | | −0.8 | | 0.2 | |
| 1963 | Age 0 | Home | 9.1 | | 429.4 | | | | | | | | |
| | Age 1 | Home | 1.6 | | | 419.2 | 1.5 | | | | | | |
| | Age 2 | Home | 0.5 | | | | | 395.4 | 11.2 | | | | |
| | | School | | | | | | | 1.4 | | | | |
| | Age 3 | Home | 0.3 | | | | | | | 284.6 | 97.7 | | |
| | | School | | | | | | | | | 12.7 | | |
| | Age 4 | Home | 0.2 | | | | | | | | | 8.9 | 264 |
| | | School | | | | | | | | | | | 102 |
| | Age 5 | Home | | | | | | | | | | | |
| | | School | 0.2 | | | | | | | | | | |
| **England and Wales** | | | | | | | | | | | | | |
| 1964 | Age 0 | Home | 8.9 | | | | | | | | | | |
| | Age 1 | Home | 1.4 | | | | | | | | | | |
| | Age 2 | Home | 0.5 | | | | | | | | | | |
| | | School | | | | | | | | | | | |
| | Age 3 | Home | 0.3 | | | | | | | | | | |
| | | School | | | | | | | | | | | |
| | Age 4 | Home | 0.2 | | | | | | | | | | |
| | | School | | | | | | | | | | | |
| | Age 5 | Home | | | | | | | | | | | |
| | | School | 0.2 | | | | | | | | | | |
| Total inflows | | | — | 451.1 | 429.4 | 419.0 | 1.5 | 394.7 | 12.6 | 283.8 | 110.4 | 9.1 | 367 |

1964 and 1964–1965, the same group which has already been shown disaggregated by years of age in Table 8.5. This example is given in Table 8.6.

The pattern of Table 8.6 is similar to that of the others. Thus, in the columns for age 2 in 1964: of the 420,700 1-year-olds who flowed in from 1963, 419,200 remained at home and 1500 went to school for the first time; net immigration was negative, that is, there were 200 net emigrants. And, in the rows for age 2 in 1964: 500 died during the year; and 420,000 survived and can be found aged 3 in 1965, when 407,400 remained at home, 11,100 went to school for the first time and 1500 continued at school. And so on for the other ages: each year, as they grow one year older, a certain number flow from home to home, a certain number flow from home to school and a certain number flow from school to school.

Wales

| | 1965 | | | | | | | | | Total outflows |
| Age 0 | Age 1 | Age 2 | | Age 3 | | Age 4 | | Age 5 | | |
| Home | Home | Home | School | Home | School | Home | School | Home | School | |
|---|---|---|---|---|---|---|---|---|---|---|
| 443.2 | | | | | | | | | | — |
| 0.0 | 0.0 | −0.5 | | 0.3 | | −0.7 | | −0.8 | | — |
| | | | | | | | | | | 438.5 |
| | | | | | | | | | | 422.3 |
| | | | | | | | | | | 407.1 |
| | | | | | | | | | | 1.4 |
| | | | | | | | | | | 382.6 |
| | | | | | | | | | | 12.7 |
| | | | | | | | | | | 273.8 |
| | | | | | | | | | | 102.4 |
| | | | | | | | | | | (5.6) |
| | | | | | | | | | | (364.6) |
| | 442.2 | | | | | | | | | 451.1 |
| | | 426.7 | 1.3 | | | | | | | 429.4 |
| | | | | 407.4 | 11.1 | | | | | 419.0 |
| | | | | | 1.5 | | | | | 1.5 |
| | | | | | | 292.4 | 99.0 | | | 394.7 |
| | | | | | | | 12.6 | | | 12.6 |
| | | | | | | | | 9.3 | 274.3 | 283.8 |
| | | | | | | | | | 110.4 | 110.4 |
| | | | | | | | | | | (9.1) |
| | | | | | | | | | | (367.1) |
| 443.2 | 442.2 | 426.2 | 1.3 | 407.7 | 12.6 | 294.7 | 111.6 | 8.5 | 384.7 | |

As with Tables 8.1 and 8.3, the inflows and outflows of 1964 in Tables 8.5 and 8.6 could be rearranged as sets of accounts, each set representing a further stage in the deconsolidation of ages 0–5 in Table 8.4 (just as Table 8.4 was itself a deconsolidation of Table 8.2). In a mainly theoretical paper, however, it seems unnnecessary to labour the point by actually drawing up these accounts. The statistically minded may find some amusement in doing it for themselves.

A word about the derivation of the figures. Births and deaths are derived from the Registrar General's estimates in U.K. General Register Office (1965–1968). The stocks of survivors at ages 0 and 1 are calculated from them. The stocks of survivors at age 2 to 5 are taken directly from the Government Actuary's estimates in U.K. Department of Education and Science (1965–1967). Net immigrations are derived from these stock estimates, with an adjustment for deaths: for example, the survivors aged 4 at the end of 1964 were 394,000 and the deaths at age 4 in 1964 were 200, giving a total outflow of 394,200; but at the beginning of 1964 the inflow of survivors from the 3-year-olds of 1963 had been 395,000; and so the net immigrations into 1964 at age 4 were $394,200 - 395,000 = -800$. The numbers going to, and continuing at, school are based on the January counts of the school population in U.K. Department of Education and Science (1965–1967); and the numbers remaining at home are residual figures which balance the inward and outward flows.

## Cohort Analysis

Each of the two submatrices in Table 8.6 provides a cross-section analysis of six groups of children born in six successive years; in the language of demography, six cohorts. Thus in the 1963–1964 submatrix the 5-year-olds of 1964 belong to the 1959 cohort, the 4-year-olds to the 1960 cohort and so on; similarly, in the 1964–1965 submatrix the 5-year-olds of 1965 belong to the 1960 cohort, the 4-year-olds to the 1961 cohort and so on. By relating successive ages from a series of such matrices it is possible to draw up a set of accounts for a single cohort. As I said above, we have constructed cross-section matrices back to 1960–1961, and these enable us to follow the 1960 cohort from birth up to the flows into and out of age 5.

This information is shown both in accounting and in matrix form in Tables 8.7 and 8.8. In these tables the outflow of survivors from one age appears as the main inflow into the next. It will be noticed that the entries for ages $\frac{3}{4}$ in 1963–1964 and $\frac{4}{5}$ in 1964–1965 are taken directly from Table 8.6.

## Demographic Input-Output

Let us now return to Table 8.6 and consider what information can be extracted from it which will be useful for purposes of model-building. The most interesting information relates to the transition coefficients, that is the proportions in which the survivors of one year distribute themselves during the succeeding year among the various activities open to them. These coefficients are obtained by taking the entries in a row of the matrix and dividing them by their sum, in the same way as economic input-output coefficients are obtained from an input-output table by dividing the entries in a column by their sum. In order to see how the transition coefficients can be used to build demographic input-output models, it is easiest to start by setting out the basic accounting framework in symbolic form. This is done in Table 8.9.

The inflow into this year is made up of the births and immigrations of this year and of the survivors from last year. Writing this inflow as a column vector, we have

$$p = b + \Lambda^{-1}S'i \tag{8-1}$$

Table 8.7.   The Accounts for the 1960 Cohort, 1960 to 1965 (*Thousands of Males*)

| 1960: Age 0 | | | |
|---|---|---|---|
| **Inflows** | | **Outflows** | |
| Births | 404.4 | Deaths | 8.9 |
| Net immigrants | 0.0 | Survivors into 1961: home | 395.3 |
| Total | 404.2 | Total | 404.2 |

| 1961: Age 1 | | | |
|---|---|---|---|
| Survivors from 1960: home | 395.3 | Deaths | 1.5 |
| Net immigrants | 0.0 | Survivors into 1962: home | 393.8 |
| Total | 395.3 | Total | 395.3 |

| 1962: Age 2 | | | | |
|---|---|---|---|---|
| Survivors from 1961: | | | Deaths | 0.5 |
|   remaining at home | 392.4 | | Survivors into 1963: | |
|   entering school | 1.4 | |   at home | 393.6 |
| | | 393.8 |   at school | 1.4 |
| Net immigrants | | 1.7 | | 395.0 |
| Total | | 395.5 | Total | 395.5 |

| 1963: Age 3 | | | | |
|---|---|---|---|---|
| Survivors from 1962: | | | Deaths | 0.3 |
|   remaining at home | 382.3 | | Survivors into 1964: | |
|   entering school | 11.3 | |   at home | 382.3 |
|   continuing at school | 1.4 | |   at school | 12.7 |
| | | 395.0 | | 395.0 |
| Net immigrants | | 0.3 | | |
| Total | | 395.3 | Total | 395.3 |

| 1964: Age 4 | | | | |
|---|---|---|---|---|
| Survivors from 1963: | | | Deaths | 0.2 |
|   remaining at home | 284.6 | | Survivors into 1965: | |
|   entering school | 97.7 | |   at home | 283.6 |
|   continuing at school | 12.7 | |   at school | 110.4 |
| | | 395.0 | | 394.0 |
| Net immigrants | | −0.8 | | |
| Total | | 394.2 | Total | 394.2 |

| 1965: Age 5 | | | | |
|---|---|---|---|---|
| Survivors from 1964: | | | Deaths | 0.2 |
|   remaining at home | 9.3 | | Survivors into 1966: | |
|   entering school | 274.3 | |   at home | 8.5 |
|   continuing at school | 110.4 | |   at school | 384.5 |
| | | 394.0 | | 393.0 |
| Net immigrants | | −0.8 | | |
| Total | | 393.2 | Total | 393.2 |

Table 8.8.   A Matrix for the 1960 Cohort, 1960 to 1965 (*Thousands of Males*)

| | | Outside world | 1960 | 1961 | 1962 | | 1963 | | 1964 | | 1965 | | Total |
|---|---|---|---|---|---|---|---|---|---|---|---|---|---|
| | | Deaths | Age 0 | Age 1 | Age 2 | | Age 3 | | Age 4 | | Age 5 | | outflows |
| | | | Home | Home | Home | School | Home | School | Home | School | Home | School | |
| **Outside world** | | | | | | | | | | | | | |
| Births | | | 404.2 | | | | | | | | | | — |
| Net immigrations | | | 0.0 | 0.0 | 1.7 | | 0.3 | | −0.8 | | −0.8 | | — |
| **England and Wales** | | | | | | | | | | | | | |
| 1960 Age 0 | Home | 8.9 | | 395.3 | | | | | | | | | 404.2 |
| 1961 Age 1 | Home | 1.5 | | | 392.4 | 1.4 | | | | | | | 395.3 |
| 1962 Age 2 | Home | 0.5 | | | | | 382.3 | 11.3 | | | | | 394.1 |
| | School | | | | | | | 1.4 | | | | | 1.4 |
| 1963 Age 3 | Home | 0.3 | | | | | | | 284.6 | 97.7 | | | 382.6 |
| | School | | | | | | | | | 12.7 | | | 12.7 |
| 1964 Age 4 | Home | 0.2 | | | | | | | | | 9.3 | 274.3 | 283.8 |
| | School | | | | | | | | | | | 110.4 | 110.4 |
| 1965 Age 5 | Home | 0.2 | | | | | | | | | | | (8.5) |
| | School | | | | | | | | | | | | (384.7) |
| Total inflows | | — | 404.2 | 395.3 | 394.1 | 1.4 | 382.6 | 12.7 | 283.8 | 110.4 | 8.5 | 384.7 | |

Table 8.9.  A Demographic Accounting Matrix in Symbolic Form[a]

| | | Outside world | Our country | | Total outflows |
| | | | This year | Next year | |
| --- | --- | --- | --- | --- | --- |
| Outside world | | | $b'$ | $\Lambda b'$ | |
| Our country | Last year | $\Lambda^{-1}d$ | $\Lambda^{-1}S$ | | $\Lambda^{-1}p$ |
| | This year | $d$ | | $S$ | $p$ |
| Total inflows | | | $p'$ | $\Lambda p'$ | |

[a] The meaning of the symbols is as follows:

$p$ = the vector of population flows

$b$ = the vector of births and immigrations

$d$ = the vector of deaths and emigrations

$S$ = the matrix of survivors; the typical element, $s_{jk}$ say, of $S$ represents the number of people in category $j$ in year $\tau$ who flow into category $k$ in year $\tau + 1$

$\Lambda$ = the lag, or shift, operator (often written as $E$); thus $\Lambda^{-1}d$, $\Lambda^{-1}S$, and $\Lambda^{-1}p$ denote, respectively, last year's deaths and emigrations, last year's survivors, and last year's population flows; and $\Lambda b$ and $\Lambda p$ denote next year's births and immigrations and next year's population flows;

(a prime superscript) indicates transposition; that is, $p'$ is a row vector formed from the elements of the column vector $p$; and $S'$ is a matrix formed by transposing the rows and the columns of $S$.

where $i$ denotes the unit vector and so $\Lambda^{-1}S'i$ denotes the column sums of $\Lambda^{-1}S$. The population vector at the beginning of this year is

$$(p - b) = \Lambda^{-1}S'i \qquad (8\text{--}2)$$

and so the total population at the beginning of this year is

$$i'(p - b) = i'\Lambda^{-1}S'i \qquad (8\text{--}3)$$

The outflow from this year is made up of the deaths and emigrations of this year and of the survivors into next year. Thus

$$p = d + Si \qquad (8\text{--}4)$$

This population vector at the end of this year is

$$(p - d) = Si \qquad (8\text{--}5)$$

and the total population at the end of this year is

$$i'(p - d) = i'Si \qquad (8\text{--}6)$$

These are the simple relationships between the flow vector $p$, the opening and closing stock vectors $(p - b)$ and $(p - d)$ and the opening and closing stock totals $i'(p - b)$ and $i'(p - d)$.

In the language of input-output, Eq. (8–1) says that in any given year total inputs, the elements of $p$, are equal to primary inputs, the elements of $b$, *plus* intermediate inputs, the elements of $\Lambda^{-1}S'i$; and Eq. (8–4) says that in the given year total outputs, the elements of $p$, are equal to final outputs, the elements of $d$, *plus* intermediate outputs, the elements of $Si$.

If we now divide the elements in each row of $\Lambda^{-1}S$ by the corresponding element of $\Lambda^{-1}p$, we obtain a matrix of transition coefficients, $C$ say, of which the typical element, $c_{jk}$ say, shows the proportion of the population in category $j$ last year which is to be found in category $k$ this year. Thus we can write

$$\Lambda^{-1}S' = C\Lambda^{-1}\hat{p} \qquad (8\text{--}7)$$

where $\Lambda^{-1}\hat{p}$ denotes a diagonal matrix formed from the vector $\Lambda^{-1}p$. By substituting for $\Lambda^{-1}S'$ from (8–7) into (8–1), we obtain

$$p = b + C\Lambda^{-1}p \qquad (8\text{--}8)$$

which shows this year's inflow vector, $p$, as composed of this year's births and immigrations, $b$, *plus* this year's rearrangement of last year's outflow vector.

The contrast between the matrix of transition coefficients, $C$, and the matrix of economic input-output coefficients, usually denoted by $A$, is now apparent. The elements of $A$ are static input coefficients which show how the intermediate inputs of one year are related to the outputs of the same year. The elements of $C$ are dynamic output coefficients which show how the intermediate outputs of one year are related to the inputs of the following year.

Let us assume that $C$ is constant over time. On this assumption, if we multiply Eq. (8–8) by successive powers of $\Lambda$ and substitute for the terms in $\Lambda^\theta p$, $\theta = 1, 2, \ldots, \tau$, we obtain

$$\Lambda p = \Lambda b + C p \qquad (8\text{--}9)$$

$$\Lambda^2 p = \Lambda^2 b + C\Lambda p$$

$$= \Lambda^2 b + C\Lambda b + C^2 p \qquad (8\text{--}10)$$

and, in general,

$$\Lambda^\tau p = \sum_{\theta=0}^{\tau-1} C^\theta \Lambda^{\tau-\theta} b + C^\tau p \qquad (8\text{--}11)$$

Equation (8–11) expresses the population flow vector $\tau$ years hence in terms of the present flow vector and of the vectors of future births and immigrations up to and including those in year $\tau$. The existing population, as it ages, goes through a fixed sequence of transitions and so do the additions to the population from births and immigrations. For values of $\tau$ exceeding the human life-span, $C^\tau$ is the null matrix and $\Lambda^\tau p$ arises entirely from the gradual transformation of successive $b$s by means of $C$. In other words, given estimates of births and immigrations up to year $\tau$, if we assume that the transition proportions are constant over time, we can forecast the number in each age and activity in year $\tau$.

However, this assumption would be unrealistic. The transition proportions are likely to change over time, and we should therefore date $C$ just as we have dated $p$ and $b$. In this case it is more convenient to define $C$ as

$$C = S'\hat{p}^{-1} \qquad (8\text{--}12)$$

and to rechristen the $C$ in (8.8), which we have used so far, $\Lambda^{-1}C$. On this basis, (8–

9) can remain as before and (8–10) becomes

$$\Lambda^2 p = \Lambda^2 b + \Lambda C \Lambda p$$
$$= \Lambda^2 b + \Lambda C \Lambda b + \Lambda C C p \qquad (8\text{–}13)$$

Further,

$$\Lambda^3 p = \Lambda^3 b + \Lambda^2 C \Lambda^2 p$$
$$= \Lambda^3 b + \Lambda^2 C \Lambda^2 b + \Lambda^2 C \Lambda C \Lambda b + \Lambda^2 C \Lambda C C p \qquad (8\text{–}14)$$

and, in general,

$$\Lambda^\tau p = \Lambda^\tau b + \sum_{\theta=1}^{\tau-1} \left[ \prod_{\lambda=\tau-\theta}^{\tau-1} (\Lambda^\lambda C) \right] \Lambda^{\tau-\theta} b + \sum_{\theta=0}^{\tau-1} (\Lambda^\theta C) p \qquad (8\text{–}15)$$

From the middle term on the right-hand side of (8–15) we can see that the multiplier of $\Lambda^{\tau-\theta} b$ is $\Lambda^{\tau-1} C \times \Lambda^{\tau-2} \times \cdots \times \Lambda^{\tau-1} C$, instead of the $C^\theta$ in (8–11).

If the matrix $C$ is expected to change in the future, we must replace (8–11) by (8–15) in making projections; and, in addition to estimating the present elements of $p$ and the future elements of $b$, we must estimate the future elements of $C$.

Thus, given $p$ and a set of future $b$s and $C$s, we can build up a series of future $p$s. Computationally this is quite straightforward and the only practical difficulty arises from the high order of the $C$ matrices. However, apart from a concentration of entries just below the leading diagonal, these matrices are entirely composed of zeros and, as shown in Slater (1967), this means that the storage space needed can be greatly reduced.

## Changing Coefficients

The transition proportions that we can derive from a demographic accounting matrix are measures of past changes from one period to the next. Like input-output coefficients in economic models, they depend on the forces of supply and demand and may represent either an equilibrium position, or a stage in an adaptive process, or simply a disequilibrium position which persists because the equilibrating mechanism is feeble and its purpose not clearly specified.

Thus the main problem encountered in building the model described in the preceding section is to find out what has been happening to the transition proportions in the past and what is likely to happen to them in the future. The steps to be taken can be described as follows.

(a) The first thing to do is simply to calculate sets of transition coefficients for a series of years and note any changes in the individual coefficients over time and any relationships there may be between different sets. For example, let us examine the coefficients that can be derived from a series of tables like Table 8.6. Since, as I have said, the series we have constructed goes back to 1960–1961, we can trace the changes in any given set of coefficients over a five-year period. In the case of small children the most interesting coefficients are those which show the proportions of

Table 8.10. Transition Proportions from Home to School, Ages 1 to 5, 1960–1961 to 1964–1965 (*Per Thousand Males*)

|  | Age 1–2 | Age 2–3 | Age 3–4 | Age 4–5 |
|---|---|---|---|---|
| 1960–1961 | 0.004 | 0.031 | 0.271 | 0.983 |
| 1961–1962 | 0.004 | 0.031 | 0.253 | 0.981 |
| 1962–1963 | 0.003 | 0.029 | 0.246 | 0.979 |
| 1963–1964 | 0.004 | 0.028 | 0.255 | 0.967 |
| 1964–1965 | 0.003 | 0.026 | 0.251 | 0.967 |

the outflow of one year who go to school for the first time in the following year. These are set out in Table 8.10.

As can be seen, about 3 or 4 per mille of the 1-year-old survivors go to school for the first time at age 2; about 3 percent of the 2-year-olds who are not at school go there for the first time at age 3; about 25 percent of the 3-year-olds who are not at school go there for the first time at age 4; and about 97 or 98 percent of the 4-year-olds who are not at school go there for the first time at age 5, the age at which school attendance becomes compulsory.

There seems to be a fairly general tendency for the proportions to become a little smaller with time, which is surprising in view of the emphasis that is coming to be placed on early education. The falling tendency might be due to an increasing pressure of numbers on facilities that have not kept pace; or it might be due to a greater willingness to have young children at home as a consequence of higher incomes and greater leisure or of some change in social mores. None of these explanations, however, should have much influence on the 4–5 ratios, since 5 is the age at which school attendance becomes compulsory. If we look at the time series of the 4–5 ratios we can see that the fall is most marked between 1962–1963 and 1963–1964. This suggests that either there was some change in the age regulations for school admittance or there was some change in the concept of ineducability which would lower the intake of the system.

(*b*) Once the proportions have been calculated and possible reasons for the changes noted, the next thing to do is to try to decide how far each of the possible forces has in fact been at work. As a rule one will not be able to reach a definite conclusion without outside help, but sometimes a change in certain ratios can confidently be attributed to a known change in circumstances. For example, in 1963 there was a change in the regulations relating to school leaving, which required that children who had their fifteenth birthday late in the year, who would previously have been allowed to leave at the end of the autumn term, should continue at school until the end of the spring term of the following year. The effect of this change of regulations on secondary modern schools is illustrated in Table 8.11.

From the first column of this table, relating to boys who leave school in the year in which they become 15, we can see that in 1962, 76.6 percent of the boys who entered the year aged 14 left school when they became 15; whereas in 1963 the

Table 8.11.  Transition Proportions from Secondary Modern Schools to Home and Employment at Ages 14–15 and 15–16, 1960–1961 to 1964–1965 (*Per Thousand Males*)

|  | Age 14–15 | Age 15–16 |
|---|---|---|
| 1960–1961 | 0.793 | 0.766 |
| 1961–1962 | 0.766 | 0.711 |
| 1962–1963 | 0.555 | 0.724 |
| 1963–1964 | 0.556 | 0.830 |
| 1964–1965 | 0.523 | 0.807 |

corresponding proportion fell to 55.5 percent. Since we know about the change in the regulations, we can be pretty sure that the change in the ratios is almost entirely due to it. What is more, since we are dealing with children who leave school as soon as possible, we can predict that the decrease in autumn-term leavers in 1963 will be matched by an increase in spring-term leavers in 1964. Indeed, when we look at the second column of the table, relating to boys who leave school in the year in which they become 16, we see that the proportion rises from 72.4 percent in 1963 to 83 percent in 1964.

If we want to make sure that our surmise is correct, we can calculate for each triad of years (1960–1962, 1961–1963, etc.) the proportion of boys aged 14 in the first year of the triad who had left school by the time they were 16, two years later. The series runs: 0.939, 0.934, 0.924, 0.913. Thus, while there was a slightly diminishing tendency to leave school at these ages, the trend is smooth, and we can conclude that the violent changes which appear in Table 8.11 result, as we had surmised, from the change in regulations.

(*c*) Many changes are the result of processes more complicated than that just described. For example, as time goes on a larger and larger proportion of students aspire to the higher rungs of the educational ladder. What controls an adaptive process like this one? As usual, supply and demand. Factors on the side of supply include the facilities available, the standard of ability required and the grants provided to students; factors on the side of demand include the cultural and economic standards of different groups in society, an awareness of the advantages of higher education, which may be generally heightened by the example of others, and costs, whether these represent charges to be met or present income to be foregone.

A basis for modelling this kind of situation might be provided by an epidemic model, as was suggested in Stone (1965). In such a model, the rate of change in the numbers of applicants for higher education would be made to depend partly on the proportion who were accepted and partly on the proportion who were rejected. The proportion of admissions would follow a logistic curve, the maximum rate of change being reached when the proportion was one half. In the simplest case, two parameters would be needed, one relating to the maximum proportion of applicants and the other relating to the speed of the adjustment process. By making these parameters functions of variables, a more complicated expression would be obtained which, if it could not be solved analytically, would perhaps yield to numerical treatment.

(d) In many cases (they may even be the majority in the early stages of this kind of model-building) we shall find it impossible to model the variations in a particular transition proportion even though we may have ideas about the factors behind these variations. In such cases there is little we can do but resort to some form of trend analysis: try to decide whether there is a significant trend and how far it seems reasonable to project it into the future, or whether a constant value would be more in accord with the data. In this field of research, trend analysis should be regarded as a last resort because, whereas in economic planning we may be content to look five or at most ten years ahead, in educational planning we may well want to look from ten to twenty-five years ahead. Twenty-five years is an uncomfortably long time over which to allow even a small linear trend to build up.

(e) Finally, as in input-output analysis, it is necessary to consider possible future changes in the coefficients which do not follow from any factor that has been at work in the past. An example is an intended change in the school leaving age: in Britain it is proposed to raise this age by one year to sixteen in 1971. One effect of such a change is clearly predictable: children will be required to spend their fifteenth year at school. But it is also to be expected that the change will bring about alterations in the transition proportions at later ages and in general will lead to more voluntary attendance at these ages. Here it would seem desirable to look at what happened when the leaving age was raised from fourteen to fifteen after the last war and at the changes in the proportions from age fifteen onwards over the last twenty years. By these means it might be possible to make some plausible allowance for the indirect effects of the next compulsory change.

**A Link with Economic Accounting**

In this paper I am concerned only with a framework into which information on human flows and stocks can be fitted. For the moment I do not see very clearly how a general system of demographic accounts can best be linked with a general system of economic accounts. In the specific case of the educational system, however, I see the connection in clearer focus.

Having obtained from the model estimates of flows and stocks classified by age and type of educational institution, we can take the estimates for the age group attending each type of institution (ages 2 to 4 at nursery school, 5 to 11 at primary school, 9 to 13 at preparatory school, and so on) and aggregate them over the individual ages in the group. This will give a measure, expressed in student numbers, of the activity levels expected of each type of institution. The economic inputs needed to sustain these activity levels can then be set out in a series of columns, one for each type of institution. These columns will form a table, each row of which will relate to a particular kind of economic input, precisely like the rows in an ordinary input-output table.

The inputs into educational processes, like those into any other process can be classified under three broad heads.

### Intermediate Inputs

Items like books and stationery, food and cleaning materials, give rise to no special problems. They must be obtained from the productive system as and when they are wanted, as is done for any other productive activity.

### Primary Inputs

As far as labour is concerned, while clerical and domestic staff should not be forgotten, the main component is teachers. And teachers, before they become inputs into the educational system, form an important part of its output. If the stock of teachers is to be maintained and expanded, enough teachers must be trained to make good the wastage due to death, retirement and, in the case of women teachers, marriage. In order to work out the flows in the feedback loop connecting activity levels in one period with teacher training in an earlier period, it is necessary to know something about the age distribution of teachers, the capacity of the teaching profession to hold its members through their working life and the pattern of exodus and return to be expected of women teachers. Future recruitment problems may well emerge from a study of this kind which may lead to an improvement in the status and pay of teachers and to a reconsideration of existing educational technology.

As far as capital is concerned, needs can largely be expressed in terms of specialised buildings and equipment. Whatever standard is adopted, buildings and equipment must be in place when they are needed, and can best be considered under a separate heading.

### Capital Goods

Just as with teachers we can trace a feedback loop to teacher training, so we can trace a similar loop connecting activity levels in education to educational building activity in an earlier period. As in the case of teachers, we need to know something about wastage (scrapping) rates as well as about activity levels. At the national level, no more can be done than assess the order of magnitude of future building programmes which would be called for by future activity levels operated according to certain standards of space and hygiene. At the regional level many more specific factors would come in, since educational building plans have usually to fit into a general programme of rural or urban development and renewal.

In conclusion, one part of educational planning consists of ensuring that resources will be available to meet the expected needs of the educational system; another part consists of deciding what these needs ought to be. Models like those described above may throw some light on the second type of questions. They cannot be expected to resolve them, however. For this purpose we must turn, as in economics, from input-output models to programming models, exemplified in the field of educational planning by the work of Armitage and Smith (1967), Armitage et al. (1968), Bénard (1967), Clough (1968), Clough and McReynolds (1966), Von Weizsäcker (1967), and many others.

In this paper I have emphasised the application of demographic accounting to educational problems simply because this is the aspect of the subject on which I have done most work. Evidently, however, many other aspects are important and we may hope that, before long, demographic accounting will come to be recognised as an essential branch of social accounting.

### References

Armitage, P. and C. Smith. 1967. The development of computable models of the British educational system and their possible uses. In *Mathematical Models in Educational Planning*, ed. Directorate for Scientific Affairs. Paris: OECD.

Armitage, P., C. Smith, and P. Alper. 1968. Models for Educational Decision-Making.

Bénard, J. 1967. General optimization model for the economy and education. In *Mathematical Models in Educational Planning*, ed. Directorate for Scientific Affairs. Paris: OECD.

Clough, D. J. 1968. A model for education-employment systems analysis.

Clough, D. J. and W. P. McReynolds. 1966. State transition model of an educational system incorporating a constraint theory of supply and demand. *Ontario Journal of Educational Research* 9: 1–18.

Hicks, J. R. 1952. *The Social Framework*. 2nd ed. Oxford: The Clarendon Press.

Slater, L. 1967. Large sparse matrices.

Stone, R. 1965. A model of the educational system. *Minerva* 3(2): 172–186.

Stone, R. 1966. Input-output and demographic accounting. *Minerva* 4(3): 365–380.

Stone, R., G. Stone, and J. Gunton. 1968. An example of demographic accounting. *Minerva* 6(2): 185–212.

U.K. Department of Education and Science. 1965–1967. *Statistics of Education 1964*. Vol. 1; *SE 1965*, Vol. 1; *SE 1966*, Vol. 1. London: H.M.S.O.

U.K. Department of Education and Science. 1965–1968. *Statistical Review of England and Wales 1963*, pts. I and II; *SREW 1964*, pts. I and II; *SREW 1965*, pts. I and II; *SREW 1966*, pts. I and II. London: H.M.S.O.

Von Weizsäcker, C. C. 1967. Training policies under conditions of technical progress: a theoretical treatment. In *Mathematical Models in Educational Planning*. Paris: OECD.

# III

## Studies on
## the Coefficients Matrix

# 9

## Alternative Proof of the Substitution Theorem for Leontief Models in the General Case

KENNETH J. ARROW

It is of some interest to state and prove, in a manner which does not involve the use of the calculus, the theorem concerning substitutability in Leontief models stated elsewhere in this volume by Professor Samuelson (Samuelson, 1951). The chief virtue of such restatement is not the generalization to nondifferentiable production functions but the greater clarity given to the importance of the special conditions of the problem. This approach has been developed by Professor Koopmans (1951b) for the case of three outputs; the present chapter seeks to generalize his results.

### The Assumptions of the Samuelson-Leontief Model

Samuelson's assumption will be restated here in the terminology of linear programming (Koopmans, 1951a). We shall let $n + 1$ be the total number of commodities involved; the first $n$ will be termed "products" and the $(n + 1)$th "labor."

ASSUMPTION I: *There is a collection of basic activities, each represented by a vector with n + 1 components, such that every possible state of production is represented by a linear combination of a finite number of the basic activities with nonnegative coefficients.*[1] *The collection of basic activities from which such combinations are formed need not itself be finite.*

ASSUMPTION II: *No basic activity has more than one output.*

ASSUMPTION III: *In every basic activity labor is a nonzero input.*

ASSUMPTION IV: *There is a given supply of labor from outside the system, but none of any product.*

This article was published in *Activity Analysis of Production and Allocation*, edited by Tjalling C. Koopmans. Originally published by John Wiley & Sons, Inc., New York, 1951. © Cowles Commission for Research in Economics, pp. 155–164. Reprinted by permission of Yale University Press.

ASSUMPTION I is that of constant returns to scale; II states the absence of joint production in the basic activities; III states that labor appears solely as a primary input; and IV states that no product is a primary input.

In the vector representation of activities, let the $(n + 1)$th component be labor. As usual, inputs will be represented by negative numbers, outputs by positive ones. By an activity of the $i$th industry we shall understand an activity in which no component other than the $i$th is positive. Clearly, any linear combination of the basic activities of the $i$th industry with nonnegative coefficients is itself an activity of the $i$th industry. Further, let $y$ be any activity. Then, by ASSUMPTION I,

$$y = \sum_k x_k b^k \tag{9-1}$$

where $x_k \geq 0$ and $b^k$ is a vector representing a basic activity.[2] Number the activities $b^k$ in such a way that those with $k = 1, \ldots, n_1$ are basic activities of the first industry, and, in general, those with $k = n_{i-1} + 1, \ldots, n_i$ are basic activities of the $i$th industry, where $n_0 = 0$. Then, from Eq. (9–1),

$$y = \sum_{i=1}^{n} \sum_{k=n_{i-1}+1}^{n_i} x_k b^k \tag{9-2}$$

As noted,

$$\sum_{k=n_{i-1}+1}^{n_i} x_k b^k$$

is an activity of the $i$th industry. Hence, every activity is expressible as a sum of $n$ activities, one from each industry.

Further, let a normalized activity be one in which the labor input is 1. From ASSUMPTION I it follows that every activity of the $i$th industry is the nonnegative multiple of a normalized activity of that industry, and conversely. Hence every activity is a linear combination of $n$ normalized activities, one from each industry, with nonnegative coefficients. The amount of labor used in any activity is, therefore, the sum of these coefficients. If, finally, we choose the units of labor so that the total supply of labor available, as guaranteed by ASSUMPTION IV, is 1, we may say that any activity $y$ is expressible in the form

$$y = \sum_{j=1}^{n} x_j a^j \tag{9-3}$$

where

$$x_j \geq 0, \qquad \sum_{j=1}^{n} x_j = 1 \tag{9-4}$$

and $a^j$ is a normalized activity of the $j$th industry. As now defined, all vectors $y$, $a^j$ have $-1$ as their $(n + 1)$th component; let us redefine them to have only their first $n$ components.

Note that the set of all normalized activities of the $j$th industry is a convex set; call it $S_j$. From ASSUMPTION II,

$$\text{if} \quad a \in S_j, \quad \text{then} \quad a_k \leq 0 \quad \text{for all} \quad k \neq j \tag{9-5}$$

(Here the symbol $\in$ means "belongs to"; $a_k$ denotes the $k$th component of $a$.)
Finally, it follows from ASSUMPTION IV that

$$y \geqslant 0 \tag{9–6}$$

(The names and symbols for various partial ordering relations among vectors will be those introduced by Koopmans [1951a].)

The set $S$ of feasible points in the product space is that satisfying (9–3), (9–4), and (9–6). The problem is to characterize the set of efficient points[3] of $S$ if the assumption contained in (9–5) is made.

The set of all points satisfying (9–3) and (9–4) will be referred to as the *convex hull of the union* of $S_1, \ldots, S_n$. $S$ is then the intersection of the nonnegative orthant (of Euclidean $n$-space) with the convex hull of the union of $S_1, \ldots, S_n$.

The following notation and terminology will be used: $A$ will denote a square matrix of order $n$, $a_i^j$ will be the element in the $i$th row and $j$th column of $A$, and $a^j$ will be the vector which is the $j$th column of $A$. $A$ will be said to be *admissible* if $a^j \in S_j$ for every $j$. A *weight vector*, $x$, has the properties $x \geqslant 0$, $\sum_{j=1}^{n} x_j = 1$. A pair $(A, x)$ is said to be a *representation* if $A$ is admissible, $x$ is a weight vector, and $Ax \geqslant 0$. A vector $y$ for which there exists a representation $(A, x)$ such that $y = Ax$ is termed *feasible*; this definition agrees with that given in the first sentence of the preceding paragraph.

In the light of (9–3)–(9–6), the economic significance of these definitions is obvious. In particular, the set of feasible points, or vectors, is precisely $S$; a representation is a mode of industrial organization which will achieve a given feasible point. Note that, in view of (9–5), $a_i^j \leqslant 0$ for all $i \neq j$.

Two forms of Samuelson's theorem will be established, corresponding to Koopmans' "strong" and "weak" assumptions, respectively. In the first case we assume that it is possible to produce a positive net output of all products; in the second we assume only that some net product is possible.

## The Substitution Theorem Under Strong Assumptions

THEOREM I: *For each* $j = 1, \ldots, n$, *let* $S_j$ *be a convex set in Euclidean* $n$-*space such that, if* $a \in S_j$, *then* $a_i \leqslant 0$ *for* $i \neq j$. *Let* $S$ *be the intersection of the convex hull of the union of* $S_1, \ldots, S_n$ *with the nonnegative orthant. If* $S$ *is a compact set[4] with at least one positive element, then the set of efficient points of* $S$ *is the intersection with the nonnegative orthant of an* $(n-1)$-*dimensional hyperplane the direction coefficients of whose outward normal are all positive.*

LEMMA 1: *If* y′ *belongs to the compact set* S, *there is an efficient point* y″ *of* S *such that* y″ $\geqslant$ y′.[5]

*Proof:* Let $U$ be the set of points $y$ such that $y \in S$, $y \geqslant y'$. $U$ is a compact set, so that the continuous function $\sum_{i=1}^{n} y_i$ attains a maximum in $U$, say at $y''$. Since $y'' \in U$, $y'' \geqslant y'$. If $y''$ were not efficient, there would be a point $\bar{y}$ of $S$ such that $\bar{y} \geqslant y''$; but then $\bar{y} \in U$, $\sum_{i=1}^{n} \bar{y}_i > \sum_{i=1}^{n} y''_i$, contrary to the construction of $y''$.[6]

LEMMA 2: *If* A *is a (square) matrix such that* $a_i^j \leqslant 0$ *for* $i \neq j$, *and* x *and* y *are vectors such that* $Ax = y$, $x \geqslant 0$, $y \geqslant 0$, $y_i > 0$, *then* $x_i > 0$.

*Proof:* By hypothesis, $a_i^j x_j \leqslant 0$ for $i \neq j$, so that $\sum_{j \neq i} a_i^j x_j \leqslant 0$. Hence $0 < y_i = a_j^i x_i + \sum_{j \neq i} a_i^j x_j \leqslant a_i^i x_i$. Since $x_i \geqslant 0$, we must have $x_i > 0$.

LEMMA 3: *Let* A *be a matrix such that* $a_i^j \leqslant 0$ *for* $i \neq j$ *and for which there exists a vector* x *such that* $Ax > 0$. *Then* (a) $Ax' \geqslant 0$ *implies* $x' \geqslant 0$; (b) $Ax' \geqslant 0$ *implies* $x' \geqslant 0$.[7]

*Proof:* By LEMMA 2, the hypothesis $Ax > 0$ implies $x > 0$. The ratios $x_j'/x_j$ are therefore defined; let

$$m = \min_j (x_j'/x_j) \tag{9-7}$$

where $j$ varies from 1 to $n$, and choose $i$ so that

$$x_i'/x_i = m \tag{9-8}$$

From (9-7) and the hypotheses,

$$a_i^j x_j' = a_i^j x_j (x_j'/x_j) \leqslant a_i^j x_j m \qquad (j \neq i) \tag{9-9}$$

Suppose $Ax' \geqslant 0$. Then, from (9-8) and (9-9),

$$0 \leqslant \sum_{j=1}^n a_i^j x_j' \leqslant a_i^i x_i m + \sum_{j \neq i} a_i^j x_j m = m \sum_{j=1}^n a_i^j x_j$$

By hypothesis, $\sum_{j=1}^n a_i^j x_j > 0$, so that $m \geqslant 0$. From (9-7), $x_j' \geqslant 0$, since $x_j > 0$ for all $j$, establishing (a).

If $Ax' \geqslant 0$, then, clearly, $x' \geqslant 0$ by (a), $x' \neq 0$, so that $x' \geqslant 0$.

LEMMA 4: *If* A *is a matrix such that* $Ax \geqslant 0$ *only if* $x \geqslant 0$, *then* A *is nonsingular.*

*Proof:* If $x$ is such that $Ax = 0$, then $A(-x) = 0$. By hypothesis, $x \geqslant 0$, $-x \geqslant 0$, so that $x = 0$. Hence $Ax = 0$ implies $x = 0$, so that $A$ must be nonsingular.

LEMMA 5: *If* $(A, x)$ *is a representation of* $y > 0$, *let* Q *be the set of points* $y' \geqslant 0$ *for which there exists a vector* x' *such that* $Ax' = y'$, $\sum_{j=1}^n x_j' = 1$. *Then every point of* Q *is feasible.*

*Proof:* By hypothesis,

$$a_i^j \leqslant 0 \quad \text{for} \quad i \neq j, \tag{9-10}$$

$$Ax = y > 0 \tag{9-11}$$

From (9-10), (9-11), and LEMMA 3, $Ax' \geqslant 0$ implies $x' \geqslant 0$. Since $\sum_{j=1}^n x_j' = 1$, $x'$ is a weight vector. Therefore $(A, x')$ is a representation, and $y'$ is a feasible point.

LEMMA 6: *If* Q *is defined as in* LEMMA 5, *there do not exist two points* y′, y″, *in* Q *such that* y′ ⩾ y″.

*Proof*: Suppose the contrary. Let $y' = Ax'$, $y'' = Ax''$, where

$$\sum_{j=1}^{n} x'_j = 1 = \sum_{j=1}^{n} x''_j \qquad (9\text{--}12)$$

$A(x' - x'') \geqslant 0$. By the proof of LEMMA 5, $A$ satisfies the hypotheses of LEMMA 3, so that $x' - x'' \geqslant 0$; but then, $\sum_{j=1}^{n} (x'_j - x''_j) > 0$, contrary to (9–12).

LEMMA 7: *If, for each* k = 1, ..., p, $y^{(k)}$ *has representation* $A^{(k)}$, $x^{(k)}$, *and* $t_k > 0$, *and if* $\sum_{k=1}^{p} t_k = 1$, *then* $y = \sum_{k=1}^{p} t_k y^{(k)}$ *is feasible and has a representation* (A, x), *where* $x = \sum_{k=1}^{p} t_k x^{(k)}$, *and* $a^j = (\sum_{k=1}^{p} t_k x_j^{(k)} a^{(k)j})/x_j$, *for all* j *for which* $x_j > 0$.

*Proof*: Define $x$ and $a^j$ as in the hypothesis; for all $j$ such that $x_j = 0$, choose $a^j$ to be any element of $S_j$. Since the sets $S_j$ are convex, it follows that $a^j \in S_j$ for each $j$, so that $A$ is admissible. It is also easy to see that $x$ is a weight vector, that $y = Ax$, and that $y \geqslant 0$, so that $(A, x)$ is a representation of $y$.

LEMMA 8: *Let* y > 0 *be an efficient point with representation* (A, x), *and let* T *be defined in terms of* y *in the same way that* Q *is defined in* LEMMA 5. *Then,* (a) A *is nonsingular;* (b) *every efficient point of* S *belongs to* T.

*Proof*: By the proof of LEMMA 5, $A$ satisfies the hypotheses of LEMMA 3 and hence is nonsingular by LEMMAS 3 and 4.

Let $y'$ be any efficient point. Since there is a positive efficient point, we cannot have $y' = 0$. Since $A$ is nonsingular, there is a vector $x'$ such that $Ax' = y' \geqslant 0$. By LEMMA 3, $x' \geqslant 0$, and therefore $\sum_{j=1}^{n} x'_j > 0$. Let $t_0 = 1/\sum_{j=1}^{n} x'_j$. Then, $A(t_0 x')$ $= t_0 y' \geqslant 0$, $\sum_{j=1}^{n} t_0 x'_j = 1$, so that

$$t_0 y' \in T \qquad (9\text{--}13)$$

By (9–13) and LEMMA 5, $t_0 y'$ is feasible. If $t_0 > 1$, then $t_0 y' \geqslant y'$, which is impossible for an efficient point $y'$. Hence

$$0 < t_0 \leqslant 1 \qquad (9\text{--}14)$$

The variable point $t t_0 y' + (1 - t)y > 0$ for $t = 0$. Hence we can choose $t_1$ so that

$$t_1 < 0 \qquad (9\text{--}15)$$

$$y'' = t_1 t_0 y' + (1 - t_1)y > 0 \qquad (9\text{--}16)$$

Let $x'' = t_1 t_0 x' + (1 - t_1)x$; then, by the definition of $t_0$ and the fact that $x$ is a weight vector, $\sum_{j=1}^{n} x''_j = 1$; also, $y'' = Ax''$. From (9–16) and the definition of $T$, $y'' \in T$. By LEMMA 5,

$$y'' \text{ is a feasible point} \qquad (9\text{--}17)$$

Let $t_2 = (t_1 t_0)/(t_1 t_0 - 1)$, $t_3 = (1 - t_1)/(1 - t_1 t_0)$. From (9–14) and (9–15),

$$0 < t_2 < 1 \tag{9–18}$$

$$t_3 \geqslant 1 \tag{9–19}$$

From (9–16),

$$t_3 y = t_2 y' + (1 - t_2) y'' \tag{9–20}$$

From (9–18), (9–20), (9–17), and LEMMA 7, $t_3 y$ is a feasible point. If $t_3 > 1$, then $t_3 y > y$, so that $y$ would not be efficient, contrary to hypothesis. Hence, from (19), $t_3 = 1$, which implies that $t_0 = 1$. From (9–13), then, $y' \in T$.

*Proof of Theorem I*: By hypothesis, there is at least one positive feasible point. By LEMMA 1, there is an efficient point $y > 0$. Let $T$ be defined as in LEMMA 8. Then every efficient point of $S$ belongs to $T$. Conversely, let $y'$ be any point of $T$. If $y'$ is not efficient, there is, by LEMMA 1, an efficient point $y'' \geqslant y'$. Since $y''$ is efficient, it belongs to $T$ by LEMMA 8; but this contradicts LEMMA 6. Hence $y'$ is efficient, so that $T$ is precisely the set of efficient points.

$T$ is the intersection with the nonnegative orthant of the hyperplane defined parametrically by the equations $Ax' = y$, $\sum_{j=1}^{n} x'_j = 1$. By LEMMA 8, $A$ is nonsingular, so that $x' = A^{-1}y$. Let $A^i_j$ be the element in the $j$th row and $i$th column of $A^{-1}$. and $A^i$ be the $i$th column. Then the equation of the hyperplane is

$$\sum_{i=1}^{n} \left( \sum_{j=1}^{n} A^i_j \right) y_i = 1$$

Hence the numbers $\sum_{j=1}^{n} A^i_j$ are the direction numbers of the outward normal to $T$. For each $i$, $AA^i$ is a vector all of whose components are zero except for the $i$th, which is 1. Therefore $AA^i \geqslant 0$; by LEMMA 3, $A^i \geqslant 0$, so that $\sum_{j=1}^{n} A^i_j > 0$ for all $i$.

## The Substitution Theorem Under Weak Assumptions

A generalization of THEOREM I in which it is assumed only that there is a feasible point $y \geqslant 0$ (instead of $y > 0$) will be developed in this section. Some new terminology and notation will be needed.

A representation $(A, x)$ will be said to be *trivial* if there is a nonnull set of integers, $I$, such that $x_i > 0$ for some $i$ in $I$, and $\sum_{j \in I} a^i_j x_j = 0$ for all $i$ in $I$. The mode of industrial organization displayed by a trivial representation has the property that there is a collection of industries in which there is some net input of labor and possibly of other commodities and such that the output of any one industry in the group is completely absorbed by the other industries in the group. This group, then, is only a drain on the net resources of the nation. The main result of this section is that any industry which can be used in any system of industrial organization not of the degenerate type just described can yield a positive net output; therefore, Samuelson's theorem applies.

LEMMA 9: *Let A be a matrix such that $a_i^j \leqslant 0$ when $i \neq j$; x and y vectors such that $x \geqslant 0$, $y \geqslant 0$, $y = Ax$; I a set of integers (between 1 and n); and i an element of I. Then, (a) $\sum_{j \in I} a_i^j x_j \geqslant y_i \geqslant 0$; (b) if $\sum_{j \in I} a_i^j x_j = 0$, then $y_i = 0$, and $a_i^j = 0$ for all $j \in -I$ such that $x_j > 0$. (By $-I$ is meant the set of integers between 1 and n not in I.)*

*Proof:* From the hypothesis,

$$a_i^j x_j \leqslant 0 \quad \text{for} \quad i \neq j \tag{9-21}$$

so that

$$\sum_{j \in -I} a_i^j x_j \leqslant 0 \tag{9-22}$$

From (9–22) and the hypotheses,

$$0 \leqslant y_i = \sum_{j \in I} a_i^j x_j + \sum_{j \in -I} a_i^j x_j \leqslant \sum_{j \in I} a_i^j x_j$$

establishing (a). If $\sum_{j \in I} a_i^j x_j = 0$, then clearly $y_i = 0$, and $\sum_{j \in -I} a_i^j x_j = 0$, so that, from (9–21), $a_i^j x_j = 0$ for $j \in -I$, from which (b) follows.

LEMMA 9 is a generalization of LEMMA 2.

LEMMA 10: *If $y \geqslant 0$ has a trivial representation $(A, x)$, then y is not efficient.*

*Proof:* By hypothesis, there is a set of integers, $I$, such that

$$x_i > 0 \quad \text{for} \quad \text{some} \quad i \in I \tag{9-23}$$

$$\sum_{j \in I} a_i^j x_j = 0 \quad \text{for all} \quad i \in I \tag{9-24}$$

From (9–24) and LEMMA 9b, $y_i = 0$ for all $i$ in $I$; since $y_k > 0$ for some $k$, we must have $k$ in $-I$. By LEMMA 2, then, $x_k > 0$ for some $k$ not in $I$. Together with (9–1), this shows that $0 < \sum_{j \in I} x_j < 1$. Let $t = 1/(1 - \sum_{j \in I} x_j)$, and define $x_j' = 0$ for $j \in I$, $x_j' = t x_j$ for $j \in -I$. Then

$$t > 1 \tag{9-25}$$

$$x' \text{ is a weight vector} \tag{9-26}$$

Let $y' = Ax'$. For $i$ in $I$, it follows from (9–24) and Lemma 9b that $a_i^j x_j = 0$ for $j$ in $-I$. Hence

$$y_i' = \sum_{j \in I} a_i^j x_j' + \sum_{j \in -I} a_i^j x_j' = 0 = t y_i \tag{9-27}$$

for $i$ in $I$. For $i \in -I$, $a_i^j x_j \leqslant 0$ for $j$ in $I$. Therefore

$$0 \leqslant y_i = \sum_{j \in I} a_i^j x_j + \sum_{j \in -I} a_i^j x_j \leqslant \sum_{j \in -I} a_i^j x_j$$

so that

$$y_i' = \sum_{j \in I} a_i^j x_j' + \sum_{j \in -I} a_i^j x_j' = t \sum_{j \in -I} a_i^j x_j \geqslant t y_i$$

for $i$ in $-I$, or, with (9–27),

$$y' \geqq ty \qquad (9\text{–}28)$$

$A$ is an admissible matrix by hypothesis; $x'$ is a weight vector, by (9–26); and from (9–28), (9–25), and the hypothesis, $y' \geq 0$, so that $y'$ is a feasible point. Furthermore, from (9–28), (9–25), and the hypothesis that $y \geq 0$, it follows that $y' \geq y$, so that $y$ is not efficient.

The proof of LEMMA 10 amounts to saying that the industrial organization represented by a trial representation can always be improved by shutting down the group of industries which yields no net aggregate output and distributing the released labor to the other industries in proportion to the numbers already employed.

We shall also need the following generalization of LEMMA 3:

LEMMA 11: *Let A be a matrix such that $a_i^j \leq 0$ for $i \neq j$ and for which there exists a vector $x > 0$ such that $(A, x)$ is a nontrivial representation. Then (a) $Ax' \geqq 0$ implies $x' \geq 0$; and (b) $Ax' \geq 0$ implies $x' \geq 0$.*

*Proof*: Since $x > 0$, the ratios $x_j'/x_j$ are defined. Let

$$m = \min_{j} (x_j'/x_j) \qquad (9\text{–}29)$$

and let $I$ be the set of integers such that $x_j'/x_j = m$; $I$ is nonnull. From the hypothesis, $a_i^j x_j < 0$ for $i$ in $I$, $j$ in $-I$, if $a_i^j \neq 0$. We then have

$$x_j'/x_j = m \quad \text{for} \quad j \in I \qquad (9\text{–}30)$$

$$a_i^j x_j' = a_i^j x_j (x_j'/x_j) < m a_i^j x_j \qquad (9\text{–}31)$$

if $i$ is in $I$, $j$ in $-I$, and $a_i^j \neq 0$. Suppose that for all $i$ in $I$, $\sum_{j \in I} a_i^j x_j = 0$; since $x_j > 0$ for all $j$, it would follow that $(A, x)$ is trivial, contrary to hypothesis. Hence, by LEMMA 9a, there is some $i$ in $I$ such that

$$\sum_{j \in I} a_i^j x_j > 0 \qquad (9\text{–}32)$$

From (9–31),

$$\sum_{j \in -I} a_i^j x_j' < m \sum_{j \in -I} a_i^j x_j \qquad (9\text{–}33)$$

if $a_i^j \neq 0$ for some $j$ in $-I$. Suppose $Ax' \geq 0$. Then, using (9–30),

$$0 \leq \sum_{j \in I} a_i^j x_j' + \sum_{j \in -I} a_i^j x_j' = m \sum_{j \in I} a_i^j x_j + \sum_{j \in -I} a_i^j x_j' \qquad (9\text{–}34)$$

If $a_i^j = 0$ for all $j$ in $-I$, then, from (9–32) and (9–34), it follows that $m \geq 0$. If $a_i^j \neq 0$ for some $j$ in $-I$, then, from (9–33) and (9–34),

$$0 < m \sum_{j=1}^{n} a_i^j x_j$$

Since $\sum_{j=1}^{n} a_i^j x_j \geq 0$ by the hypothesis that $(A, x)$ is a representation, we must have

$m > 0$. Hence, in either case, it follows from (9–29) that $x' \geqslant 0$. Part (b) follows from (a) as in LEMMA 3.

An integer, $i$, between 1 and $n$ will be said to denote a *useful industry* if there is some nontrivial representation $(A, x)$ in which $x_i > 0$. LEMMA 10 guarantees us that, in the search for efficient points, industries which are not useful can be regarded as nonexistent, so there is no loss of generality in assuming that all numbers denote useful industries.

It is possible that the set of feasible points is empty, in which case Samuelson's theorem naturally has no particular content. Hence we shall assume that there is at least one useful industry.

THEOREM II: *For each* $j = 1, \ldots, n$, *let* $S_j$ *be a convex set in Euclidean* n-*space such that if* $a \in S_j$, *then* $a_i \leqslant 0$ *for* $i \neq j$. *Let* S *be the intersection of the nonnegative orthant with the convex hull of the union of* $S_1, \ldots, S_n$. *If* S *is a compact set, and if every number from 1 to* n *denotes a useful industry, then the set of efficient points of* S *is the intersection with the nonnegative orthant of a hyperplane the direction coefficients of whose outward normal are all positive.*

*Proof:* For each $k$, let $y^{(k)}$ be a feasible point with a nontrivial representation $(A^{(k)}, x^{(k)})$ such that $x_k^{(k)} > 0$ for each $k$; the existence of these points follows from the hypothesis that every number from 1 to $n$ denotes a useful industry. Let $y = (\sum_{k=1}^{n} y^{(k)})/n$; by LEMMA 7, $y$ is a feasible point with representation $(A, x)$, where $x = (\sum_{k=1}^{n} x^{(k)})/n$, so that $x > 0$, and $a^j = (\sum_{k=1}^{n} x_j^{(k)} a^{(k)j})/nx_j$. Suppose $(A, x)$ is trivial; then, for some set of integers $I$, $\sum_{j \in I} a_i^j x_j = 0$ for all $i$ in $I$. From this, it follows that

$$\sum_{k=1}^{n} \left( \sum_{j \in I} a_i^{(k)j} x_j^{(k)} \right) = 0$$

for all $i$ in $I$. From LEMMA 9a, then $\sum_{j \in I} a_i^{(k)j} x_j^{(k)} = 0$ for each $k$ and all $i$ in $I$; in particular, the equation holds for any $k$ in $I$. Since $x_k^{(k)} > 0$, and therefore $x_i^{(k)} > 0$ for at least one $i$ in $I$, we would have $(A^{(k)}, x^{(k)})$, a trivial representation, contrary to hypothesis. Hence $(A, x)$ is a nontrivial representation with $x > 0$. All the conditions of LEMMA 11 are satisfied, so that, by LEMMAS 11 and 4, $A$ is nonsingular.

Let $y'$ be any positive vector. Then there is a vector $x'$ such that $Ax' = y' > 0$. By LEMMA 11, $x' \geqslant 0$; let $t = 1/(\sum_{j=1}^{n} x_j') > 0$. Then $tx'$ is a weight vector, and $ty' = A(tx')$ is a positive feasible point with representation $(A, tx')$. All the hypotheses of THEOREM I are then fulfilled, and the conclusion follows.

## NOTES

1. The restriction to linear combinations of a *finite* number of basic activities is unnecessary. The generalization of a set of nonnegative weights is a *measure* over the space of basic activities. (For the definition of a measure, see Saks [1937, pp. 7–17].) If $b$ stands for a variable basic activity and $\mu$ is a measure over the space of basic activities, then any state of production is of the form $\int b \, d\mu$. All subsequent results apply equally well to this more general case, with completely analogous proofs.

2. In this chapter all vectors are column vectors. For future reference, note that the prime symbol will not denote transposition but will serve to distinguish different column vectors.
3. For the relevant definition of an efficient point see Koopmans (1951), considering labor as the only primary commodity.
4. That is, closed and bounded.
5. This lemma has been proved by von Neumann and Morgenstern (1947, p. 593) for the case where $S$ has a finite number of elements.
6. See Koopmans (1951a), for definitions of "$\geq$" and "$\leq$".
7. Recall that in this chapter the prime is not used as a transposition sign.

**References**

Koopmans, Tjalling C. 1951a. Analysis of production as an efficient combination of activities. In *Activity Analysis of Production and Allocation*, ed. Tjalling C. Koopmans. New York: John Wiley & Sons, Ch. III, pp. 33–97.

Koopmans, Tjalling C. 1951b. Alternative proof of the substitution theorem for Leontief models in the case of three industries. In *Activity Analysis of Production and Allocation*, ed. Tjalling C. Koopmans. New York: John Wiley & Sons, Ch. VIII, pp. 147–154.

Neumann, John von, and Oskar Morgenstern. 1947. *Theory of Games and Economic Behavior*, 2nd ed. Princeton: Princeton University Press.

Saks, Stanislaw. 1937. *Theory of the Integral*, 2nd ed. New York: G. H. E. Stechert and Company.

Samuelson, Paul A. 1951. Abstract theorem concerning substitutability in open Leontief models. In *Activity Analysis of Production and Allocation*, ed. Tjalling C. Koopmans. New York: John Wiley & Sons, Inc., Ch. VII, pp. 142–146.

# 10

## Alternative Proof of the Substitution Theorem for Leontief Models in the Case of Three Industries

TJALLING C. KOOPMANS

Samuelson (1951) arrives at an important theorem which shows that Leontief's model of interindustry relationships has a greater generality than a literal reading of its assumptions suggests. In this theorem it is assumed that:

(I) each industry produces only one commodity; and
(II) each industry consumes, besides the commodities produced by other industries, only one scarce primary factor (labor), and that factor is the same for all industries.

Assuming, further, constant returns to scale in each industry, Samuelson finds that, even if each industry has a choice of many alternative processes for the production of its commodity, it is compatible with efficiency of production as a whole that each industry uses only one of the processes available to it, and this same process can be used regardless of the commodity composition of the net output of all industries taken together and regardless of the amount of labor available.

In this chapter a proof of this theorem is given which does not require that the alternative processes available to each industry can be subsumed in a production function possessing derivatives. We shall merely assume that the (finite or infinite number of) processes between which choice can be made by each industry have the properties associated with the notion of an activity:

(III) all inputs and outputs of a process can be multiplied by a nonnegative scale factor (*divisibility*); and
(IV) net outputs of different processes available to an industry can be added together to make a new available process (*additivity*).

Assumptions (III) and (IV) are also made by Leontief for the (unique) processes characterizing the several industries. Assumption (III) is implied in Samuelson's assumption of homogeneous production functions (constant returns to scale), but

This article was published in *Activity Analysis of Production and Allocation*, edited by Tjalling C. Koopmans. Originally published by John Wiley & Sons, Inc., New York, 1951. © Cowles Commission for Research in Economics, pp. 147–154. Reprinted by permission of Yale University Press.

his counterpart of (IV) is the more restrictive assumption of differentiable production functions.

We shall explicitly use two further assumptions:

(V) it is possible for each industry to select a process from among those available to it, and a scalar level of its operation, such that the total net output of all industries is positive for each commodity (except, of course, labor); and

(VI) the net output vectors of the alternative processes available to any one industry with a given labor input form a closed and bounded set in the commodity space.

ASSUMPTIONS I, II, III, and IV have no economic meaning as a model of production unless ASSUMPTION V is satisfied, and ASSUMPTION V is implied in Leontief's model and in Samuelson's discussion of it. An explicit criterion for its validity has been given by Hawkins and Simon (1949) for the present model. In Koopmans (1951a, Ch. III, Sect. 3.6) the same criterion, Postulate $C_1$, is explored for a more general model. Arrow has proved (Arrow, 1951) that a slightly weaker assumption, Postulate $C_2$, together with a further assumption excluding certain degenerate types of technology matrices, implies ASSUMPTION V.

ASSUMPTION VI is sufficient but not necessary for the validity of the theorem. A refinement of this assumption is also given by Arrow (1951).

In the present chapter only the case of three industries will be considered. The proof given here is generalized by Arrow (1951) to $n$ industries. It appears in that generalization that the "visual" elements, intentionally employed in the present proof because they aid intuitive understanding, are not essential to the mathematical argument. We now formulate in mathematical terms the theorem to be proved.[1]

THEOREM: *Let there be three closed and bounded convex sets,* $S_1, S_2, S_3$, *of points,* $y \equiv (y_1, y_2, y_3)$, *in three-dimensional space with the following properties:*

$$\begin{cases} \text{if} \quad a_{(1)} \equiv (a_{11}, a_{21}, a_{31}) \in S_1, \quad \text{then} \quad a_{11} > 0, \quad a_{21} \lessgtr 0, \quad a_{31} \lessgtr 0 \\ \text{if} \quad a_{(2)} \equiv (a_{12}, a_{22}, a_{32}) \in S_2, \quad \text{then} \quad a_{21} \lessgtr 0, \quad a_{22} > 0 \quad a_{32} \lessgtr 0 \quad (10\text{--}1) \\ \text{if} \quad a_{(3)} \equiv (a_{13}, a_{23}, a_{33}) \in S_3, \quad \text{then} \quad a_{31} \lessgtr 0, \quad a_{32} \lessgtr 0, \quad a_{33} > 0 \end{cases}$$

*Let there exist three points,*[2] $a'_{(1)} \in S_1$, $a'_{(2)} \in S_2$, $a'_{(3)} \in S_3$, *and three scalar weights,* $x'_1, x'_2, x''_3$, *such that*

$$\begin{cases} y'_1 \equiv a'_{11}x'_1 + a'_{12}x'_2 + a'_{13}x'_3 > 0 \\ y'_2 \equiv a'_{21}x'_1 + a'_{22}x'_2 + a'_{23}x'_3 > 0 \\ y'_3 \equiv a'_{31}x'_1 + a'_{32}x'_2 + a'_{33}x'_3 > 0 \end{cases} \qquad (10\text{--}2)$$

*and*

$$1 = x'_1 + x'_2 + x'_3, \qquad x'_1 \geqq 0, \qquad x'_2 \geqq 0, \qquad x'_3 \geqq 0 \qquad (10\text{--}3)$$

*Denote by* T *the set of those points* y *of the convex hull* $\bar{S}$ *(as defined below) of* $S_1, S_2$, *and* $S_3$ *which* (a) *belong to the closed positive octant* **P** *defined by*

$$y_1 \geqslant 0, \qquad y_2 \geqslant 0, \qquad y_3 \geqslant 0 \qquad (10\text{–}4)$$

*and* (b) *are such that no different point,* $y^* \equiv (y_1^*, y_2^*, y_3^*)$, *exists in* $\bar{S}$ *which satisfies*

$$y_1^* \geqslant y_1, \qquad y_2^* \geqslant y_2, \qquad y_3^* \geqslant y_3 \qquad (10\text{–}5)$$

*Then* T *is a plane triangle with one vertex on each of the positive coordinate axes, and such that all its points* y *can be obtained through linear combination*

$$\begin{cases} y_1 = a_{11}''x_1 + a_{12}''x_2 + a_{13}''x_3 \\ y_2 = a_{21}''x_1 + a_{22}''x_2 + a_{23}''x_3 \\ y_3 = a_{31}''x_1 + a_{32}''x_2 + a_{33}''x_3 \end{cases} \qquad (10\text{–}6)$$

*of the same three points*

$$a_{(1)}'' \in S_1, \qquad a_{(2)}'' \in S_2, \qquad a_{(3)}'' \in S_3 \qquad (10\text{–}7)$$

*with scalar weights,* $x_1, x_2, x_3$, *satisfying*

$$1 = x_1 + x_2 + x_3, \qquad x_1 \geqslant 0, \qquad x_2 \geqslant 0, \qquad x_3 \geqslant 0 \qquad (10\text{–}8)$$

The interpretation of the theorem is as follows. The three sets, $S_1, S_2, S_3$, incorporate the alternative modes of production available to each industry *at a labor input equal to unity*. The coordinates $a_{11}, a_{21}, a_{31}$ of a "point" $a_{(1)} \in S_1$ specify the positive output ($a_{11}$) of commodity "1" and the nonnegative inputs ($-a_{21}$, $-a_{31}$) of commodities "2" and "3" arising from the choice of the process $a_{(1)}$. The sign restrictions in (10–1) are imposed to satisfy ASSUMPTIONS I and II. The sets $S_1$, $S_2$, $S_3$ are made convex to satisfy ASSUMPTIONS III and IV, bounded to satisfy ASSUMPTION VI, and closed because productive processes cannot be measured with the absolute accuracy needed to give meaning to the distinction between closed and not closed sets, whereas without the assumption of closedness the theorem would not be valid. The existence of a solution of (10–2) expresses ASSUMPTION V.

The convex hull $\bar{S}$ of $S_1, S_2, S_3$ is the set of all points $(y_1, y_2, y_3)$ such that

$$\begin{cases} y_1 = a_{11}x_1 + a_{12}x_2 + a_{13}x_3 \\ y_2 = a_{21}x_1 + a_{22}x_2 + a_{23}x_3 \\ y_3 = a_{31}x_1 + a_{32}x_2 + a_{33}x_3 \end{cases} \qquad (10\text{–}9)$$

for some choice of processes, one for each industry,[3]

$$a_{(1)} \in S_1, \qquad a_{(2)} \in S_2, \qquad a_{(3)} \in S_3 \qquad (10\text{–}10)$$

and some set of levels of operation $x_1, x_2, x_3$ satisfying (10–8) so as to absorb all available labor. Since we assume no net inflow of any commodity except labor, the *attainable*[4] point set is the intersection S of $\bar{S}$ and P. The theorem concentrates on the set T of *efficient*[5] points $y$ of S, i.e., those points that cannot be improved upon, in the sense of (10–5), by any other attainable point $y^*$. The theorem says that this set T is a plane triangle, and that all its points can be obtained as combinations of

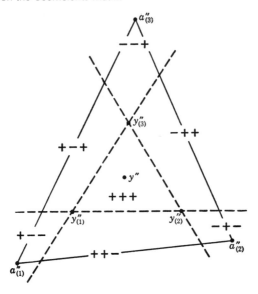

Fig. 10.1.

the same three processes, $a''_{(1)}$, $a''_{(2)}$, $a''_{(3)}$, one for each industry.

We proceed to the proof of the theorem. The point $y'$ defined by (10–2) is a point of $\bar{S}$. Now consider the point $y'' = \lambda'' y'$, where $\lambda''$ is the algebraically largest value of $\lambda$ for which

$$\lambda y' = (\lambda y'_1, \lambda y'_2, \lambda y'_3) \tag{10–11}$$

is contained in $\bar{S}$. ($\lambda''$ is finite because $\bar{S}$, as the convex hull of bounded sets, is itself bounded.) Then $y''$ is on the boundary of $\bar{S}$ and can be made to satisfy

$$y''_1 = a''_{11} x''_1 + a''_{12} x''_2 + a''_{13} x''_3 > 0$$

$$y''_2 = a''_{21} x''_1 + a''_{22} x''_2 + a''_{23} x''_3 > 0 \tag{10–12}$$

$$y''_3 = a''_{31} x''_1 + a''_{32} x''_2 + a''_{33} x''_3 > 0$$

where

$$a''_{(1)} \in S_1, \qquad a''_{(2)} \in S_2, \qquad a''_{(3)} \in S_3 \tag{10–13}$$

for some $x''$ satisfying (10–8) if $x''$ is substituted for $x$.

Table 10.1.   Sign Configuration of the Coordinates of Various Points or Point Sets in $\bar{T}$

| | Vertices | | | Edges | | Internal point |
|---|---|---|---|---|---|---|
| $a''_{(1)}$ | $a''_{(2)}$ | $a''_{(3)}$ | $\{a''_{(1)}, a''_{(2)}\}$ | $\{a''_{(2)}, a''_{(3)}\}$ | $\{a''_{(3)}, a''_{(1)}\}$ | $y''$ |
| + | − | − | + | − | + | + |
| − | + | − | + | + | − | + |
| − | − | + | − | + | + | + |

Consider the triangle $\bar{T}$ spanned by $a''_{(1)}, a''_{(2)}, a''_{(3)}$, i.e., the set of all points (10–6) for which $x$ satisfies (10–8). These points belong to $\bar{S}$. Table 10.1 indicates that what can be said about the sign configurations of the coordinates of various points or point sets in $\bar{T}$. Here $+$ stands for $>0$, $-$ for $\leqslant 0$. Since $y''$ has positive coordinates and is contained in the triangle $\bar{T}$, the plane $L$ of $\bar{T}$ does not coincide with a coordinate plane. It follows from the table that the intersections of this triangle with the coordinate side planes must run as indicated by Figure 10.1 (see dotted lines, with $-$ meaning $\leqslant 0$): within $L$ the intersection of $L$ with

$$y_1 = 0 \text{ separates } y'' \text{ from } \{a''_{(2)}, a''_{(3)}\}$$

$$y_2 = 0 \text{ separates } y'' \text{ from } \{a''_{(3)}, a''_{(1)}\} \qquad (10\text{–}14)$$

$$y_3 = 0 \text{ separates } y'' \text{ from } \{a''_{(1)}, a''_{(2)}\}$$

separation meaning strictly that $y''$ is not on the line $y_1 = 0$ and that no point of $\{a''_{(2)}, a''_{(3)}\}$ is on the same side of $y_1 = 0$ as $y''$ is, etc.

Denote by

$$y''_{(1)} \equiv (y''_{11}, 0, 0), \qquad y''_{(2)} \equiv (0, y''_{22}, 0), \qquad y''_{(3)} \equiv (0, 0, y''_{33}) \qquad (10\text{–}15)$$

the three points at which the lines of separation intersect. Now suppose (see Figure 10.2) that $\bar{S}$ contains any point $y'''$ of $P$ (hence also of $S$) separated from the origin by $L$. Then, since $S$ is convex, the entire tetrahedron constructed on the vertices $y''_{(1)}$, $y''_{(2)}, y''_{(3)}, y'''$ belongs to $S$, has $y''$ on its boundary facing the origin, and hence contains a point (10–3) with $\lambda > \lambda''$, in contradiction to the definition of $\lambda''$. It follows that the triangle $T = \{y''_{(1)}, y''_{(2)}, y''_{(3)}\}$ is a part of the boundary of $S$. We also read from Figure 10.1 that in (10–15)

$$y''_{11} > 0, \qquad y''_{22} > 0, \qquad y''_{33} > 0 \qquad (10\text{–}16)$$

Figure 10.2 suggests that all points $y$ of $T$ are efficient points. An algebraic argument runs as follows. Let

$$p_1 y_1 + p_2 y_2 + p_3 y_3 = p_1 y''_1 + p_2 y''_2 + p_3 y''_3 = p_0 \qquad (10\text{–}17)$$

say, be the equation of the plane $L$ of $\bar{T}$ (and of $T$), where $p_1, p_2, p_3$ do not all vanish. Since each of the vertices (10.15) of $T$ must satisfy this equation if substituted for $y$, we have

$$p_1 y''_{11} = p_2 y''_{22} = p_3 y''_{33} = p_0 \qquad (10\text{–}18)$$

It follows from (10–16) that we can choose the sign of $p_1$ such that

$$p_0 > 0, \qquad p_1 > 0, \qquad p_2 > 0, \qquad p_3 > 0 \qquad (10\text{–}19)$$

Then $S$ contains no point $y^*$ satisfying

$$p_1 y_1^* + p_2 y_2^* + p_3 y_3^* > p_0 \qquad (10\text{–}20)$$

because such a point would be separated from the origin by $L$.

Now let $y$ be a point of $T$ and therefore of $P$. Any point $y^*$ in $\bar{S}$ satisfying (10–5)

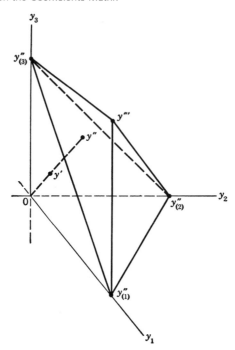

Fig. 10.2.

would also have to be in $P$ and therefore in $S$. But (10–5) would imply, in view of (10–19)

$$p_1 y_1^* + p_2 y_2^* + p_3 y_3^* > p_1 y_1 + p_2 y_2 + p_3 y_3 = p_0 \qquad (10–21)$$

[unless all three equality signs hold in (10–5), in which case $y^*$ is not different from $y$], and we have already seen that $S$ contains no such point $y^*$. Hence all points $y$ of $T$ satisfy the requirements of the theorem.

Conversely let $y'$ be in $S$ but not in $T$, and therefore on the same side of $L$ as the origin. The construction (10–11) already indicated will then lead to a point $y''$ in $\bar{S}$ such that (10–5) is satisfied if $y''$ and $y'$ are substituted for $y^*$ and $y$, respectively. Hence only points $y$ of $T$ satisfy the requirements of the theorem. This completes the proof of the theorem.

It can be shown that the theorem proved ceases to be true when either (i) more than one scarce primary input is required, or (ii) one of the three industries has joint production (e.g., $a_{11} > 0$, $a_{21} > 0$). Any proof should therefore make use of the restrictions (10–1) on the signs of the coefficients $a_{nk}$.

It might be thought that, whenever each of the sets $S_1, S_2, S_3$ is the convex hull of a smaller set of points representing more "elementary" technological processes, the three points $a''_{(1)}, a''_{(2)}, a''_{(3)}$ from which the efficient set $T$ is combined might have to be combinations of elementary processes. This, however, need not be the case. To be precise, let an extreme point of a convex set $S_k$ be defined as a point which cannot be represented as a linear combination of other points of $S_k$ with positive weights whose sum is unity. Then it will be possible to choose as basis points $a''_{(1)}, a''_{(2)}, a''_{(3)}$ of $T$ three extreme points of $S_1, S_2, S_3$, respectively. A proof of this statement can be based on the first paragraph of Ch. 3, Sect. 10, of Bonnesen and Fenchel (1934).

It is further easily seen that, while three processes suffice as a basis from which to combine all efficient points, the sets $S_1$, $S_2$, $S_3$ may be such that one or more of these processes can be chosen in more than one way.

## NOTES

1. I am indebted to Saunders MacLane for valuable discussions concerning this theorem.
2. Since no transposition signs are needed in this chapter, the symbols ' and " are used to denote different points.
3. There is no need to select more than one process from each $S_i$ because the $S_i$ are themselves assumed to be convex, and therefore the output of any linear combination of processes taken from some $S_i$ is the output of one single process contained in that $S_i$.
4. See Samuelson (1951), Definition 5.1. If we allowed some labor to go unused, it would be necessary to replace $S$ by the convex hull $\bar{S}_0$ of $\bar{S}$ and the origin 0. Since this would not add any efficient points, we need not consider this possibility.
5. See Samuelson (1951), Definition 5.2.

### References

Arrow, Kenneth, J. 1951. Alternative proof of the substitution theorem for Leontief models in the general case. In *Activity Analysis of Production and Allocations*, ed. Tjalling C. Koopmans. New York: John Wiley & Sons, Ch. IX.

Bonnesen, T., and W. Fenchel. 1934. *Theorie der convexen Körper*, Ergebnisse der Mathematik und ihrer Grenzgebiete, Vol. 3, No. 1, Berlin: Julius Springer, 1934; New York: Chelsea Publishing Company, 1948.

Hawkins, David, and Herbert Simon. 1949. Some conditions of macro-economic stability. *Econometrica* 17, 245–248.

Koopmans, Tjalling C. 1951. Analysis of production as an efficient combination of activities. In *Activity Analysis of Production and Allocation*, ed. Tjalling C. Koopmans. New York: John Wiley & Sons, Ch. III, pp. 33–97.

Samuelson, Paul A. 1951. Abstract theorem concerning substitutability in open Leontief models. In *Activity Analysis of Production and Allocation*, ed. Tjalling C. Koopmans. New York: John Wiley & Sons, Ch. VII, pp. 142–146.

# 11

## Abstract of a Theorem Concerning Substitutability in Open Leontief Models

PAUL A. SAMUELSON

Leontief (1941, 1946) assumes that total production of each of $n$ outputs, $x_1, \ldots, x_n$, is divided up into final outputs, $C_1, \ldots, C_n$, and into inputs used to help produce (with labor) all the inputs. Hence, for all $i$,

$$x_i = C_i + \sum_{j=1}^{n} x_{ji} \qquad (i = 1, 2, \ldots, n)$$

Labor, the $(n + 1)$th good, can be thought of as the sole "primary factor" or "nonproduced good," and its given total is allocated among all the different industries as follows:

$$x_{n+1} = 0 + \sum_{j=1}^{n} x_{j,n+1}$$

Note that joint products are ruled out, so the $x_{ji}$'s are functionally independent.

Since Leontief works with so-called "fixed" coefficients of production, it is usually thought that he must try to approximate reality by a production function of the form shown in Figure 11.1a, rather than of the more general form admitting of substitution as shown in Figure 11.1b. Actually, *all* his theory in its present form is compatible with the more general case of substitutability. With labor the only primary factor, *all desirable substitutions have already been made by the competitive market*. and no variation in the composition of final output or in the total quantity of labor will give rise to price change or substitution. Only the circled points in Figure 11.1b will ever be observed. The following discussion shows that this is a property of the efficiency frontier always reached under competition.

1. Let each good be subject to a production function, homogeneous of the first order.

$$x_i = F_i(x_{i1}, x_{i2}, \ldots, x_{i,n+1}) = mF_i\left(\frac{x_{i1}}{m}, \ldots, \frac{x_{i,n+1}}{m}\right) \qquad (11\text{--}1)$$

This article was published in *Activity Analysis of Production and Allocation*, edited by Tjalling C. Koopmans. Originally published by John Wiley & Sons, Inc., New York, 1951. © Cowles Commission for Research in Economics, pp. 142–146. Reprinted by permission of Yale University Press.

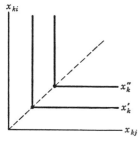

Fig. 11.1a. Equal output curves for $x_k$ with fixed coefficients.

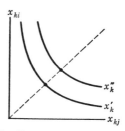

Fig. 11.1b. General equal output curves for $x_k$.

Our equilibrium requires that any $C$, such as $C_1$, be at a maximum subject to fixed values of total labor, $x_{n+1}$, and all other $C$s; that is,

$$C_1 = F_1(x_{11}, x_{12}, \ldots, x_{1,n+1}) - \sum_{j=1}^{n} x_{j1}$$

is to be a maximum subject to

$$F_i(x_{i1}, x_{i2}, \ldots, x_{i,n+1}) - \sum_{j=1}^{n} x_{ji} = C_i \qquad (i = 2, 3, \ldots, n) \qquad (11\text{--}2)$$

$$0 - \sum_{j=1}^{n} x_{j,n+1} = -x_{n+1}$$

where $F_{n+1}$, the amount of labor produced, can be set equal to zero. We have the $n(n + 1)$ variables of the form $x_{ij}$ to determine. We assume that with a finite amount of labor some finite quantity of each good is producible.

2. Because of homogeneity or constant returns to scale, the coefficients of production, $a_{ij} = x_{ij}/x_i$, are not constants but are connected by a relation of the form $F_i(a_{i1}, a_{i2}, \ldots, a_{i,n+1}) = 1$. Except for scale, this is shown in Figure 11.1b. Nevertheless, the following remarkable theorem holds:

THEOREM: *Regardless of the assigned values of* $C_2, C_3, \ldots, C_n, x_{n-1}$, *the optimal coefficients of production will always assume the same constant values, and the resulting production-possibility schedule for society will be of the simple linear form*

$$K_1 C_1 + K_2 C_2 + \cdots + K_n C_n = x_{n+1}$$

*where the* $K$s *are constants independent of the* $C$s *and* $x_{n+1}$. *It is also true that relative prices of the form* $P_i/P_j$ *will be similar constants.*

*Proof*: Form the Lagrangian expression

$$\lambda_1 C_1 + \lambda_2\left(F_2 - \sum_{j=1}^{n} x_{j2} - C_2\right) + \lambda_3\left(F_3 - \sum_{j=1}^{n} x_{j3} - C_3\right) + \cdots \qquad (11\text{--}3)$$

and differentiate it with respect to each $x_{jk}$, treating the $\lambda$s as undetermined multipliers with $\lambda_1 = 1$. This gives us

$$\lambda_i \frac{\partial F_i}{\partial x_{ij}} - \lambda_j = 0 \quad (i = 1, 2, \ldots, n; \, j = 1, 2, \ldots, n + 1) \tag{11–4}$$

We can eliminate the $\lambda$s to get the equations[1]

$$\frac{\partial F_1}{\partial x_{11}} = 1, \quad \frac{\partial F_1}{\partial x_{1i}} \frac{\partial F_i}{\partial x_{ij}} - \frac{\partial F_1}{\partial x_{1j}} = 0 \quad (i = 2, \ldots, n; \, j = 1, 2, \ldots, n + 1) \tag{11–5}$$

There are, by Eq. (11–5), $1 + (n - 1)(n + 1)$ equations to determine our $n(n + 1)$ variables $x_{ij}$. Their economic significance in terms of prices or equivalent marginal rates of substitution is easily expressed. The missing $n$ equations are supplied by the specified $C$s and $x_{n+1}$. It may be added that, if we admitted the case of joint production, this simple elimination of the $\lambda$s would not be possible.

Since each of the $F$-functions is homogeneous of order one, each of our partial derivatives must be homogeneous of order zero (i.e., the economic assumption of constant returns to scale implies that all marginal productivities depend on the *proportions* of the inputs alone). Hence the set of equations in (11–5) may be written so that instead of their involving $(n^2 + n)$ $x_{ij}$s they involve only the $n^2$ *proportions* of inputs of the form $b_{ij} = x_{ij}/x_{i,n+1}$, where $i$ and $j$ now range only from 1 to $n$.

*Equation (11–5) determines all the proportions,* $b_{ij}$, *independently of the $C$s and* $x_{n+1}$. With proportions always being invariant, it follows that we observe only one invariant set of "coefficients of production," $a_{ij}$, and the remaining assertions of the theorem are clearly implied.[2]

3. All the above is valid on the assumption that the partial derivatives of Eqs. (11–5) exist everywhere and define a unique interior solution to (11–5). In the usual problems of linear programming, where only a finite number of activities are considered, the functions have corners at which the partial derivatives are undefined, and the optimum solution is defined by boundary inequalities rather than interior equalities of the partial derivatives. Also, we must consider the possibility that more than one set of values satisfy Eqs. (11–5).

Nonetheless, the theorem remains true; a change in the bill of goods, $C_1, \ldots, C_n$, cannot make substitution profitable, and the frontier of efficiency points remains linear. A sketch of a brief but rigorous proof is as follows:[3]

First, we accept the easily proved fact that the efficiency frontier defined by our maximum problem must be a convex set in consequence of our strong homogeneity assumptions. We then show that through *any* efficient point there passes a linear hyperplane of feasible points. It must follow that the frontier locus is itself a linear hyperplane, for, if it anywhere had a corner or a curved surface, it would be impossible for us to find a hyperplane of feasible points going on all sides of the efficient point in question.

The only problem is to show that through any efficient point, $(C_1^0, C_2^0, \ldots, C_n^0, x_{n+1}^0)$, there does go a hyperplane of feasible points,

$$\sum_1^n \alpha_i(C_i - C_i^0) = x_{n+1} - x_{n+1}^0$$

for some constant $\alpha$s. Suppose that there really were absolute constant $a_{ij}$s. Then it is a well-known property of Leontief systems (Leontief, 1941, 1946a) that the bill of goods is constrained to follow a linear hyperplane by the equations

$$C_i = x_i - \sum_1^n a_{ji} x_j \qquad (i = 1, 2, \ldots, n)$$

$$-x_{n+1} = -\sum_1^n a_{j,n+1} x_j$$

Consider now an actual efficient point $(C_1^0, C_2^0, \ldots, C_n^0, x_{n+1}^0)$ being produced in the general case of Figure 11.1b by $(x_{ij}^0)$. These quantities implicitly define a set $(a_{ij}^0)$. Although it may not be obvious that it is *efficient* to stick to these fixed coefficients, the result will certainly be *feasible*. Hence there does exist a set of feasible points along a hyperplane through $(C^0, x_{n+1}^0)$, and the theorem follows.

Less heuristically, we can easily show that

$$C_i = F_i(t_i x_{i1}, t_i x_{i2}, \ldots) - \sum_1^n x_{ki} t_k$$

$$-x_{n+1} = -\sum_1^n x_{k,n+1}^0 t_k$$

define, because of the homogeneity property of the $F_i$, linear parametric equations in terms of the $t$s; for all the $t$s equal to one we get $(C_1^0, \ldots, C_n^0, x_{n+1}^0)$, and for all nearby $t$s we get feasible points on a linear hyperplane.

**NOTES**

1. Because of the necessary convexity of the $F$s, each of whose Hessian is required to be negative semidefinite, these necessary first order equations for a maximum are also sufficient. If some $x_{ij}$ does not appear at all in $F_i$, then we drop the corresponding equation in (11–4), replacing it by $x_{ij} = 0$. We also make obvious modifications in (11–5). If a good uses no labor, we must modify our use of $b$s in a simple and inessential fashion.
2. I have assumed that the price ratios are the same thing as (marginal) cost ratios, as indeed they will be if something of both goods in question is being produced.
3. I believe this argument is closely related to the more elaborate argument of Koopmans (1951).

**References**

Koopmans, Tjalling C. 1951. Alternative proof of the substitution theorem for Leontief models in the case of three industries. In *Activity Analysis of Production and Allocation*, ed. Tjalling C. Koopmans. New York: John Wiley & Sons, Ch. VIII.

Leontief, Wassily W. 1941. *The Structure of the American Economy, 1919–1929*. Cambridge, Mass.: Harvard University Press.

Leontief, Wassily, W. 1946. Exports, imports, domestic output, and employment. *Quarterly Journal of Economics* 60 (February): pp. 171–193.

# 12

## Note: Some Conditions of Macroeconomic Stability

DAVID HAWKINS AND HERBERT A. SIMON

In a recent paper by one of us[1] there is an error in the statement of a supposed sufficient condition that a system of linear homogeneous equations should have solutions all of the same sign. The present note is intended to correct that error, to state and prove an apparently new, necessary, and sufficient condition that the stated consequence should hold, and finally to interpret the significance of this condition in economic terms.

Two preliminary remarks are in order. First, the error in the theorem originally stated does not affect the substance of the paper to which reference is being made. That paper sets forth theorems on the stability of systems which *do* have stationary solutions with all variables positive. The lemma under consideration here gives necessary and sufficient conditions that a system *will* have stationary solutions with all variables positive. That is, it gives a criterion to test whether the theorems in the body of the paper are applicable to any particular system of equation.

Second, the conditions under which the variables satisfying a system of linear equations will be all positive are of economic interest in their own right. They are, in fact, the conditions determining whether a system of linear production functions is capable, given a sufficient supply of the "fixed" factors of production, of producing any desired schedule of consumption goods.

The system of equations[2] is the following:[3]

$$\sum_{j=1}^{n} b_{ij}x_j = 0 \qquad (i = 1, \ldots, n) \tag{12-1}$$

with $\Delta = 0$ and of rank $n - 1$; $b_{ij} > 0$ for all $i \neq j$; $b_{ii} = 0$ for all $i$.

Instead of dealing directly with the system (12–1), it will be more convenient to consider the associated nonhomogeneous system:

$$\sum_{j=1}^{m} a_{ij}x_j - k_i = 0 \qquad (i = 1, \ldots, m) \tag{12-2}$$

with

$$m = n - 1; \; k_i = b_{in}; \; a_{ij} = -b_{ij}, \quad (i, j = 1, \ldots, m); \; |A| = |a_{ij}| \neq 0$$

This article was published in *Econometrica*, Volume 17, Number 3–4, July–October, 1949, pp. 245–248. © University of Chicago Press. Reprinted by permission.

It is clear that, for $x_n = 1$, the solution of (12–1) is identical with that of (12–2), and the $[x_i]$ satisfying (12–1) will all be of the same sign if and only if the $[x_i]$ satisfying (12–2) are all positive. Further, without loss of generality, we can take $a_{ii} = -b_{ii} = 1$.

In Eq. (12–2), $x_i$ is the total quantity of the $i$th commodity produced; $k_i$ is the quantity of the $i$th commodity consumed; $-a_{ij}x_j$ is the quantity of the $i$th commodity used in producing the $j$th commodity. The $n$th equation in system (12–1), which is linearly dependent upon the first $m$ equations, may be interpreted as a consumption function. Alternatively, the vector $[k_i]$ in (12–2), which gives the relative quantities of the various commodities consumed, may be considered the schedule of consumption goods.

The production system (12–2) is economically meaningful only if the $[x_i]$ satisfying it are all positive. Conceivably, the signs of the $[x_i]$ may depend upon the magnitude of the $[k_i]$—that is, upon the schedule of consumption goods. Hence we will be interested in knowing under what conditions the $[x_i]$ will be positive for some given set $[k_i]$, and under what conditions the $[x_i]$ will be positive for *any* set $[k_i \geqslant 0]$.

The defective theorem is the following:

LEMMA: *The system of equations (12–1) is satisfied only for* $x_i$ *all of the same sign.*

This theorem is true only for $n \leqslant 3$, as shown by the following counter example

$$-2x_1 + 4x_2 + x_3 + x_4 = 0$$

$$4x_1 - 2x_2 + x_3 + x_4 = 0$$

$$x_1 + x_2 - 2x_3 + 4x_4 = 0$$

$$x_1 + x_2 + 4x_3 - 2x_4 = 0$$

We verify immediately that $\Delta = 0$, and is of rank $n - 1$, and that $b_{ij} > 0$ for all $i \neq j$, while $b_{ii} < 0$ for all $i$. But the general solution of this system is: $x_1 = K; x_2 = K; x_3 = -K; x_4 = -K$; where $K$ is an arbitrary constant.

The fallacy in the proof offered for the lemma lies in paragraph III of Hawkins' paper. Specifically, it is not correct that: if all members of a set of hyperplanes intersect in a common line through the origin, and if each member of the set has points lying in the first quadrant, then the common line of intersection must lie in the first quadrant.

We now proceed to a valid, necessary, and sufficient condition that the Eq. (12–2) be satisfied only for $[x_i]$ all positive.

THEOREM: *A necessary and sufficient condition that the* $x_i$ *satisfying (12–2) be all positive is that all principal minors of the matrix* $\|a_{ij}\|$ *be positive.*

To prove this theorem we first consider the augmented $m \cdot n$ matrix $\|a_{ij} - k_i\|$ and proceed to reduce this matrix, row by row, to triangular form. That is, by adding to each row an appropriate linear combination of the preceding rows, we obtain a matrix in which all elements to the left of the main diagonal are zero. This procedure can always be carried out step by step until a row (say the $j$th) is reached with a nonpositive diagonal term. It does not alter the solution of the system and does not alter the values of the principal minors consisting of the first $i$ rows and columns ($i = 1, \ldots, m$).

Because of the arrangement of signs in our particular matrix, all elements in the first column except the first can be made zero by adding to each row an appropriate *positive* multiple of the first row. The signs of all other elements off the main diagonal will remain negative. The sign of $a_{22}$ may remain positive or become negative. In the former case, the third and all following elements in the second column can be made zero by adding to the corresponding rows an appropriate *positive* multiple of the second row. In general, if the first $i$ elements on the main diagonal remain positive after the first $i - 1$ steps in the triangularization, then the $i$th step in triangularization can be carried out by adding to the remaining rows a *positive* multiple of the $i$th row; otherwise by adding a *negative* multiple of the $i$th row. We carry out the triangularization until we reach a row with a nonpositive diagonal term.

For the matrix finally obtained, we distinguish two cases: (A) all the diagonal terms in the triangular matrix are positive, (B) at least one term on the main diagonal is nonpositive (and the $j$th term, say, is the first nonpositive one). We now prove that in case (A) all the principal minors are positive and all the $x_i$ are positive; while in case (B) at least one principal minor is nonpositive and at least one of the $x_i$ is negative—a statement equivalent to our theorem.

A. In case (A) we solve the corresponding system of equations successively for $x_m, x_{m-1}, \ldots, x_1$ in terms of the $k_i$. Since $k_m > 0$, we must have $x_m > 0$. Since $k_{m-j} > 0$, it follows that if all $x_{m-i} > 0$ ($i < j$), then $x_{m-j} > 0$. Hence by induction, all the $x_i$ must be positive. But, since a triangular determinant equals the product of the elements on its main diagonal, all principal minors consisting of the first $k$ rows and columns of the triangular matrix are positive ($k = 1, \ldots, m$). But these minors are equal to the corresponding minors of the original matrix $\|a_{ij}\|$.

B. In case (B), all elements to the right of the main diagonal in the $j$th row of the diagonalized matrix are negative, and the diagonal term is nonpositive. Suppose now that all $x_i$ for $i > j$ are positive. Then, since $k_j$ is positive, $x_j$ must be negative.[4] But the principal minor of the first $j$ rows and columns of the triangularized matrix will be negative or zero, since the $j$th element in the principal diagonal is nonpositive, the others positive. Hence the corresponding minor in $\|a_{ij}\|$ will be nonpositive.

Since the signs of the $x_i$ obviously do not depend on the order in which the equations are arranged before triangularization, in case (A) *all* the principal minors of $\|a_{ij}\|$ must be positive.

This completes the proof of the theorem. Moreover, our proof gives a direct method of testing whether the $x_i$ satisfying a given matrix are all positive.

COROLLARY: *A necessary and sufficient condition that the* $x_i$ *satisfying* (2) *be all positive for any set* [$k_i > 0$] *is that all principal minors of the matrix* $\|a_{ij}\|$ *are positive.*

This corollary follows immediately from the theorem, and from the consideration that the elements of the matrix $\|a_{ij}\|$ are independent of the $[k_i]$.

*Economic Interpretations.* From the corollary, we see that if the production equations are internally consistent in permitting the production of some fixed schedule of consumption goods, then these consumption goods can be obtained in any desired proportion from this production system. Hence the system will be consistent with *any* schedule of consumption goods.

The condition that all principal minors must be positive means, in economic terms, that the group of industries corresponding to each minor must be capable of supplying more than its own needs for the group of products produced by this group of industries. If this is true, and if the condition $\Delta = 0$ for Eq. (12–1) is satisfied, then we can say that each group of industries must be just capable of supplying its own demands upon itself *and* the demands of the other industries in the economy. For example, if the principal minor involving the $i$th and $j$th commodities is negative, this means that the quantity of the $i$th commodity required to produce one unit of the $j$th commodity is greater than the quantity of the $i$th commodity that can be produced with an input of one unit of the $j$th commodity. Under these circumstances, the production of these two commodities could not be continued, for they would exhaust each other in their joint production.

## NOTES

1. David Hawkins, "Some Conditions of Macroeconomic Stability," *Econometrica* 16: 309–322 (1948).
2. The system (12–1) is essentially that introduced by W. W. Leontief in *The Structure of American Economy*, 1919–1929. Cambridge: Harvard University Press, 1941, p. 48.
3. In Hawkins' original system we require only that $b_{ij} \geqslant 0$ for $i \neq j$. The stronger condition $b_{ij} > 0$ employed here simplifies the statement of the theorem and its proof and, because of the continuity of solutions of these equations with respect to variations of these coefficients, does not involve any essential loss of generality.
4. Or, if the diagonal term is zero, we have a contradiction—i.e. all $x$ for $i > j$ cannot be positive.

# 13

## Nonnegative Square Matrices

GERARD DEBREU AND I. N. HERSTEIN

### Introduction

Square matrices, all of whose elements are nonnegative, have played an important role in the probabilistic theory of finite Markov chains (see Fréchet, 1938 and the references there given) and, more recently, in the study of linear models in economics (Arrow, 1951; Bray, 1922; Chipman, 1950, 1951; Georgescu-Roegen, 1951; Goodwin, 1950; Hawkins and Simon, 1949; Metzler, 1945, 1951a, 1951b; Morishima, 1952; Mosak, 1944).

The properties of such matrices were first investigated by Perron (1907a, 1907b), and then very thoroughly by Frobenius (1908; 1909; 1912). Lately Wielandt (1950) has given notably more simple proofs for the results of Frobenius.

In the section entitled "Nonnegative Indecomposable Matrices," we study nonnegative indecomposable matrices from a different point of view (that of the Brouwer fixed point theorem); a concise proof of their basic properties is thus obtained. In the sections that follow we study, respectively: properties of a general nonnegative square matrix $A$ are derived from those of nonnegative indecomposable matrices; theorems about the matrix $sI - A$ are proved—they cover in a unified manner a number of results recurringly used in economics; and a systematic study of the convergence of $A^p$ when $p$ tends to infinity ($A$ is a general complex matrix) is linked to combinatorial properties of nonnegative square matrices.

Unless otherwise specified, all matrices considered will have *real* elements. We define for $A = (a_{ij})$, $B = (b_{ij})$:

$$A \leqq B \quad \text{if} \quad a_{ij} \leqq b_{ij} \quad \text{for all} \quad i, j$$

$$A \leq B \quad \text{if} \quad A \leqq B \quad \text{and} \quad A \neq B,$$

$$A < B \quad \text{if} \quad a_{ij} < b_{ij} \quad \text{for all} \quad ij$$

Primed letters denote transposes.

When $A$ is an $n \cdot n$ matrix, $A_T = TAT^{-1}$ denotes the transform of $A$ by the nonsingular $n \cdot n$ matrix $T$.

This article was published in *Econometrica*, Volume 21, Number 4, October 1953, pp. 597–607. © University of Chicago Press. Reprinted by permission.

## Nonnegative Indecomposable Matrices

An $n \cdot n$ matrix $A$ ($n \geqslant 2$) is said to be *indecomposable* if for no permutative matrix[1]
$\Pi$ does $A_\pi = \Pi A \Pi^{-1} = \begin{pmatrix} A_{11} & A_{12} \\ 0 & A_{22} \end{pmatrix}$ where $A_{11}, A_{22}$ are square.

THEOREM I: *Let $A \geqslant 0$ be indecomposable. Then*
1. *$A$ has a characteristic root $r > 0$ such that*
2. *to $r$ can be associated an eigen-vector $x_0 > 0$;*
3. *if $\alpha$ is any characteristic root of $A$, $|\alpha| \leqslant r$;*
4. *$r$ increases when any element of $A$ increases;*
5. *$r$ is a simple root.*

PROOF: 1. (a) *If $x \geqslant 0$, then $Ax \geqslant 0$.* For if $Ax = 0$, $A$ would have a column of zeros, and so would not be indecomposable.

1. (b) *$A$ has a characteristic root $r > 0$.*
Let $S = \{x \in R^n | x \geqslant 0, \sum x_i = 1\}$ be the fundamental simplex in the Euclidean $n$-space, $R^n$. If $x \in S$, we define $T(x) = [1/\rho(x)]Ax$ where $\rho(x) > 0$ is so determined that $T(x) \in S$ (by (1.a) such a $\rho$ exists for every $x \in S$). Clearly $T(x)$ is a continuous transformation of $S$ into itself, so, by the Brouwer fixed-point theorem (see for example Lefschetz, 1949), there is an $x_0 \in S$ with $x_0 = T(x_0) = [1/\rho(x_0)]Ax_0$. Put $r = \rho(x_0)$.

2. $x_0 > 0$. Suppose that after applying a proper $\Pi$, $\tilde{x}_0 = \begin{pmatrix} \xi \\ 0 \end{pmatrix}$, $\xi > 0$. Partition $A_\pi$
accordingly. $A_\pi \tilde{x}_0 = r \tilde{x}_0$ yields $\begin{pmatrix} A_{11} & A_{12} \\ A_{21} & A_{22} \end{pmatrix} \begin{pmatrix} \xi \\ 0 \end{pmatrix} = \begin{pmatrix} r\xi \\ 0 \end{pmatrix}$, thus $A_{21}\xi = 0$, so $A_{21}$
$= 0$, violating the indecomposability of $A$.

If $M = (m_{ij})$ is a matrix, we henceforth denote by $M^*$ the matrix $M^* = (|m_{ij}|)$.
3–4. *If $0 \leqslant B \leqslant A$, and if $\beta$ is a characteristic root of $B$, then $|\beta| \leqslant r$. Moreover, $|\beta| = r$ implies $B = A$.*
$A'$ is indecomposable and therefore has a characteristic root $r_1 > 0$ with an eigen-vector $x_1 > 0$: $A'x_1 = r_1 x_1$. Moreover $\beta y = By$. Taking absolute values and using the triangle inequality, we obtain

(i) $|\beta| y^* \leqslant By^* \leqslant Ay^*$

So

(ii) $|\beta| x_1' y^* \leqslant x_1' Ay^* = r_1 x_1' y^*$

Since $x_1 > 0$, $x_1' y^* > 0$, thus $|\beta| \leqslant r_1$.
Putting $B = A$ one obtains $|\alpha| \leqslant r_1$. In particular $r \leqslant r_1$ and since, similarly, $r_1 \leqslant r$, $r_1$ is equal to $r$.
Going back to the comparison of $B$ and $A$ and assuming that $|\beta| = r$ one gets from (i) and (ii)

$$ry^* = By^* = Ay^*.$$

From $ry^* = Ay^*$, application of 2 gives $y^* > 0$. Thus $By^* = Ay^*$ together with $B \leqslant A$ yields $B = A$.

5. (a) *If* B *is a principal submatrix of* A *and* $\beta$ *a characteristic root of* B, $|\beta| < $ r. $\beta$ is also a characteristic root of the $n \cdot n$ matrix $\bar{B} = \begin{pmatrix} B & 0 \\ 0 & 0 \end{pmatrix}$. Since $A$ is indecomposable, $\bar{B} \leqslant A_\pi$ for a proper $\Pi$ and $|\beta| < r$ (by 3–4).

5. (b) r *is a simple root of* $\Phi(t) = \det(tI - A) = 0$.

$\Phi'(r)$ is the sum of the principal $(n-1) \cdot (n-1)$ minors of $\det(rI - A)$. Let $A_i$ be one of the principal $(n-1) \cdot (n-1)$ submatrices of $A$. By 5(a) $\det(tI - A_i)$ cannot vanish for $t \geqslant r$, whence $\det(rI - A_i) > 0$ and $\Phi'(r) > 0$.[2]

With a proof practically identical to that of 3–4, one obtains the more general result:

*If* B *is a complex matrix such that* $B^* \leqslant A$, A *indecomposable, and if* $\beta$ *is a characteristic root of* B, *then* $|\beta| \leqslant $ r. *Moreover* $|\beta| = $ r *implies* $B^* = A$.

*More precisely if* $\beta = re^{i\varphi}$, $B = e^{i\varphi} DAD^{-1}$ *where* D *is a diagonal matrix such that* $D^* = I$. A proof of this last fact is given in (Wielandt, 1950, p. 646 lines 4–11).

From this can be derived

THEOREM II: *Let* $A \geqslant 0$ *be indecomposable. If the characteristic equation* $\det(tI - A) = 0$ *has altogether* k *roots of absolute value* r, *the set of* n *roots (with their orders of multiplicity) is invariant under a rotation about the origin through an angle of* $2\pi/k$, *but not under rotations through smaller angles. Moreover there is a permutation matrix* $\Pi$ *such that*

$$\Pi A \Pi^{-1} = \begin{bmatrix} 0 & A_{12} & 0 & \cdot & 0 \\ 0 & 0 & A_{23} & \cdot & 0 \\ \cdot & \cdot & \cdot & \cdot & \cdot \\ 0 & 0 & 0 & \cdot & A_{k-1,k} \\ A_{k1} & 0 & 0 & \cdot & 0 \end{bmatrix} \qquad (13–1)$$

*with square submatrices on the diagonal.*

Again the reader is referred to the excellent proof of Wielandt (1950, pp. 646–647).

If $k = 1$, the indecomposable matrix $A \geqslant 0$ is said to be *primitive*.

## Nonnegative Square Matrices

If $A$ is an $n \cdot n$ matrix, there clearly exists a permutation matrix $\Pi$ such that

$$\Pi A \Pi^{-1} = \begin{bmatrix} A_1 & & & * \\ & A_2 & & \\ & & \cdot & \\ 0 & & & \cdot \\ & & & A_H \end{bmatrix}$$

where the $A_h$ are square submatrices on the diagonal and every $A_h$ is either indecomposable or a $1 \cdot 1$ matrix.

The properties of $A$ will therefore be easily derived from those of the $A_h$. For example $\det(tI - A) = \Pi^H_{h=1} \det(tI - A_h)$ and THEOREM I gives

THEOREM I*: *If* $A \geqslant 0$ *is a square matrix, then*
1. $A$ *has a characteristic root* $r \geqslant 0$ *such that*
2. *to* $r$ *can be associated an eigen-vector* $x_0 \geqslant 0$;
3. *if* $\alpha$ *is any characteristic root of* $A$, $|\alpha| \leqslant r$;
4. $r$ *does not decrease when an element of* $A$ *increases.*

Let $r_h$ be the maximal nonnegative characteristic root of $A_h$, we take $r = \text{Max}_h \, r_h$; 1–3–4 are then immediate. To prove 2 we consider a sequence $A_\iota$ of $n \cdot n$ matrices converging to $A$ such that for all $\iota A_\iota > 0$. Let $r_\iota$ be the maximal positive characteristic root of $A_\iota$, $x_\iota > 0$ its associated eigen-vector so chosen that $x \in S$, the fundamental simplex of $R^n$. Clearly $r_\iota$ tends to $r$. Let us then select $x_0 \in S$ a limit point of the set $(x_\iota)$; thus there is a subsequence $x_{\iota'}$ converging to $x_0 \geqslant 0$ and for every $\iota'$, $A_{\iota'} x_{\iota'} = r_{\iota'} x_{\iota'}$, therefore $Ax_0 = rx_0$.

Statement 5 of THEOREM I no longer holds, but 5(a) becomes:

*If* B *is a principal submatrix of* A *and* $\beta$ *a characteristic root of* B, $|\beta| \leqslant r$.

The proof, almost identical, now rests on 4 of THEOREM I*.[3]

## Properties of $sI - A$ for $s > r$

*In this section* $A \geqslant 0$ *is an* $n \cdot n$ *matrix, and* $r$ *is its maximal nonnegative characteristic root.*

LEMMA*: *If for an* $x > 0$, $Ax \leqslant sx$ (*resp.* $\geqslant$), *then* $r \leqslant s$ (*resp.* $\geqslant$).
*If for an* $x \geqslant 0$, $Ax < sx$ (*resp.* $>$), *then* $r < s$ *resp.* $>$).

The proofs of the four statements being practically identical, we present only the first one. Let $x_0 \geqslant 0$ be a characteristic vector of $A'$ associated with $r$ (2 of THEOREM I*): $A'x_0 = rx_0 \cdot Ax \leqslant sx$ with $x > 0$, therefore $x'_0 Ax \leqslant sx'_0 x$, i.e., $rx'_0 x \leqslant sx'_0 x$ and, since $x'_0 x > 0$, $r \leqslant s$.

We now derive two theorems (III* and III) from the study of the equation

$$(sI - A)x = y \tag{13–2}$$

THEOREM III*: $(sI - A)^{-1} \geqslant 0$ *if and only if* $s > r$.

*Sufficiency.* Since $s > r$, Eq. (13–2) has a unique solution $x = (sI - A)^{-1}y$ for every $y$; we show that $y \geqslant 0$ implies $x \geqslant 0$.

If $x$ had negative components Eq. (13–2) could be given the form [by proper (identical) permutations of the rows and columns and partition]

$$\begin{bmatrix} sI - A_1 & -A_{12} \\ -A_{21} & sI - A_2 \end{bmatrix} \begin{bmatrix} -x_1 \\ x_2 \end{bmatrix} = y$$

where $x_1 > 0$, $x_2 \geqq 0$, $y \geqq 0$. Therefore $-(sI - A_1)x_1 - A_{12}x_2 \geqq 0$, i.e., $-(sI - A_1)x_1 \geqq 0$, i.e., $A_1x_1 \geqq sx_1$. From the LEMMA* the maximal nonnegative characteristic root of $A_1$, $r_1 \geqq s$, a contradiction to the fact that $r \geqq r_1$ (see end of section "Nonnegative Square Matrices"), and $s > r$.

*Necessity.* Since $(sI - A)^{-1} \geqq 0$, to a $y > 0$ corresponds an $x \geqq 0$. Therefore from $sx - Ax = y$ follows $Ax < sx$ and, by the LEMMA*, $r < s$.

If $A$ is indecomposable these results can be sharpened to the

LEMMA: *Let* A *be indecomposable.*
*If for an* x $\geqq 0$, Ax $\lessgtr$ sx *(resp.* $\geqq$*), then* r $\lessgtr$ s *(resp.* $\geqq$*).*
*If for an* x $\geqq 0$, Ax $\leqslant$ sx *(resp.* $\geqq$*), then* r $<$ s *(resp.* $>$*).*

The proofs, practically identical to those of the LEMMA*, use a *positive* characteristic vector of $A'$ associated with $r$. One of these statements indeed has already been proved in 3–4 of THEOREM I.

THEOREM III: *Let* A *be indecomposable.* (sI − A)$^{-1}$ > 0 *if and only if* s > r.

*Sufficiency.* We show that $y \geqq 0$ implies $x > 0$. It is already known (from the proof of sufficiency of THEOREM III*) that $x \geqq 0$. If $x$ had zero components, (2) could be given the form

$$\begin{bmatrix} sI - A_1 & -A_{12} \\ -A_{21} & sI - A_2 \end{bmatrix} \begin{bmatrix} x_1 \\ x_2 \end{bmatrix} = y$$

where $x_1 = 0$, $x_2 > 0$, $y \geqq 0$. Therefore $-A_{12}x_2 \geqq 0$, and, since $x_2 > 0$, $A_{12} = 0$ violating the indecomposability of $A$.

The *Necessity* has already been proved since $(sI - A)^{-1} > 0$ implies $(sI - A)^{-1} \geqq 0$.[4]

THEOREM IV: *The principal minors of* sI − A *of orders* 1, $\cdots$, n *are all positive if and only if* s > r.

*Sufficiency.* $\det(tI - A)$ cannot vanish for $t > r$, thus $\det(sI - A) > 0$ for $s > r$. Similarly, the maximal nonnegative characteristic root of a principal submatrix of $A$ is not larger than $r$ (see end of section "Nonnegative Square Matrices"); it is therefore smaller than $s$, and the corresponding minor of $sI - A$ is positive.

*Necessity.* The derivative of order $m(<n)$ of $\det(tI - A)$ with respect to $t$, for $t = s$, is a sum of principal minors of order $(n - m) \cdot (n - m)$ of $sI - A$ and thus is positive. As its derivatives of all orders $(0, 1, \cdots, n - 1, n)$ are positive for $t = s$, the polynomial $\det(tI - A)$ can vanish for no $t \geqq s$, i.e., $s > r$.[5,6]

Since a square matrix with nonpositive (resp. negative) off-diagonal elements can always be given the form $sI - A$ where $A \geqq 0$ (resp. $> 0$), many of the results of

Arrow (1951), Bray (1922), Chipman (1950, 1951), Georgescu-Roegen (1951), Goodwin (1950), Hawkins and Simon (1949), Metzler (1945, 1950, 1951a, 1951b), Morishima[7] (1952), Mosak (1944), Solow (1952) are contained in the above.

## Convergence[8] of $A^p$

THEOREM V: *Let* A *be a* n·n *complex matrix. The sequence* A, $A^2$, $\cdots$, $A^p$, $\cdots$ *of its powers converges if and only if*
  1. *each characteristic root* $\alpha$ *of* A *satisfies either* $|\alpha| < 1$ *or* $\alpha = 1$;
  2. *when the second case occurs the order of multiplicity of the root* 1 *equals the dimension of the eigen-vector space associated with that root.*
There is a nonsingular complex matrix $T$ such that

$$A_T = TAT^{-1} = \begin{bmatrix} J_1 & & & \\ & \ddots & & 0 \\ & & J_i & \\ 0 & & & \ddots \\ & & & & J_q \end{bmatrix} \quad \text{where}$$

$$J_i = \begin{bmatrix} \alpha_\iota & 1 & & & & \\ & \ddots & \ddots & & & 0 \\ & & \ddots & 1 & & \\ & & & \alpha_\iota & 1 & \\ 0 & & & & \ddots & \ddots \\ & & & & & \ddots & 1 \\ & & & & & & \alpha_\iota \end{bmatrix}$$

is a square matrix on the diagonal and $\alpha_\iota$ a characteristic root of $A$. To every root $\alpha_\iota$ corresponds at least one $J_i$ (for this reduction of $A$ to its Jordan canonical form see for example Van der Waerden, 1950).
    Since

$$TA^pT^{-1} = \begin{bmatrix} J_1^p & & & \\ & \ddots & & 0 \\ & & J_i^p & \\ 0 & & & \ddots \\ & & & & J_q^p \end{bmatrix},$$

$A^p$ converges if and only if every one of the $J_i^p$ converges. Let us therefore study one of them; for this purpose we drop the subscripts $i$ and $\iota$.
    $J$ is a $k \cdot k$ matrix of the form $J = \alpha I + M$ where $M = (m_{st})$: $m_{st} = 1$ if $t = s + 1$,

$m_{st} = 0$, otherwise.

$$J^p = \alpha^p I + \binom{p}{1}\alpha^{p-1}M + \cdots \binom{p}{k-1}\alpha^{p-k+1}M^{k-1}.$$

It is easily seen that for $M^h$, $m_{st}^{(h)} = 1$ if $t = s + h$ and $m_{st}^{(h)} = 0$ otherwise. Thus $M^h = 0$ if $h \geqslant k$; also the nonzero elements of $M^h$ and $M^{h'}$ ($h \neq h'$) never occur in the same place so $J^p$ converges if and only if every term of the right-hand sum does.

The first term shows that necessarily either $|\alpha| < 1$ or $\alpha = 1$.

If $|\alpha| < 1$, every term tends to zero and $J^p$ converges.

If $\alpha = 1$ no term other than the first one converges and necessarily $k = 1$, i.e., $J = [1]$; clearly $J^p$ converges in this case.

We wish, however, to obtain for this necessary and sufficient condition of convergence an expression independent of a reduction to Jordan canonical form.

Consider then an arbitrary $n \cdot n$ complex matrix $A$ and let $\mathscr{I}$ be the set of $i$ for which $J_i$ corresponds to the root 1. The equation $A_T x = x$, in which $x$ is partitioned in the same way as $A_T$, yields $J_i x_i = x_i$ for all $i$, i.e.,

$$\text{if } i \notin \mathscr{I}, \, x_i = 0$$

if $i \in \mathscr{I}$, all components of $x_i$ but the first one equal zero.

Thus the dimension of the eigen-vector space associated with the root 1 equals the number of elements of $\mathscr{I}$. This number, in turn, equals the order of multiplicity of the root 1 if and only if $J_i = [1]$ for all $i \in \mathscr{I}$.

The above theorem and method of proof were first given by Oldenburger (1940).

We now assume that the limit $C$ exists and give its expression. If 1 is not a characteristic root of $A$, $C = 0$. Let therefore 1 be a root of $A$ of order $\mu$. Thus $x$ (resp. $y$), an eigen-vector of $A$ (resp. $A'$) associated with the root 1, has the form $x = X\xi$ (resp. $y = Y\eta$) where $X$ (resp. $Y$) is a $n \cdot \mu$ matrix of rank $\mu$ and $\xi$ (resp. $\eta$) is a $\mu \cdot 1$ matrix. For an arbitrary $x$ the relation $AA^p x = A^{p+1}x$ gives in the limit $AC_x = Cx$, i.e., $Cx = X\xi(x)$. To determine $\xi(x)$ we remark that $Y' = Y'A$, i.e., by iteration $Y' = Y'A^p$, and therefore $Y' = Y'C$; thus $Y'x = Y'Cx = Y'X\xi(x)$. $Y'X$ is a nonsingular[9] $\mu \cdot \mu$ matrix, i.e., $\xi(x) = (Y'X)^{-1}Y'x$. Finally for all $x$, $Cx = X(Y'X)^{-1}Y'x$, i.e., $C = X(Y'X)^{-1}Y'$.

COROLLARY: *Let* $A \geqslant 0$ *be indecomposable and* 1 *be its maximal positive characteristic root. The sequence* $A^p$ *converges if and only if* $A$ *is primitive.*

The necessity is obvious. The sufficiency follows from the fact that 1 is a simple root.

Let then $x_0 > 0$ (resp. $y_0 > 0$) be an eigen-vector of $A$ (resp. $A'$) associated with the root 1, the limit $C$ of $A^p$ has the simple expression $C = x_0 y_0'/y_0' x_0$.

Clearly $C > 0$, thus if the indecomposable matrix $A \geqslant 0$ is primitive, there is a positive integer $m$ such that $A^p > 0$ when $p \geqslant m$. The converse is an immediate consequence of the decomposition (1) of THEOREM II.[10]

**NOTES**

1. A permutation matrix is obtained by permuting the columns of an identity matrix. $\Pi A \Pi^{-1}$ is obtained by performing the same permutation on the rows and on the columns of $A$.

2. As an immediate consequence of 4 one obtains:

$$\text{Min}_i \sum_j a_{ij} \leqslant r \leqslant \text{Max}_i \sum_j a_{ij}$$

*and one equality holds only if all row sums are equal (then they both hold).*

This is proved by increasing (resp. decreasing) some elements of $A$ so as to make all row sums equal to

$$\text{Max}_i \sum_j a_{ij} \quad (\text{resp. Min}_i \sum_j a_{ij}).$$

A similar result naturally holds for column sums.

3. *A stochastic $n \cdot n$ matrix $P$* is defined by $p_{ij} > 0$ for all $i, j$ and $\sum_j p_{ij} = 1$ for all $i$. Clearly 1 is a characteristic root of $P$ (take an eigen-vector with all components equal). 1 is therefore a root of some of the indecomposable matrices $P_1, P_2, \cdots, P_H$. Suppose that 1 is a root of $P_h$, it follows from footnote (2) that all row sums of $P_h$ are equal to 1, i.e.,

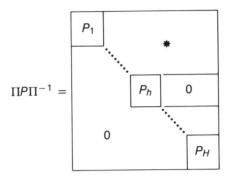

This remark makes many properties of stochastic matrices (the subject of the theory of finite Markov chains; see (Fréchet, 1938) and its references) ready consequences of the results of this article.

4. It is worth (Hawkins and Simon, 1949) emphasizing a result obtained in the proof of necessity of THEOREM III*.

*Remark.* Let $A \geqslant 0$ (resp. $A \geqslant 0$ indecomposable) be a square matrix. If for a $y > 0$ (resp. $y \geqslant 0$), $x \geqslant 0$, then $(sI - A)^{-1} \geqslant 0$ (resp. $(sI - A)^{-1} > 0$).

The proof for indecomposable matrices uses the LEMMA instead of the LEMMA*.

5. Georgescu-Roegen (1951) stated a result whose counterpart here would be the following theorem (stronger than IV): *The $n$ northwest principal minors of $sI - A$ of orders $1, \cdots, n$ are all positive if and only if $s > r$.*

6. We give a last property useful in economics (Metzler, 1951a, 1951).

THEOREM: *Let $A > 0$ be a square matrix and let $C_{ij}$ be the cofactor of the $i^{th}$ row, $j^{th}$ column element of $sI - A$. If $s > \sum_j a_{ij}$ for all $i$, then $i \neq j$ implies $C_{ii} > C_{ij}$.*

Let us define the matrix $B = (b_{pq})$ as follows:

$$b_{pq} = a_{pq} \quad \text{if} \quad p \neq i; \; b_{iq} = 0 \quad \text{if} \quad i \neq q \neq j; \; b_{ii} = s/2 = b_{ij}$$

$B$ is indecomposable, moreover $\sum_q b_{iq} = s$, $\sum_q b_{pq} < s$ for $p \neq i$. Therefore (see footnote 3) the maximal positive characteristic root of $B$, $r(B) < s$. Thus det $(sI - B) > 0$; a development according to the $i^{th}$ row yields:

$$s/2 C_{ii} - s/2 C_{ij} > 0.$$

7. Morishima studies square matrices $A$ such that for a permutation matrix $\Pi$,

$$\Pi A \Pi^{-1} = A_\pi = \begin{bmatrix} A_{11} & A_{12} \\ A_{21} & A_{22} \end{bmatrix},$$

where $A_{11} \geqq 0$ and $A_{22} \geqq 0$ are square, $A_{12} \leqq 0$, $A_{21} \leqq 0$. The relation

$$\begin{bmatrix} I & 0 \\ 0 & -I \end{bmatrix}\begin{bmatrix} A_{11} & A_{12} \\ A_{21} & A_{22} \end{bmatrix}\begin{bmatrix} I & 0 \\ 0 & -I \end{bmatrix} = \begin{bmatrix} A_{11} & -A_{12} \\ -A_{21} & A_{22} \end{bmatrix}$$

shows how properties of $A_\pi$ can be immediately derived from those of the *nonnegative* matrix

$$A_\pi^S = \begin{bmatrix} A_{11} & -A_{12} \\ -A_{21} & A_{22} \end{bmatrix}$$

In particular $A_\pi$ and $A_\pi^S$ have the same characteristic roots.

8. The Cesaro convergence of $A^p$, i.e., the convergence of $\frac{1}{p}(A + A^2 + \cdots + A^p)$ can be studied in exactly the same fashion.

9. $X_T = TX$ (resp. $Y'_T = Y'T^{-1}$) plays for $A_T$ the same role as $X$ (resp. $Y'$) does for $A$. Moreover $Y'X = Y'_T X_T$. The right-hand matrix is nonsingular for the form taken by the Jordan matrix $A_T$ in the convergence case implies that the eigen-vector space $U$ generated by $X_T$ is identical with the eigen-vector space $V$ generated by $Y_T$. Thus $Y'_T X_T \xi = 0$ implies $X_T \xi = 0$ (there is no vector different from zero in $U$ perpendicular to $V$, i.e., to $U$) therefore $\xi = 0$ since the rank of $X_T$ is $\mu$.

10. This characterization of a primitive matrix, due to Frobenius, is typical of the purely combinatorial properties of the nonnegative square matrix $A$ (used for example in the theory of communication networks): the smallest $m$ satisfying the above condition is independent of the values of the nonzero elements of $A$ as long as they stay positive.

The development of combinatorial techniques adapted to the treatment of such properties is the subject of König, 1936.

In the references, we have tried to cover the economic literature with reasonable completeness. No such attempt has been made for the mathematical literature, of which only a few essential papers have been quoted.

**References**

Alexandroff, P., and H. Hope. 1935. *Topologie*. Berlin: J. Springer. p. 480–481.

Arrow, K. J. 1951. Alternative proof of the substitution theorem for Leontief models in the general case, Chap. 9 In *Activity Analysis of Production and Allocation*, ed. T. C. Koopmans. New York: John Wiley & Sons. pp. 155–164.

Bray, H. E. 1922. Rates of exchange. *The American Mathematical Monthly* 29 (November-December): pp. 365–371.

Chipman, J. S. 1950. The multi-sector multiplier. *Econometrica* 18 (October): pp. 355–374.

Chipman, J. S. 1951. *The Theory of Inter-Sectoral Money Flows and Income Formation*. Part III. Baltimore: The Johns Hopkins Press.

Fréchet, M. 1938. *Traité du Calcul des Probabilités et de ses Applications*. (E. Borel ed.). Tome I, Fascicule III, Second livre. Paris: Gauthier-Villars.

Frobenius, G. 1908. Über Matrizen aus positiven Elementen. *Sitzungsberichte der königlich preussichen Akademie der Wissenschaften* 1: pp. 471–476.

Frobenius, G. 1909. Über Matrizen aus positiven Elementen II. *Sitzungsberichte* 1: pp. 514–518.

Frobenius, G. 1912. Über Matrizen aus nicht negativen Elementen. *Sitzungsberichte*, 1: pp. 456–477.

Georgescu-Roegen, N. 1951. Some properties of a generalized Leontief model. Chap. 10, In *Activity Analysis of Production and Allocation*, ed. T. C. Koopmans. New York: John Wiley & Sons. pp. 165–173.

Goodwin, R. M. 1950. Does the matrix multiplier oscillate. *The Economic Journal* 60 (December): pp. 764–770.

Hawkins, D., and H. A. Simon. 1949. Note: Some conditions of macroeconomic stability. *Econometrica.* 17 (July-October): pp. 245–248.

König, D. 1936. *Theorie der endlichen und unendlichen Graphen.* Leipzig: Akademische Verlagsgesellschaft M. B. H.

Lefschetz, S. 1949. *Introduction to Topology.* Princeton: Princeton University Press. Ch. 5.

Metzler, L. A. 1945. Stability of multiple markets: The Hicks conditions. *Econometrica* 13 (October): pp. 277–292.

Metzler, L. A. 1950. A multiple-region theory of income and trade. *Econometrica* 18 (October): pp. 329–354.

Metzler, L. A. 1951a. A multiple-county theory of income transfers. *The Journal of Political Economy* 59 (February): pp. 14–29.

Metzler, L. A. 1951b. Taxes and subsidies in Leontief's input-output model. *The Quarterly Journal of Economics* 65 (August): pp. 433–438.

Morishima, M. 1952. On the laws of change of the price-system in an economy which contains complementary commodities. *Osaka Economic Papers* 1: pp. 101–113.

Mosak, J. L. 1944. *General Equilibrium Theory in International Trade.* Bloomington, Indiana: The Principia Press. pp. 49–51.

Oldenburger, R. 1940. Infinite powers of matrices and characteristic roots. *Duke Mathematical Journal* 6 (June): pp. 357–361.

Perron, O. 1907a. Grundlagen für eine Theorie des Jacobischen Kettenbruchalgorithmus. *Mathematische Annalen* 64 (May): pp. 1–76.

Perron, O. 1907b. Zur Theorie der Matrices. *Mathematische Annalen* (July): pp. 248–263.

Solow, R. 1952. On the structure of linear models. *Econometrica* 20 (January): pp. 29–46.

van der Waerden, B. L. 1950. *Modern Algebra.* Vol. 2, Section 111, New York: Frederick Ungar Publishing Co.

Wielandt, H. 1950. Unzerlegbare, nicht negative Matrizen. *Mathematische Zeitschrift* 52 (March): pp. 642–648.

# 14

## Criteria for Aggregation in Input-Output Analysis

WALTER D. FISHER

From the early days of input-output analysis its practitioners have recognized the importance of the aggregation problem and the fact that the results of the analysis depend upon the particular aggregation procedures used to combine industries. Recently, an increasing number of writers have become interested in theoretical as well as practical aspects of the problem.[1]

The potential user of input-output analysis often needs to reduce a given table to smaller size. A large table—say 200 by 200 or even 50 by 50—is cumbersome for many purposes. The theorist finds it too detailed to understand, and the forecaster finds it too extensive to use in making further numerical computations. Of course some information is always lost by aggregation. But the same considerations that lead to the computations of averages and index numbers also lead to the consolidation of input-output tables: ease of comprehension and economy of manipulation.

The following question has been posed in the literature:[2] given a detailed input-output table in which many "small" industries appear, and given the desire of the research worker to have only a few "large" industries to deal with, when is it possible to consolidate the small industries into large industries, and still obtain the same results of the analysis or predictions as would have been obtained by using the many small industries in the first place? The answer to this question has been shown to be "almost never." The special cases where it is possible have indicated the type of conditions needed for what might be called "good" aggregation procedures—meaning procedures which give "approximately the same" results as given by the detailed model. The need to develop criteria for "good" aggregation has been recognized, but few specific suggestions have been made.[3]

The purpose of this paper is to suggest specific criteria and procedures for "good" aggregation of a given input-output table. First, reasonable criteria are developed, based on the usual objectives of input-output analysis and on some special assumptions concerning final demand. Some approximations to these criteria are then proposed that are closely related to previous suggestions to aggregate on the basis of "similarity of coefficients" or "homogeneity of input

This article was published in *The Review of Economics and Statistics*, Volume 40, Number 3, August, 1958, pp. 250–260. © By The President and Fellows of Harvard College. Reprinted by permission.

structure." The criteria are then illustrated and the usefulness of the approximations tested by numerical experiments involving a number of aggregations of an $18 \times 18$ matrix into $8 \times 8$ matrices.

Emphasis is placed on the problem of selecting the industries of the original input-output table that are to be grouped together, given the degree of aggregation desired and the manner of aggregating given industries. Questions of optimal degree of aggregation and optimal form of the aggregating function are not considered. The modest scale of the numerical experiments undertaken puts this study in the class of a pilot study.

## A Criterion of Predictive Accuracy

It is assumed that an economy is adequately represented by a static, open Leontief input-output model of $n$ industries, for which the input coefficients are known exactly. The usual assumptions are assumed to hold strictly: e.g., just one commodity produced by each industry, strict complementarity among the commodities used as inputs by an industry, and constant returns to scale. This model will be called the *detailed model* and the industry classifications used will be called *small industries*. It is further assumed that an investigator, having certain prediction objectives, and employing a consolidation procedure to be described, desires to reduce the detailed model to a smaller *aggregate model of m large industries*, where $m$ is a preassigned number smaller than $n$. This formulation admittedly bypasses many problems connected with the construction of the detailed model and the validity of that model, but has the advantage of concentrating attention on other matters.[4]

One frequent important objective of input-output analysis is the prediction of output of certain industries, given a set of final demands. In some applications it is desired to predict the output of only a single industry. For example, with given estimates of final demand for the various industrial products of the United States, the iron and steel industry may desire to know what its output will be. Put in other words, it may desire to know what its ouptut should be, if the projected final demands are to be realized. Such an application of input-output analysis has been recommended by Evans.[5] A prediction of an output of one industry will be called a *special-purpose prediction* and the industry concerned will be called a *key industry*. The aggregation problem related to a special purpose prediction would be to perform combinations of small industries other than the key industry in such a manner that error in the output prediction for the key industry is minimized. If the method of aggregation that gave this result happened to involve large errors in predicting the output of the other industries, this would not matter.

In other applications it may be important to consider more than one industry, or even all industries in the model. This would probably be the case in general economic planning or mobilization. When it is desired to have output predictions of all intermediate industries of the model, it will be said that a *general-purpose prediction* is desired. In choosing an aggregation procedure for a general-purpose prediction, a criterion of minimizing errors in all industries would be used. This

212 Studies on the Coefficients Matrix

paper will be concerned with these two types of prediction objectives. We are now ignoring other criteria for aggregation, such as conventional distinctions between industries.

It is to be expected that, for various special-purpose predictions, the optimal method of aggregation would not be the same—that is, the method of aggregation that minimizes prediction errors in Industry No. 1 would not be the same as the method of aggregation that minimizes prediction errors in Industry No. 2.[6] Hence, in order to develop a criterion for general-purpose predictions, some sort of compromise would have to be made among the various criteria used for each industry separately, on the assumption that it were a key industry. This compromise will involve the assigning of an a priori weight to each separate industry, indicating the relative importance of minimizing errors in that industry. This idea will be developed after the specific consolidation procedure is described.

Following previous authors,[7] the rules of consolidation involve simple summation of money flows in a particular base period, for which a detailed input-output table is available. The flows will be assumed to be represented on a gross basis[8] in money units. Let $x_{ij}$ denote the flow from small industry $i$ to small industry $j$; let $y_i$ denote the final demand for small industry $i$; and let $x_i$ denote the total output of small industry $i$ (so that $x_i = \sum_{j=1}^{n} x_{ij} + y_i$). Let $x_{IJ}$ denote the flow from large industry $I$ to large industry $J$; let $y_I$ denote the final demand for large industry $I$; and let $x_I$ denote the total output of large industry $I$ (so that $x_I = \sum_{J=1}^{m} x_{IJ} + y_I$). Let a bar over a letter (such as $\bar{x}_{ij}$) denote its value in the base period. Define an *aggregation partition* or an *aggregation pattern* as the system of partitioning the $n$ small industries into $m$ mutually exclusive groups and the assignment of each group to a large industry. Then, for some particular aggregation partition, the large-industry flows and outputs in the base period are defined by the following rules of consolidation:

$$\bar{x}_{IJ} = \sum_{i \in I} \sum_{j \in J} \bar{x}_{ij}, \tag{14-1}$$

$$\bar{y}_I = \sum_{i \in I} \bar{y}_i, \tag{14-2}$$

$$\bar{x}_I = \sum_{i \in I} \bar{x}_i \tag{14-3}$$

where the symbol $\sum_{i \in I}$ means "the sum for all those small industries that are assigned to large industry $I$ according to the aggregation partition."

Generalizing now to any period, let the input coefficient $a_{ij}$ denote the amount of input from small industry $i$ that is needed by small industry $j$ in order to produce one unit of output. Then $a_{ij} = x_{ij}/x_j$. Although these coefficients are computed from the data of the base period, they are assumed to remain fixed for all periods; therefore bars are not shown in the definition just given. Small industry final demands, $y_i$, are assumed to be exogenous variables that may differ from their base period values, $\bar{y}_i$.

Input coefficients for the large industries defined by the aggregation partition may also be computed from the base period data: $a_{IJ} = \bar{x}_{IJ}/\bar{x}_J$. These are called *aggregate coefficients*. It is also useful to define a *semiaggregate coefficient*: $a_{Ij} = \sum_{i \in I} a_{ij}$. This is a sum of the original coefficients pertaining to output of small industry $j$ that originate from a single *large* industry $I$ according to the partition. (That is, rows, but not columns, of the original table of coefficients are combined.) Then it can be shown, by using definition (1) and juggling summation signs that $a_{IJ} = \sum_{j \in J} \bar{x}_j a_{Ij}/\bar{x}_J$. That is, the aggregate coefficient is a weighted average of the semiaggregate coefficients belonging to its large industry of destination, the weights being proportional to base-year outputs of the included small industries. The aggregate coefficients therefore depend on base-period flows.

Two alternative procedures may be used for predicting the output of a large industry $I$ in any period, given a set of final demands. First, all of the small-industry outputs $x_i$ may be calculated by solving a set of $n$ simultaneous linear equations. Then these outputs may be consolidated into large-industry outputs, using an equation like (14–3) above only without bars. Let $x_I^*$ denote the output of industry $I$ as determined by this procedure. Since exact knowledge of all of the $a_{ij}$ and of the $y_i$ is assumed, there is no error in $x_I^*$.

Another procedure would be first to aggregate the small final demands $y_i$ into large final demands $y_I$ by means of Eq. (14–2) without bars, and then solve a set of only $m$ simultaneous linear equations containing the $y_I$ and the aggregate coefficients $a_{IJ}$. Let $x_I^0$ denote the output of industry $I$ so determined.

It has been shown by others[9] that in general $x_I^0$ will differ from $x_I^*$, and that the discrepancy, or aggregation bias, is

$$x_I^0 - x_I^* = \sum_{j=1}^{n} (b_{IJ} - b_{Ij})y_j \qquad (14\text{–}4)$$

where $b_{IJ}$ is the element of the aggregate Leontief inverse matrix $[I - (a_{IJ})]^{-1}$ that corresponds in position with $a_{IJ}$, $b_{Ij} = \sum_{i \in I} b_{ij}$, where $b_{ij}$ is the element of the Leontief inverse matrix $[I - (a_{ij})]^{-1}$ corresponding to $a_{ij}$, $y_j$ is the final demand for industry $j$ in the current period, and $j$ belongs to $J$ according to the aggregation partition. For convenience the quantities $b_{Ij}$ will be called *semiaggregated inverse coefficients*.

It has also been shown by these same authors that when the final demands in any year are proportional to their base-year values, the aggregation bias is zero—that is,

$$\sum_{j=1}^{n} (b_{IJ} - b_{Ij})\lambda \bar{y}_j = 0 \qquad (14\text{–}5)$$

where $\lambda$ is a constant.

If an investigator in practice uses the prediction $x_I^0$ rather than $x_I^*$, his prediction will be more accurate when the bias $x_I^0 - x_I^*$ is smaller. It would seem reasonable to select aggregation patterns that make the biases as small as possible. In the following section this idea will be made more precise.

## The Criterion with Uncorrelated Final Demands

The discrepancies that we desire to minimize are shown by Eq. (14–4) to be functions of the input coefficients as well as of the final demands in the current year. Since the input coefficients are regarded, at least in theory, as being stable over a number of years, while the final demands are regarded as autonomous variables that vary from year to year, and since we desire to develop criteria for aggregation that will have some properties of invariance over individual years, we shall at this point adopt a randomization or statistical approach by assuming that the final demands are random variables with certain assumed probability distributions, and then seek to minimize certain *expectations* involving the biases.

It is not claimed that final demands are in reality determined by a random or stochastic process. A stochastic model is being proposed as an artifact to approximate the average situation produced by a series of final demands.

For special-purpose predictions the following criterion is proposed:

$$C_I = E(x_I^0 - x_I^*)^2 \tag{14-6}$$

where industry $I$ is the key industry of interest. A lower value of $C_I$ indicates a better aggregation pattern. The use of the squared value of the discrepancy is justified by the following reasons: it gives equal recognition to positive or negative discrepancies, it is algebraically convenient, and it is somewhat conventional in statistical treatment of discrepancies or errors.[10]

For general-purpose predictions, the following criterion is proposed:

$$C = \sum_{I=1}^{m} C_I = \sum_{I=1}^{m} E(x_I^0 - x_I^*)^2 \tag{14-7}$$

where a lower value of $C$ indicates a better aggregation pattern. Here the a priori weights for the various large industries mentioned in the previous section are each unity. This is justified, partially, by the selection of money values as units for measurement of input-output flows, of total outputs, and consequently of the discrepancies $x_I^0 - x_I^*$. Assuming that the relative importance of errors of prediction is adequately expressed for each large industry separately as the square of its discrepancy in dollar terms and that the combined importance is adequately expressed by simple addition, then unit weights for the various $C_I$ are appropriate in forming $C$.

Assume that the final demand for small industry $j$ is a random variable, $y_j$. Let $\text{var}_j$ denote the variance of $y_j$ and $\text{cov}_{jk}$ the covariance of $y_j$ with $y_k$. Then, by squaring Eq. (14-4) and taking expectations, Eq. (14-6) becomes

$$C_I = \sum_{j=1}^{n} (b_{IJ} - b_{Ij})^2 \, \text{var}_j + \sum_{j=1}^{n} \sum_{k \neq j} (b_{IJ} - b_{Ij})(b_{IK} - b_{Ik}) \, \text{cov}_{jk} \tag{14-8}$$

where small industry $j$ belongs to large industry $J$, and small industry $k$ belongs to large industry $K$.

Now the following additional specific assumptions will be made about the distribution of the final demands $y_j$. It is assumed that $y_j$ has an expected value equal to its base year value, $\bar{y}_j$, has variance proportional to this base-year value, and has covariance with all other final demands equal to zero. That is, it is assumed:

$$E(y_j) = \bar{y}_j \tag{14-9a}$$

$$\text{var}_j = v\bar{y}_j \tag{14-9b}$$

$$\text{cov}_{jk} = 0 \text{ for all } j \text{ and } k \qquad (14\text{–}9c)$$

where $E$ denotes expected value and $v$ is a positive constant.

Assumption (14–9a) implies that the base year is somewhat typical of all prospective years, in terms of expected final demands; assumption (14–9b) implies that the average variation above and below the base-year value will be greater when the base-year value itself is greater (and, specifically, will be zero when the base-year value is zero); assumption (14–9c) implies that there is no interrelationship of final demands—that is, a high demand for one industry is associated with neither a high nor a low demand for any other industry. Assumptions (14–9a) and (14–9b) seem reasonable approximations to reality, or at least could be, if the base year were well chosen. Assumption (14–9c) would not seem reasonable if we were to regard final demand as determined by individual behavior patterns and utility or preference schedules, since complementarities in final demand are denied. Even so, it may be useful as a first rough approximation.[11] This assumption is more reasonable if we regard final demands as controllable and if it is desired to discover best aggregations over a time span when *all possible combinations* of final demands are to be considered (subject to assumptions 14–9a and 14–9b). Such a viewpoint may be of use when government demand is a large part of the final bill of goods.

By making use of assumptions (14–9a), (14–9b), and (14–9c), Eq. (14–8) becomes

$$C_I = v \sum_{j=1}^{n} (b_{IJ} - b_{Ij})^2 \bar{y}_j \qquad (14\text{–}10)$$

and Eq. (14–7) becomes:

$$C = v \sum_{I=1}^{m} \sum_{j=1}^{n} (b_{IJ} - b_{Ij})^2 \bar{y}_j \qquad (14\text{–}11)$$

The suggested criteria are now expressed as weighted sums of squares, each squared term involving a difference between a certain semiaggregated inverse coefficient and a corresponding aggregate inverse coefficient, the weights being final demands in the base period.

## Minimal Distance Criteria

Criteria (14–10) and (14–11) suffer from at least two disadvantages. First, since the semiaggregated inverse coefficients $b_{Ij}$ depend on the $b_{ij}$ by the relations $b_{Ij} = \sum_{i \in I} b_{ij}$; and since $b_{ij}$ is an element of a detailed Leontief inverse matrix, it is necessary to have inverted a matrix of size $n \times n$ (i.e., solved a system of $n$ simultaneous linear equations) before the criteria can be applied. But to avoid such a procedure is one main objective of the aggregation. This objective would not then be served by going through a computation of the individual $b_{ij}$, except in situations when the optimal aggregation arrived at in one problem could be extrapolated to further problems of similar nature.

Second, to use the criteria, it is necessary also to determine a set of $m \times m$ aggregate inverse coefficients $b_{IJ}$ for each aggregation partition that is to be compared. This involves, for each aggregation partition, consolidating the original

$a_{ij}$s into $a_{IJ}$s, and then inverting an $m \times m$ matrix. Since usually a large number of aggregation partitions are to be compared, the computations will be quite extensive.

Two approximate criteria will now be defined. One of these will get rid of the second difficulty just mentioned, and one of them will get rid of both difficulties. First, we define for the special-purpose problem the approximate criterion

$$C'_I = \sum_{j=1}^{n} (\bar{b}_{IJ} - b_{Ij})^2 \bar{y}_j \qquad (14\text{-}12)$$

and for the general-purpose problem the approximate criterion

$$C' = \sum_{I=1}^{m} C'_I = \sum_{I=1}^{m} \sum_{j=1}^{n} (\bar{b}_{IJ} - b_{Ij})^2 \bar{y}_j \qquad (14\text{-}13)$$

where $\bar{b}_{IJ}$ is the weighted mean $\sum_{j \in J} \bar{y}_j b_{Ij} / \sum_{j \in J} \bar{y}_j$. We state that a lower value of $C'_I$ (or $C'$) indicates a better aggregation pattern.

By comparing Eqs. (14–12) and (14–13) with Eqs. (14–10) and (14–11) it is seen that $C'_I$ and $C'$ differ from $C_I$ and $C$ only in that weighted means $\bar{b}_{IJ}$ are used in place of the aggregate inverse coefficients $b_{IJ}$. Since the constant $v$ in Eq. (14–10) is the same for all partitions, it may be ignored. Looking at Eq. (14–12), it is seen that the problem of minimizing $C'_I$ is that of finding a partition of the $n$ given numbers, $\bar{b}_{IJ}$, into $m$ mutually exclusive groups so that the weighted sum of the squared deviations of the individual numbers from their group means is as small as possible.

From geometrical analogy the quantity $C'_I$ may be called "squared distance," and the problem of minimizing it may be called a "minimal distance problem" in one dimension.[12] This problem, for moderate sized $n$ and $m$, has been found to be amenable to solution by automatic digital computer.[13] The structure of the general-purpose problem is, however, more complicated.[14]

The term 'approximate criteria" for $C'_I$ and $C'$ in the sense that these two quantities are approximations to $C_I$ and $C$ would be justified only if the differences $(C'_I - C_I)$ and $(C' - C)$ were small in comparison with their absolute values, or if, for different aggregation partitions, $C'_I$ were approximately proportional to $C_I$. In these situations one would not go far wrong in using the approximate criteria as a substitute for $C_I$ or $C$ in judging the relative merit of different aggregation partitions. One of the objectives of the numerical experiments to be described in the next section was to determine the effect of using this approximation.

A somewhat cruder, but more useful, approximate criterion may be defined as

$$C''_I = \sum_{j=1}^{n} (\bar{a}_{IJ} - a_{Ij})^2 \qquad (14\text{-}14)$$

for the special-purpose problem, and as

$$C'' = \sum_{I=1}^{m} C''_I = \sum_{I=1}^{m} \sum_{j=1}^{n} (\bar{a}_{IJ} - a_{Ij})^2 \qquad (14\text{-}15)$$

for the general-purpose problem, where $a_{Ij}$ is the semiaggregated coefficient $a_{Ij}$ $= \sum_{i \in I} a_{ij}$, and $\bar{a}_{IJ}$ is the unweighted mean $\sum_{j=1}^{n} a_{Ij} / n_J$. This criterion employs the input coefficients in the detailed model without inversion. To minimize $C''_I$ in Eq.

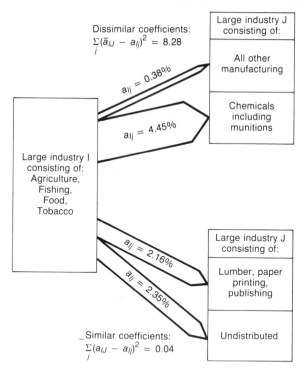

Fig. 14.1. A criterion of similarity of semiaggregated input coefficients.

(14–14) is a one-dimensional minimal distance problem with all weights equal to unity. The use of the original semiaggregated coefficients $a_{Ij}$ in substitution of inverse coefficients $b_{Ij}$ in an "approximation" of this type is justified by the mathematical representation of the operation of inversion of a Leontief matrix as a power series.[15] In fact, Theil, making use of this power series representation and dropping off certain terms, develops an expression which he calls "first order aggregation bias,"[16] which may be written as an approximation to our equation (14–4) as

$$x_I^0 - x_I^* \cong \sum_{j=1}^{n} (a_{IJ} - a_{Ij})y_j. \qquad (14\text{–}16)$$

Using this equation (14–16) as a starting point, instead of (14–4), and making the randomizing assumptions regarding final demands that were made above, one could develop a criterion similar to $C$, except that $a$'s are used instead of $b$'s. Our $C''$ would still differ from such a criterion in that $\bar{a}_{IJ}$ is used instead of $a_{IJ}$, and all the weights are equal. These discrepancies are the result of shortcuts made in the interest of computational simplicity.

The meaning of criterion $C''$ is illustrated in Figure 14.1. The figure also illustrates varying degrees of similarity of semi-aggregated input coefficients that are assigned to the same large industry of destination $J$; $C_I''$ consists of a sum of many squared terms associated with large industry $I$, two of which are illustrated in the figure; $C''$ consists of a larger sum of such terms, including all of the large industries $I$. Similar interpretations apply to $C$ and $C'$ if inverse coefficients are substituted for direct coefficients.

The approximations $C_I''$ and $C''$ will also be used in the numerical experiments described in the next section.

**Numerical Experiments**

Experiments were undertaken in which an $18 \times 18$ matrix representing the United States economy in 1939 was aggregated in 13 different ways into $8 \times 8$ matrices, and the various criteria for each of the different aggregation partitions were computed and compared. With some differences, these experiments continue a type of research initiated by Morgenstern and Whitin and by Balderston and Whitin.[17] The $18 \times 18$ matrix, published by Balderston and Whitin,[18] represents already a considerable aggregation from much larger tables previously compiled by the Bureau of Labor Statistics. It is used here because of its availability and the availability of its inverse, because a certain amount of experimentation has already been performed on it, and because its small size makes it suitable for pilot studies in the methodology of aggregation. The small industries of this matrix are shown in Table 14.1.

Three of the aggregated $8 \times 8$ matrices are those defined by Balderston and Whitin[19] and used by them to make certain comparisons. Apparently the chief purpose of these aggregations was to illustrate the point that the manner of aggregation affects the magnitude of a certain inverse coefficient relating two particular small industries, Ferrous metals and Rubber. These authors state that,

Table 14.1.   Small Industries in 18 × 18 Matrix

1. Ferrous metals
2. Rubber
3. Agriculture and fishing
4. Food, tobacco, and kindred products
5. Motor vehicles, industrial and heating equipment
6. Metal fabricating
7. Nonferrous metals and products
8. Nonmetallic minerals and products
9. Fuel and power
10. Chemicals, including munitions
11. Lumber, paper, and their products; printing and publishing
12. Textiles and leather
13. All other manufacturing
14. Construction
15. Transportation
16. Trade
17. Business and personal services
18. Undistributed

aside from keeping these two small industries separate, "it was difficult to choose the grouping to be followed in the aggregations on any basis other than common sense, because the criterion for aggregation (e.g. proportionality of output between industries in an aggregate) could hardly be applied when the whole economy is represented by such a small number of industries."[20]

The remaining ten aggregated $8 \times 8$ matrices were selected by means of minimal distance criteria with the intention that they be good aggregations in terms of the criteria developed in this paper. Eight of them were selected as good for special-purpose predictions in four different key industries, and two as being good for general-purpose predictions. The definitions of these ten aggregation partitions are shown in Table 14.2.

Of the eight special-purpose aggregations, four (nos. 4, 5, 6, 7) are chosen to give minimal $C_I'$ and four (nos. 9, 10, 11, 12) are chosen to give minimal $C_I''$. The four key industries considered are: Case 1 (Ferrous metals), Case 2 (Rubber), Case 3, Small industries 3 and 4 combined (Agriculture and Food), and Case 4, a conglomerate of small industries 10, 13, 16, 17, and 18. Since, in a special-purpose aggregation, the key industry is specified in advance, the problem of finding the best aggregation partition becomes that of partitioning the small industries not included in the key industry into 7 large industries such that the squared distance of the semiaggregated coefficients $b_{Ij}$ (or $a_{Ij}$) within large industries is minimized, where $I$ is the key industry. The four aggregations giving minimal $C_I'$ involved a weighted sum of squares and were obtained by using the program on the Illiac mentioned in footnote 13. The four aggregations giving minimal $C_I''$ involved only an unweighted sum of squares and were easily obtained by plotting the relevant $a_{Ij}$ as points on a line, and making the groupings of nearest points by visual inspection.

Of the two general-purpose aggregations, one (no. 8) is intended to give a low value of $C'$, and one (no. 13) is intended to give a low value of $C''$. A procedure has not yet been found that guarantees selection of the optimal partition in these cases. The problem is more complicated than the one-dimensional minimal distance problem described above. For one thing, the relevant "distance" is in $m$ dimensions instead of one dimension; for another, the locations of the points in this space (obtained from the values of the semiaggregated coefficients) are not given in advance but vary for the different possible aggregation partitions.

After inspection of the $18 \times 18$ Leontief coefficient matrix revealed that the diagonal entries were quite large in relation to most of the nondiagonal entries,[21] a procedure was used for selecting the optimal partitions that is based on the idea that this matrix approximates a diagonal matrix. Aggregations 8 and 13 were chosen as the solutions to one-dimensional minimal distance problems in which the given points are diagonal elements of the matrices $(b_{ij})$ and $(a_{ij})$, respectively. Aggregation 8 was chosen so as to minimize $\sum_{i=1}^{n} (b_{ii} - \bar{b}_{II})^2 \bar{y}_i$, and Aggregation 13 was chosen so as to minimize $\sum_{i=1}^{n} (a_{ii} - \bar{a}_{II})^2$.[22] The resulting values of $C'$ and $C''$ are probably near the minimal values.

For each of the four key industries used, the Balderston-Whitin aggregations[23] were compared with the two minimal distance aggregations that were intended to be optimal for that industry by computing the special-purpose criteria $C_I$, $C_I'$, and $C_I''$ (where $I$ denotes the key industry considered). The results are shown in Table

Table 14.2.   Definitions of Aggregation Partitions Obtained from Minimal Distance Criteria

**Aggregations for Minimal $C_I'$**

| Large industry | No. 4 | No. 5 | No. 6 | No. 7 | No. 8[a] |
|---|---|---|---|---|---|
| | | | Small industries[b] included in large industry | | |
| 1 | 1 | 2 | 3, 4 | 10, 13, 16, 17, 18 | 7 |
| 2 | 6 | 5 | 12 | 12 | 1, 11, 12 |
| 3 | 5 | 18 | 10 | 6, 4 | 9, 5 |
| 4 | 14 | 12, 6, 3 | 11, 18 | 2, 5, 8 | 10, 18 |
| 5 | 18, 7 | 10, 13, 4, 8 | 14 | 11 | 6, 8 |
| 6 | 15, 11, 9, 3 | 16, 11, 9, 1 | 13 | 3, 9, 1 | 3, 4, 13, 16 |
| 7 | 4, 8, 10, 12, 13, 16, 2 | 15, 14, 7 | 2, 6, 8, 5, 1, 16, 9 | 14, 7 | 15, 17 |
| 8 | 17 | 17 | 7, 17, 15 | 15 | 2, 14 |

**Aggregations for Minimal $C_I''$**

| Large industry | No. 9 | No. 10 | No. 11 | No. 12 | No. 13[c] |
|---|---|---|---|---|---|
| | | | Small industries[b] included in large industry | | |
| 1 | 1 | 2 | 3, 4 | 10, 13, 16, 17, 18 | 7 |
| 2 | 6 | 5 | 12 | 2, 12 | 1 |
| 3 | 5 | 18 | 10 | 6, 8 | 12, 11 |
| 4 | 14 | 3, 15 | 18 | 4 | 5 |
| 5 | 18 | 6, 12 | 11 | 3, 5, 11, 9 | 9, 10 |
| 6 | 7 | 17 | 14 | 1 | 8 |
| 7 | 15, 11 | 16, 10, 13 | 13 | 7 | 6, 4, 3, 13 |
| 8 | { 3, 2, 4, 8, 9, 10, 12, 13, 16, 17 | 14, 4, 9, 11, 1, 7, 8 | 1, 2, 5, 6, 7, 8, 9, 15, 16, 17 | 14, 15 | 17, 2, 15, 18, 14, 16 |

[a] Selected to minimize $\sum_{i=1}^{N} (b_{ii} - \bar{b}_{II})^2 \bar{y}_i$.

[b] For names of the small industries, see Table 14-1.

[c] Selected to minimize $\sum_{i=1}^{N} (a_{ii} - \bar{a}_{II})^2$.

14.3. The three Balderston-Whitin aggregations were then compared with minimal distance aggregations 8 and 13 by means of each of the general-purpose criteria $C$, $C'$, and $C''$. These results are shown in Table 14.4. The computations perform two functions: first, they compare the different aggregation partitions in terms of criterion $C$; second, they compare the approximate criteria $C'$ and $C''$ with criterion $C$ as indicators of good aggregation partitions.[24]

Tables 14.3 and 14.4 show that in every case studied (each of the four separate key industries and all industries combined) in terms of criterion $C$ an aggregation pattern better than Nos. 1, 2, or 3 was found by the use of minimum-distance ideas. That is, one of the aggregation patterns 4 to 13 always gave a lower value of $C$ than did any of patterns 1, 2, or 3.[25]

The relative degree of improvement over patterns 1, 2, or 3 attained by the optimal pattern, judging by ratios of $C$ values, was greatest for the smallest key industries (Ferrous metals and Rubber); was moderate for the all-industries case; and was least for the two largest key industries (cases 3 and 4). In absolute terms

Table 14.3. Comparisons of Aggregations by Special-Purpose Criteria

| Aggregation No.[a] | $C_I$ | $C_I'$ | $C_I''$ |
|---|---|---|---|
| Case 1: Large Industry $I$ comprises Small Industry 1 (ferrous metals) | | | |
| 1 | 0.015731 | 0.008440 | 0.003458 |
| 2 | 0.008605 | 0.006988 | 0.007927 |
| 3 | 0.024065 | 0.023731 | 0.007637 |
| 4 | 0.000071[b] | 0.000056[b] | 0.000059 |
| 9 | 0.000809 | 0.000804 | 0.000005[b] |
| Case 2: Large Industry $I$ comprises Small Industry 2 (rubber) | | | |
| 1 | 0.000734 | 0.000386 | 0.000462 |
| 2 | 0.000652 | 0.000461 | 0.000664 |
| 3 | 0.000708 | 0.000698 | 0.000559 |
| 5 | 0.000002[b] | 0.000002[b] | 0.000016 |
| 10 | 0.000015 | 0.000015 | 0.000000[b] |
| Case 3: Large Industry $I$ comprises Small Industries 3 and 4 (agriculture and food) | | | |
| 1 | 0.074958 | 0.072567 | 0.026403 |
| 2 | 0.071124 | 0.070164 | 0.025640 |
| 6 | 0.068625[b] | 0.067914[b] | 0.024114 |
| 11 | 0.068704 | 0.067981 | 0.024112[b] |
| Case 4: Large Industry $I$ comprises Small Industries 10, 13, 16, 17, 18 | | | |
| 2 | 0.488579 | 0.423380 | 0.212058 |
| 7 | 0.368363[b] | 0.289765[b] | 0.129019 |
| 12 | 0.418621 | 0.329722 | 0.108443[b] |

[a] For definitions of aggregations 1, 2, 3, see Balderston and Whitin, op. cit., 119, 121, and 123. For names of small industries see Table 14–1. For definitions of remaining aggregations, see Table 14–2.

[b] Minimum value of the criterion for the aggregations considered.

Table 14.4. Comparisons of Aggregations by General-Purpose Criteria

| Aggregation No.[a] | $C$ | $C'$ | $C''$ |
|---|---|---|---|
| 1 | 1.1448 | 0.8438 | 0.5270 |
| 2 | 0.8381 | 0.7085 | 0.4632 |
| 3 | 1.0471 | 0.9668 | 0.5418 |
| 8 | 0.5876 | 0.3163[b] | 0.4037 |
| 13 | 0.5670[b] | 0.3806 | 0.2538[b] |

[a] For definitions of aggregations 1, 2, 3, see Balderston and Whitin, op. cit., 119, 121, and 123. For names of small industries see Table 14.1. For definitions of remaining aggregations, see Table 14.2.

[b] Minimum value of the criterion for the aggregations considered.

(arithmetic differences of $C$ values) the greatest improvement was attained in the all-industries case; next in special cases 1, 3, and 4; and least in special case 2.

The approximate distance criteria $C'$ and $C''$ showed up well as indicators of criterion $C$, and hence of good aggregation partitions. In each of the four special industry cases the aggregation pattern having minimal $C'$ also had minimal $C$. In cases 1, 2, and 3 the loss incurred by basing the selection of a partition on the value of $C''$ rather than on $C'$ would have been extremely small. In case 4 (in which the given key industry was a rather large and heterogeneous conglomerate) the loss by using $C''$ would have been somewhat larger. In the all-industries case the aggregation pattern having minimal $C''$ also had minimal $C$. The loss would have been slight if minimal $C'$ had been used.

In the preceding paragraph the expression "minimal $C$" is applied to the set of aggregation patterns listed in Tables 14.3 and 14.4. The attempt was not made in any case to find the partition with the absolute minimum value of $C$ out of all possibilities. It would be an interesting project to find a procedure for determining this partition. It is possible that some improvement could be made by such a determination.

In summary, these results indicate that, given the prediction objective, substantially lower errors may be obtained by deliberate aggregation procedures based on minimal distance ideas than on haphazard procedures.

## Conclusions

The approach of this paper has been shown to have a close relationship with previous suggestions that small industries in input-output tables be aggregated into large industries on the basis of "similarity of coefficients," "stability of the inverse," "homogeneity of industries," "similarity of production functions," "similarity of cost structure." The minimum distance idea described in the preceding pages seems to encompass all of these suggestions.

It is recognized that the special statistical assumptions made in this paper concerning final demand limit the applicability of the numerical experiments reported. Formulas (14–10) and (14–11) for $C_I$ and $C$ hold only under the special assumptions that the variance is a constant multiple of the mean, and that all covariances are zero. By making the assumption of zero covariances the possibility is denied of recognizing tendencies toward complementarity of final demands of certain small industries (sometimes included in the term "stability of product mix"), and including such complementarities in the determination of our criterion of predictive accuracy, $C$.[26]

Some effort was made to test crudely the variance assumption (14–9b) as applied to "adjusted" final demands (see footnote 11) by noting changes that took place in the final demands for the 18 industrial categories from 1939 to 1947, thus estimating a variance on the basis of one degree of freedom. The data showed some slight indication that either assumption (14–9b) or an assumption that the standard deviation is a constant multiple of the mean would be acceptable. No attempt was made to test the covariance assumption. It should be noted that the usefulness of formulas (14–10) and (14–11) does not depend upon the assumption of strict zero value for all covariances; it is sufficient that the cross-product terms in formula (14–8) be small in relation to the squared terms.

If further research should demonstrate that the special statistical assumptions are not tenable or useful, and if one can make good a priori assumptions about the values of the variances and covariances, then an approach could be followed of using criterion $C$ based on formula (14–8) as it stands and thus allowing for correlated demands.

Another theoretical limitation of the approach of this paper, and of previous approaches by others, results from basing the definition of $C$ on outputs of large industries rather than small. In some applications the ultimate objective will be the prediction or determination of outputs of small industries, and in these cases the large industry outputs would have to be allocated among the component small industries by some procedure of disaggregation. Fei has emphasized the importance of this disaggregation problem.[27]

The approach suggested in this paper, of grouping industries together on the basis of the accuracy of predictions of output, has led to the neglect of other considerations, such as conventional grouping of industries, legal connections, bottleneck situations not recognized in the Leontief assumptions, and so on. This restricted approach has the merit of conceptual simplicity by concentrating attention on one aspect of the problem, but neglects the probability that the investigator must in fact consider some of these aspects also in his aggregation decision. By glancing at some of the groupings shown in Table 14–2, it can be seen that there are some strange bedfellows. We might desire at times to sacrifice some accuracy of prediction if we could attain more familiar large industries.

In principle it is possible to introduce modifying criteria formally into the statement of the aggregation problem by means of side conditions or constraints. It is possible to state and to solve the minimum-distance problem when the partitions to be considered are limited in some manner in advance. If workable procedures can be found for finding optimal aggregation partitions for input-output analysis that incorporate relevant restrictions of the type just mentioned, then the approach suggested in this paper can be extended further in the direction of realism.

**NOTES**

1. J. B. Balderston and T. M. Whitin, "Aggregation in the Input-Output Model" in Oskar Morgenstern, ed., *Economic Activity Analysis* (New York, 1954), 79–128; Tibor Barna, "Classification and Aggregation in Input-Output Analysis" in Tibor Barna, ed., *The Structural Interdependence of the Economy* (Proceedings of an International Conference on Input-Output Analysis, Varenna, 1954), ch. 7; John C. H. Fei, "A Fundamental Theorem for the Aggregation Problem of Input-Output Analysis," *Econometrica*, XXIV (October 1956), 400–412; M. Hatanaka, "Note on Consolidation Within a Leontief System," *Econometrica*, XX (April 1952), 301–303; Mathilda Holzman, "Problems of Classification and Aggregation" in Wassily Leontief, ed., *Studies in the Structure of the American Economy* (New York, 1953), ch. 9; John McCarthy, "Aggregation in the Open Leontief Model" (paper presented to the Econometric Society at Cleveland, Ohio, December 1956); M. McManus, "General Consistent Aggregation in Leontief Models," *Yorkshire Bulletin*, VIII (June 1956), 28–48; Edmond Malinvaud, "Aggregation Problems in Input-Output Models" in Tibor Barna ed., op. cit., ch. 8; Oskar Morgenstern and Thomson M. Whitin, "Comments" in National Bureau of Economic Research ed., *Input-Output Analysis: An Appraisal* (Studies in Income and Wealth, Vol. 18, Princeton, 1955), 128–135; Herbert A. Simon and Albert Ando, "Aggregation of Variables in Dynamic Systems" (paper presented to the Econometric Society at Cleveland, Ohio, December 1956); H. Theil, "Linear Aggregation in Input-Output Analysis," *Econometrica*, XXV (January 1957), 111–122.

2. See works cited of Balderston and Whitin, Hatanaka, Malinvaud, McManus, Theil.

3. Some preliminary suggestions have been made by McCarthy and by Simon and Ando in their papers cited.

4. Here we are following Malinvaud (op. cit., 190), who has drawn a distinction between aggregation problems associated with empirical construction of input-output tables, and problems associated with using the tables.

5. W. Duane Evans, "Marketing Uses of Input-Output Data," *Journal of Marketing*, XVII (July 1952), 11–21.

6. Cf. Balderston and Whitin, op. cit., 88, n. 11, where a similar point is made with regard to particular coefficients of the inverse matrix, based on a previous comment of Leontief.

7. E.g., Wassily W. Leontief, *The Structure of American Economy, 1919–1939* (2nd ed., New York, 1951), 14–15.

8. I.e., allowing for intra-industry flows.

9. Balderston and Whitin, op. cit., equation (7), p. 108; Theil, op. cit., the term $Px_w$ in equation (6.7), p. 114.

10. L. J. Savage, in his recent book *The Foundations of Statistics* (New York, 1954), ch. 15, presents some theoretical basis for the squared error criterion from the viewpoint of decision theory.

11. The above remarks are partly based on the fact that no loss of generality occurs by restricting consideration to bills of goods that represent deviations from a uniform proportional movement of the base year bill of goods. More specifically, if $y_j$ is an arbitrary final demand in the current year, let $z_j = y_j - \lambda \bar{y}_j$, where $\bar{y}_j$ is the base year final demand and $\lambda$ is a constant so chosen that $\sum_{j=1}^{n} z_j = 0$. Then no change in the aggregation bias results if $y_j$ is replaced by $z_j$ [as shown by, e.g., Balderston and Whitin, op. cit., equation (8), page 108, and can also be shown by multiplying our Eq. (14–5) above by $\lambda$ and subtracting same from Eq. (14–4)]. Assumptions (14–9) are more reasonable when applied to the "adjusted" final demands $z_j$ than to the original final demands $y_j$, since common proportional movements have been removed from the latter. See also the discussion in the concluding section of this paper.

12. If the numbers $b_{ij}$ are pictured as $n$ points on a line, then $C'_I$ represents a sum of squared distances of the individual points from their group centers of gravity. When the points are bunched closely around their group centers of gravity (the $b_{Ij}$ belonging to the same group $J$ having similar magnitudes), $C'_I$ becomes small. In analysis of variance terminology the problem is to define groups so that the "sum of squares within groups" is minimized. This problem is discussed in a previous article. Walter D. Fisher, "On a Pooling Problem from the Statistical Decision Viewpoint," *Econometrica*, XXI (October 1953), 567–85.

13. The author wrote and checked a program that finds an optimal partition and its associated squared

distance for $n$ up to 200 and $m$ up to 10 on the University of Illinois Illiac. Details were to be published elsewhere, but acknowledgment was given to the University of Illinois and particularly Professor William A. Neiswanger of the Department of Economics, and Kern Dickman, Computer Consultant, for aid and facilities in preparing this program during the summer of 1956.

14. See below, "Numerical Experiments."
15. See Frederick V. Waugh, "Inversion of the Leontief Matrix by a Power Series," *Econometrica*, XVIII (April 1950), 142–154.
16. Theil, op. cit., equation (8.5), page 117. Notation converted from Theil's to ours.
17. Op. cit.
18. Op. cit. 116–117. Its inverse has been computed and is shown on page 118.
19. *Ibid.*, 119, 121, 123.
20. *Ibid.*, 104.
21. Some nondiagonal $a_{ij}$ in rows 16 and 18 and in a few other scattered locations were fairly large.
22. For a diagonal matrix $(a_{ij})$ it can be shown that $C'$ becomes $\sum_{i=1}^{n} (b_{ii} - \bar{b}_{II})^2 \bar{y}_i$, and $C''$ becomes

$$\sum_{i=1}^{n} (a_{ii} - \bar{a}_{II})^2.$$

23. Only aggregations 1 and 2 could be used for case 3, and only aggregation 2 could be used for case 4, since the other aggregations did not specify the key industry as a large industry.
24. In the remaining portion of this paper $C$, $C'$, and $C''$ will be understood to stand also for $C_I$, $C'_I$, and $C''_I$, respectively. For computation of criterion $C$ for the minimal distance aggregations the inversion of ten $8 \times 8$ aggregated matrices was required. These computations were programmed for the Ferut automatic digital computer at Toronto by Mr. Klaus Hoecksmann under the supervision of Professor T. E. Hull of the Department of Mathematics at the University of British Columbia.
25. Of course no criticism of the work of Balderston and Whitin is implied by these results, since their aggregations were constructed for a different purpose.
26. Cf. Malinvaud, op. cit., section 8; Theil, op. cit., section 10.
27. Fei, op. cit., 408. Fei's "adjustment problem" (page 407) may be regarded as a type of minimal distance problem.

# 15

## The Stability of Input-Output Coefficients

PER SEVALDSON

We still know relatively little about the fluctuations over time in input-output coefficients and the causes of these fluctuations. This is the subject of a research program carried out by the Central Bureau of Statistics of Norway. One aspect of this program has been the study of time series of input-output accounts at constant prices. The presentation of some results from this study is the topic of the present paper.

It is generally believed that the stability of input-output coefficients depends on the level of sector specification in the data. This problem is discussed, the data are presented, and some results of computations in a relatively detailed sector specification are presented. In the final section of this paper these results are compared to the results of computations in aggregated sector specifications.

### The Effects of Aggregation on Variability in Input-Output Coefficients

It is generally recognized that aggregation in input-output tables entails tendencies both to make the coefficients more stable and to make them more variable. A cause of increased stability can be found in the fact that when sectors are aggregated, sectors which produce raw materials that are close substitutes will frequently be combined, and the coefficients for use of the sum of their products may reasonably be expected to be more stable than the coefficients for the use of each of them. A cause of increased instability is the fact that when two or more sectors with different coefficients for the same input are combined, the aggregate coefficient will be an average of the coefficients of the individual sectors, and the average will depend on the relative weight of the production in each sector. When these weights vary, the average coefficient will vary, even if the individual coefficients are constant.

Let us examine the possibilities a little more closely. We shall use the following notation:

$x_j$      = production in sector $j$ in the detailed specification, measured in value at constant prices ($j = 1, 2, \ldots, n$).

$x_{ij}$      = input from sector $i$ to sector $j$ ($i = 1, 2, \ldots, n; j = 1, 2, \ldots, n$).

This article was published in *Applications of Input-Output Analysis*, edited by Anne P. Carter and Andrew Bródy, © North-Holland Publishing Company, 1970, pp. 207–237. Reprinted by permission.

$a_{ij}$ = $x_{ij}/x_j$ = input coefficient for use in detailed sector $j$ of products from detailed sector $i$ ($i = 1, 2, ..., n; j = 1, 2, ..., n$).

$a_{Ij}$ = $\sum_{i \in I} a_{ij}$ = input coefficient for use in detailed sector $j$ of products from aggregate sector $I(I = 1, 2, ..., N; j = 1, 2, ..., n; n > N)$. $\sum_{i \in I}$ means the sum over all those sectors $i$, which belong to the aggregate sector $I$.

$A_{IJ}$ = $\sum_{i \in I} \sum_{j \in J} x_{ij}/(\sum_{j \in J} (x_j - \sum_{i \in J} x_{ij}))$ = input coefficient for use in aggregate sector $J$ of products from aggregate sector $I$ ($I = 1, 2, ..., N; J = 1, 2, ..., N; I \neq J$).

$A_{II}$ = $0$ ($I = 1, 2, ..., N$).

$r_j$ = $x_j/(\sum_{j \in J} (x_j - \sum_{i \in J} x_{ij}))$ = relative share of total production in detailed sector $j$ in (net) production in aggregate sector $J$ ($j = 1, 2, ..., n; J = 1, 2, ..., N$).

var $a_{ij}$ = variance of $a_{ij}$
var $a_{Ij}$ = variance of $a_{Ij}$
var $A_{IJ}$ = variance of $A_{IJ}$
var $r_j$ = variance of $r_j$

We may now write (for $I \neq J$):

$$A_{IJ} = \sum_{i \in I} \sum_{j \in J} \frac{x_{ij}}{\sum_{j \in J} (x_j - \sum_{i \in J} x_{ij})} = \sum_{j \in J} \frac{\sum_{i \in I} x_{ij}}{x_j} \cdot \frac{x_j}{\sum_{j \in J} (x_j - \sum_{i \in J} x_{ji})} = \sum_{j \in J} r_j \cdot a_{Ij}$$

In principle there may exist interdependency between $r_j$ and $a_{Ij}$, since they are both derived from items in an input-output table, and the items of such a table are bound together through definitional relationships (identity of column sums with corresponding row sums). However, this is likely to impose only a mild restriction on the free variability of $r_j$ in relation to $a_{Ij}$. More serious is the possibility of an interdependence due to the fact that $x_j$ occurs in the definition of both $r_j$ and $a_{Ij}$, and that $x_{Ij}$ is part of $x_j$ (through the relationship $x_j = \sum_i x_{ij}$ + value added in sector $j$). However, even through the possibility of interdependency cannot be ruled out, we will disregard it here.

If $r_j$ and $a_{Ij}$ are independent, we have for the variance of the product ($r_j a_{Ij}$):

$$\text{var} (r_j a_{Ij}) = \text{var } r_j \cdot \text{var } a_{Ij} + \bar{a}_{Ij}^2 \text{ var } r_j + \bar{r}_j^2 \text{ var } a_{Ij}$$

where $\bar{a}_{Ij}$ and $\bar{r}_j$ are the average (expected) values of $a_{Ij}$ and $r_j$, respectively. Furthermore, if, the individual products ($r_j a_{Ij}$) are independent for $j \in J$, then

$$\text{var } A_{IJ} = \sum_{j \in J} \text{var} (r_j a_{Ij}) = \sum_{j \in J} \text{var } r_j \text{ var } a_{Ij} + \sum_{j \in J} \bar{a}_{Ij}^2 \text{ var } r_j + \sum_{j \in J} \bar{r}_j^2 \text{ var } a_{Ij}$$

or

$$\text{var } A_{IJ} = \sum_{j \in J} (\bar{r}_j^2 + \text{var } r_j) \text{ var } a_{Ij} + \sum_{j \in J} \bar{a}_{Ij}^2 \text{ var } r_j$$

The variance of the aggregate coefficient is, in this case, a weighted sum of the variances of the coefficients and the variances of the relative shares of the individual sectors in production in the aggregate sector.

Since both $\sum_{j \in J} (\bar{r}_j^2 + \text{var } r_j)$ and $\sum_{j \in J} \bar{a}_{Ij}^2$ may easily be considerably less than 1, and even their sum may be below 1, we should not be surprised to find that var $A_{IJ}$ will frequently be less than the typical variances of both the $r_j$ and the $a_{Ij}$. Moreover, since $A_{IJ}$ will tend to be bigger than the average of the $a_{Ij}$ coefficients (because the sum of weights in $A_{IJ} = \sum_{j \in J} r_j a_{Ij}$ will normally exceed 1) we may expect var $A_{IJ}$ quite often to be less than the variance of coefficients $a_{Ij}$ or $a_{ij}$ of the same magnitude. These conclusions rest on the previous assumptions about independence but since there are few reasons to expect positive covariation between $r_j$ and the corresponding $a_{Ij}$ or between the various $(r_j a_{Ij})$ in the same aggregate sector, this does not appear to be a serious restriction.

Let us consider the weights of var $a_{Ij}$ and var $r_j$ a little more closely. When production is very unevenly distributed among the detailed sectors in an aggregate sector, one $\bar{r}_j$ may be large (near 1) and the others must be limited in number and small. In this case all the var $r_j$ may be expected to be small since we may generally expect the $r_j$ with small $\bar{r}_j$ to have small variances and the variance of the one $r_j$ with great $\bar{r}_j$ will be approximately equal to the variance of the (small) sum of the small $r_j$'s. The weight sum $\sum_{j \in J} (\bar{r}_j^2 + \text{var } r_j)$ will be dominated by the square of the largest $\bar{r}_j$ $(j \in J)$. We may therefore expect the sum of weights for the coefficient variances to be less than 1, even if the dominating sector has an $\bar{r}_j$ as big as 0.8 or 0.9. With such large fractions for one sector, the number of detailed sectors in the aggregate sector will also normally be limited and the weight sum $\sum_{j \in J} \bar{a}_{Ij}^2$ for the terms var $r_j$ will be small, even for values of $\bar{a}_{Ij}$ up to 0.2 or 0.3, which must be considered to be large. At the same time var $r_j$ may be expected to be small, as already mentioned. Consequently, we conclude that, in this case, we can normally expect var $A_{IJ}$ to be less than the typical var $a_{Ij}$ for $j \in J$.

If the $\bar{r}_j$ are small or moderate in size, then their squares and sum of squares will be well below 1, and even if the var $r_j$ are added to the squares, their sum is not likely to reach the neighbourhood of 1. Now, if the number of detailed sectors to be aggregated is large, and if the coefficients $\bar{a}_{Ij}$ are large, the sum $\sum_{j \in J} \bar{a}_{Ij}^2$ may become large too, and the sizes of the variances of the production shares gain increasing influence on var $A_{IJ}$. However, with an increasing number of detailed sectors to be aggregated, their individual shares tend to decrease, and so do the variances of these shares (in absolute terms). Consequently, the weight sum for var $a_{Ij}$ decreases and the var $r_j$ themselves decrease, and var $A_{IJ}$ may still well be less than the typical var $a_{Ij}$.

We have assumed that the $r_j$ and $a_{Ij}$ are independent for any given $j$ in $J$. By and large this appears to be a plausible assumption. One possible cause for a negative correlation might appear if there were a random component in the productivity of sector $j$, which acted so that, in certain periods, the output was high in relation to inputs, and that high output also led to a higher than normal share in production, $r_j$, while in other periods output in relation to inputs—and thus share in production—was low. It is hard to imagine that such tendencies should dominate the figures, but if they existed we would expect them to *diminish* and not to increase the variance of the products $(r_j a_{Ij})$.

Our other previous assumption, that the individual products in the sum $\sum_{j \in J} r_j a_{Ij}$ are independent, is probably not realistic, since the $r_j$ are likely to be negatively correlated, at least when the number of detailed sectors in each aggregate sector is small. However, this should tend to diminish the variance of the sum.

If we compare the disaggregated coefficients, $a_{ij}$, and the aggregated coefficients $A_{IJ}$, we arrive at the following conclusion: the variance of the "semi-aggregated" coefficients $a_{Ij}$ will be in general less than the variances of the component detailed coefficients $a_{ij}$ ($i \in I$) because of the substitution effect. But it does not necessarily follow that in general the variance of semi-aggregated coefficients will be smaller than the variance of detailed coefficients of the *same* size. We shall, therefore, assume that the variance of semi-aggregated coefficients is of the same order of magnitude as the variance of detailed coefficients of the same size. The coefficients $A_{IJ}$ will tend to be greater than the average of the corresponding semi-aggregated coefficients, $a_{Ij}$ ($j \in J$), since the sum of the weights ($\sum_{j \in J} r_j$) in general will exceed 1.

Now we have demonstrated that the variance of $A_{IJ}$, under quite plausible assumptions, will be less than the variance of the "semi-aggregated" coefficients $a_{Ij}$. Consequently, we expect the variance of $A_{IJ}$ in general to be smaller than the variance of detailed coefficients $a_{ij}$ of the same size. We have not proved that var $A_{IJ}$ will always be less than the variance of detailed coefficients $a_{ij}$ of the same size, but we have demonstrated that the opposite will occur only for rather extreme values of the product shares $r_j$. Let us consider a few numerical examples with an aggregation of four detailed sectors into one aggregate sector. We make the following assumptions:

$$0.1 \leqslant r_j \leqslant r$$

$$1 \leqslant \sum_{j \in J} r_j \leqslant 1.1$$

$$\text{var } r_j \leqslant \tfrac{1}{9}\bar{r}_j^2, \text{ i.e., (the standard deviation of } r_j) \leqslant \tfrac{1}{3}\bar{r}_j$$

$$a_{Ij} \leqslant a_I$$

$$\text{var } a_{Ij} \leqslant \tfrac{1}{4}a_{Ij}^2 (\leqslant \tfrac{1}{4}a_I^2)$$

where $r$ and $a_I$ are maximum values for $r_j$ and $a_{Ij}$, respectively. We now get

$$\text{var } A_{IJ} = \sum_{j \in J} ((\bar{r}_j^2 + \text{var } r_j) \text{ var } r_j) \text{ var } a_{Ij} + \bar{a}_{Ij}^2 \text{ var } r_j)$$

$$\leqslant \sum_{j \in J} (\tfrac{10}{9}\bar{r}_j^2 \cdot \tfrac{1}{4}a_I^2 + a_I^2 \tfrac{1}{9}\bar{r}_j^2) = \tfrac{14}{36}a_I^2 \sum_{j \in J} \bar{r}_j^2$$

and

| for $a_I$ | = | 0.05 | 0.10 | 0.20 | 0.30 |
|---|---|---|---|---|---|
| var $a_{Ij}$ | $\leqslant$ | 0.00062 | 0.00250 | 0.01000 | 0.02250 |
| var $A_{IJ} \leqslant$ | for $r = 0.6$ | 0.00046 | 0.00180 | 0.00731 | 0.01645 |
| | $r = 0.5$ | 0.00042 | 0.00167 | 0.00669 | 0.01506 |
| | $r = 0.4$ | 0.00036 | 0.00144 | 0.00576 | 0.01295 |
| | $r = 0.3$ | 0.00030 | 0.00121 | 0.00482 | 0.01080 |

In all these cases we have established an upper bound for var $A_{IJ}$ that is below the upper bound for var $a_{ij}$, whereas the upper bound for $A_{IJ}$ is 1.1 $a_I$ (i.e., 1.1 times the upper bound for $a_{ij}$).

Obviously, greater stability in the coefficients of an aggregated table is no argument for the preference of an aggregated model to a more detailed one. A reasonable interpretation is that we should expect greater precision in the estimates of aggregate production levels—even when derived from a model with an aggregate input-output table—than in the estimates of detailed production levels, estimated from a model with a detailed input-output table. But we should expect to get even more accurate estimates of aggregate production levels when they are taken as sums of detailed production levels, estimated from a model with a detailed input-output table.

## The Data

### The Detailed Input-Output Accounts

Annual input-output accounts for Norway for the period 1949–1960, in fixed 1955 prices, are specified for 89 production sectors. Ten of these sectors have been excluded from the investigation of input-output relationships, either because they are only dummy sectors with no real counterparts in the productive activity of the economy, or because they have no inputs of raw materials, so that their value added is identically 100 percent of gross production in all years. The accounts are given at purchasers' market prices, but in order to eliminate a possibly unstable element gross trade margins have been deducted from the value of production of each sector in this investigation. Correspondingly, the inputs from the trade sector covering these margins have been eliminated from the input accounts. With deliveries from each of 83 production sectors and each of 60 import sectors to each of the 79 sectors of the investigation, and one item for value added for each of these 79 sectors, the number of possible items each year is 11,297. Actually, we had only about 1500 items, the rest being zeroes in all years.

Since the very small coefficients were considered to be of limited interest for the analysis, all input items which were less than 2 percent of gross production in a sector in all the years, and less than 1 percent in at least one year, were lumped together into one item for each sector, called "small unspecified." These inputs (75 altogether) are not analysed in the same detail as the others. The specified input items were classified into five main groups, namely:

Norwegian competitive (161 items)
Norwegian non-competitive (153 items)
Imports, competitive (137 items)
Imports, noncompetitive (26 items)
Gross value added (79 items).

For each sector the following sums were taken and treated as separate inputs in the analysis: the sum of each item of Norwegian competitive and the corresponding competitive imports (225 items); the sum of electricity and all fuel inputs (53 items);

the sum of the principal input and all inputs which could be expected to be relatively close substitutes for it (53 items); and the sum of all imported inputs (68 items).

The inputs were classified according to whether the receiving sector was an extractive or a service-producing industry (37 sectors), or a commodity-processing industry (42 sectors). The inputs were also classified by type into the following categories: direct materials (455 items), auxiliary materials (233 items), service inputs (79 items), and packaging materials (41 items).

## *The Aggregated Input-Output Accounts*

The aggregated accounts are based on the detailed accounts and cover the same series of annual data. The aggregations give a 14-sector and a 5-sector specification, and were designed for purposes other than the present investigation. (The figures have been published in *National Accounts Classified by Fourteen and Five Industrial Sectors 1949–1961, Vol. 1*, Central Bureau of Statistics of Norway, Oslo, 1965). Since coefficients had already been calculated on the basis of values at purchasers' prices, including trade margins, no deduction from the value of gross production was made in these series.

## Characteristics of the Ordinary Proportional Input-output Coefficient

Our computations give estimates of this coefficient in two alternative forms, namely

$$\frac{x_{ij}(t)}{x_j(t)} = a_{ij} + u_{ij}(t) \qquad (15\text{--}1)$$

$$\frac{x_{ij}(t)}{x_j(t)} = b_{ij} + d_{ij}t + v_{ij}(t) \qquad (15\text{--}2)$$

Here

$x_{ij}(t)$ = amount of item $i$ absorbed in sector $j$ in absolute, constant (1955) prices (kroner) in year $t$ (purchasers' prices);

$x_j(t)$ = total production in sector $j$ in absolute, constant (1955) prices (kroner) in year $t$ (producers' prices);

$t$ = year $t$, $t = 1$ for 1949, $t = 2$ for 1950, ..., $t = 12$ for 1960.

$a_{ij}$, $b_{ij}$, and $d_{ij}$ = constants estimated over the period.

$u_{ij}(t)$ and $v_{ij}(t)$ = residual terms, when the constants are determined so that

$$\sum_{t=1}^{12} u_{ij}(t) = \sum_{t=1}^{12} v_{ij}(t) = 0$$

$$\sum_{t=1}^{12} [u_{ij}(t)]^2 = \text{minimum}$$

$$\sum_{t=1}^{12} [v_{ij}(t)]^2 = \text{minimum}$$

We have $0 \leqslant a_{ij} < 1$ and $0 \leqslant b_{ij} < 1$ for all $i$ and $j$. The standard deviations of $u_{ij}(t)$, "standard deviation about the average coefficient," and of $v_{ij}(t)$, "standard deviation about the trend in the coefficients," are given in Table 15.1.[1]

The standard deviation about the average coefficient is of the order of 1 to 3 percent of production and the standard deviation about a linear trend through the period is roughly three quarters of the standard deviation about the average. Since the specified coefficients vary in average size from around 2 percent up to 100 percent (and in a special case even above 100) it will be important to ascertain whether the standard deviation varies with the size of the coefficient. This is done in Table 15.2. Table 15.2 indicates a clear correlation between the average size of the coefficients and their variance. The average standard deviation about the average coefficient varies from less than 1 percent, for coefficients of 2 percent or less, to 5 percent and more for the largest coefficients. The notable exception is the coefficient for gross value added. This coefficient is by definition equal to 1 minus the sum of all other coefficients, and its standard deviations might, as well, be expected to vary in inverse proportion to its size. This appears to be the case for the three upper size groups. In the lower size groups there are only two observations (one with high standard deviations, 0.052 and 0.056, respectively, and one with low, 0.014 and 0.009).

The frequency distributions of the standard deviations about the average coefficient for the 477 specified input items show that, particularly for the small coefficients, the distributions are quite concentrated, with a few cases of extreme variability. Table 15.3 illustrates how peaked these distributions are in comparison to the normal distribution.

If we compare the first four categories of specified input coefficients (Table 15.2), the variance for the noncompetitive items appears to be slightly less than for the competitive items, but it does not seem to reduce the variance appreciably to combine corresponding domestic and foreign competitive items.

Table 15.1.  Standard Deviations About Average Coefficients and Coefficient Trends for Main Categories of Inputs

| Type of inputs | Number of coefficients | Standard deviation | |
| --- | --- | --- | --- |
| | | About average | About trend |
| Norwegian competitive | 161 | 0.017 | 0.013 |
| Norwegian noncompetitive | 153 | 0.008 | 0.006 |
| Imports, competitive | 137 | 0.017 | 0.012 |
| Imports, noncompetitive | 26 | 0.014 | 0.011 |
| All specified inputs | 477 | 0.014 | 0.010 |
| Competitive inputs combined | 225 | 0.018 | 0.012 |
| Fuels combined | 53 | 0.006 | 0.005 |
| Substitution groups | 53 | 0.031 | 0.023 |
| Import sums | 68 | 0.025 | 0.020 |
| Small unspecified | 75 | 0.008 | 0.006 |
| Gross value added | 79 | 0.032 | 0.026 |

Table 15.2.   Standard Deviations About Average Coefficients and Coefficient Trends for Main Categories of Inputs, Classified by Average Size of the Coefficient

| | 0.02 and less | | | 0.02–0.05 | | | Average size of coefficient 0.05–0.10 | | | 0.10–0.25 | | | 0.25–0.50 | | | Over 0.50 | | |
|---|---|---|---|---|---|---|---|---|---|---|---|---|---|---|---|---|---|---|
| | Number | Average[a] | Trend[a] | Number | Average[a] | Trend[a] | Number | Average[a] | Trend[a] | Number | Average[a] | Trend[a] | Number | Average[a] | Trend[a] | Number | Average[a] | Trend[a] |
| Norwegian competitive | 55 | 0.007 | 0.006 | 53 | 0.013 | 0.009 | 23 | 0.016 | 0.011 | 17 | 0.032 | 0.025 | 8 | 0.064 | 0.054 | 5 | 0.050 | 0.034 |
| Norwegian noncompetitive | 64 | 0.005 | 0.004 | 55 | 0.006 | 0.004 | 23 | 0.015 | 0.011 | 9 | 0.018 | 0.015 | 1 | 0.023 | 0.025 | 1 | 0.029 | 0.031 |
| Imports, competitive | 54 | 0.009 | 0.007 | 39 | 0.011 | 0.009 | 23 | 0.020 | 0.015 | 14 | 0.032 | 0.024 | 5 | 0.054 | 0.036 | 2 | 0.084 | 0.050 |
| Imports, noncompetitive | 9 | 0.007 | 0.006 | 9 | 0.008 | 0.006 | 1 | 0.019 | 0.020 | 4 | 0.014 | 0.007 | 2 | 0.031 | 0.023 | 1 | 0.090 | 0.089 |
| All specified inputs | 182 | 0.007 | 0.006 | 156 | 0.010 | 0.007 | 70 | 0.017 | 0.012 | 44 | 0.028 | 0.021 | 16 | 0.054 | 0.043 | 9 | 0.060 | 0.044 |
| Competitive inputs combined | 57 | 0.008 | 0.006 | 77 | 0.011 | 0.008 | 36 | 0.015 | 0.011 | 32 | 0.027 | 0.018 | 14 | 0.060 | 0.040 | 9 | 0.043 | 0.033 |
| Fuels combined | 38 | 0.002 | 0.002 | 9 | 0.008 | 0.005 | 3 | 0.017 | 0.011 | 2 | 0.021 | 0.019 | — | — | — | 1 | 0.058 | 0.057 |
| Substitution groups | 3 | 0.002 | 0.002 | 4 | 0.008 | 0.004 | 4 | 0.010 | 0.006 | 12 | 0.023 | 0.017 | 14 | 0.041 | 0.023 | 16 | 0.045 | 0.039 |
| Import sums | 8 | 0.006 | 0.006 | 13 | 0.010 | 0.009 | 6 | 0.017 | 0.014 | 26 | 0.028 | 0.022 | 11 | 0.037 | 0.031 | 4 | 0.074 | 0.061 |
| Small unspecified | 22 | 0.004 | 0.003 | 30 | 0.009 | 0.006 | 21 | 0.009 | 0.008 | 1 | 0.010 | 0.008 | — | — | — | — | — | — |
| Gross value added | — | — | — | — | — | — | 2 | 0.033 | 0.033 | 6 | 0.058 | 0.048 | 27 | 0.042 | 0.036 | 44 | 0.022 | 0.016 |

[a] Standard deviation.

Table 15.3.   Characteristics of Size Distributions of Standard Deviations About the Average Coefficient for the 477 Specified Input Items

| Group average coefficient (percent) | Percentage of values of more than the average plus or minus | | |
| | 3 times | 2 times | 1 times |
| | the standard deviation of the distribution | | |
|---|---|---|---|
| 0– 2.00 | 2.2 | 3.3 | 5.0 |
| 2.01– 5.00 | 2.5 | 2.5 | 12.3 |
| 5.01–10.00 | 2.9 | 2.9 | 14.3 |
| 10.01–25.00 | 2.3 | 6.8 | 27.3 |
| 25.01 and over | 0.0 | 4.0 | 12.0 |
| The normal distribution | 0.3 | 4.6 | 31.7 |

The combination of all fuels (energy sources) into one item appears to reduce the variability slightly as compared to that of other items. If we compare the combined fuels items with the specified fuels items, we get what is shown in Table 15.4.

There appears to be a definite reduction in variability when the fuels items are combined instead of being treated as separate items.

If we group the coefficients according to the size of delivery in absolute value (kroner) we obtain the figures in Table 15.5. The standard deviation is still in general larger for big items than for small. However, the correlation with coefficient size is far less pronounced.

Table 15.6 gives a joint distribution by size of coefficient and by input size. There appears to be a slight tendency for the standard deviations to be somewhat smaller when the coefficient is calculated for an input which is large in absolute (kroner) value than when the input is small in absolute value. The important difference, however, is between the standard deviations for small and big coefficients.

In this study the 79 production sectors were divided into two groups: Group 0, extractive industries and service industries; and Group 1, commodity processing

Table 15.4.   Variance of Coefficients

| | Average size of coefficient | | | | | | | | | | | |
| | 0.02 and less | | | 0.02–0.05 | | | 0.05–0.10 | | | Greater than 0.50 | | |
| | | Standard deviation about | | | Standard deviation about | | | Standard deviation about | | | Standard deviation about | |
| | Number | Average | Trend | Number | Average | Trend | Number | Average | Trend | Number | Average | Trend |
|---|---|---|---|---|---|---|---|---|---|---|---|---|
| Fuels combined | 38 | 0.002 | 0.002 | 9 | 0.008 | 0.005 | 3 | 0.017 | 0.011 | 1 | 0.058 | 0.057 |
| Specified fuel items | 14 | 0.008 | 0.006 | 7 | 0.016 | 0.011 | 6 | 0.013 | 0.010 | 1 | 0.073 | 0.064 |

Table 15.5.  Standard Deviations About Average Coefficients and Coefficient Trends for Main Categories of Inputs, Classified by Average Size of the Input Item

| | Average size of input item in millions of 1955-kroner | | | | | | | | | | | | | | | | | | | | |
| | 0–10.0 | | | 10.1–50.0 | | | 50.1–100.0 | | | 100.1–250.0 | | | 250.1–500.0 | | | 500.1–1000.0 | | | 1000.1 and over | | |
| | Number | Average[a] | Trend[a] | Number | Average[a] | Trend[a] | Number | Average[a] | Trend[a] | Number | Average[a] | Trend[a] | Number | Average[a] | Trend[a] | Number | Average[a] | Trend[a] | Number | Average[a] | Trend[a] |
|---|---|---|---|---|---|---|---|---|---|---|---|---|---|---|---|---|---|---|---|---|---|
| Norwegian competitive | 85 | 0.011 | 0.009 | 43 | 0.018 | 0.013 | 13 | 0.026 | 0.020 | 10 | 0.048 | 0.033 | 9 | 0.018 | 0.013 | 1 | 0.044 | 0.032 | — | — | — |
| Norwegian non-competitive | 108 | 0.008 | 0.006 | 36 | 0.008 | 0.007 | 3 | 0.005 | 0.005 | 4 | 0.003 | 0.002 | 2 | 0.016 | 0.017 | — | — | — | — | — | — |
| Imports, competitive | 69 | 0.011 | 0.008 | 50 | 0.020 | 0.015 | 9 | 0.031 | 0.024 | 7 | 0.012 | 0.009 | 2 | 0.082 | 0.034 | — | — | — | — | — | — |
| Imports, non-competitive | 15 | 0.006 | 0.006 | 6 | 0.017 | 0.010 | 3 | 0.017 | 0.014 | 1 | 0.090 | 0.089 | — | — | — | — | — | — | 1 | 0.018 | 0.010 |
| All specified inputs | 277 | 0.010 | 0.008 | 135 | 0.016 | 0.012 | 28 | 0.024 | 0.019 | 22 | 0.030 | 0.023 | 13 | 0.028 | 0.017 | 1 | 0.044 | 0.032 | 1 | 0.018 | 0.010 |
| Competitive inputs combined | 93 | 0.011 | 0.009 | 75 | 0.018 | 0.012 | 24 | 0.023 | 0.014 | 19 | 0.034 | 0.024 | 14 | 0.026 | 0.017 | — | — | — | — | — | — |
| Fuels combined | 39 | 0.003 | 0.003 | 14 | 0.014 | 0.011 | — | — | — | — | — | — | — | — | — | — | — | — | — | — | — |
| Substitution groups | 3 | 0.053 | 0.045 | 14 | 0.031 | 0.021 | 9 | 0.030 | 0.022 | 9 | 0.035 | 0.022 | 12 | 0.031 | 0.025 | 5 | 0.017 | 0.013 | 1 | 0.013 | 0.014 |
| Import sums | 16 | 0.013 | 0.010 | 25 | 0.027 | 0.023 | 10 | 0.024 | 0.021 | 12 | 0.033 | 0.028 | 3 | 0.054 | 0.031 | 1 | 0.009 | 0.006 | 1 | 0.018 | 0.011 |
| Gross value added | 3 | 0.082 | 0.073 | 13 | 0.043 | 0.038 | 13 | 0.035 | 0.028 | 31 | 0.023 | 0.019 | 10 | 0.025 | 0.015 | 1 | 0.035 | 0.030 | 4 | 0.025 | 0.013 |

[a] Standard deviation.

Table 15.6.  Standard Deviations About Average Coefficients and Coefficient Trends for Main Categories of Inputs, Classified by Average Size of the Coefficients and Average Size of the Input Item (*in Kroner*)

| Coefficient, % Input, mill. kr. | 0–10.0 0–50 | | | Over 10.0 0–50 | | | 0–10.0 Over–50 | | | Over 10.0 Over 50 | | |
|---|---|---|---|---|---|---|---|---|---|---|---|---|
| | Number | Average[a] | Trend[a] | Number | Average[a] | Trend[a] | Number | Average[a] | Trend[a] | Number | Average[a] | Trend[a] |
| Norwegian competitive | 119 | 0.012 | 0.008 | 9 | 0.045 | 0.041 | 12 | 0.010 | 0.006 | 21 | 0.043 | 0.032 |
| Norwegian noncompetitive | 135 | 0.007 | 0.005 | 9 | 0.019 | 0.016 | 7 | 0.002 | 0.002 | 2 | 0.021 | 0.022 |
| Imports, competitive | 110 | 0.012 | 0.010 | 9 | 0.045 | 0.035 | 7 | 0.014 | 0.011 | 11 | 0.039 | 0.025 |
| Imports, noncompetitive | 18 | 0.008 | 0.007 | 3 | 0.017 | 0.008 | 1 | 0.002 | 0.002 | 4 | 0.040 | 0.035 |
| All specified items | 382 | 0.010 | 0.007 | 30 | 0.035 | 0.029 | 27 | 0.009 | 0.006 | 38 | 0.040 | 0.030 |
| Competitive inputs combined | 153 | 0.011 | 0.008 | 15 | 0.043 | 0.032 | 18 | 0.008 | 0.005 | 39 | 0.037 | 0.024 |
| Fuels combined | 50 | 0.004 | 0.003 | 3 | 0.033 | 0.031 | — | — | — | — | — | — |
| Substitution groups | 6 | 0.008 | 0.005 | 11 | 0.049 | 0.036 | 5 | 0.006 | 0.002 | 31 | 0.033 | 0.025 |
| Import sums | 26 | 0.012 | 0.011 | 15 | 0.038 | 0.030 | 2 | 0.009 | 0.006 | 25 | 0.032 | 0.026 |
| Gross value added | 2 | 0.033 | 0.033 | 14 | 0.053 | 0.046 | — | — | — | 63 | 0.027 | 0.021 |

[a] Standard deviation.

Table 15.7. Standard Deviations About Average Coefficients and About Coefficient Trends for Main Categories of Inputs, Classified by Average Size of Coefficient and Type of Sector

| | 0.02 and less | | | 0.02–0.05 | | | 0.05–0.10 | | | 0.10–0.25 | | | 0.25–0.50 | | | Over 0.50 | | |
|---|---|---|---|---|---|---|---|---|---|---|---|---|---|---|---|---|---|---|
| | Number | Average[a] | Trend[a] | Number | Average[a] | Trend[a] | Number | Average[a] | Trend[a] | Number | Average[a] | Trend[a] | Number | Average[a] | Trend[a] | Number | Average[a] | Trend[a] |
| **Extractive and service sectors** | | | | | | | | | | | | | | | | | | |
| Norwegian competitive | 13 | 0.004 | 0.003 | 6 | 0.012 | 0.008 | 4 | 0.008 | 0.004 | 2 | 0.021 | 0.007 | — | — | — | 1 | 0.033 | 0.029 |
| Norwegian noncompetitive | 25 | 0.003 | 0.003 | 40 | 0.006 | 0.004 | 16 | 0.012 | 0.007 | 8 | 0.018 | 0.014 | 1 | 0.023 | 0.025 | — | — | — |
| Imports, competitive | 9 | 0.008 | 0.005 | 4 | 0.007 | 0.005 | 3 | 0.019 | 0.011 | 2 | 0.032 | 0.020 | — | — | — | — | — | — |
| Imports, noncompetitive | 3 | 0.005 | 0.005 | 5 | 0.005 | 0.005 | 1 | 0.019 | 0.020 | 2 | 0.016 | 0.003 | 2 | 0.031 | 0.023 | — | — | — |
| All specified inputs | 50 | 0.004 | 0.003 | 55 | 0.006 | 0.004 | 24 | 0.012 | 0.008 | 14 | 0.020 | 0.012 | 3 | 0.028 | 0.024 | 1 | 0.033 | 0.029 |
| Competitive inputs combined | 9 | 0.007 | 0.005 | 12 | 0.011 | 0.007 | 7 | 0.012 | 0.006 | 6 | 0.021 | 0.014 | — | — | — | 1 | 0.031 | 0.029 |
| Fuels combined | 8 | 0.002 | 0.002 | 2 | 0.014 | 0.008 | 1 | 0.024 | 0.011 | — | — | — | — | — | — | — | — | — |
| Substitution groups | 3 | 0.002 | 0.002 | 2 | 0.006 | 0.002 | 2 | 0.009 | 0.004 | 4 | 0.012 | 0.011 | — | — | — | — | — | — |
| Import sums | 8 | 0.006 | 0.006 | 8 | 0.009 | 0.006 | 2 | 0.018 | 0.010 | 6 | 0.019 | 0.014 | 2 | 0.032 | 0.029 | — | — | — |
| Gross value added | — | — | — | — | — | — | — | — | — | — | — | — | 4 | 0.030 | 0.024 | 33 | 0.019 | 0.014 |
| **Commodity processing sectors** | | | | | | | | | | | | | | | | | | |
| Norwegian competitive | 42 | 0.008 | 0.007 | 47 | 0.013 | 0.009 | 19 | 0.017 | 0.013 | 15 | 0.034 | 0.028 | 8 | 0.064 | 0.054 | 4 | 0.055 | 0.036 |
| Norwegian noncompetitive | 39 | 0.006 | 0.005 | 15 | 0.008 | 0.007 | 7 | 0.021 | 0.019 | 1 | 0.020 | 0.022 | — | — | — | 1 | 0.029 | 0.031 |
| Imports, competitive | 45 | 0.009 | 0.007 | 35 | 0.011 | 0.009 | 20 | 0.021 | 0.016 | 12 | 0.032 | 0.025 | 5 | 0.054 | 0.036 | 2 | 0.084 | 0.050 |
| Imports, noncompetitive | 6 | 0.008 | 0.006 | 4 | 0.011 | 0.007 | — | — | — | 2 | 0.013 | 0.012 | — | — | — | 1 | 0.090 | 0.089 |
| All specified inputs | 132 | 0.008 | 0.006 | 101 | 0.011 | 0.009 | 46 | 0.019 | 0.015 | 31 | 0.030 | 0.025 | 13 | 0.060 | 0.047 | 8 | 0.063 | 0.045 |
| Competitive inputs combined | 48 | 0.008 | 0.006 | 65 | 0.011 | 0.008 | 29 | 0.016 | 0.012 | 26 | 0.028 | 0.019 | 14 | 0.060 | 0.040 | 8 | 0.045 | 0.033 |
| Fuels combined | 30 | 0.002 | 0.002 | 7 | 0.007 | 0.005 | 2 | 0.014 | 0.011 | 2 | 0.021 | 0.019 | — | — | — | 1 | 0.058 | 0.057 |
| Substitution groups | — | — | — | 2 | 0.011 | 0.006 | 2 | 0.012 | 0.008 | 8 | 0.029 | 0.020 | 13 | 0.043 | 0.025 | 16 | 0.045 | 0.039 |
| Import sums | — | — | — | 5 | 0.012 | 0.013 | 4 | 0.016 | 0.016 | 20 | 0.031 | 0.024 | 9 | 0.038 | 0.031 | 4 | 0.074 | 0.061 |
| Gross value added | — | — | — | — | — | — | 2 | 0.033 | 0.033 | 6 | 0.058 | 0.049 | 23 | 0.044 | 0.038 | 11 | 0.030 | 0.020 |

[a] Standard deviation.

industries. Table 15.7 gives the same information as Table 15.2 for each of the two groups separately. Surprisingly, the coefficients appear to be more stable in the extractive and service sectors than in the commodity processing sectors, over the period investigated.

## Trends in the Input-Output Coefficients

Tables 15.1–15.7 all indicate that the standard deviation about a trend in the coefficients is somewhat smaller than the standard deviation about the average coefficient for the whole period, and we shall now examine more closely the existence of trends in the coefficients. Trends can appear quite easily in series of only 12 items. Technical change may be expected to register as coefficient trends, since the change will usually take effect more quickly in some establishments than in others, and even within an establishment the switch to new techniques will often be gradual.

In our computations the trend effect in the coefficients was tested in the form

$$\frac{x_{ij}(t)}{x_j(t)} = b_{ij} + d_{ij}t + v_{ij}(t)$$

We then computed the standard deviation of $d_{ij}$, $\sigma_{ij}{}^2$ and $d_{ij}/\sigma_{ij}$. If $d_{ij}$ is normally distributed about "true" value of 0, with standard deviation $\sigma_{ij}$, then $d_{ij}/\sigma_{ij}$ will be distributed according to the $t$-distribution with 10 degrees of freedom.

The distributions for the specified input items are given in Table 15.8. The corresponding $t$-distributions are also shown. The deviations from the $t$-distributions are very marked and indicate the existence of negative as well as positive trends in the coefficients. For the $t$-distribution, 5 percent of the observations will deviate from zero by more than 2.2 times the standard deviation and 1 percent will deviate by at least 3.2 times the standard deviation, if the true value of the coefficient is zero (no trend). Accordingly, we have grouped the coefficients into the following groups (Tables 15.9 and 15.10):

1. Clear positive trend $\qquad\qquad \dfrac{d_{ij}}{\sigma_{ij}} > 3.2$

2. Moderate positive trend $\qquad 3.2 \geqslant \dfrac{d_{ij}}{\sigma_{ij}} > 2.2$

3. No trend $\qquad\qquad\qquad\quad \geqslant 2.2 \dfrac{d_{ij}}{\sigma_{ij}} \geqslant -2.2$

4. Moderate negative trend $\quad -2.2 > \dfrac{d_{ij}}{\sigma_{ij}} \geqslant -3.2$

5. Clear negative trend $\qquad -3.2 > \dfrac{d_{ij}}{\sigma_{ij}}$

For all coefficient groups, about 40 percent fall in the no trend category, 10 percent in each of the categories of moderate trends, and 20 percent in each of the extreme

Table 15.8.   Distribution of Ratios of Trend Coefficients to Standard Deviations[a]

| Ratio | Norwegian competitive | Norwegian noncompetitive | Imports, competitive | Imports noncompetitive | All specified inputs | Competitive inputs combined | Fuels combined | Substitution groups | Import sums | Gross value added |
|---|---|---|---|---|---|---|---|---|---|---|
| −16.01 to −17.00 | — | 1 | — | — | 1 | — | — | — | — | — |
| −15.01 to −16.00 | — | — | — | — | — | — | — | — | — | — |
| −14.01 to −15.00 | — | — | — | — | — | 1 | — | — | — | — |
| −13.01 to −14.00 | — | 1 | — | — | 1 | — | — | 1 | — | 1 |
| −12.01 to −13.00 | — | 1 | — | — | 1 | — | — | — | — | — |
| −11.01 to −12.00 | — | — | — | — | — | 1 | — | — | — | — |
| −10.01 to −11.00 | — | 2 | — | — | 2 | — | — | — | — | — |
| −9.01 to −10.00 | — | 3 | — | — | 3 | — | — | — | — | — |
| −8.01 to −9.00 | 2 | 3 | 2 | — | 7 | 1 | — | 1 | — | 2 |
| −7.01 to −8.00 | 2 | 4 | 1 | 1 | 8 | 2 | 1 | 1 | 1 | — |
| −6.01 to −7.00 | 3 | 7 | 4 | — | 15 | 3 | 4 | 2 | — | 6 |
| −5.01 to −6.00 | 6 | 2 | 1 | 1 | 9 | 6 | 1 | 3 | 2 | 1 |
| −4.01 to −5.00 | 15 | 6 | 11 | 4 | 36 | 9 | 3 | 1 | 3 | 2 |
| −3.01 to −4.00 | 13 (1) | 5 (1) | 6 (1) | 1 | 25 (3) | 16 | 3 | 5 | 2 | 6 (1) |
| −2.01 to −3.00 | 11 (5) | 8 (5) | 10 (4) | 1 | 30 (14) | 11 (2) | 4 (2) | 3 (2) | 5 (3) | 7 (2) |
| −1.01 to −2.00 | 16 (22) | 16 (20) | 6 (18) | 4 (4) | 42 (64) | 15 (6) | 8 (7) | 6 (7) | 5 (9) | 9 (10) |
| −0.01 to −1.00 | 12 (52) | 16 (51) | 13 (45) | 4 (9) | 45 (157) | 16 (30) | 8 (18) | 5 (17) | 7 (22) | 7 (26) |
| 0.00 to 0.99 | 17 (53) | 9 (52) | 18 (46) | 2 (9) | 46 (158) | 17 (73) | (17) | 6 (18) | 11 (22) | 6 (27) |
| 1.00 to 1.99 | 12 (22) | 14 (20) | 14 (18) | 3 (4) | 43 (64) | 18 (74) | 8 (18) | 3 (7) | 9 (9) | 7 (10) |
| 2.00 to 2.99 | 16 (5) | 15 (5) | 11 (4) | 3 | 45 (14) | 28 (30) | 8 (7) | 3 (2) | 4 (3) | 12 (2) |
| 3.00 to 3.99 | 12 (1) | 17 (1) | 13 (1) | 1 | 43 (3) | 20 (6) | 4 (2) | 3 | 8 | 4 (1) |
| 4.00 to 4.99 | 6 | 11 | 11 | — | 28 | 16 (2) | 2 | 2 | 3 | 2 |
| 5.00 to 5.99 | 6 | 8 | 6 | 1 | 21 | 7 | 4 | 1 | 6 | 2 |
| 6.00 to 6.99 | 6 | 1 | 4 | — | 11 | 12 | 3 | 1 | — | 5 |
| 7.00 to 7.99 | 2 | 1 | 3 | — | 6 | 11 | — | — | 1 | — |
| 8.00 to 8.99 | 1 | — | 2 | — | 4 | 3 | — | 1 | 1 | — |
| 9.00 to 9.99 | 2 | 1 | — | — | 1 | 5 | — | 1 | — | — |
| 10.00 to 10.99 | — | — | 1 | — | 4 | 2 | — | 2 | — | — |
| 11.00 to 11.99 | — | — | — | — | — | 2 | — | — | — | — |
| 12.00 to 12.99 | — | — | — | — | — | 1 | — | 2 | — | — |
| 13.00 to 13.99 | — | — | — | — | — | 1 | — | — | — | — |
| 14.00 to 14.99 | — | — | — | — | — | — | — | — | — | — |
| Total | 161 (161) | 153 (153) | 137 (137) | 26 (26) | 477 (477) | 225 (225) | 53 (53) | 53 (53) | 68 (68) | 79 (79) |

[a] Figures in parentheses are the figures that would be expected if the distributions were $t$-distributions with 10 degrees of freedom.

Table 15.9.  Trend Characteristics of Coefficients by Size Groups (*Number of Coefficients*)

| | Norwegian competitive | Norwegian noncompetitive | Imports competitive | Imports noncompetitive | All specified inputs | Competitive inputs combined | Fuels combined | Substitution groups | Import sums | Gross value added |
|---|---|---|---|---|---|---|---|---|---|---|
| **Coefficient size 0.02 and less** | | | | | | | | | | |
| Clear positive trend | 8 | 12 | 15 | — | 35 | 15 | 7 | — | 2 | — |
| Moderate positive trend | 6 | 10 | 4 | 1 | 21 | 5 | 2 | — | — | — |
| No trend | 26 | 28 | 26 | 5 | 85 | 27 | 19 | 3 | 6 | — |
| Moderate negative trend | 3 | 2 | 5 | — | 10 | 4 | 3 | — | — | — |
| Clear negative trend | 12 | 12 | 4 | 3 | 31 | 6 | 7 | — | — | — |
| Total | 55 | 64 | 54 | 9 | 182 | 57 | 38 | 3 | 8 | — |
| **Coefficient size 0.02–0.05** | | | | | | | | | | |
| Clear positive trend | 11 | 12 | 11 | 1 | 35 | 18 | 1 | 3 | 3 | — |
| Moderate positive trend | 5 | 7 | 2 | 2 | 16 | 5 | 1 | — | 2 | — |
| No trend | 19 | 20 | 18 | 5 | 62 | 29 | 3 | — | 7 | — |
| Moderate negative trend | 3 | 2 | — | — | 5 | 7 | — | — | 1 | — |
| Clear negative trend | 15 | 14 | 8 | 1 | 38 | 18 | 4 | 1 | — | — |
| Total | 53 | 55 | 39 | 9 | 156 | 77 | 9 | 4 | 13 | — |
| **Coefficient size 0.05 and over** | | | | | | | | | | |
| Clear positive trend | 12 | 7 | 12 | — | 31 | 24 | 1 | 9 | 11 | 11 |
| Moderate positive trend | 5 | 2 | 6 | — | 13 | 8 | — | 3 | 4 | 9 |
| No trend | 22 | 16 | 10 | 4 | 52 | 31 | 3 | 20 | 20 | 36 |
| Moderate negative trend | 2 | 2 | 7 | 1 | 12 | 5 | 1 | 3 | 5 | 7 |
| Clear negative trend | 12 | 7 | 9 | 3 | 31 | 23 | 1 | 11 | 7 | 16 |
| Total | 53 | 34 | 44 | 8 | 139 | 91 | 6 | 46 | 47 | 79 |
| **All coefficients** | | | | | | | | | | |
| Clear positive trend | 31 | 31 | 38 | 1 | 101 | 57 | 9 | 12 | 16 | 11 |
| Moderate positive trend | 16 | 19 | 12 | 3 | 50 | 18 | 3 | 3 | 6 | 9 |
| No trend | 67 | 64 | 54 | 14 | 199 | 87 | 25 | 23 | 33 | 36 |
| Moderate negative trend | 8 | 6 | 12 | 1 | 27 | 16 | 4 | 3 | 6 | 7 |
| Clear negative trend | 39 | 33 | 21 | 7 | 100 | 47 | 12 | 12 | 7 | 16 |
| Total | 161 | 153 | 137 | 26 | 477 | 225 | 53 | 53 | 68 | 79 |

groups. As might be expected, the no trend category is slightly greater for the smallest coefficients than for the other size groups.

The notable exceptions are the various import coefficients. The larger competitive import coefficients had relatively fewer cases of no trend and smaller competitive import coefficients fewer cases of negative trends than other input categories, with a corresponding overrepresentation of positive trends for all size groups. Noncompetitive import coefficients had relatively more cases of no trend and negative trends than usual, and relatively fewer cases of positive trends, for all

Table 15.10.   Trend Characteristics of Coefficients (*Percentage Distribution*)

| | Norwegian competitive | Norwegian noncompetitive | Imports competitive | Imports noncompetitive | All specified inputs | Competitive inputs combined | Fuels combined | Substitution groups | Import sums | Gross value added |
|---|---|---|---|---|---|---|---|---|---|---|
| **Coefficient size 0.02 and less** | | | | | | | | | | |
| Clear positive trend | 15 | 19 | 28 | — | 19 | 26 | 18 | — | 25 | — |
| Moderate positive trend | 11 | 15 | 8 | 11 | 12 | 9 | 6 | — | — | — |
| No trend | 47 | 44 | 48 | 56 | 47 | 47 | 50 | 100 | 75 | — |
| Moderate negative trend | 5 | 3 | 9 | — | 5 | 7 | 8 | — | — | — |
| Clear negative trend | 22 | 19 | 7 | 33 | 17 | 11 | 18 | — | — | — |
| Total | 100 | 100 | 100 | 100 | 100 | 100 | 100 | 100 | 100 | — |
| Number of coefficients | 55 | 64 | 54 | 9 | 182 | 57 | 38 | 3 | 8 | — |
| **Coefficient size 0.02–0.05** | | | | | | | | | | |
| Clear positive trend | 21 | 22 | 28 | 11 | 23 | 23 | 11 | 75 | 23 | — |
| Moderate positive trend | 9 | 13 | 5 | 22 | 10 | 7 | 11 | — | 15 | — |
| No trend | 36 | 36 | 46 | 56 | 40 | 38 | 33 | — | 54 | — |
| Moderate negative trend | 6 | 4 | — | — | 3 | 9 | — | — | 8 | — |
| Clear negative trend | 28 | 25 | 21 | 11 | 24 | 23 | 45 | 25 | — | — |
| Total | 100 | 100 | 100 | 100 | 100 | 100 | 100 | 100 | 100 | — |
| Number of coefficients | 53 | 55 | 39 | 9 | 156 | 77 | 9 | 4 | 13 | — |
| **Coefficient size 0.05 and over** | | | | | | | | | | |
| Clear positive trend | 23 | 21 | 27 | — | 22 | 26 | 17 | 19 | 23 | 14 |
| Moderate positive trend | 9 | 6 | 14 | — | 9 | 9 | — | 7 | 9 | 11 |
| No trend | 41 | 47 | 23 | 50 | 38 | 34 | 50 | 43 | 42 | 46 |
| Moderate negative trend | 4 | 6 | 16 | 12 | 9 | 6 | 16 | 7 | 11 | 9 |
| Clear negative trend | 23 | 20 | 20 | 38 | 22 | 25 | 17 | 24 | 15 | 20 |
| Total | 100 | 100 | 100 | 100 | 100 | 100 | 100 | 100 | 100 | 100 |
| Number of coefficients | 53 | 34 | 44 | 8 | 139 | 91 | 6 | 46 | 47 | 79 |
| **All coefficients** | | | | | | | | | | |
| Clear positive trend | 19 | 20 | 28 | 4 | 21 | 25 | 17 | 23 | 23 | 14 |
| Moderate positive trend | 10 | 12 | 9 | 11 | 10 | 8 | 6 | 6 | 9 | 11 |
| No trend | 42 | 42 | 39 | 54 | 42 | 39 | 47 | 43 | 49 | 46 |
| Moderate negative trend | 5 | 4 | 9 | 4 | 6 | 7 | 7 | 6 | 9 | 9 |
| Clear negative trend | 24 | 22 | 15 | 27 | 21 | 21 | 23 | 22 | 10 | 20 |
| Total | 100 | 100 | 100 | 100 | 100 | 100 | 100 | 100 | 100 | 100 |
| Number of coefficients | 161 | 153 | 137 | 26 | 477 | 225 | 53 | 53 | 68 | 79 |

size groups. The sum coefficients for imports had relatively more positive and less negative trends than normal.

The standard deviation about the average coefficient is generally greater for the coefficients with a moderate trend than for those with no trend and greater for those with a clear trend than for those with no trend. But when the standard

Table 15.11. Standard Deviations About Average Coefficients and About Trends for Input Coefficients, Classified by Trend Characteristics

| | Average size of coefficient | | | | | | | | |
| | 0.02 and less | | | 0.02–0.05 | | | 0.05–0.10 | | |
| | Number | Average[a] | Trend[a] | Number | Average[a] | Trend[a] | Number | Average[a] | Trend[a] |
|---|---|---|---|---|---|---|---|---|---|
| **All specified inputs** | | | | | | | | | |
| No trend | 85 | 0.006 | — | 62 | 0.008 | — | 21 | 0.015 | — |
| Moderate trend | 31 | 0.008 | 0.006 | 21 | 0.009 | 0.007 | 14 | 0.016 | 0.013 |
| Clear trend | 66 | 0.007 | 0.004 | 73 | 0.012 | 0.007 | 35 | 0.019 | 0.011 |
| **Competitive inputs combined** | | | | | | | | | |
| No trend | 27 | 0.007 | — | 29 | 0.008 | — | 14 | 0.010 | — |
| Moderate trend | 9 | 0.010 | 0.007 | 12 | 0.012 | 0.010 | 6 | 0.011 | 0.009 |
| Clear trend | 21 | 0.008 | 0.005 | 36 | 0.013 | 0.009 | 16 | 0.022 | 0.011 |
| **Gross value added** | | | | | | | | | |
| No trend | — | — | — | — | — | — | 1 | 0.052 | — |
| Moderate trend | — | — | — | — | — | — | — | — | — |
| Clear trend | — | — | — | — | — | — | 1 | 0.014 | 0.009 |

[a] Standard deviation.

| | Average size of coefficient | | | | | | | | |
| | 0.10–0.25 | | | 0.25–0.50 | | | Greater than 0.50 | | |
| | Number | Average[a] | Trend[a] | Number | Average[a] | Trend[a] | Number | Average[a] | Trend[a] |
|---|---|---|---|---|---|---|---|---|---|
| **All specified inputs** | | | | | | | | | |
| No trend | 21 | 0.018 | — | 6 | 0.049 | — | 4 | 0.045 | — |
| Moderate trend | 8 | 0.026 | 0.020 | 2 | 0.042 | 0.036 | 1 | 0.073 | 0.064 |
| Clear trend | 15 | 0.041 | 0.025 | 8 | 0.061 | 0.038 | 4 | 0.072 | 0.038 |
| **Competitive inputs combined** | | | | | | | | | |
| No trend | 9 | 0.019 | — | 5 | 0.045 | — | 3 | 0.036 | — |
| Moderate trend | 5 | 0.017 | 0.014 | — | — | — | 2 | 0.035 | 0.028 |
| Clear trend | 18 | 0.034 | 0.019 | 9 | 0.067 | 0.037 | 4 | 0.053 | 0.033 |
| **Gross value added** | | | | | | | | | |
| No trend | 2 | 0.061 | — | 15 | 0.036 | — | 17 | 0.016 | — |
| Moderate trend | 2 | 0.069 | 0.055 | 6 | 0.058 | 0.050 | 8 | 0.022 | 0.017 |
| Clear trend | 1 | 0.029 | 0.011 | 6 | 0.042 | 0.025 | 19 | 0.027 | 0.014 |

[a] Standard deviation.

deviation is taken about the trend value for coefficients with trend, the size of the standard deviation does not appear to increase as we move from no trend coefficients to moderate trend and clear trend coefficients (Table 15.11).

A comparison of manufacturing sectors with primary and service sectors with respect to trend characteristics of input coefficients does not indicate strong systematic differences. There is some indication of a greater proportion of no trend coefficients for inputs in manufacturing (Tables 15.12 and 15.13). Also when we compare the trend characteristics of the various types of input coefficients, there do not seem to be systematic differences between direct materials, auxiliary materials, service inputs, and packaging materials (Tables 15.14 and 15.15).

We shall finally consider the slopes of the trend lines (Tables 15.16 and 15.17). We have grouped the coefficients with clear or moderate negative or positive trends according to the numerical size of the estimated annual change in trend value. The coefficients classified as having no trend have been kept apart. We find that trends which change a coefficient as much as 1 percentage point[3] per year (i.e., by 10 percentage points or more over a period of 10 years) are quite rare. In the entire 79 sector input-output matrix, with potentially more than 11,000 coefficients, and with 1500 registered nonzero coefficients, only 6 intermediate input coefficients, 11 import coefficients, and 11 gross value-added coefficients showed changes in trend values of 1 percentage point or more per annum. For changes of $\frac{1}{2}$ or more precentage points per annum, the figures are 32 intermediate input coefficients, 25 import coefficients, and 27 gross value-added coefficients. (The latter figure is of course more than $\frac{1}{3}$ of all the value-added coefficients and not negligible.) Considering that 87 of the intermediate input coefficients, 52 of the import coefficients, and all the 79 gross value-added coefficients were in size groups with average standard deviations about the trend value of more than 1 percentage point, trends in the coefficients do not appear to be a major source of instability in input-output coefficients over moderate time intervals. This conclusion is confirmed by the relatively moderate reduction in average standard deviation for the coefficients, when it is taken about the trend value instead of about the arithmetic average.

From the data for Table 15.2 we obtain the figures for Table 15.18.

## Stability of Coefficients in Detailed and Aggregated Tables

We have previously demonstrated the plausibility of having the variance of the coefficients decrease with progressive aggregation in the data. Comparing the standard deviations in Tables 15.2 and 15.19 makes it possible to confront this hypothesis with the data. We find that, as we move from the 79-sector table to the 14-sector table, there is a drastic reduction in the standard deviations about the average coefficient, and for most of the coefficient classes there are further reductions as we move on from 14 to 5 sectors. Since all tables are derived from the same set of data, the stability in the aggregate tables must be due to substantial stability in the shares of individual detailed sectors' contributions to the aggregate sectors. Also since most coefficients are quite small, each detailed sector is normally a small fraction of each aggregate sector.

Table 15.12. Trend Characteristics of Coefficients by Size Groups and Type of Sector (*Number of Coefficients*)

| | Norwegian competitive | Norwegian noncompetitive | Imports competitive | Imports noncompetitive | All specified inputs | Competitive inputs combined | Fuels combined | Substitution groups | Import sums | Gross value added |
|---|---|---|---|---|---|---|---|---|---|---|
| **Coefficient size 0.02 and less** | | | | | | | | | | |
| *Extractive and service sectors* | | | | | | | | | | |
| Clear positive trend | 2 | 4 | 4 | — | 10 | 3 | 1 | — | 2 | — |
| Moderate positive trend | — | 6 | 1 | — | 7 | 1 | — | — | — | — |
| No trend | 7 | 11 | 4 | 3 | 25 | 5 | 4 | 3 | 6 | — |
| Moderate negative trend | 1 | — | — | — | 1 | — | 1 | — | — | — |
| Clear negative trend | 3 | 4 | — | — | 7 | — | 2 | — | — | — |
| Total | 13 | 25 | 9 | 3 | 50 | 9 | 8 | 3 | 8 | — |
| *Commodity processing sectors* | | | | | | | | | | |
| Clear positive trend | 6 | 8 | 11 | — | 25 | 12 | 6 | — | — | — |
| Moderate positive trend | 6 | 4 | 3 | 1 | 14 | 4 | 2 | — | — | — |
| No trend | 19 | 17 | 22 | 2 | 60 | 22 | 15 | — | — | — |
| Moderate negative trend | 2 | 2 | 5 | — | 9 | 4 | 2 | — | — | — |
| Clear negative trend | 9 | 8 | 4 | 3 | 24 | 6 | 5 | — | — | — |
| Total | 42 | 39 | 45 | 6 | 132 | 48 | 30 | — | — | — |
| **Coefficient size 0.02–0.05** | | | | | | | | | | |
| *Extractive and service sectors* | | | | | | | | | | |
| Clear positive trend | 3 | 7 | — | — | 10 | 5 | — | 2 | 3 | — |
| Moderate positive trend | 1 | 6 | — | 2 | 9 | — | 1 | — | 1 | — |
| No trend | 1 | 12 | 2 | 3 | 18 | 5 | — | — | 4 | — |
| Moderate negative trend | — | 1 | — | — | 1 | 1 | — | — | — | — |
| Clear negative trend | 1 | 14 | 2 | — | 17 | 1 | 1 | — | — | — |
| Total | 6 | 40 | 4 | 5 | 55 | 12 | 2 | 2 | 8 | — |
| *Commodity processing sectors* | | | | | | | | | | |
| Clear positive trend | 8 | 5 | 11 | 1 | 22 | 13 | 1 | 1 | — | — |
| Moderate positive trend | 4 | 1 | 2 | — | 6 | 5 | — | — | 1 | — |
| No trend | 18 | 8 | 16 | 2 | 34 | 24 | 3 | — | 3 | — |
| Moderate negative trend | 3 | 1 | — | — | 1 | 6 | — | — | 1 | — |
| Clear negative trend | 14 | — | 6 | 1 | 10 | 17 | 3 | 1 | — | — |
| Total | 47 | 15 | 35 | 4 | 73 | 65 | 7 | 2 | 5 | — |

| | Norwegian competitive | Norwegian noncompetitive | Imports competitive | Imports noncompetitive | All specified inputs | Competitive inputs combined | Fuels combined | Substitution groups | Import sums | Gross value added |
|---|---|---|---|---|---|---|---|---|---|---|
| **Coefficient size 0.05 and over** | | | | | | | | | | |
| | | | *Extractive and service sectors* | | | | | | | |
| Clear positive trend | 4 | 6 | 3 | — | 13 | 5 | — | 2 | 2 | 5 |
| Moderate positive trend | 1 | — | — | — | 1 | 3 | — | 1 | 1 | 5 |
| No trend | 1 | 12 | — | 2 | 15 | 2 | — | 4 | 2 | 14 |
| Moderate negative trend | — | — | 1 | 1 | 2 | 1 | — | — | 1 | 3 |
| Clear negative trend | 1 | 7 | 1 | 2 | 11 | 3 | 1 | — | 4 | 10 |
| Total | 7 | 25 | 5 | 5 | 42 | 14 | 1 | 7 | 10 | 37 |
| | | | *Commodity processing sectors* | | | | | | | |
| Clear positive trend | 8 | 1 | 9 | — | 11 | 19 | 1 | 7 | 9 | 6 |
| Moderate positive trend | 4 | 2 | 6 | — | 8 | 5 | — | 2 | 3 | 4 |
| No trend | 21 | 4 | 10 | 2 | 21 | 29 | 3 | 16 | 18 | 22 |
| Moderate negative trend | 2 | 2 | 6 | — | 9 | 4 | 1 | 3 | 4 | 4 |
| Clear negative trend | 11 | — | 8 | 1 | 14 | 20 | — | 11 | 3 | 6 |
| Total | 46 | 9 | 39 | 3 | 63 | 77 | 5 | 39 | 37 | 42 |
| **All coefficients** | | | | | | | | | | |
| | | | *Extractive and service sectors* | | | | | | | |
| Clear positive trend | 9 | 17 | 7 | — | 33 | 13 | 1 | 4 | 7 | 5 |
| Moderate positive trend | 2 | 12 | 1 | 2 | 17 | 4 | 1 | 1 | 2 | 5 |
| No trend | 9 | 35 | 6 | 8 | 58 | 12 | 4 | 7 | 12 | 14 |
| Moderate negative trend | 1 | 1 | 1 | 1 | 4 | 2 | 1 | — | 1 | 3 |
| Clear negative trend | 5 | 25 | 3 | 2 | 35 | 4 | 4 | — | 4 | 10 |
| Total | 26 | 90 | 18 | 13 | 147 | 35 | 11 | 12 | 26 | 37 |
| | | | *Commodity processing sectors* | | | | | | | |
| Clear positive trend | 22 | 14 | 31 | 1 | 58 | 44 | 8 | 8 | 9 | 6 |
| Moderate positive trend | 14 | 7 | 11 | 1 | 28 | 14 | 2 | 2 | 4 | 4 |
| No trend | 58 | 29 | 48 | 6 | 115 | 75 | 21 | 16 | 21 | 22 |
| Moderate negative trend | 7 | 5 | 11 | — | 19 | 14 | 3 | 3 | 5 | 4 |
| Clear negative trend | 34 | 8 | 18 | 5 | 48 | 43 | 8 | 12 | 3 | 6 |
| Total | 135 | 63 | 119 | 13 | 268 | 190 | 42 | 41 | 42 | 42 |

Table 15.13.  Trend Characteristics of Coefficients by Size Groups and Type of Sector (*Percentage Distributions*)

| | Norwegian competitive | Norwegian noncompetitive | Imports competitive | Imports noncompetitive | All specified inputs | Competitive inputs combined | Fuels combined | Substitution groups | Import sums | Gross value added |
|---|---|---|---|---|---|---|---|---|---|---|
| **Coefficient size 0.02 and less** | | | | | | | | | | |
| *Extractive and service sectors* | | | | | | | | | | |
| Clear positive trend | 15 | 16 | 45 | — | 20 | 33 | 13 | — | 25 | — |
| Moderate positive trend | — | 24 | 10 | — | 14 | 11 | — | — | — | — |
| No trend | 54 | 44 | 45 | 100 | 50 | 56 | 50 | 100 | 75 | — |
| Moderate negative trend | 8 | — | — | — | 2 | — | 12 | — | — | — |
| Clear negative trend | 23 | 16 | — | — | 14 | — | 25 | — | — | — |
| Total | 100 | 100 | 100 | 100 | 100 | 100 | 100 | 100 | 100 | |
| Number of coefficients | 13 | 25 | 9 | 3 | 50 | 9 | 8 | 3 | 8 | — |
| *Commodity processing sectors* | | | | | | | | | | |
| Clear positive trend | 14 | 20 | 24 | — | 19 | 25 | 20 | — | — | — |
| Moderate positive trend | 14 | 10 | 7 | 17 | 11 | 8 | 7 | — | — | — |
| No trend | 45 | 44 | 49 | 33 | 45 | 46 | 50 | — | — | — |
| Moderate negative trend | 5 | 5 | 11 | — | 7 | 8 | 7 | — | — | — |
| Clear negative trend | 22 | 21 | 9 | 50 | 18 | 13 | 16 | — | — | — |
| Total | 100 | 100 | 100 | 100 | 100 | 100 | 100 | — | — | — |
| Number of coefficients | 42 | 39 | 45 | 6 | 132 | 48 | 30 | — | — | — |
| **Coefficient size 0.02–0.05** | | | | | | | | | | |
| *Extractive and service sectors* | | | | | | | | | | |
| Clear positive trend | 50 | 17 | — | — | 18 | 42 | — | 100 | 37 | — |
| Moderate positive trend | 17 | 15 | — | 40 | 16 | — | 50 | — | 13 | — |
| No trend | 17 | 30 | 50 | 60 | 33 | 42 | — | — | 50 | — |
| Moderate negative trend | — | 3 | — | — | 2 | 8 | — | — | — | — |
| Clear negative trend | 16 | 35 | 50 | — | 31 | 8 | 50 | — | — | — |
| Total | 100 | 100 | 100 | 100 | 100 | 100 | 100 | 100 | 100 | — |
| Number of coefficients | 6 | 40 | 4 | 5 | 55 | 12 | 2 | 2 | 8 | — |
| *Commodity processing sectors* | | | | | | | | | | |
| Clear positive trend | 17 | 33 | 31 | 25 | 30 | 20 | 14 | 50 | — | — |
| Moderate positive trend | 9 | 7 | 6 | — | 8 | 8 | — | — | 20 | — |
| No trend | 38 | 53 | 46 | 50 | 47 | 37 | 43 | — | 60 | — |
| Moderate negative trend | 6 | 7 | — | — | 1 | 9 | — | — | 20 | — |
| Clear negative trend | 30 | — | 17 | 25 | 14 | 26 | 43 | 50 | — | — |
| Total | 100 | 100 | 100 | 100 | 100 | 100 | 100 | 100 | 100 | — |
| Number of coefficients | 47 | 15 | 35 | 4 | 73 | 65 | 7 | 2 | 5 | — |

|  | Norwegian competitive | Norwegian noncompetitive | Imports competitive | Imports noncompetitive | All specified inputs | Competitive inputs combined | Fuels combined | Substitution groups | Import sums | Gross value added |
|---|---|---|---|---|---|---|---|---|---|---|
| **Coefficient size 0.05 and over** | | | | | | | | | | |
| *Extractive and service sectors* | | | | | | | | | | |
| Clear positive trend | 58 | 24 | 60 | — | 31 | 36 | — | 39 | 20 | 14 |
| Moderate positive trend | 14 | — | — | — | 2 | 21 | — | 14 | 10 | 13 |
| No trend | 14 | 48 | — | 40 | 36 | 15 | — | 57 | 20 | 38 |
| Moderate negative trend | — | — | 20 | 20 | 5 | 7 | — | — | 10 | 8 |
| Clear negative trend | 14 | 28 | 20 | 40 | 26 | 21 | 100 | — | 40 | 27 |
| Total | 100 | 100 | 100 | 100 | 100 | 100 | 100 | 100 | 100 | 100 |
| Number of coefficients | 7 | 25 | 5 | 5 | 42 | 14 | 1 | 7 | 10 | 37 |
| *Commodity processing sectors* | | | | | | | | | | |
| Clear positive trend | 17 | 11 | 23 | — | 18 | 25 | 20 | 18 | 24 | 14 |
| Moderate positive trend | 9 | 22 | 15 | — | 13 | 6 | — | 5 | 8 | 10 |
| No trend | 46 | 45 | 26 | 67 | 33 | 38 | 60 | 41 | 49 | 52 |
| Moderate negative trend | 4 | 22 | 15 | — | 14 | 5 | 20 | 8 | 11 | 10 |
| Clear negative trend | 24 | — | 21 | 33 | 22 | 26 | — | 28 | 8 | 14 |
| Total | 100 | 100 | 100 | 100 | 100 | 100 | 100 | 100 | 100 | 100 |
| Number of coefficients | 46 | 9 | 39 | 3 | 63 | 77 | 5 | 39 | 37 | 42 |
| **All coefficients** | | | | | | | | | | |
| *Extractive and service sectors* | | | | | | | | | | |
| Clear positive trend | 35 | 19 | 39 | — | 22 | 37 | 9 | 33 | 27 | 14 |
| Moderate positive trend | 7 | 13 | 6 | 15 | 12 | 11 | 9 | 8 | 8 | 13 |
| No trend | 35 | 39 | 33 | 62 | 39 | 34 | 37 | 59 | 46 | 38 |
| Moderate negative trend | 4 | 1 | 6 | 8 | 3 | 6 | 9 | — | 4 | 8 |
| Clear negative trend | 19 | 28 | 16 | 15 | 24 | 12 | 36 | — | 15 | 27 |
| Total | 100 | 100 | 100 | 100 | 100 | 100 | 100 | 100 | 100 | 100 |
| Number of coefficients | 26 | 90 | 18 | 13 | 147 | 35 | 11 | 12 | 26 | 37 |
| *Commodity processing sectors* | | | | | | | | | | |
| Clear positive trend | 16 | 22 | 26 | 8 | 22 | 23 | 19 | 20 | 21 | 14 |
| Moderate positive trend | 11 | 11 | 9 | 8 | 10 | 7 | 5 | 5 | 10 | 10 |
| No trend | 43 | 46 | 41 | 46 | 43 | 40 | 50 | 39 | 50 | 52 |
| Moderate negative trend | 5 | 8 | 9 | — | 7 | 7 | 7 | 7 | 12 | 10 |
| Clear negative trend | 25 | 13 | 15 | 38 | 18 | 23 | 19 | 29 | 7 | 14 |
| Total | 100 | 100 | 100 | 100 | 100 | 100 | 100 | 100 | 100 | 100 |
| Number of coefficients | 135 | 63 | 119 | 13 | 268 | 190 | 42 | 41 | 42 | 42 |

Table 15.14.   Trend Characteristics of Coefficients by Input Types (*Number of Coefficients*)

| | Norwegian competitive | Norwegian noncompetitive | Imports competitive | Imports noncompetitive | All specified inputs | Competitive inputs combined | Fuels combined | Substitution groups |
|---|---|---|---|---|---|---|---|---|
| **Direct materials** | | | | | | | | |
| Clear positive trend | 26 | 6 | 26 | — | 58 | 35 | — | 10 |
| Moderate positive trend | 13 | 3 | 12 | — | 28 | 17 | — | 3 |
| No trend | 52 | 8 | 34 | 5 | 99 | 58 | — | 18 |
| Moderate negative trend | 7 | 2 | 10 | — | 19 | 11 | — | 3 |
| Clear negative trend | 30 | 1 | 16 | 3 | 50 | 34 | — | 12 |
| Total | 128 | 20 | 98 | 8 | 254 | 155 | — | 46 |
| **Auxiliary materials** | | | | | | | | |
| Clear positive trend | 5 | 12 | 12 | 1 | 30 | 20 | 9 | 2 |
| Moderate positive trend | 1 | 7 | — | 1 | 9 | 1 | 3 | — |
| No trend | 7 | 23 | 20 | 3 | 53 | 22 | 25 | 2 |
| Moderate negative trend | 1 | 1 | 2 | — | 4 | 3 | 4 | — |
| Clear negative trend | 5 | 14 | 5 | 3 | 27 | 7 | 12 | — |
| Total | 19 | 57 | 39 | 8 | 123 | 53 | 53 | 4 |
| **Service inputs** | | | | | | | | |
| Clear positive trend | — | 12 | — | — | 12 | — | — | — |
| Moderate positive trend | — | 9 | — | 2 | 11 | — | — | — |
| No trend | — | 28 | — | 6 | 34 | — | — | — |
| Moderate negative trend | — | 1 | — | 1 | 2 | — | — | — |
| Clear negative trend | — | 16 | — | 1 | 17 | — | — | — |
| Total | — | 66 | — | 10 | 76 | — | — | — |
| **Packaging materials** | | | | | | | | |
| Clear positive trend | — | 1 | — | — | 1 | 2 | — | — |
| Moderate positive trend | 2 | — | — | — | 2 | — | — | — |
| No trend | 8 | 5 | — | — | 13 | 7 | — | — |
| Moderate negative trend | — | 2 | — | — | 2 | 2 | — | — |
| Clear negative trend | 4 | 2 | — | — | 6 | 6 | — | — |
| Total | 14 | 10 | — | — | 24 | 17 | — | — |

Table 15.15.  Trend Characteristics of Coefficients by Input Types (*Percentage Distribution*)

| | Norwegian competitive | Norwegian noncompetitive | Imports competitive | Imports noncompetitive | All specified inputs | Competitive inputs combined | Fuels combined | Substitution groups |
|---|---|---|---|---|---|---|---|---|
| **Direct materials** | | | | | | | | |
| Clear positive trend | 20 | 30 | 27 | — | 23 | 23 | — | 22 |
| Moderate positive trend | 10 | 15 | 12 | — | 11 | 11 | — | 7 |
| No trend | 41 | 40 | 35 | 63 | 39 | 37 | — | 39 |
| Moderate negative trend | 6 | 10 | 10 | — | 7 | 7 | — | 6 |
| Clear negative trend | 23 | 5 | 16 | 37 | 20 | 22 | — | 26 |
| Total | 100 | 100 | 100 | 100 | 100 | 100 | — | 100 |
| Number of coefficients | 128 | 20 | 98 | 8 | 254 | 155 | — | 46 |
| **Auxiliary materials** | | | | | | | | |
| Clear positive trend | 27 | 21 | 31 | 13 | 25 | 38 | 17 | 50 |
| Moderate positive trend | 5 | 12 | — | 13 | 7 | 2 | 6 | — |
| No trend | 37 | 40 | 51 | 37 | 43 | 41 | 47 | 50 |
| Moderate negative trend | 5 | 2 | 5 | — | 3 | 6 | 7 | — |
| Clear negative trend | 26 | 25 | 13 | 37 | 22 | 13 | 23 | — |
| Total | 100 | 100 | 100 | 100 | 100 | 100 | 100 | 100 |
| Number of coefficients | 19 | 57 | 39 | 8 | 123 | 53 | 53 | 4 |
| **Service inputs** | | | | | | | | |
| Clear positive trend | — | 18 | — | — | 16 | — | — | — |
| Moderate positive trend | — | 14 | — | 20 | 14 | — | — | — |
| No trend | — | 42 | — | 60 | 45 | — | — | 100 |
| Moderate negative trend | — | 2 | — | 10 | 3 | — | — | — |
| Clear negative trend | — | 24 | — | 10 | 22 | — | — | — |
| Total | — | 100 | — | 100 | 100 | — | — | 100 |
| Number of coefficients | — | 66 | — | 10 | 76 | — | — | 3 |
| **Packaging materials** | | | | | | | | |
| Clear positive trend | — | 10 | — | — | 4 | 12 | — | — |
| Moderate positive trend | 14 | — | — | — | 9 | — | — | — |
| No trend | 57 | 50 | — | — | 54 | 41 | — | — |
| Moderate negative trend | — | 20 | — | — | 8 | 12 | — | — |
| Clear negative trend | 29 | 20 | — | — | 25 | 35 | — | — |
| Total | 100 | 100 | — | — | 100 | 100 | — | — |
| Number of coefficients | 14 | 10 | — | — | 24 | 17 | — | — |

Table 15.16.   Size Distribution of Trend Coefficients (*Number of Coefficients*)

| | No trend | \multicolumn{6}{c}{Annual change in trend value[a]} | | | | | |
| --- | --- | --- | --- | --- | --- | --- | --- |
| | | 0–0.49 | 0.50–0.99 | 1.00–1.99 | 2.00–2.99 | 3.00–3.99 | Total |
| Norwegian competitive | 67 | 69 | 19 | 4 | 1 | 1 | 161 |
| Norwegian noncompetitive | 65 | 81 | 7 | — | — | — | 153 |
| Imports, competitive | 54 | 61 | 11 | 10 | 1 | — | 137 |
| Imports, noncompetitive | 14 | 9 | 3 | — | — | — | 26 |
| All specified inputs | 200 | 220 | 40 | 14 | 2 | 1 | 477 |
| Competitive inputs combined | 88 | 97 | 21 | 16 | 2 | 1 | 225 |
| Fuels combined | 25 | 25 | 3 | — | — | — | 53 |
| Substitution groups | 23 | 12 | 8 | 9 | 1 | — | 53 |
| Import sums | 34 | 15 | 12 | 4 | 3 | — | 68 |
| Gross value added | 36 | 16 | 16 | 9 | 2 | — | 79 |

[a] Percent of output in receiving sector.

Table 15.17.   Size Distribution of Trend Coefficient (*Percentage Distribution*)

| | No trend | \multicolumn{6}{c}{Annual change in trend value[a]} | | | | | |
| --- | --- | --- | --- | --- | --- | --- | --- |
| | | 0–0.49 | 0.50–0.99 | 1.00–1.99 | 2.00–2.99 | 3.00–3.99 | Total |
| Norwegian competitive | 41.7 | 42.8 | 11.8 | 2.5 | 0.6 | 0.6 | 100.0 |
| Norwegian noncompetitive | 42.5 | 52.9 | 4.6 | — | — | — | 100.0 |
| Imports, competitive | 39.5 | 44.5 | 8.0 | 7.3 | 0.7 | — | 100.0 |
| Imports, noncompetitive | 53.8 | 34.6 | 11.6 | — | — | — | 100.0 |
| All specified inputs | 41.9 | 46.2 | 8.4 | 2.9 | 0.4 | 0.2 | 100.0 |
| Competitive inputs combined | 39.1 | 43.1 | 9.3 | 7.1 | 0.9 | 0.5 | 100.0 |
| Fuels combined | 47.2 | 47.2 | 5.6 | — | — | — | 100.0 |
| Substitution groups | 43.4 | 22.6 | 15.1 | 17.0 | 1.9 | — | 100.0 |
| Import sums | 50.0 | 22.1 | 17.6 | 5.9 | 4.4 | — | 100.0 |
| Gross value added | 45.6 | 20.3 | 20.2 | 11.4 | 2.5 | — | 100.0 |

[a] Percent of output in receiving sector.

Table 15.18.   Average Standard Deviations About Trend as Percents of Average Standard Deviation About the Mean[a]

|  | Average size of coefficients | | | | | |
|---|---|---|---|---|---|---|
|  | 0.02 and less | 0.02–0.05 | 0.05–0.10 | 0.10–0.25 | 0.25–0.50 | Over 0.50 |
| Norwegian competitive | 78 | 68 | 69 | 74 | 82 | (65) |
| Norwegian noncompetitive | 75 | 70 | 70 | 80 | (104) | (103) |
| Imports, competitive | 72 | 75 | 70 | 73 | 64 | (57) |
| Imports, noncompetitive | 80 | 70 | (101) | (48) | (71) | (95) |
| All specified inputs | 72 | 71 | 70 | 73 | 76 | 70 |
| Competitive inputs, combined | 75 | 67 | 66 | 64 | 64 | 72 |
| Fuels, combined | 85 | 62 | (59) | (86) | — | (94) |
| Substitution groups | (96) | (43) | (53) | 70 | 55 | 83 |
| Import sums | 92 | 84 | 81 | 74 | 80 | (79) |
| Small unspecified | 82 | 63 | 80 | (77) | — | — |
| Gross value added | — | — | (94) | 80 | 82 | 69 |

[a] Here both the standard deviation about the trend and the standard deviation about the mean have been adjusted for degrees of freedom. Figures based on five observations or less have been put in parentheses.

Table 15.19.   Standard Deviations About Average Coefficients for Main Categories of Inputs, Classified by Average Size of Coefficients for 14- and 5-Sector Aggregations

| | Average size of coefficient | | | | | |
| --- | --- | --- | --- | --- | --- | --- |
| | 0.02 and less | | 0.02–0.05 | | 0.05–0.10 | |
| | Number | Standard deviation | Number | Standard deviation | Number | Standard deviation |
| **A. 14-sector aggregation** | | | | | | |
| Norwegian competitive | 12 | 0.003 | 10 | 0.005 | 1 | 0.003 |
| Norwegian noncompetitive | 7 | 0.002 | 9 | 0.004 | 3 | 0.004 |
| Imports, competitive | 2 | 0.002 | 12 | 0.006 | 5 | 0.010 |
| Imports, noncompetitive | — | — | 1 | 0.008 | — | — |
| All specified inputs | 21 | 0.002 | 32 | 0.005 | 9 | 0.007 |
| Import sums | 2 | 0.003 | — | — | 2 | 0.008 |
| Gross value added | — | — | — | — | — | — |
| **B. 5-sector aggregation** | | | | | | |
| Norwegian competitive | 2 | 0.002 | 2 | 0.004 | 3 | 0.004 |
| Norwegian noncompetitive | 1 | 0.002 | 3 | 0.002 | — | — |
| Imports, competitive | — | — | 4 | 0.003 | 1 | 0.003 |
| Imports, noncompetitive | — | — | — | — | — | — |
| All specified inputs | 3 | 0.002 | 9 | 0.003 | 4 | 0.004 |
| Import sums | — | — | 1 | 0.003 | — | — |
| Gross value added | — | — | — | — | — | — |

| | 0.10–0.25 | | 0.25–0.50 | | Greater than 0.50 | |
| --- | --- | --- | --- | --- | --- | --- |
| **A. 14-sector aggregation** | | | | | | |
| Norwegian competitive | 5 | 0.014 | — | — | — | — |
| Norwegian noncompetitive | 7 | 0.009 | — | — | — | — |
| Imports, competitive | 3 | 0.014 | 1 | 0.031 | — | — |
| Imports, noncompetitive | — | — | 1 | 0.014 | — | — |
| All specified inputs | 15 | 0.012 | 2 | 0.022 | — | — |
| Import sums | 6 | 0.013 | 3 | 0.018 | — | — |
| Gross value added | — | — | 5 | 0.014 | 9 | 0.016 |
| **B. 5-sector aggregation** | | | | | | |
| Norwegian competitive | 2 | 0.005 | 1 | 0.006 | — | — |
| Norwegian noncompetitive | — | — | — | — | — | — |
| Imports, competitive | 3 | 0.012 | — | — | — | — |
| Imports, noncompetitive | — | — | 1 | 0.014 | — | — |
| All specified inputs | 5 | 0.009 | 2 | 0.010 | — | — |
| Import sums | 3 | 0.013 | 1 | 0.014 | — | — |
| Gross value added | — | — | 2 | 0.011 | 3 | 0.015 |

**NOTES**

1. In comparing the standard deviation about the average and about the trend, note that the estimates of standard deviation about the trend have been adjusted for degrees of freedom, whereas no such adjustment has been made in the estimates of standard deviations about the average. The adjustment would increase the estimated standard deviations about the average by 4.45 percent.
2. With $12 - 2 = 10$ degrees of freedom.
3. Percent of output in receiving sector.

# 16

## Changes Over Time in Input-Output Coefficients for the United States

BEATRICE N. VACCARA

The construction of an input-output table for a complex, roundabout economy is of necessity a difficult and time consuming process. As a consequence, it is not unusual for an input-output table to appear in print about five years after the year it describes. For example, the United States table for the year 1947, prepared by the Bureau of Labor Statistics, was published in a summary form in May 1952 (Evans and Hoffenberg, 1952). The 1958 United States table was published in November 1964 (Goldman et al., 1964), and it is anticipated that the 1963 table will be published by the fall of 1969. Although increasing reliance upon computers may shorten this process, it is doubtful, given the resources likely to be devoted to the construction of input-output tables in the United States, that production can be completed in less than four years.[1] In addition, the fact that the United States program for input-output work calls for the complete reconstruction of input-output tables only every 5 years—to correspond with the quinquennial economic censuses—makes it likely that input-output analysis of current economic problems rest on basic input relationships that are almost ten years old. This in itself is not necessarily a cause for alarm. Much depends upon the degree of stability, over time, in the basic technical relationships measured by the input-output coefficient tables.

In an industrial economy as large and long established as that of the United States, it is not unreasonable to assume that changes in the technical relationships for an *entire* industry occur slowly and orderly. Of course, changes in production processes *do* occur. New products are introduced; new materials are substituted for old; new technologies alter the production process; and changes in relative prices induce long-run substitution of one basic material for another. But these changes do not affect the total capacity of an industry at once—existing capacity in good working order is rarely scrapped because a newer, different, or more efficient production process has been introduced. Rather, it is the newly-created capacity that may utilize newer processes. Thus, while the newer industrial process for producing steel ingots, which involves the basic oxygen furnace, was perfected in

This article was published in *Applications of Input-Output Analysis*, edited by Anne P. Carter and Andrew Bródy, © North-Holland Publishing Company, 1970, pp. 238–260, Volume 2. Reprinted by permission.

1954, as of the end of 1966, 12 years later, only 25 percent of the steel making capacity in the United States relied upon this new process. (American Iron and Steel Institute, 1967.) Although changes that affect the basic input structure of an industry may occur slowly, they nevertheless do occur and cannot be ignored. It is, therefore, imperative to measure the impact on input-output coefficients of those changes that have occurred in the past, and even more important, to try to examine some of the factors contributing to these changes. For only through an understanding and separation of the *many* factors that contribute to changes in technical relationships, reflected in various input-output tables, can one hope to project these changes—either to actual, but more current years, or to future years—and thus base analysis on a table of input-output relationships that is more representative of the time period at hand. (U.S. Department of Labor, 1966, 1967.)

### Causes of Coefficient Change

It is important to remember that changes in input coefficients, as reflected in an input-output table, can be caused by many factors, only one of which may be technological change—change in the physical requirements for the specific goods and services used in producing a given basket of goods. Perhaps one of the most important factors that could cause changes over time in the input coefficients for a given industry is changing product mix. In a highly aggregative picture of the U.S. economy, such as the 86-order classification system of the 1958 input-output table, an individual industry cannot represent a single, or even a homogeneous, set of commodities. Thus, shifts over time in the product composition of an individual industry could cause associated shifts in input coefficients for that industry.

Divergence of actual technical relationships from a linear homogeneous function, and the strictly proportional relationship between changes in inputs and outputs that it assumes, could cause differences in input relationships between two time periods. For some inputs, particularly those that reflect overhead costs, this proportionality assumption is undoubtedly unrealistic. Thus, coefficients for a given year might differ from those of another year merely because the scale of operation or degree of capacity utilization was much greater in one year than the other. The problem of nonproportionality of some, if not most, inputs could be particularly important in the 1947, 1958, and 1961 comparisons in this paper since we are comparing technical relationships in time periods that reflect different phases of the business cycle. The year 1947 was one of overall expansion of the business economy, while the years 1958 and 1961 were years of business cycle troughs.

It should also be noted that the particular conventions relating to secondary production and competitive imports that are adopted for the construction of the basic input-output table can have considerable influence on the stability over time of individual input coefficients. In the United States tables, this subsidiary production is generally handled via the "transfer" approach, a procedure that builds into the table a set of base year relationships that are completely unrelated to actual production requirements, and which may thus vary quite randomly over time.[2]

The level of aggregation of the input-output tables can also affect the stability over time of the various input coefficients. In general, one would expect a relatively aggregated table to show more coefficient instability than a less aggregated table because the higher the level of aggregation, the more severe the probem of changing product mix. On the other hand, in some cases, aggregation may contribute to coefficient stability, since it can cancel out the impact of substitutions among related materials. Thus, for example, the 1958 table, which combines all nonferrous metals into one producing sector,[3] masks the shift over time between copper and aluminium. Finally, it should be pointed out that some of the differences in technical coefficients, and far too many I suspect, may reflect random factors, such as differences between the various input-output tables in the data sources and statistical methods for estimating the technical relationships.

We have noted several important factors that could contribute to changes over time in the technical input-output coefficients, factors that are in no way related to changes in the technological requirements for producing a fixed basket of goods. Rather, they may be related to changes in the size of the basket, the mix of commodities in the basket, the relative importance of the various domestic and foreign producers of this basket of goods, or the methods used to measure and value the commodities in the basket and/or the ingredients used to make these commodities. A complete analysis of coefficient change should attempt to factor out these various causes and to relate them to the underlying forces of changes in consumer tastes, changes in relative prices, and technological innovation.

In recognition of the importance of a careful analysis of changes over time in the United States input-output relationships, the Federal Government has undertaken a long-range, continuing study of such changes as revealed by the United States tables for 1947, 1958, a 1961 updated version of the 1958 table, and the forthcoming 1963 table. This work is going on both in the Office of Business Economics of the United States Department of Commerce and at the Harvard Economic Research Project (under a grant from the Interagency Growth Study). (Carter, 1967; Vaccara and Simon, 1968.) Unfortunately, this task has turned out to be more difficult and complex than anticipated and, as a result, this paper is much more of a progress report on the work of the Office of Business Economics than it is a definitive accounting and explanation of coefficient changes in the United States. While this paper provides various measures of the degree and direction of coefficient change over time, it does not "factor out" any of the causes of coefficient change.

A good deal of the difficulty in connection with our effort to compare input-output coefficients over time arises from the fact that the 1947 input-output table, as originally prepared by the Bureau of Labor Statistics, was not statistically and conceptually consistent with the national accounts, as was the 1958 table, and the subsequent input-output table prepared by the Office of Business Economics. It was thus necessary, first, to complete the complex task of reworking the 1947 table so that it was consistent with the 1958 table in terms of concepts, statistics, and aggregation level. Furthermore, before the analysis of changes in input coefficients over time, it was necessary to express the various flow tables in a single set of base year prices, those of 1958.[4]

### Deflation of Input-Output Tables

The task of deflating an input-output table is a complex and difficult one. It is not simply a process of developing a set of industry gross output deflators, which are used uniformly across the row. In many cases, it requires the development of price deflators appropriate to each cell of the table. In an input-output system as highly aggregative as that of the 1958 table, there is considerable room for variation in the mix of products that a given industry sells to its various customers. It therefore becomes important to utilize unpublished, detailed information to develop a price deflator for each cell, which represents, as far as possible, the particular mix of the given cell.

Moreover, these cell-by-cell deflators must be consistent with the deflators that have been developed elsewhere in the national accounts. When one sums the current and constant dollar values for the various cells of the final demand columns, he derives implicit deflators for gross national product, and its major components, personal consumption expenditures, fixed business investment, exports, etc. These must be consistent with deflators used for the same aggregates in the summary income and product accounts. Furthermore, when one aggregates the current and constant dollar values for the various intermediate columns, these imply residual value-added deflators that, in aggregate, must correspond to the implicit GNP deflator, developed on the product side. Since these two sets of deflators are often from diverse statistical sources, serious problems of reconciliation can result.

### Changes in Total Output Requirements as Measured by Inverse Coefficients

Although our work on coefficient analysis has not yet progressed to the point where we can factor out the causes of these changes in each of the industries, we have learned a great deal about the extent, direction, and degree of variability of the coefficient changes. A convenient way of summarizing the impact on production of the many plus-and-minus changes in individual input coefficients that occurred during the eleven-year period 1947–1958 is to compare the outputs required from each industry to produce a fixed set of final demands with the 1947 as compared to the 1958 input-output inverse table. By doing this, we learn that, on the average (ignoring signs), there was a 16 percent difference in the output required from each industry to produce the 1958 bill of goods with the 1958 technology, as compared to the 1947 technology. Although the *average* impact on an individual industry's production requirement was 16 percent, for a few industries, such as plastics and synthetic materials and machine shop products, these differences in output

Table 16.1. Change Over Time in Intermediate Requirements to Deliver 1958 Final Demand

| Input-output industry | Derived 1958 intermediate output[b] (millions of 1958 dollars) | | | Percent change | | Average annual rate of change[c] | |
|---|---|---|---|---|---|---|---|
| | 1947 Inverse | 1958 Inverse | 1961 Inverse | 1958 tech. 1947 tech. Col. (2) ÷ col. (1) | 1961 tech. 1958 tech. Col. (3) ÷ col. (2) | 1947–58 | 1958–61 |
| | (1) | (2) | (3) | (4) | (5) | (6) | (7) |
| 1 | 26,340 | 23,564 | 22,817 | −10.5 | −3.2 | −1.0 | −1.1 |
| 2 | 20,752 | 17,624 | 17,359 | −15.1 | −1.5 | −1.5 | −0.5 |
| 3 | 1,133 | 1,257 | 1,237 | +10.9 | −1.6 | +0.9 | −0.5 |
| 4 | 2,220 | 1,564 | 1,555 | −29.5 | −0.6 | −3.1 | +0.2 |
| 5 | 1,206 | 1,227 | 1,204 | +1.7 | −1.9 | +0.2 | −0.6 |
| 6 | 1,324 | 1,156 | 1,041 | −12.7 | −9.9 | −1.2 | −3.5 |
| 7 | 4,235 | 2,121 | 1,924 | −49.9 | −9.3 | +6.1 | −3.2 |
| 8 | 11,313 | 10,865 | 10,636 | −4.0 | −2.1 | −0.4 | −0.7 |
| 9 | 1,258 | 1,583 | 1,665 | +25.8 | +5.2 | +2.1 | +1.7 |
| 10 | 408 | 485 | 490 | +18.9 | +1.0 | +1.6 | +0.3 |
| 11 | 0 | 0 | 0 | — | — | — | — |
| 12 | 18,492 | 12,455 | 11,667 | −32.6 | −6.3 | −3.6 | −2.2 |
| 14 | 15,809 | 17,597 | 17,214 | +11.3 | −2.2 | +1.0 | −0.7 |
| 15 | 1,745 | 1,284 | 1,164 | −26.4 | −9.3 | −2.7 | −3.2 |
| 16, 17, 19 | 11,171 | 12,768 | 12,809 | +14.3 | +0.3 | +1.2 | +0.1 |
| 18 | 2,289 | 2,982 | 2,855 | +30.3 | −4.3 | +2.4 | −1.5 |
| 20 | 9,565 | 8,117 | 8,201 | −15.1 | +1.0 | −1.5 | +0.3 |
| 21 | 834 | 455 | 438 | −45.4 | −3.7 | −5.4 | −1.3 |
| 22 | 731 | 666 | 592 | −8.9 | −11.1 | −0.8 | −3.8 |
| 23 | 490 | 396 | 353 | −19.2 | −10.9 | −2.0 | −3.8 |
| 24, 25 | 12,478 | 12,908 | 12,970 | +3.4 | +0.5 | +0.3 | +0.2 |
| 26 | 10,334 | 9,840 | 9,753 | −4.8 | −0.9 | −0.4 | −0.3 |
| 27 | 7,457 | 10,260 | 10,910 | +37.6 | +6.3 | +2.9 | +2.1 |
| 28 | 2,087 | 3,961 | 4,369 | +89.8 | +10.3 | +6.0 | +3.3 |
| 29 | 1,509 | 2,234 | 2,318 | +48.0 | +3.8 | +3.6 | +1.3 |
| 30 | 2,124 | 1,832 | 1,829 | −13.7 | −0.2 | −1.4 | −0.1 |
| 31 | 9,223 | 9,162 | 9,151 | −0.7 | −0.1 | −0.1 | 0 |
| 32 | 4,364 | 5,143 | 5,917 | +17.9 | +15.0 | +1.5 | +4.8 |
| 33 | 1,085 | 892 | 842 | −17.8 | −5.6 | −1.8 | −1.9 |
| 34 | 377 | 415 | 394 | +10.1 | −5.1 | +0.9 | −1.7 |
| 35 | 2,144 | 2,011 | 2,131 | −6.2 | +6.0 | −0.6 | +2.0 |
| 36 | 6,338 | 7,259 | 7,404 | +14.5 | +2.0 | +1.2 | +0.7 |
| 37 | 26,091 | 18,882 | 18,808 | −27.6 | −0.4 | −2.9 | −0.1 |
| 38 | 10,989 | 9,459 | 9,628 | −13.9 | +1.8 | −1.4 | +0.6 |
| 39 | 1,892 | 2,029 | 2,046 | +7.2 | +0.8 | +0.6 | +0.3 |
| 40 | 7,174 | 7,119 | 7,041 | −0.8 | −1.1 | −0.1 | −0.4 |
| 41 | 4,253 | 3,406 | 3,143 | −19.9 | −7.7 | −2.0 | −2.6 |
| 42 | 5,182 | 5,637 | 5,531 | +8.8 | −1.9 | +0.8 | −0.6 |
| 43 | 1,177 | 1,111 | 953 | −5.6 | −14.2 | −0.5 | −4.9 |
| 44 | 579 | 699 | 704 | +20.7 | +0.7 | +1.7 | +0.2 |
| 45 | 1,216 | 1,029 | 948 | −15.4 | −7.9 | −1.5 | −2.7 |
| 46 | 631 | 506 | 555 | −19.8 | +9.7 | −2.0 | +3.1 |
| 47 | 2,077 | 2,106 | 1,934 | +1.4 | −8.2 | +0.1 | −2.8 |
| 48 | 876 | 731 | 770 | −16.6 | +5.3 | −1.6 | +1.7 |
| 49 | 2,546 | 2,305 | 2,491 | −9.5 | +8.1 | −0.9 | +2.6 |

Table 16.1 (*continued*).

| Input-output industry | Derived 1958 intermediate output[b] (millions of 1958 dollars) | | | Percent change | | Average annual rate of change[c] | |
|---|---|---|---|---|---|---|---|
| | 1947 Inverse | 1958 Inverse | 1961 Inverse | 1958 tech. 1947 tech. Col. (2) ÷ col. (1) | 1961 tech. 1958 tech. Col. (3) ÷ col. (2) | 1947–58 | 1958–61 |
| | (1) | (2) | (3) | (4) | (5) | (6) | (7) |
| 50 | 756 | 1,523 | 1,734 | +101.5 | +13.9 | +6.6 | +4.4 |
| 51 | 198 | 900 | 1,111 | +354.5 | +23.4 | +14.8 | +7.3 |
| 52 | 860 | 866 | 916 | +0.7 | +5.8 | +0.1 | +1.9 |
| 53 | 2,474 | 3,193 | 3,301 | +29.1 | +3.4 | +2.4 | +1.1 |
| 54 | 545 | 965 | 1,032 | +77.1 | +6.9 | +5.3 | +2.3 |
| 55 | 1,929 | 1,899 | 1,934 | −1.6 | +1.8 | −0.1 | +0.6 |
| 56, 57 | 2,008 | 4,287 | 4,609 | +113.5 | +7.5 | +7.1 | +2.4 |
| 58 | 1,214 | 1,041 | 1,098 | −14.3 | +5.5 | −1.4 | +1.8 |
| 59 | 9,410 | 9,560 | 9,782 | +1.6 | +2.3 | +0.1 | +0.8 |
| 60 (incl. 13) | 6,241 | 7,706 | 6,969 | +23.5 | −9.6 | +1.9 | −3.3 |
| 61 | 1,089 | 962 | 974 | −11.7 | +1.2 | −1.1 | +0.4 |
| 62 | 1,354 | 1,920 | 2,055 | +41.8 | +7.0 | +3.2 | +2.3 |
| 63 | 555 | 740 | 741 | +33.3 | +0.1 | +2.6 | 0 |
| 64 | 2,038 | 2,375 | 2,577 | +16.5 | +8.5 | +1.4 | +2.8 |
| 65 | 26,775 | 19,809 | 20,695 | −26.0 | +4.5 | −2.7 | +1.5 |
| 66 | 2,852 | 4,598 | 4,669 | +61.2 | +1.5 | +4.4 | +0.5 |
| 67 | 767 | 1,540 | 1,535 | +100.8 | −0.3 | +6.5 | −0.1 |
| 68 | 8,002 | 11,360 | 12,376 | +42.0 | +8.9 | +3.2 | +2.9 |
| 69 | 23,014 | 26,991 | 27,662 | +17.3 | +2.5 | +1.5 | +0.8 |
| 70 | 15,171 | 14,390 | 13,561 | −5.1 | −5.8 | −0.5 | −2.0 |
| 71 | 18,266 | 20,162 | 20,756 | +10.4 | +2.9 | +0.9 | +1.0 |
| 72 | 3,277 | 2,383 | 2,258 | −27.3 | −5.2 | −2.8 | −1.8 |
| 73 | 15,775 | 21,257 | 21,576 | +34.8 | +1.5 | +2.8 | +0.5 |
| 74 | — | — | — | — | — | — | — |
| 75 | 4,487 | 3,314 | 3,407 | −26.1 | +2.8 | −2.7 | +0.9 |
| 76 | 1,246 | 2,104 | 2,008 | +68.9 | −4.6 | +4.9 | −1.6 |
| 77 | 2,494 | 1,823 | 1,940 | −26.9 | +6.4 | −2.8 | +2.1 |
| 81 | 7,590 | 6,611 | 6,680 | −12.9 | +1.0 | −1.3 | +0.3 |
| Total[e] | 415,905 | 413,381 | 415,717 | 2,094.6 | 353.8 | 160.6 | 177.6 |
| Average | | | | 29.5 | 5.0 | 2.3 | 1.7 |
| Total[f] | 286,189 | 277,039 | 276,594 | 16,345.9 | 305.9 | 123.6 | 101.6 |
| Average | | | | 28.2 | 5.3 | 2.1 | 1.8 |

[a] In general, the numbering scheme compares to that of the published 1958 table. In a few cases, however, because of extensive revisions between 1947 and 1958 in the Standard Industrial Classification (SIC), it was necessary to combine industries to insure comparability in the two years. In addition, some unresolved problems of conceptual comparability between the 1947 and 1958 tables necessitated the omission of a few industries. For the name and SIC industry composition of each 1958 input-output industry see Table 16.4.

[b] Columns 1 through 3 were derived by deducting the 1958 final demand for each industry from the output results obtained by multiplying the 1958 final demands by the 1947, 1958, and 1961 inverse matrices.

[c] Computed by the compound interest formula.

[d] Industry 74 (Research and development) has been omitted from this table because of conceptual differences in intermediate output in 1947 and 1958 which could not be resolved.

[e] Based on 71 industries. In deriving totals and averages for columns 4 through 7, signs were ignored.

requirements exceeded 75 percent. On the other hand, for many important industries, such as paper and petroleum products, the change in output requirements was less than 5 percent. Because of the procedures employed—the application of the same set of final demands to different inverse matrices—it is generally to be expected that industries that sell primarily to final demand will show smaller percentage changes over time in total output requirements than industries that sell primarily to other producing industries. The most extreme case is industry 11 (new construction), which sells its entire output directly to final demand. As is to be expected, the output requirements for this industry are the same whether one uses the 1947, 1958 or 1961 inverse matrices.

### Changes in Intermediate Output Requirements as Measured by Inverse Coefficients

Perhaps a more meaningful way to summarize the impact on production of changes over time in the various input coefficients, is to concentrate on changes in *intermediate* rather than in total output requirements. Table 16.1 presents just such an analysis. For the purposes of computing this table we deducted the final demands from the outputs derived by the use of the various inverse matrices. An examination of Table 16.1 reveals an average change between 1947 and 1958 in intermediate output requirements of almost 30 percent, in contrast to the overall average change of 16 percent in total output requirements.

As might be expected, change over the three-year period 1958–1961 in the intermediate output required to produce the 1958 bill of goods was considerably smaller than over the eleven-year period 1947–1958. Moreover, even after one makes an adjustment for the difference in the length of the two time periods, by computing the average percentage change per year rather than the total for the entire period, some slowing down in the rate of change in output requirements is noted. During the period 1947–1958, the average annual rate of change in intermediate output requirements (ignoring signs) was 2.3 percent. Over the 1958–1961 period, the difference in intermediate output requirements for a fixed bill of goods averaged 1.7 percent per year. Forty-four of the 71 industry comparisons indicated a slower annual rate of change in the 1958 to 1961 period. The slowing down in the pace of coefficient change was particularly marked for large changes. Thus, for example, to supply the 1958 set of final demands with the 1958 input-output table required 90 percent (or 6 percent per year) more output from the plastics and synthetic materials industry than with the 1947 input-output table. To supply these demands with the 1961, rather than the 1958 table, required only 3.3 percent more output per year.

In general, there was a systematic relationship between the *directions* of change in the industry intermediate output requirements for the two time periods examined, 1947 to 1958 and 1958 to 1961. Of the 36 industries that showed positive increases in output requirements between 1947 and 1958, 26 showed positive

---

*f* Based on 58 industries (industries 1 through 64). In deriving totals and averages for columns 4 through 7, signs were ignored.

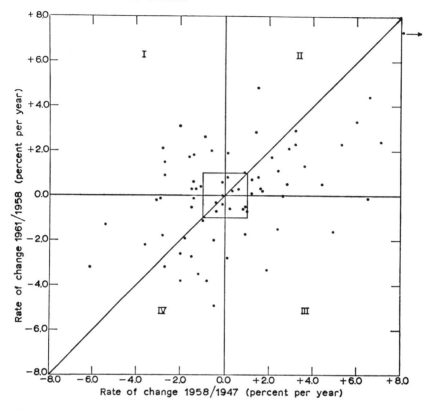

Fig. 16.1. Average annual rate of change in intermediate output requirements to deliver 1958 final demand.

increase between 1958 and 1961; 22 of the 35 industries with negative changes between 1947 and 1958 also showed negative changes between 1958 and 1961. Moreover, the majority of the 23 cases with changes in the opposite direction in the two time periods occurred in industries where the changes in output requirements were small in both time periods.

The relationship between the average annual rates of change in industry outputs in the two time periods can be observed by examination of Figure 16.1. There is considerable variation in the observed annual rates of change for the 71 industries. Large negative, as well as large positive, changes are evident: 58 observations fall outside of the range of what might be considered insignificant change, that is, 1.0 percent or less per year, in both of the time periods examined. In the 1947 to 1958 period, 49, and in the 1958 to 1961 period, 40, of the 71 industries showed average annual rates of change of over 1.0 percent. On the other hand, very few industries—only 4—showed average annual rates of change of over 3 percent in both time periods. These were: the coal mining industry, with negative changes; and the plastics and synthetic materials, machine shop products, and office and computing machines industries, with positive changes.

All the observations falling in the 2nd and 4th quadrants indicate the same direction of movement from 1947 to 1958 and 1958 to 1961; those falling in the 1st and 3rd quadrants indicate movements in the opposite direction. In quadrant 2, the observations that fall below the diagonal line have the same direction of change in the two time periods but smaller change in the later time period. The same is true, in quadrant 4, of observations that fall above the diagonal line.

The above analysis, based on the *combined* direct and indirect effects of coefficient changes, is a convenient way of summarizing the impact on a given industry's production of the various changes in input coefficients for *all* industries. It may not be a true guide to the degree or direction of change in the direct requirements for the output of the *given* industry. In some instances, the direct and indirect impacts on the output of a given industry may offset each other and thus give an impression of little coefficient change over time. Tables 16.1 and 16.2 show that this is actually the case in the iron ore industry. The combined direct and indirect impacts on the intermediate output requirements for iron ore change very little between 1947 and 1958. In reality there was a substantial increase (23 percent) in the direct requirements for iron ore.[5] This increase was almost entirely offset by the sizable decrease in requirements for steel. In other cases, the direct and indirect impacts may reinforce one another, and thus the combined impact of changes may be considerably larger than the changes in the direct coefficients.

## Changes in Intermediate Output Requirements as Measured by Direct Coefficients

To get at the underlying causes of coefficient changes, it is thus important to measure the separate impact on an industry's output requirements of changes over time in the direct input coefficients. Table 16.2 shows the percentage change in intermediate output requirements for individual industries based on direct rather than inverse coefficients. To derive these measures, the actual 1958 total domestic output for each industry was multiplied, in turn, by the 1947, 1958, and 1961 direct coefficients matrices (each matrix in 1958 prices). These multiplications yield the intermediate output that would have been required from each industry to produce the actual 1958 output levels with 1947, 1958, and 1961 technologies.

Compare Tables 16.1 and 16.2. *On the average,* percentage changes in intermediate output requirements based on inverse and direct coefficients are quite similar. For the period 1947 to 1958, changes in intermediate output requirements for 58 commodity producing industries[6] average 28 and 26 percent, respectively. For the 1958 to 1961 period, the average change in intermediate output requirements was a little over 5 percent by both methods.

Although the averages for all industries differ very little the two measures differ more for individual industries.[7] There were sizable differences between the two methods. The average annual rate of change in intermediate output requirements differed by 1 percent or more for 20 of the 58 cases examined in the 1947–1958 period, and for 16 of the 58 cases in the 1958–1961 period.

Although measures of output change based on inverse coefficients may not be an adequate guide to the *degree* of change in direct input coefficients, they are

Table 16.2.   Change Over Time in Primary Intermediate Output Requirements

| Input-output industry[a] | Derived 1958 primary intermediate output[b] (millions of 1958 dollars) | | | Percentage change | | Average annual rate of exchange[c] | |
|---|---|---|---|---|---|---|---|
| | 1947 Direct co-efficient | 1958 Direct co-efficient | 1961 Direct co-efficient | 1958 tech. 1947 tech. Col. (2) ÷ col. (1) | 1961 tech. 1958 tech. Col. (2) ÷ col. (1) | 1947–58 | 1958–61 |
| | (1) | (2) | (3) | (4) | (5) | (6) | (7) |
| 1 | 23,448 | 21,288 | 20,961 | −9.2 | −1.5 | −0.9 | −0.5 |
| 2 | 15,313 | 15,250 | 15,461 | −0.4 | +1.4 | 0 | +0.5 |
| 3 | 980 | 1,240 | 1,203 | +26.5 | −3.0 | +2.2 | −1.0 |
| 4 | 1,977 | 1,558 | 1,560 | −21.2 | +0.1 | −2.2 | 0 |
| 5 | 974 | 1,197 | 1,181 | +22.9 | −1.3 | +1.9 | −0.4 |
| 6 | 1,157 | 1,127 | 1,021 | −2.6 | −9.4 | −0.2 | −3.2 |
| 7 | 3,627 | 2,102 | 1,910 | −42.0 | −9.1 | −4.9 | −3.1 |
| 8 | 11,147 | 10,330 | 9,936 | −7.3 | −3.8 | −0.7 | −1.3 |
| 9 | 1,317 | 1,506 | 1,576 | +14.4 | +4.6 | +1.2 | +1.5 |
| 10 | 416 | 325 | 310 | −21.9 | −4.6 | −2.2 | −1.6 |
| 11 | 0 | 0 | 0 | — | — | — | — |
| 12 | 18,241 | 12,455 | 11,488 | −31.7 | −7.8 | −3.4 | −2.6 |
| 14 | 13,313 | 14,856 | 14,709 | +11.6 | +1.0 | +1.0 | −0.3 |
| 15 | 1,451 | 1,133 | 1,036 | −21.9 | −8.6 | −2.2 | −2.9 |
| 16, 17, 19 | 11,760 | 11,921 | 11,876 | +1.4 | −0.4 | +0.1 | −0.1 |
| 18 | 2,368 | 2,813 | 2,711 | +18.8 | −3.6 | +1.6 | −1.2 |
| 20 | 8,668 | 7,901 | 8,025 | −8.8 | +1.6 | −0.8 | +0.5 |
| 21 | 802 | 423 | 398 | −47.3 | −5.9 | −5.7 | −2.1 |
| 22 | 699 | 541 | 461 | −22.6 | −14.8 | −2.3 | −5.2 |
| 23 | 403 | 291 | 253 | −27.8 | −13.1 | −2.9 | −4.6 |
| 24, 25 | 13,161 | 12,518 | 12,516 | −4.9 | 0 | −0.5 | 0 |
| 26 | 5,455 | 4,141 | 4,244 | −24.1 | +2.5 | −2.4 | +0.8 |
| 27 | 7,659 | 9,281 | 9,567 | +21.2 | +3.1 | +1.8 | +1.0 |
| 28 | 2,137 | 3,602 | 3,837 | +68.6 | +6.5 | +4.9 | +2.1 |
| 29 | 1,287 | 1,794 | 1,874 | +39.4 | +4.5 | +3.1 | +1.5 |
| 30 | 1,752 | 1,719 | 1,759 | −1.9 | +2.3 | −0.2 | +0.8 |
| 31 | 8,685 | 8,305 | 8,121 | −4.4 | −2.2 | −0.4 | −0.7 |
| 32 | 4,134 | 4,674 | 5,406 | +13.1 | +15.7 | +1.1 | +5.0 |
| 33 | 1,042 | 881 | 842 | −15.5 | −4.4 | −1.5 | −1.5 |
| 34 | 310 | 326 | 311 | +5.2 | −4.6 | +0.5 | −1.6 |
| 35 | 2,209 | 1,976 | 2,052 | −10,5 | +3.8 | −1.0 | +1.3 |
| 36 | 6,132 | 6,906 | 7,053 | +12.6 | +2.1 | +1.1 | +0.7 |
| 37 | 23,649 | 17,672 | 17,744 | −25.3 | +0.4 | −2.6 | +0.1 |
| 38 | 10,028 | 8,907 | 9,068 | −11.2 | +1.8 | −1.1 | +0.6 |
| 39 | 1,840 | 1,974 | 1,989 | +7.3 | +0.8 | +0.6 | +0.3 |
| 40 | 6,092 | 6,512 | 6,467 | +6.9 | −0.7 | +0.6 | −0.2 |
| 41 | 3,992 | 2,991 | 2,709 | −25.1 | −9.4 | −2.6 | −3.3 |
| 42 | 4,602 | 5,111 | 4,952 | +11.1 | −3.1 | +1.0 | −1.1 |
| 43 | 793 | 804 | 709 | +1.4 | −11.8 | +0.1 | −4.1 |
| 44 | 261 | 434 | 447 | +66.3 | +3.0 | +4.7 | +1.0 |
| 45 | 771 | 687 | 578 | −10.9 | −15.9 | −1.1 | −5.6 |
| 46 | 424 | 378 | 433 | −10.8 | +14.6 | −1.1 | +4.7 |
| 47 | 1,806 | 1,772 | 1,605 | −1.9 | −9.4 | −0.2 | −3.3 |
| 48 | 540 | 438 | 437 | −18.9 | −0.2 | −1.9 | −0.1 |

| Input-output industry[a] | Derived 1958 primary intermediate output[b] (millions of 1958 dollars) | | | Percentage change | | Average annual rate of exchange[c] | |
|---|---|---|---|---|---|---|---|
| | 1947 Direct co-efficient | 1958 Direct co-efficient | 1961 Direct co-efficient | 1958 tech. 1947 tech. Col. (2) ÷ col. (1) | 1961 tech. 1958 tech. Col. (2) ÷ col. (1) | 1947–58 | 1958–61 |
| | (1) | (2) | (3) | (4) | (5) | (6) | (7) |
| 49 | 1,960 | 1,919 | 2,096 | −2.1 | +9.2 | −0.2 | +3.0 |
| 50 | 689 | 1,376 | 1,573 | +99.7 | +14.3 | +6.5 | +4.6 |
| 51 | 144 | 328 | 352 | +127.8 | +7.3 | +7.8 | +2.4 |
| 52 | 605 | 584 | 665 | −3.5 | +13.9 | −0.3 | +4.4 |
| 53 | 2,142 | 2,498 | 2,618 | +16.6 | +4.8 | +1.4 | +1.6 |
| 54 | 138 | 447 | 495 | +223.9 | +10.7 | +11.3 | +3.5 |
| 55 | 1,690 | 1,631 | 1,663 | −3.5 | +2.0 | −0.3 | +0.7 |
| 56, 57 | 2,004 | 3,056 | 3,182 | +52.5 | +4.1 | +3.9 | +1.4 |
| 58 | 1,059 | 841 | 897 | −20.6 | +6.7 | −2.1 | +2.2 |
| 59 | 8,700 | 8,386 | 8,582 | −3.6 | +2.3 | −0.3 | +0.8 |
| 60 (incl. 13) | 1,701 | 3,488 | 3,050 | +105.1 | −12.6 | +6.8 | −4.4 |
| 61 | 837 | 683 | 706 | −18.4 | +3.4 | −1.9 | +1.1 |
| 62 | 1,184 | 1,413 | 1,443 | +19.3 | +2.1 | +1.6 | +0.7 |
| 63 | 499 | 583 | 564 | +16.8 | −3.3 | +1.4 | −1.1 |
| 64 | 1,535 | 1,771 | 1,927 | +15.4 | +8.8 | +1.3 | +2.9 |
| Total[d] | 251,709 | 241,064 | 240,609 | 1,507.6 | 323.9 | 119.6 | 108.8 |
| Average[d] | | | | 26.0 | 5.6 | 2.1 | 1.9 |

[a] In general, the numbering scheme is that of the published 1958 table. In a few cases, however, because of extensive revisions between 1947 and 1958 in the Standard Industrial Classification (SIC), it was necessary to combine industries to insure comparability in the two years. For the name and SIC industry composition of each 1958 input-output industry, see Table 16.4.

[b] The results in columns 1–3 were obtained by the multiplication of actual 1958 total domestic outputs by the matrices of primary direct coefficients (on a domestic output base) for the years 1947, 1958, and 1961. Each of these matrices was in 1958 prices.

[c] Computed by the compound interest formula.

[d] Based on 58 industries (industries 1 through 64). In deriving totals and averages for columns 4 through 7, signs were ignored.

generally a good guide to their *direction* of change. In the 1947 to 1958 period, only 7 of the 58 industries examined showed opposite direction of change for intermediate output requirements based on inverse, as compared to direct, coefficients. For the 1958 to 1961 period, this number was 9. However, in both periods opposite direction of movement occurred mostly in industries showing very small average annual rates of change based on both direct and inverse requirements.

## Analysis of Variability of Direct Coefficient Changes

The analysis up to this point has dealt exclusively with *average* changes over time in requirements for the outputs of individual industries. We turn next to an

Table 16.3.   Analysis of Variability of Primary Direct Coefficients, 1947 to 1958

| Producing industry[a] | Consuming industry[a] | Percent change in primary direct coefficient[b] | Number of coefficients analyzed[c] | Percent of coefficients conforming[d] | Producing industry[a] | Consuming industry[a] | Percent change in primary direct coefficient[b] | Number of coefficients analyzed[c] | Percent of coefficients conforming[d] |
|---|---|---|---|---|---|---|---|---|---|
| (1) | (2) | (3) | (4) | (5) | (1) | (2) | (3) | (4) | (5) |
| 1 | ALL | −9.2 | 7 | 57 | 16, 17, 19 | ALL | +1.4 | 13 | 38 |
|   | 14 | −10.6 |   |   |   | 16, 17, 19 | +5.3 |   |   |
|   | 1 | +24.3 |   |   |   | 18 | −4.7 |   |   |
| 2 | ALL | −0.4 | 7 | 71 | 18 | ALL | +18.8 | 1 | 100 |
|   | 1 | +3.5 |   |   |   | 18 | +10.0 |   |   |
|   | 14 | −4.7 |   |   |   | 16, 17, 19 | −98.1 |   |   |
| 3 | ALL | +26.5 | 2 | 50 | 20 | ALL | −8.8 | 20 | 70 |
|   | 20 | +56.8 |   |   |   | 11 | −21.7 |   |   |
|   | 14 | +20.1 |   |   |   | 20 | +56.3 |   |   |
| 4 | ALL | −21.2 | 3 | 67 | 21 | ALL | −47.3 | 6 | 83 |
|   | 2 | −39.7 |   |   |   | 2 | −48.3 |   |   |
|   | 1 | +10.7 |   |   |   | 14 | −39.9 |   |   |
| 5 | ALL | +22.9 | 2 | 100 | 22 | ALL | −22.6 | 3 | 33 |
|   | 37 | +16.2 |   |   |   | 56, 57 | −56.7 |   |   |
|   | 5 | +144.4 |   |   |   | 11 | +21.6 |   |   |
| 6 | ALL | −2.6 | 3 | 67 | 23 | ALL | −27.8 | 2 | 50 |
|   | 38 | −5.3 |   |   |   | 71 | −100.0 |   |   |
|   | 6 | +20.9 |   |   |   | 70 | −100.0 |   |   |
| 7 | ALL | −42.0 | 11 | 82 | 24, 25 | ALL | −4.9 | 42 | 62 |
|   | 68 | −49.6 |   |   |   | 24, 25 | −2.7 |   |   |
|   | 37 | −1.6 |   |   |   | 26 | +1.7 |   |   |
| 8 | ALL | −7.3 | 4 | 50 | 26 | ALL | −24.1 | 27 | 89 |
|   | 31 | −9.9 |   |   |   | 26 | −0.4 |   |   |
|   | 68 | +17.0 |   |   |   | 70 | −31.7 |   |   |
| 9 | ALL | +14.4 | 6 | 50 | 27 | ALL | +21.2 | 32 | 56 |
|   | 36 | −16.7 |   |   |   | 27 | +24.0 |   |   |
|   | 11 | +68.5 |   |   |   | 28 | +23.6 |   |   |
| 10 | ALL | −21.9 | 2 | 100 | 28 | ALL | +68.6 | 12 | 67 |
|   | 27 | −31.2 |   |   |   | 16, 17, 19 | +92.6 |   |   |
|   | 10 | −20.0 |   |   |   | 32 | +127.4 |   |   |
| 12 | ALL | −31.7 | 19 | 84 | 29 | ALL | +39.4 | 6 | 67 |
|   | 71 | −39.4 |   |   |   | 29 | +10.5 |   |   |
|   | 65 | −35.7 |   |   |   | 77 | +196.1 |   |   |
| 14 | ALL | +11.6 | 13 | 31 | 30 | ALL | −1.9 | 12 | 67 |
|   | 14 | +25.5 |   |   |   | 12 | +36.8 |   |   |
|   | 1 | +30.9 |   |   |   | 11 | −29.6 |   |   |
| 15 | ALL | −21.9 | 1 | 100 | 31 | ALL | −4.4 | 31 | 77 |
|   | 15 | −21.9 |   |   |   | 31 | −24.3 |   |   |
|   | 65 | −100.0 |   |   |   | 65 | +44.1 |   |   |

| Producing industry[a] | Consuming industry[a] | Percent change in primary direct coefficient[b] | Number of coefficients analyzed[c] | Percent of coefficients conforming[d] | Producing industry[a] | Consuming industry[a] | Percent change in primary direct coefficient[b] | Number of coefficients analyzed[c] | Percent of coefficients conforming[d] |
|---|---|---|---|---|---|---|---|---|---|
| (1) | (2) | (3) | (4) | (5) | (1) | (2) | (3) | (4) | (5) |
| 32 | ALL | +13.1 | 32 | 50 | 45 | ALL | −10.9 | 8 | 38 |
|  | 59 | −42.1 |  |  |  | 11 | −38.9 |  |  |
|  | 65 | +4.9 |  |  |  | 73 | −82.4 |  |  |
| 33 | ALL | −15.5 | 2 | 50 | 46 | ALL | −10.8 | 3 | 67 |
|  | 34 | −15.5 |  |  |  | 11 | +24.1 |  |  |
|  | 18 | −37.2 |  |  |  | 46 | +41.6 |  |  |
| 34 | ALL | +5.2 | 1 | 100 | 47 | ALL | −1.9 | 20 | 40 |
|  | 34 | +27.3 |  |  |  | 59 | −57.3 |  |  |
|  | 72 | −76.5 |  |  |  | 47 | +24.5 |  |  |
| 35 | ALL | −10.5 | 11 | 73 | 48 | ALL | −18.9 | 5 | 80 |
|  | 14 | −5.1 |  |  |  | 26 | −65.7 |  |  |
|  | 29 | −47.5 |  |  |  | 16, 17, 19 | −5.4 |  |  |
| 36 | ALL | +12.6 | 18 | 50 | 49 | ALL | −2.1 | 20 | 55 |
|  | 11 | +14.8 |  |  |  | 59 | −47.5 |  |  |
|  | 36 | +48.0 |  |  |  | 11 | +63.2 |  |  |
| 37 | ALL | −25.3 | 50 | 86 | 50 | ALL | +99.7 | 11 | 36 |
|  | 37 | −3.1 |  |  |  | 59 | +31.7 |  |  |
|  | 11 | −18.1 |  |  |  | 61 | −81.1 |  |  |
| 38 | ALL | −11.2 | 35 | 54 | 51 | ALL | +127.8 | 1 | 100 |
|  | 38 | +22.8 |  |  |  | 51 | +323.6 |  |  |
|  | 11 | −37.5 |  |  |  | 70 | −84.2 |  |  |
| 39 | ALL | +7.3 | 6 | 33 | 52 | ALL | −3.5 | 4 | 75 |
|  | 14 | +48.8 |  |  |  | 52 | +4.6 |  |  |
|  | 31 | −47.9 |  |  |  | 54 | −26.7 |  |  |
| 40 | ALL | +6.9 | 9 | 22 | 53 | ALL | +16.6 | 19 | 47 |
|  | 11 | +32.0 |  |  |  | 53 | +23.0 |  |  |
|  | 12 | −24.7 |  |  |  | 54 | −45.8 |  |  |
| 41 | ALL | −25.1 | 26 | 92 | 54 | ALL | +223.9 | 2 | 50 |
|  | 59 | −30.5 |  |  |  | 11 | +409.3 |  |  |
|  | 60 | −18.5 |  |  |  | 12 | +67.2 |  |  |
| 42 | ALL | +11.1 | 42 | 33 | 55 | ALL | −3.5 | 7 | 86 |
|  | 11 | +3.9 |  |  |  | 11 | +24.0 |  |  |
|  | 59 | +31.8 |  |  |  | 12 | −47.3 |  |  |
| 43 | ALL | +1.4 | 9 | 11 | 56, 57 | ALL | +52.5 | 4 | 75 |
|  | 44 | −17.9 |  |  |  | 56, 57 | +2.2 |  |  |
|  | 61 | −10.3 |  |  |  | 66 | −28.4 |  |  |
| 44 | ALL | +66.3 | 2 | 0 | 58 | ALL | −20.6 | 7 | 43 |
|  | 44 | −28.2 |  |  |  | 59 | +9.2 |  |  |
|  | 2 | +122.8 |  |  |  | 75 | −40.6 |  |  |

Table 16.3 (*continued*).

| Producing industry[a] | Consuming industry[a] | Percent change in primary direct coefficient[b] | Number of coefficients analyzed[c] | Percent of coefficients conforming[d] | Producing industry[a] | Consuming industry[a] | Percent change in primary direct coefficient[b] | Number of coefficients analyzed[c] | Percent of coefficients conforming[d] |
|---|---|---|---|---|---|---|---|---|---|
| (1) | (2) | (3) | (4) | (5) | (1) | (2) | (3) | (4) | (5) |
| 59 | ALL | −3.6 | 11 | 64 | 62 | ALL | +19.3 | 6 | 67 |
|  | 59 | +11.9 |  |  |  | 62 | −32.4 |  |  |
|  | 75 | +9.4 |  |  |  | 77 | +10.4 |  |  |
| 60 | ALL | +105.1 | 2 | 50 | 63 | ALL | +16.8 | 4 | 25 |
| (incl. 13) | 60 | +93.8 |  |  |  | 77 | −42.2 |  |  |
|  | (incl. 13) |  |  |  |  | 63 | −21.9 |  |  |
|  | 65 | +162.5 |  |  |  |  |  |  |  |
| 61 | ALL | −18.4 | 3 | 100 | 64 | ALL | +15.4 | 6 | 33 |
|  | 65 | −32.3 |  |  |  | 64 | −25.6 |  |  |
|  | 61 | −15.5 |  |  |  | 72 | +7.4 |  |  |

[a] In general, the numbering scheme follows that of the published 1958 table. In a few cases, because of extensive revisions between 1947 and 1958 in the Standard Industrial Classification (SIC), it was necessary to combine industries to insure comparability in the two years. For the name and SIC industry composition of each 1958 input-output industry see Table 16.4.

[b] For any given producing industry, the first entry in this column is a weighted average of the percent change in the primary direct coefficients for all of its intermediate customers. The next two entries indicate the percent change in the primary direct coefficients for the 1st and 2nd largest customers of the given producing industry. The ranking of customer size is not determined by the size of the input coefficients, but by the level of the dollar flow that results from the multiplication of the 1958 actual domestic output by the 1947 primary direct coefficient.

[c] In computing a measure of the extent to which individual coefficient changes correspond to the direction of change of the average coefficient, only those coefficients which were at least 0.005 in 1947 were included. This column shows the number of consuming industries which meet this criterion. The 1st and 2nd largest customers of an industry (in value terms) were not necessarily included in this group.

[d] The figure in this column shows the percentage of consuming industries (with coefficients of at least 0.005) that showed changes between 1947 and 1958 in the same direction as the change in the weighted average of the coefficients for all the customers of a given producing industry.

examination of cell-by-cell coefficients to determine whether or not the input coefficients for the various direct customers of a given producing industry move in the same way as the overall average.[8] An examination of the data on *individual* input coefficients indicates a very marked degree of variability in the extent and direction of change over time. Industries that showed positive changes, between 1947 and 1958, in overall intermediate requirements for their output, showed frequent decreases in coefficients into individual consuming industries. Likewise industries with overall negative changes in the direct requirements for their outputs showed numerous instances of increasing requirements for particular industrial consumers. Moreover, this opposite direction of movement was not necessarily confined to the smaller customers of a given industry's output. Often coefficients for

Table 16.4.[10]  Industry Numbering for the 1958 Input-Output Study

| Industry no. and industry title | Industry no. and industry title |
|---|---|

**Agriculture, forestry and fisheries**
1   Livestock and livestock products
2   Other agricultural products
3   Forestry and fishery products
4   Agricultural, forestry and fisheries services

**Mining**
5   Iron and ferroalloy ores mining
6   Nonferrous metal ores mining
7   Coal mining
8   Crude petroleum and natural gas
9   Stone and clay mining and quarrying
10   Chemical and fertilizer mineral mining

**Construction**
11   New construction
12   Maintenance and repair construction

**Manufacturing**
13   Ordnance and accessories
14   Food and kindred products
15   Tobacco manufactures
16   Broad and narrow fabrics, yarn, and thread mills
17   Miscellaneous textile goods and floor coverings
18   Apparel
19   Miscellaneous fabricated textile products
20   Lumber and wood products, except containers
21   Wooden containers
22   Household furniture
23   Other furniture and fixtures
24   Paper and allied products, except containers and boxes
25   Paperboard containers and boxes
26   Printing and publishing
27   Chemicals and selected chemical products
28   Plastics and synthetic materials
29   Drugs, cleaning, and toilet preparations
30   Paints and allied products
31   Petroleum refining and related industries
32   Rubber and miscellaneous plastics products
33   Leather tanning and industrial leather products
34   Footwear and other leather products
35   Glass and glass products
36   Stone and clay products
37   Primary iron and steel manufacturing
38   Primary nonferrous metals manufacturing
39   Metal containers

40   Heating, plumbing and fabricated structural metal products
41   Screw machine products, bolts, nuts, etc., and metal stampings
42   Other fabricated metal products
43   Engines and turbines
44   Farm machinery and equipment
45   Construction, mining, oil field machinery and equipment
46   Materials handling machinery and equipment
47   Metalworking machinery and equipment
48   Special industry machinery and equipment
49   General industrial machinery and equipment
50   Machine shop products
51   Office, computing and accounting machines
52   Service industry machines
53   Electric transmission and distribution equipment, and electrical industrial apparatus
54   Household appliances
55   Electric lighting and wiring equipment
56   Radio, television, and communication equipment
57   Electronic components and accessories
58   Miscellaneous electrical machinery, equipment and supplies
59   Motor vehicles and equipment
60   Aircraft and parts
61   Other transportation equipment
62   Professional, scientific, and controlling instruments and supplies
63   Optical, ophthalmic, and photographic equipment and supplies
64   Miscellaneous manufacturing

**Transportation, communication, electric, gas, and sanitary services**
65   Transportation and warehousing
66   Communications, except radio and television broadcasting
67   Radio and T.V. broadcasting
68   Electric, gas, water, and sanitary services

**Wholesale and retail trade**
69   Wholesale and retail trade

**Finance insurance and real estate**
70   Finance and insurance
71   Real estate and rental

Table 16.4 (*continued*).

| Industry no. and industry title | Industry no. and industry title |
|---|---|
| **Services** | **Imports** |
| 72  Hotels and lodging places; personal and repair services, except automobile repair | 80  Gross imports of goods and services |
| 73  Business services | **Dummy industries** |
| 74  Research and development | 81  Business travel, entertainment, and gifts |
| 75  Automobile repair and services | 82  Office supplies |
| 76  Amusements | 83  Scrap, used, and secondhand goods |
| 77  Medical, educational services, and non-profit organizations | **Special industries** |
|  | 84  Government industry |
| **Government enterprises** | 85  Rest of the world industry |
| 78  Federal Government enterprises | 86  Household industry |
| 79  State and local government enterprises |  |

[10] For further details, see Goldman et al. (1964).

the two largest intermediate customers of an industry's output change was in a direction opposite to that of the overall average.

While there was a tendency for the majority of the customers of a given industry to show direct coefficient changes of the same direction as the overall change, this tendency was by no means marked. As can be seen from column 5 of Table 16.3, only 34 of the 58 industries examined showed coefficient changes of the same direction as the average for the majority of their individual consuming industries.[9] It might be expected that the conformity of individual coefficient changes to the industry average would be most marked for those industries that showed large average changes between 1947 and 1958 in the direct requirements for their output, and that the cases of opposite direction of movement would be confined to those industries with small average changes over the period. This is not generally the case. The attempt to correlate average percentage change in direct coefficients with the percentage of an industry's customers showing coefficient changes in the same direction as the average (columns 3 and 5 of Table 16.3) yields a correlation coefficient of virtually zero.

This seemingly unpatterned behavior of changes between 1947 and 1958 in the various individual input coefficients is indeed disconcerting. One would have hoped to observe more regularity in the pattern of change. How much easier would be the task of updating and projecting coefficients if the average pattern for an industry was characteristic of all the customers of an industry! Of course, there are many possible explanations for this erratic behavior, not the least important of which may be errors of statistical estimation. It is also likely that some of the factors mentioned earlier in this paper, particularly changing product mix, are working to disguise a well-defined, orderly pattern of technological change.

The answers to the many unresolved questions about the reasons for coefficient changes cannot, of course, be found via an analysis such as the present one, which looks at changes in input coefficients along the row—that is, from the point of view of the producing industry. Such answers may come at the next stage of the analysis, which will focus on coefficient changes from the point of view of the individual columns. Such an analysis alone can provide the knowledge of what substitutions among the various inputs of a given consuming industry have taken place. And it is through this knowledge that the underlying causes of such substitutions can be unearthed.

**NOTES**

1. One important element of delay in producing an input-output table is the amount of time it takes before the detailed product statistics from the general economic censuses become available to input-output analysts. For example, the bulk of the necessary data from 1963 Census of Manufactures, the collection of which could not, of course, begin before 1963 was completed, did not become available until the spring of 1966. Every effort is being made to speed up this process and it is hoped that for the 1967 economic census, there will be direct access to the data on computer tapes rather than through published, printed documents.
2. Because it has been recognized that these transfers, both domestic and foreign, can seriously affect the comparisons of individual industry input coefficients over time, our analysis is based on inverse and direct coefficients that are computed on a domestic rather than a total output base. Moreover, our analysis of direct coefficients is based on primary flows only. Secondary flows have been excluded. By these means we have minimized the influence of 'fictitious transfers' on the industry input coefficients.
3. Subsequent to the original publication of the 1958 table, additional detail on non-ferrous metals was provided in Department of Commerce (1966). There is, however, no 1958 inverse matrix that contains this additional detail.
4. It is generally believed that input coefficients will be more stable over time if they are expressed in constant, rather than current, prices. This belief rests on the concept that input coefficients reflect basic technological relationships which, in turn, reflect the physical quantity requirements of production. Expressing coefficients in constant prices is thus a way of adjusting value data to reflect changes in quantity inputs. For numerous reasons, it is, however, possible that input coefficients expressed in current dollars would be equally, if not more, stable over time than coefficients expressed in constant dollars. Subsequent work at OBE will analyze coefficient changes in current dollars as well.
5. This increase in the requirements for iron ore was associated with a decrease in the requirements for iron and steel scrap materials.
6. At the time this paper was written, the cell-by-cell repricing of the 1947 input-output table had not been completed. Some problems in the service areas still remained. It was therefore necessary to confine this portion of the analysis to the goods producing sectors of the economy (industries 1–64).
7. There is another, minor, potential source of difference between the two measures of change in intermediate output requirements. Output requirements derived by use of the inverse matrices include requirements for both secondary and primary production. Output requirements derived by the use of direct coefficients include primary production only. Secondary output requirements can either augment or offset changes in primary output requirements.
8. The analysis relating to the variability of the individual primary direct input coefficients is confined to the time period 1947–1958. It would not be meaningful to include the 1958–1961 time period in this portion of the analysis since, in general, the procedure for developing the 1961 input-output table was one of adjusting all 1958 coefficients from a given producing industry by the average change for that industry.
9. This analysis of the direction of change in individual input coefficients was confined to coefficients which were at least 0.005 in 1947.
10. For further details, see Goldman et al. (1964).

**References**

American Iron and Steel Institute. 1967. Steel facts. No. 198; October-November.

Carter, A. P. 1967. Changes in the structure of the American economy, 1947–1948 and 1962. *Review of Economics and Statistics 49*; 209–224.

Evans, W. D. and M. Hoffenberg. 1952. The interindustry relations study for 1947. *Review of Economics and Statistics 34*: 97–142.

Goldman, M. R., M. L. Marimont, and B. N. Vaccara. 1964. Survey of current business. U.S. Department of Commerce. Office of Business Economics. Washington.

U.S. Department of Commerce. Office of Business Economics. 1965. Survey of Current Business. Washington.

U.S. Department of Commerce. Office of Business Economics. 1966. Survey of Current Business. Washington.

U.S. Department of Labor. Bureau of Labor Statistics. 1966. Projections of 1970 Interindustry Relationships, Potential Demand, Employment. Bulletin No. 1536. Washington.

U.S. Department of Labor. Bureau of Labor Statistics. 1967. 1970 Input-Output Coefficients Report No. 327. Washington.

Vaccara, B. N. and N. W. Simon. 1968. Factors affecting the post-war industrial composition of real product. In: *The Industrial Composition of Income and Product, Studies in Income and Wealth*, ed. John W. Kendrick. No. 32. New York: National Bureau of Economic Research.

# 17

## An Assessment of the RAS Method for Updating Input-Output Tables

R. G. LYNCH

The United Kingdom Central Statistical Office has published official input-output studies for the years 1954, 1963, and 1968, based upon comprehensive Censuses of Production. These enquiries provided detailed data on industry inputs and outputs analysed by commodity, allowing firmly-based input-output tables to be constructed. Tables have also been produced for 1970, 1971, and 1972, years for which no detailed information on industry inputs was available. For these years, the industry input structures as shown by Table B, the absorption matrix in the published tables, were estimated by updating the 1968 input-output structure using a modified version of a mechanical method known as the RAS method. This is a technique in which the base year absorption matrix is adjusted to sum to given row and column totals for the update year, by successive prorating of the rows and columns until consistency is achieved.

Since being developed in the Department of Applied Economics, Cambridge (Lecomber, 1964), the RAS method has been examined in various studies (Paelink and Waelbroeck, 1965; Schneider, 1965; Tilanus, 1966; Allen, 1974; Barker, 1975; Lecomber, 1975; Allen and Lecomber, 1975). These studies have increasingly emphasised the potential inaccuracies arising out of the RAS method. This paper examines an updating of the 1963 United Kingdom absorption flow matrix to 1968 by the RAS method, and compares the updated matrix and derived statistics with the corresponding ones for the firmly based 1968 tables, at a 69-industry level of disaggregation.

### The RAS Method

The basic method is described by the Department of Applied Economics, Cambridge (1963, pp. 27–30) and in a publication of the United Nations on input-output tables and analyses (1973, pp. 65–74). In terms of the input-output coefficient matrix $A$ obtained from the absorption flow matrix by dividing each cell

This paper was presented at the Seventh International Conference on Input-Output Techniques held in Innsbruck, Austria, April 9–13, 1979, pp. 1–29. © British Crown Copyright 1985. Reprinted by permission of the Controller of Her Majesty's Britannic Stationary Office.

271

value by the total corresponding industrial input (identically equal to industrial output), each $a_{ij}$ can be subject to two effects over time:

(*i*) The effect of substitution, measured by the extent to which commodity *i* has been replaced by, or used as a substitute for, other commodities in industrial production.

(*ii*) the effect of fabrication, measured by the extent of which industry *j* has come to absorb a greater or smaller ratio of intermediate to total inputs in its production.

It is assumed that each effect works uniformly (e.g., that commodity *i* is increasing or decreasing as an input into all industries at the same rate, and that any change in the ratio of intermediate to total inputs into an industry has the same effect on all commodities used as inputs). The substitution multipliers which operate along the rows are denoted as the vector *r* and the fabrication multiplier operating on the columns as the vector *s*. Each cell in the base matrix $A_0$ will be subject to these two effects and the new coefficient matrix $A_1$ can thus be written as $A_1 = rA_0s$, *r* and *s* being matrices with the vectors *r* and *s* in the diagonals and zeros elsewhere.

In terms of the absorption flow matrix, the RAS method consists of finding a set of multipliers to adjust the rows of the existing matrix and a set of multipliers to adjust the columns so that the cells in the adjusted matrix will sum to the required row and column totals relating to the update year.

In order to find the updated absorption matrix $X_1$, knowing the row and column totals and the base matrix $X_0$, then the estimation process of obtaining $X_1$ from $X_0$ by the RAS method is no more than proportional adjustment of the base matrix successively along its rows and columns until convergence is reached, and $X_1 = rX_0s$. In this case the *r* and *s* are no longer simple measures of substitution and fabrication. A description of the mathematical properties of the method can be found in Bacharach (1970).

This method can be modified to allow the fixing of selected rows and columns and individual cells to known values in the update year. This is accomplished by setting the value of the original cells corresponding to the fixed ones to zero and then applying the normal RAS procedure to this modified matrix so that the new row and column totals less the fixed cell values are satisfied, and finally inserting the fixed cell values for the update year. A fuller description is given in the United Nations publication (1973, pp. 65–74) on input-output tables.

## United Kingdom Methodology

For United Kingdom input-output studies, three basic tables are constructed. They are:

a. An analysis of commodities made by industries—the Make matrix
b. An analysis of domestically produced commodities purchased by industries and by final demand—the Absorption matrix

*c.* An analysis of the imported commodities purchased by industries and by final demand—the Imports matrix.

Input-output tables for 1968 have been published based upon the comprehensive Census of Production for 1968 which provided data on industry inputs and industry outputs analysed by commodity. These formed a firm base for the detailed analysis of commodity outputs and commodity inputs (items *a* and *b* above) and also provided some basis for the detailed analysis of the purchases of final demand by commodity.

Tables for 1970, 1971, and 1972 have also been published and their construction is described in articles in *Economic Trends* (May 1974; April 1975). The major difference between the 1968 and later input-output tables is that no detailed information on industry inputs was available for these later years. The annual Census of Production only provided figures of industry total outputs and purchases of materials and fuels, wages and salaries, and stocks held by census industries. Hence the industry input structures for 1970, 1971, and 1972 were based upon an updating of the 1968 Absorption matrix using the RAS method. This requires that the intermediate row and column totals of the Absorption matrix for the update years be calculated.

The first step is to derive the overall row and column totals of the absorption matrix, i.e., the total commodity and industry outputs, respectively. This is done via the Make matrix where the industries are considered in two groups. The first of these consists of all manufacturing industries plus construction. For these industries the commodity outputs (rows) are first revalued at the update year prices. This revaluation leads to a new set of industry outputs (columns) which are adjusted pro rata to equal the actual industry outputs taken from the Annual Census of Production. This in turn gives revised row totals which are now estimates of update year commodity outputs at update year prices. The entries in the Make matrix for the remaining industries are estimated directly from other sources.

The overall row and column totals thus estimated for the Make matrix provide the overall row and column totals for the Absorption matrix. The intermediate row totals are calculated by subtracting final demand from total commodity outputs and the intermediate column totals are similarly obtained by subtracting primary inputs from total industry outputs. Much of the data for both final demand and primary inputs is available directly, although 1968 patterns are used for some items. The estimation of the elements of the intermediate transactions matrix can then be undertaken. Six rows and columns of the 1970 Absorption matrix were estimated from data provided by government departments (agriculture, forestry, and fishing) or from data provided by the Digest of United Kingdom Energy Statistics (coal mining, mineral oil refining, gas, and electricity). There remained 84 rows and columns of the Absorption matrix for which no firmly based information was available. To provide estimates of purchases of commodities by these domestic industries, the RAS method was used. A similar procedure was carried out for 1971 and 1972, when only 59 industries were identified compared with 90 for 1970.

**The Data Used in This Study**

This study was made possible by the existence of firmly based input-output tables for the years 1963 and 1968. The 1963 tables were originally compiled on the 1958 Standard Industrial Classification and had first to be reclassified to the 1968 Standard Industrial Classification. In order to produce comparable tables, it was necessary to reduce the 70 industry tables of 1963 to 69 industries, and so the 90 industry tables of 1968 were also reduced to 69 industries. It was then possible to assess RAS updates of the 1963 Absorption matrix to 1968 by comparison with the firmly based 1968 matrix. It was found helpful in assessing the magnitude of errors in the updated tables to compare them with the errors which occur using the most recent firmly based tables, without updating—in this case the 1963 tables.

In the early stages of this study, several discrepancies between the updated and firmly based 1968 absorption matrices had obviously arisen because of improvements in the allocation of unidentified purchases in 1968. Whilst it is a perfectly valid exercise to assess the updating procedure without allowing for such improvements, specifically in order to test the RAS method the large discrepancies were resolved as far as possible where justification existed for changes in allocations in the 1963 table. This was done by adjusting the particular cells which could be reasonably changed in 1963 by a comparison of 1963 and 1968 procedures, and then consistency obtained by a series of row and column pro rata adjustments. This is similar to the method used by Allen (1974) in an assessment of the RAS method using 1954 and 1963 input-output tables.

It was, therefore, possible to produce Make, Absorption, Primary Inputs, and Final Demand matrices for 1963 on a 1968 SIC basis and as far as possible adjust for differences in procedures for the allocation of unidentified purchases by an industry.

**The Updating of the Absorption Matrix**

This exercise is a test of the RAS method under near ideal conditions of data availability, i.e., the row and column intermediate totals from the firmly based 1968 absorption matrix were used as the constraints in updating the 1963 matrix to 1968. Table 17.1 shows in absolute and percentage terms the number and total value of elements in the absorption matrix for 1968 above various lower bounds.

Table 17.1.  The Number and Value of Elements in the 1968 Absorption Matrix in Absolute and Percentage Terms Above Various Lower Bounds

| Category | Lower bound | Number of elements | Percentage of total | Value of elements (£m) | Percentage of total |
|---|---|---|---|---|---|
| A | nonzero | 2,578 | 54 | 27,738 | 100 |
| B | £10m | 471 | 10 | 23,578 | 85 |
| C | 1% of intermediate input of individual industry | 1,028 | 22 | 25,609 | 92 |

Table 17.2. Root Mean Square Error (£m) in Updated Absorption Matrix for 1968

| Category | standard RAS | Modified RAS excluding fixed elements | Modified RAS including fixed elements |
|----------|--------------|---------------------------------------|---------------------------------------|
| A | 9.0 | 8.9 | 8.2 |
| B | 20.6 | 20.3 | 18.6 |
| C | 14.0 | 13.8 | 12.7 |

In the results which follow, zero elements occuring in both base and update matrices are not included in the error measures since the RAS method ensures that a zero element in the base year matrix will be zero in the update. The percentage error is defined as $100((a_u - a_f)/a_f)\%$ for $a_f \neq 0$. $a_u$ is an element of the updated matrix, $a_f$, an element of the firmly based matrix. This conforms with the natural idea that if £75m is the update estimate for a firmly based element of £100m, the percentage error is $-25\%$. This definition yields a skew distribution, the largest negative error possible being $-100\%$ with no limit to the size of the positive errors.

The updating procedure was performed both with and without exogenous information, i.e., the modified and standard RAS methods. The same six rows and columns were used as in the update of 1968 to 1970 to produce the 1970 input-output tables (i.e., Agriculture, Forestry and Fishing, Coal Mining, Mineral Oil Refining, Gas, and Electricity). The differences between the updated and firmly based 1968 absorption matrices are shown in Table 17.2 for the standard and modified RAS, both including and excluding the fixed elements in the error measure.

The distribution of percentage errors as percentages of their own population is shown is Table 17.3 for categories B and C of the lower bounds. The distribution for category A is similar in pattern, apart from a bias to high percentage errors due to small elements of the 1968 absorption matrix occuring in the denominator of the error measure.

These results suggest that updating by the RAS method yields errors in the absorption matrix estimate for the later year which are disturbingly large. For entries in the 1968 absorption matrix greater than £10m (85% of total by value), in

Table 17.3. The Distribution of % Errors in the Updated Absorption Matrix of the Standard and Modified RAS Procedures

| Method | Lower bound category | −100 to −75% | −75 to −50% | −50 to −25% | −25 to 0% | 0 to 25% | 25 to 50% | 50 to 75% | 75 to 100% | 0ver 100% |
|--------|---------|------|------|------|------|------|------|------|------|------|
| Standard RAS | B | 4.9 | 9.2 | 17.3 | 24.8 | 23.5 | 10.0 | 5.11 | 2.4 | 2.8 |
| | C | 6.0 | 9.1 | 15.6 | 22.0 | 22.0 | 11.2 | 4.9 | 2.4 | 5.8 |
| Modified RAS excl. | B | 4.5 | 8.5 | 16.3 | 25.5 | 24.8 | 11.3 | 5.0 | 1.0 | 3.1 |
| fixed elements | C | 6.0 | 8.9 | 15.8 | 22.4 | 20.6 | 12.1 | 6.0 | 2.7 | 5.5 |
| Modified RAS incl. | B | 3.8 | 7.3 | 13.9 | 29.1 | 28.4 | 9.6 | 4.3 | 0.8 | 2.8 |
| fixed elements | C | 4.6 | 7.9 | 12.8 | 27.4 | 26.8 | 8.6 | 5.5 | 1.7 | 4.7 |

the standard RAS case approximately 50% of the update elements are in error by more than 25%, and 25% are in error by more than 50%. The inclusion of exogenous information improves the overall accuracy slightly, but has surprisingly little effect on the accuracy with which the nonfixed elements are estimated.

It should be borne in mind that the size of the errors shown is due in part to three factors:

a. The length of time (five years) over which the update is made.
b. No reclassification exercise such as that carried out on the 1963 absorption matrix to change from the 1958 SIC to the 1968 SIC can be wholly satisfactory.
c. A general improvement in coverage of data sources, quality of data, and techniques of data-processing occurs with time, resulting in differences between the allocation of purchases of commodities amongst industries for 1963 and 1968.

**Updated Input-Output Tables as Aids in Economic Analysis**

Despite the considerable errors in the updated absorption matrix, the tables derived from this matrix could still show a significant improvement over those obtained from the original 1963 matrix. To test this possibility, comparisons were made of derived tables for 1968 obtained by three different means:

i. By using the firmly based 1968 absorption matrix and final demand.
ii. By using the updated matrix and the 1968 final demand.
iii. By using the 1963 matrix and the 1968 final demand.

For many analytical purposes, the fundamental input-output matrix is the industry-by-industry coefficient matrix $E$, where

$$g = Eg + f \qquad (17\text{-}1)$$

$g$ is the gross industry output vector and $f$ the total final demand vector. It is this matrix $E$ which has been most subject to tests in previous assessments of the RAS method. It was necessary, therefore, to construct the matrix $E$ for 1963, 1968, and the update to 1968 from the corresponding absorption matrices using the Central Statistical Office's normal procedures as described in the 1968 input-output study.

The 1963 coefficient matrix $E$ was changed to flow form by multiplying each element by the appropriate 1968 industry output. This allowed all the matrices to be compared in flow form, at the same time allowing the lower bound criteria to be applied but the percentage errors apply equally well to the coefficient form. Both the 1963 version and the 1968 updated version (shown in the tables as "1963–1968") were compared with the firmly based 1968 one, and the results are shown in Table 17.4. The figures in brackets are the corresponding errors for the 34-sector version of the 69 sector tables (the summary tables published regularly in *Economic Trends* are 34 sector). The 34-sector version was obtained by reducing the 69-sector flow matrices as a last step.

Using the three different versions of $E$, the inverses and associated derived

Table 17.4. Errors in the Estimates of Coefficient Matrix E for 1968.[a]

| Lower bound (£m) estimate | | Root mean square error for flow form (£m) | Mean absolute % error |
|---|---|---|---|
| 0 | 1963 | 7.9 (17.8) | 456.0 (369.8) |
| | 1963–68 | 6.6 (15.4) | 496.8 (294.0) |
| 10 | 1963 | 22.0 (28.9) | 45.4 (46.4) |
| | 1963–68 | 17.7 (24.7) | 40.6 (41.4) |
| 20 | 1963 | 28.4 (33.0) | 38.9 (37.9) |
| | 1963–68 | 23.0 (28.5) | 34.3 (33.3) |

[a] The figures for the 34-sector versions are shown in brackets.

matrices of the 1968 published tables were then calculated using the same methodology as for the 1968 study. At first glance it might seem that a good test of the adequacy of the Leontief inverse $(I - E)^{-1}$ (Table E of the 1968 publication) would be how successfully total output for 1968 could be predicted using

$$g = (I - E)^{-1}f \tag{17-2}$$

However, the RAS method explicity uses the relationship between gross output and final demand to derive intermediate outputs and so this equation is satisfied identically by an RAS updated matrix E. Another way of measuring the adequacy of the inverse is to consider how well each element of the inverse is estimated. In order to apply a lower bound criterion to exclude smaller elements, each element of the inverse was weighted by the appropriate final demand for 1968. As each element $e_{ij}$ of E is defined as the amount of industry $i$ output required to produce one unit of final output for industry $j$, the flow version represents the allocation of industry gross output to final demand both directly and indirectly. The results are shown in Table 17.5 and, as before, the percentage figures apply equally to the coefficient form of the inverses.

The Central Statistical Office's usual practice is to calculate the inverse excluding intraindustry transaction, the diagonal elements of **Eg**. However, because of differences in the size of the intraindustry transactions between the 1968 flow matrix **Eg** and the RAS update, this results in the measure of total domestic output net of intraindustry transactions being different for the 1968 firmly based and updated versions of the matrices, as reflected in

$$G_n = (I - En)^{-1}f \tag{17-3}$$

Table 17.5. Errors in the Estimates of the Inverse Matrix $(I-E)^{-1}$ for 1968[a]

| Lower bound (£m) estimate | | Root mean square error for flow form (£m) | Mean absolute percent error |
|---|---|---|---|
| 5 | 1963 | 13.3 (19.6) | 33.4 (31.5) |
| | 1963–68 | 9.5 (13.5) | 28.0 (24.2) |
| 10 | 1963 | 17.0 (22.9) | 26.7 (27.1) |
| | 1963–68 | 12.0 (15.7) | 21.0 (20.6) |

[a] The figures for the 34-sector versions are shown in brackets.

where $n$ denotes net of intraindustry transactions. Similarly, this measure of domestic gross output will change as the level of aggregation changes, so complicating comparisons of updated and firmly based versions of the 1968 matrices at different levels of aggregation. For these reasons the intraindustry transactions are included when calculating the inverse for the purpose of this study.

Although a test on the adequacy of the updated matrix $E$ in its role in predicting gross output from final demand for the update year is rendered trivial by the nature of the RAS method, it is nevertheless of interest to compare the allocations of gross or net output amongst the various sectors of final demand such as consumers' expenditure or exports for individual industries, as demonstrated in Table H of the 1968 input-output study. Taking $F$ to be the matrix of final demand, Eq. (17–2) becomes

$$G_f = (I - E)^{-1} F \qquad (17\text{–}4)$$

where $G_f$ is a matrix showing the allocation of gross output to the sectors of final demand by industry. It is of course possible to premultiply rows of this matrix by the ratio of net to gross output for each industry and so express it in net output terms, but as percentage errors are perhaps the most illuminating this was not felt necessary.

Table 17.6 summarises the results of the comparison between the 1968 flow version of $G_f$ in gross output terms and the alternative estimates. The 1963 version is obtained by applying the 1963 inverse to the 1968 final demand matrix on an industry basis and, similarly, the 1963–1968 update inverse provides the updated version. It must be remembered that the RAS method assures that the total final demand allocation of an industry's output in the updated version is identical with the 1968 total. As it is straightforward to adjust the rows of the 1963 matrix $G_f$ to conform to these industry output totals also by simple prorating, a more rigorous test of the RAS method, in as far as it accomplished more than just consistency with 1968 totals, is to constrain the 1963 estimates of $G_f$ to 1968 totals and compare the errors in these matrices of final demand in flow form, as shown in Table 17.6. This is

Table 17.6.  Errors in the Estimates of the 1968 Matrix Showing Allocation of Total Output to Final Demand (1968 *Study: Table H*)[a]

| Lower bound (£m) estimate | | Root mean square error in flow form (£m) | Mean absolute percentage error |
|---|---|---|---|
| 5 | 1963 | 14.6 (20.3) | 11.7 (10.8) |
| | 1963(c)[b] | 8.4 (11.3) | 7.0 (6.6) |
| | 1963–68 | 5.0 (6.4) | 5.6 (4.4) |
| 10 | 1963 | 15.5 (21.4) | 11.4 (10.8) |
| | 1963(c)[b] | 8.9 (11.9) | 6.6 (6.1) |
| | 1963–68 | 5.3 (6.7) | 5.2 (4.0) |
| 20 | 1963 | 17.0 (22.5) | 10.2 (9.0) |
| | 1963(c)[b] | 9.7 (12.6) | 5.2 (4.4) |
| | 1963–68 | 5.7 (7.1) | 4.6 (3.9) |

[a] 34-sector version is shown in brackets.
[b] Constrained.

equivalent to considering the estimated matrices as a means of indicating the proportional allocation of total output between the sectors of final demand by industry, the form of Table I of the 1968 input-output study. The 1963 estimates constrained to 1968 totals are denoted by 1963 (c) in Tables 17.6 and 17.7.

As each estimate of the flow matrix $G_f$ includes the same direct components of final demand for 1968, a truer assessment of the adequacy of the inverse estimates is achieved by considering only the indirect flow part of $G_f$. The differences are shown in Table 17.7.

The figures shown in Tables 17.5, 17.6, and 17.7 prompt the following deductions.

1. The updated versions of $E$, $(I - E)^{-1}$ and $G_f$ are better estimates of the 1968 equivalents than the 1963 versions, although the improvement is relatively modest.

2. The results of the constrained 1963 table comparison in Tables 17.6 and 17.7 do, however, suggest that simple prorating procedures yield almost as good results as the standard RAS method in determining allocations of output to final demand, an important use of input-output methods.

3. The aggregation of the tables for 69 to 34 sectors as a last step does little to improve the percentage errors occurring in the basic and derived input-output tables.

This last, rather surprising result can be rationalized as follows. The aggregation of 69 sectors to 34 is not the simple 2:1 reduction that it seems. Of the 34 sectors created by the aggregation, 20 are unchanged from the 69-sector classification and so many errors revealed in the 69-sector comparison are unaffected by aggregation. Also, if errors in cells at the 69-sector level which are aggregated are positively correlated, then percentage errors will not decrease to the extent that would be expected if the errors were independent. Where a cell value at the 69-sector level constitutes most of the value of a 34-sector aggregated cell, then positively correlated errors can result in an increase of the average absolute percentage error on aggregation.

Table 17.7. Errors in the Estimates of the 1968 Matrix Showing Indirect Allocation of Outputs to Final Demand (1968 *Study: Table I*)[a]

| Lower bound (£m) | estimate | Root mean square error in flow form (£m) | Mean absolute percentage error |
|---|---|---|---|
| 5 | 1963 | 15.5 (21.3) | 22.0 (21.3) |
| | 1963(c)[b] | 6.2 (8.1) | 10.0 (8.1) |
| | 1963–68 | 5.3 (6.6) | 9.5 (7.5) |
| 10 | 1963 | 17.1 (22.5) | 21.4 (20.2) |
| | 1963(c)[b] | 6.8 (8.7) | 9.4 (7.8) |
| | 1963–68 | 5.9 (7.1) | 8.6 (6.9) |
| 20 | 1963 | 19.6 (24.5) | 19.7 (17.3) |
| | 1963(c)[b] | 7.7 (9.5) | 8.2 (6.4) |
| | 1963–68 | 6.6 (7.8) | 7.3 (6.0) |

[a] 34-sector version is shown in brackets.
[b] Constrained.

## The Effect of Different Aggregation and Exogenous Information Levels on the Adequacy of RAS Updated Input-Output Tables

It is trivially true that either increasing the exogenous information level to 100% or reducing the level of aggregation to 1 will produce updated tables with no estimation errors in the intermediate sector. These tables will have the disadvantage of either costing much more (in that more exogenous information is required) or revealing less of the economic structure in the update year due to the high level of aggregation. The question naturally arises whether updated tables can be produced using an RAS with exogenous information which are significantly less costly to produce in terms of time and resources (including the resources required to conduct the Census of Production and purchases inquiry) than firmly based tables, in which the errors do not exceed acceptable levels and yet the level of aggregation allows meaningful deductions and analytical exercises highlighting economic structure to take place.

Four different levels of aggregation were chosen: 69, 34, 13, and 6. In the progressive aggregation from 69 to 6, homogeneous industries were chosen to be combined in as far as this was possible. Four different levels of exogenous information were chosen which in terms of fixed rows and columns at the 69-industry level were 0, 6, 10, and 20. In choosing the rows and columns to be determined exogenously, the criterion used was the likelihood of such information becoming available in the UK in the foreseeable future short of a full-scale purchases inquiry (e.g., industry sectors with a large content of nationalised industries were chosen first and industries such as distribution were not determined exogenously).

It should be emphasised here that complete rows and columns were chosen at the 69-industry level for convenience. It would have been possible to choose individual cells instead. Also, the criterion used is not the only possible one for selecting exogenous information. One equally valid criterion is whether the elements in question are likely to be estimated badly by RAS (e.g., a heterogeneous commodity group) and so most require exogenous determination, another is whether the elements to be fixed are important in terms of uses to which the input-output table can be put (Allen, 1974). Methods of incorporating not fully reliable

Table 17.8.  Errors in the Estimates of the 1968 Symmetric Flow Absorption Matrix (1968 *Study: Table D*)

| Aggregation levels | 1963 | | 1963–68 Exogenous information levels in terms of fixed rows and columns | | | | | | | |
|---|---|---|---|---|---|---|---|---|---|---|
| | | | 0 | | 6 | | 10 | | 20 | |
| | rmse | % | rmse | % | rmse | % | rmse | % | rmse | % |
| 69 | 17.6 | 51.9 | 14.2 | 48.3 | 12.3 | 40.3 | 11.0 | 32.1 | 9.3 | 23.8 |
| 34 | 25.0 | 51.9 | 21.6 | 47.0 | 19.3 | 36.1 | 15.2 | 26.2 | 12.6 | 19.9 |
| 13 | 48.3 | 31.3 | 37.6 | 28.9 | 25.4 | 18.9 | 20.6 | 11.7 | 13.8 | 9.6 |
| 6 | 75.8 | 19.9 | 55.8 | 16.9 | 23.6 | 7.8 | 28.1 | 6.9 | 27.0 | 6.5 |

Table 17.9.  Alternative Measures of the Level of Exogenous Information
Available

| Number of fixed rows/columns | | | | |
|---|---|---|---|---|
| at 69-sector level | 0 | 6 | 10 | 20 |
| % of nonzero cells fixed | 0 | 17.1 | 30.8 | 52.6 |
| % of value fixed | 0 | 19.4 | 38.6 | 59.7 |

exogenous information and the results of tests on these methods are given in a
paper by Allen and Lecomber (1975).

The details of the method used in this study are as follows. Cells in the 1963
absorption matrix corresponding to the fixed rows and columns were set to zero.
The RAS procedure was then applied to this modified absorption matrix and the
exogenous information added to produce a 1968 update. At higher levels of
aggregation, the endogenous and exogenous matrices were treated separately. The
1963 absorption matrix at the 69-industry level with cells for which information
existed set to zero was aggregated and then the RAS procedure applied, and then
the aggregated form of the matrix of exogenous information was added back to
produce the 1968 update. This allowed the level of exogenous information to be
held constant in terms of value and ease of procurement whilst the aggregation
level was varied.

The absorption matrices were then transformed by the Central Statistical
Office's usual procedures to industry-by-industry matrices. To provide a 1963
estimate of 1968, the 1963 flow matrix was put into coefficient form by dividing
intermediate industry inputs by the appropriate gross industry output for 1963,
and then converted to 1968 flow form by multiplying by the corresponding 1968
gross industry outputs. This allowed the matrices to be compared using lower
bound criteria for inclusion, but still yielding percentage errors which applied
equally well to the coefficient forms of this industry-by-industry matrix. Table 17.8
shows the results of this exercise, for a lower bound of £5m. The percent errors
shown in the table are the mean absolute percent errors.

Note that the exogenous information levels denoted by the number of fixed rows
and columns in Table 17.8 can be represented by alternative measures, and this is
illustrated in Table 17.9.

Table 17.10.  Percentage in the Sample Using £5m Lower Bound
for Different Sector Levels

| Sector level | percent of nonzero cells by number | percent of nonzero cells by value |
|---|---|---|
| 69 | 21 | 92 |
| 34 | 51 | 98 |
| 13 | 87 | 99.8 |
| 6 | 100 | 100 |

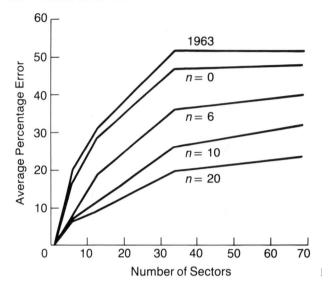

Fig. 17.1. Number of sectors.

Table 17.10 shows the percentage coverage of the £5m lower bound for different levels of aggregation.

The error measures of Table 17.8 are of the updated matrix including the exogenous information; thus, the errors are a measure of the RAS method with extra information. The results of Table 17.8 are illustrated in Figures 17.1 and 17.2. Figure 17.1 demonstrates that increasing the number of rows and columns increases the mean absolute percentage error for the various levels of exogenous information, but at a decreasing rate.

Figure 17.2 shows that increasing the level of exogenous information reduces the average absolute percentage error as would be expected, but the effect is greater at higher levels of disaggregation.

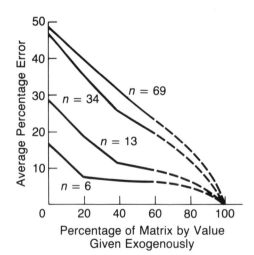

Fig. 17.2. Percentage of matrix by value given exogeneously.

## Conclusions

This study has shown that large percentage errors exist in an update of the 1963 absorption flow matrix to 1968 using the RAS method. The estimate is improved by the inclusion of six fixed rows and columns but the errors remain large. It has also demonstrated that the 1963 firmly based absorption matrix can be used to examine the economic structure in 1968 by operating on a firmly based 1968 final demand matrix, almost as effectively as an updated version of the absorption matrix. In order to bring percentage errors down to the order of 10% in the updated absorption matrix, either the number of industries must be reduced to six or lower, or the exogenous information by either value or percentage of nonzero calls must be about 60% (i.e., about 18 rows and columns in terms of this study of 69 sectors).

There can be no doubt that the reclassification exercise and improvements in data sources and methodology between 1963 and 1968 are responsible to some extent for the differences between the updated and firmly based tables for 1968. These factors are of course difficult to measure in quantitative terms. However, it is unlikely that differences arising from these factors significantly affect either the assessment of the RAS method described in this paper, or the conclusions drawn.

It would seem, therefore, that the RAS method can be used as a convenient means of constraining matrices to given row and column totals, but that little should be expected of it as a means of forecasting absorption matrices over a period as long as five years at the 69-industry level. This in turn suggests that publishing input-output absorption matrices for years between full-scale censuses is of dubious merit unless a considerable amount of exogenous information is available; in terms of illustrating the contemporary economic structure of the United Kingdom, the latest firmly based table will do almost as well.

The Central Statistical Office is very much aware of these points, and efforts are being made to incorporate as much exogenous information as possible into the input-output tables for years for which there was no purchases inquiry.

### References

Allen, R. I. G. 1974. Some experiments with the RAS method of input-output coefficients. *Oxf. Bull. Econ Statist.*, 36: 215–228.

Allen, R. I. G., and J. R. C. Lecomber. 1975. A critique of methods of adjusting, updating and projecting matrices. In *Estimating and Projecting Input-output Coefficients*, ed. R. I. G. Allen and W. F. Gossling. pp. 43–56. London: Input-output Publishing Company.

Bacharach, M. 1970. *Biproportional Matrices and Input-output Change.* Cambridge: Cambridge University Press.

Barker, T. S. 1975. An analysis of the updated 1963 input-output transactions table. In *Estimating and Projecting Input-output Coefficients*, ed. R. I. G. Allen and W. F. Gossling. pp. 57–66. London: Input-output Publishing Company.

Central Statistical Office. 1961. Input-output tables for the United Kingdom 1954. London: H.M.S.O.

———. 1970. Input-output tables for the United Kingdom 1963. London: H.M.S.O.

———. 1973. Input-output tables for the United Kingdom 1968. London: H.M.S.O.

———. 1974a. Input-output tables for the United Kingdom 1970. London: H.M.S.O.

———. 1974b. Summary input-output tables for 1970: Economic Trends. No. 247 (May): pp. 13–26. London: H.M.S.O.

————. 1975a. Input-output tables for the United Kingdom 1971. London: H.M.S.O.

————. 1975b. Summary input-output tables for 1971. Economic Trends. No. 258 (April) pp. 105–119. London: H.M.S.O.

————. 1976. Input-output tables for the United Kingdom 1972. London: H.M.S.O.

Department of Applied Economics. University of Cambridge. 1963. *Input-output Relationships, 1954–1966.* London: Chapman and Hall.

Lecomber, J. R. C. 1975. A critique of methods of adjusting, updating and projecting matrices. In Estimating and Projecting Input-output Coefficients, ed. R. I. G. Allen and W. F. Gossling. pp. 1–25. London: Input-output Publishing Company.

Miernyk, W. H. 1975. The projection of technical coefficients for medium-term forecasting. Paper presented at the Input-output Research Association Conference (February). London.

Paelinck, J. and Waelbroeck, J. 1963. Etude empirique sur l'evolution de coefficients input-output. *Economie Appliqueé* 16(1): pp. 81–111.

Schneider, J. 1965. An Evaluation of two alternative methods for updating input-output tables. B.A. dissertation. Harvard College.

Tilanus, C. B. 1966. *Input-output Experiments, The Netherlands 1948–1961.* Rotterdam: Rotterdam University Press.

United Nations. 1973. Input-output Tables and Analysis. Studies in Methods. Series F, No. 14, Rev 1. New York: United Nations.

# 18

# Investment in Input-Output Models and the Treatment of Secondary Products

CLOPPER ALMON

## The Promise of Input-Output

Forecasting demand for investment goods ought to be one of the most fruitful and practical applications of input-output, for the value of input-output in studying the demand for a good grows with the indirectness of that demand. It helps but little in forecasting demand for food, but can serve us well in forecasting the demand for copper. When a model derives investment from the growth of output within itself, then the demand for capital goods becomes indirect. The demand for electric generators then depends upon the growth of all industries that use electricity. Moreover, the demand for the ingredients of capital items, such things as metal and plastics, becomes doubly indirect—and doubly amenable to input-output. Besides these benefits, there are opportunities. Only input-output can provide the detail by purchasing industry necessary to check the forecasts against direct knowledge of what is going on in the world. Only with this much detail can we help the makers of machines used in only few industries to foresee their demands.

Despite these attractions, only a few input-output studies have generated investment within the model. Indeed, there are problems that arise, or become acute, only when investment becomes endogenous. This paper discusses three of these problems. The first concerns the techniques used to generate investment; the second and third concern, respectively, the flow coefficients and the capital investment coefficients. On the first problem, I shall have to argue that, after all, no special technique is needed. Any good way to explain investment can be built into an input-output model. The second problem deals with defining flow coefficients when secondary products are numerous, as they are when the machinery sectors are divided sufficiently finely to take advantage of the endogenous investment. The third deals with evidence on the stability of coefficients in the investment matrix, specifically with the shares of six types of metal cutting machine tools in the investment spending of metalworking industries.

This article was published in *Applications of Input-Output Analysis*, edited by Anne P. Carter and Andrew Bródy, © North-Holland Publishing Company, 1970, pp. 103–116, Volume 2. Reprinted by permission.

**Ways to Generate Investment**

Leontief's (1953) original exposition of the "dynamic" input-output system set off a now extensive literature on the determination of investment in an input-output growth model. Much of this work centered on the "switching" problem, or as it is sometimes called, the "stability" of this model. Simply put, the question was, "Will the model generate negative outputs or otherwise come to an impasse?" And the answer in most of the studies has been "Yes". I believe that this literature concerned only mathematical curiosities that follow from various formulations of the investment process, and, which do not match the way *any* economy invests. Basically, such formulations suppose that the growth of industries is determined so that the investment required for that growth exactly uses up whatever remains from current output after subtraction of current consumption, export, and government use. This formulation, of course, has the cart before the horse, for industries invest to grow, not grow to invest. Naturally, such models produced some rather bizarre growth paths. It is these results that have, unfortunately, created the impression (under which I long labored) that there is some "problem" about investment peculiar to input-output models. Some of the devices described below reflect this impression. Our present work, however, shows this view to be quite mistaken; any good investment function can be put into an input-output model that is patterned on the way the economy works.

The simplest device for including investment in an input-output model is intended for making a rough projection of a single future year $T$ years ahead. $T$ is apt to be 5 or 10. If we assume linear growth between now and then, the rate of growth of output $x(t)$ will be $(x(T) - x(0))/T$ and investment and output are found simultaneously from the equation

$$x(T) = Ax(T) + B(x(T) - x(0))/T + y(T) \qquad (18\text{--}1)$$

where $A$ is the input-output matrix, $y$ is the vector of final demands, and $B$ is the investment matrix (i.e., $b_{ij}$ shows the amount of capital goods $i$ needed by the $j^{th}$ industry to make one unit of its output per year). Since growth may be more nearly exponential than linear, $(x(T) - x(0))/T$ may understate growth in year $T$. Alan Manne (1964) has suggested, therefore, that we "mark up" the linear growth rate $(x(T) - x(0))/T$ by the ratio of the slope of an exponential curve to the slope of a straight line which has risen by the same amount, that is, by $c = ae^{at}/((e^{at} - 1)/T)$ where the $a$ is a growth rate, say 0.04, likely to be typical of the industries of the economy in question. This $c$ depends upon $a$, but $c$ changes slowly with respect to $a$, and the correction is likely to help in all but exceptionally slowly growing industries.

This approach implies a simultaneity between output and investment decisions. In economies like that of the United States, output generally precedes investment. Only over a long period, say five or ten years, do they appear to be simultaneous. Hence, only when $T$ is about five or more can this formulation yield a realistic, if rough, estimate of investment.

A method of finding investment for equilibrium growth—i.e., growth in which capacity is always kept at exactly its desired level—was developed and applied in

Almon (1963). In this method, the equilibrium growth path, $x(t)$, must satisfy the equations

$$x(t) = A(t)x(t) + \mathbf{D}B(t)x(t) + y(t) \qquad (18\text{--}2)$$

where $\mathbf{D}$ is the differential operator, and the $A$ and $B$ matrices are explicitly made to depend on time $t$. (It must be emphasized that Eq. (18–2) holds only on the equilibrium growth path. Since it does not hold when starting from arbitrary initial conditions, the study of its "stability" is meaningless.) We may rewrite Eq. (18–2) as

$$x(t) = (I - A(t))^{-1}\mathbf{D}B(t)x(t) + (I - A(t))^{-1}y(t)$$

or

$$x(t) = (I - (I - A(t))^{-1}\mathbf{D}B(t))^{-1}(I - A(t))^{-1}y(t)$$

or

$$x(t) = (I - Q(t)\mathbf{D}B(t))^{-1}Q(t)y(t)$$

where $Q(t) = (I - A(t))^{-1}$. Then we expand the differential operator on the right in power series

$$x(t) = Q(t)y(t) + Q(t)\mathbf{D}B(t)Q(y)y(t) + (Q(t)\mathbf{D}B(t))^2 Q(t)y(t) + \cdots \quad (18\text{--}3)$$

If the series on the right converges, substitution into Eq. (18–2) shows that the power series expansion is valid. The first term shows the output that would be necessary for producing final demands if capital investment could be ignored. The second term shows the investment necessary for the growth in the first term, ignoring, however, the investment necessary to support the growth in the second term. The third term gives the investment for the growth in the second term, and so on to higher terms. This layer-upon-layer interpretation insures that the solution will be sensible and provides a convenient method of numerically calculating $x(t)$.

I used this series in a model that produced the whole time path of $x(t)$ from the present out to the terminal year. Stone (1962) had previously discovered and applied a special case of this series to the problem of finding investment in a single target year. In this special case, $A$ and $B$ are constant and each element of $y(t)$ is an exponential, so that $\mathbf{D}f(t) = \hat{r}y(t)$, where $\hat{r}$ is a diagonal matrix made up of the growth rates of the elements of $y(t)$. The series then becomes

$$x(t) = Q(I + BQ\hat{r} + (BQ)^2(\hat{r}^2) + \cdots)y(t)$$

and the matrix $V = \sum_{n=1}^{\infty}(BQ)^n\hat{r}^n$, when multiplied by $y(t)$, gives investment necessary for exponential growth. Consequently, if when we specify $y(t)$ we also specify the rates of growth of its elements—the $\hat{r}$ matrix—then the balance equations for the target year can be written

$$x = Ax + (I + V)y \qquad (18\text{--}4)$$

This is the form used by Stone. Equation (18–4) makes investment depend upon the growth in output in the same year; Eq. (18–1) makes the terminal year's investment depend upon the average growth over the period from the present to the target year. Thus, Eq. (18–1) ties down the target year by connecting it with the present, while Eq. (18–4) lets it float in the air. But if we can safely assume that the intervening years will take care of their own investment needs, then the connection of year $T$ with the present is out-of-place and the floating formula is appropriate. The proper choice depends upon the context.

The series solution (18–3) suffers from the disadvantage that proof of its convergence depends (a) upon the constancy of the $A$ and $B$ matrices and (b) upon the form of the function $y(t)$. A functional space approximation enables us to prove convergence without the second condition, and, at the same time, to achieve faster convergence. In this method, used in Almon (1966), we begin wtih a trial projection of the path of $x(t)$. Then, by least squares, we fit a polynomial to this path and use this polynomial, call it $x^0(t)$, for the investment term, $DB(t)x^0(t)$, on the right of Eq. (18–2). With this term fixed, we then solve Eq. (18–2) for $x(t)$, approximate this $x(t)$ by a polynomial, put it back into Eq. (18–2) and repeat the process. Provided the degree of the polynomials used to fit to the paths is low enough—fourth or fifth degree is safe for a twelve-year forecast—the process can be proved to converge. If the initial (zeroth) approximation of the output series is that output is always zero, then the first approximation will be equal to the first term in the series (18–3), the second approximation will be equal to the sum of the first two terms and so on. Obviously, by taking a better first guess, the functional space approximation method will get to the answer faster than does the series.

The functional approximation method also permits any sort of lag relationship between output and investment. Investment may depend on future output as in the aerospace industry, or on past output as in the automobile industry. This capability may make this method useful also for planning of less developed countries.

My present view, however, is that the functional approximation method is needlessly cumbersome for ordinary forecasting of the U.S. economy. In our present work, based on a 93-sector $A$ matrix, with the 93 aggregated to 69 for investment functions, we use a simulation approach. In most industries, investment in a given year can be explained in terms of sales, cash flow, and other variables of the preceding year. Consequently, the scheme of our model is very simple: each year's investment depends on variables calculated in the preceding year. (In a few industries, this year's output helps explain this year's investment; for these industries, we go back and make a correction after the initial calculation of outputs.) This simple scheme has allowed us to increase the complexity of the investment equation in other directions. We have several different types of equations for different industries. In the most usual type, we assume that in each industry equipment investment in year $t$, $I_t$, is proportional to the gap between the desired stock $K_t^*$, and the stock that would be available at the end of the year if no investment was made $(1 - d) K_t$; where $K_t$ is actual stock at the beginning of the year and $d$ is the depreciation rate. Thus, we assume

$$I_t = a_t[K_t^* - (1 - d)K_t].$$    (18–5)

We next suppose that $K_t^*$ depends on output $X_t$,

$$K_t^* = c_0 + c_1 X_{t-1}$$    (18–6)

and that the rate of adjustment $a_t$ depends on the adequacy of cash flow $CF$ (retained earnings plus depreciation), relative to the size of the job to be done. Thus

$$a_t = c_2 + c_3 CF_t / K_t^* - (1 - d)K_t)\tag{18–7}$$

Substitution of Eqs. (18–6) and (18–7) into Eq. (18–5) yields

$$I_t = b_1 + b_2 X_{t-1} - b_3 K_t + b_4 CF_t\tag{18–8}$$

where the $b$s are constants to be estimated by regression. We create the capital stock by accumulating past equipment investment not yet retired. The cash flow variable must be approximated because cash flow data are always on a company basis, rather than on an establishment basis. Consequently, we calculate our establishment-base cash flow by multiplying the sales of these establishments by the ratio of cash flow to sales in the most appropriate company-based industry. In forecasting, we project this ratio exogenously. By changing it, we could study, for example, the effects of tax changes.

We have fit this equation for all 69 of our equipment investment series. About half of them give entirely satisfactory results: good fits, reasonable coefficients with small standard errors, and sensible forecasts. The other half suffer from a variety of maladies. The most common is a positive coefficient on the stock term. Such equations say that simple exponential growth explains investment better than does our stock adjustment mechanism. We found that they always gave forecasts that did not respond to changes in the growth rates of output. For these sectors, we had to specify a priori the coefficient on the stock term. The resulting fits were usually just about as good as the least-squares fit. However, for one group of sectors, including transportation, petroleum refining, electrical generation, and steel, no one equation could both explain the past and forecast the future well. For these, we had to recognize past or impending technological changes—and past miscalculations. In several of these industries, input-ouput studies contribute to the investment forecasts. For example, in petroleum refining the volume of investment depends very much on the difference between the growth in gasoline sales and the growth in fuel oil sales. In railroading, growth in traffic that moves on flat cars would lead to investment; growth in cattle traffic would not, for cattle cars are in excess supply. In electrical generation, investment depends on the growth in demand; but nuclear power has radically changed the incremental capital-output ratio.

The ability of an input-output model to accommodate even these exceptional cases clearly demonstrates that input-output poses no artificial obstacles to building a good investment model but rather provides the detail necessary in order to benefit from specific industry studies.

## Product-to-Product Matrix

One of the perennial trouble makers in input-output is the secondary product; the pump made in the electric motor establishment, refrigerator made in the pump plant, and the motor made in the construction machinery establishment. As these examples indicate, this problem becomes serious as soon as the machinery sectors are divided finely enough to take advantage of the endogenous generation of investment.

There are two standard methods of handling secondary products. One is the transfer along the row; the other, transfer along the column. In the row transfer, a pump made in a motor plant appears as a sale from motors to pumps. In the column transfer, this pump appears as a negative pump input into motors. These entries are not only confusing when looked at in the table; they can also distort the outcome of its use. If we use the row transfer system and ask what would be the impact of an increase in pump exports, this motor-to-pump secondary entry would lead to an immediate increase in the output of the motor industry and thereby to higher demand for copper wire and armature steel. Moreover, because the establishments in the construction machinery sector make some motors, which are transferred to the motor sector, the demand for pumps also generates demand for construction machinery to "go into" the pumps, a sort of secondary secondary. If we use the column transfer system, an increase in the demand for electrical motors will have the apparent effect of reducing the output of pumps, for the system insists that motor establishments must make motors and pumps in fixed proportions.

These problems led to a desire to "purify" the matrix so as to create a table showing the inputs of products into products, the matrix of the elementary discussions of input-output. We have devised a purification rite which seems to provide a workable solution.

The simplest approach postulates that each product is made by the same process, no matter what kind of establishment makes it. The vector of inputs into any establishment is therefore just a linear combination of the pure processes for the items it makes. In symbols

$$f_i = Mp_i \qquad (i = 1, \ldots, n), \tag{18–9}$$

where:

$f_i$ a column vector, is the transpose of the $i$th row of the primary flow matrix; that is, $f_{ij}$ equals the purchases of product $i$ by establishments in industry $j$.

$M$ is the product mix matrix: $m_{ij}$ = the fraction of product $j$ made in establishments in industry $i$. (The columns of $M$ sum to 1.0).

$p_i$ is the transpose of the $i$th row of the pure flow matrix; $p_{ij}$ = inputs of product $i$ into project $j$.

Equation (18–9) can be solved for $p_i$ by

$$p_i = M^{-1}f_i \qquad (i = 1, \ldots, n) \tag{18–10}$$

The $p_i$ vectors found by this formula will generally have some small negative elements and some small positive entries where the corresponding $f_i$ vectors have zeros. The former are clearly nonsense; the latter are at best dubious and certainly inconvenient, especially as we move toward larger matrices for which we will wish to store only nonzero entries in our computers. By analysing the solution of Eq. (18–10), we shall find a way to put a stop to this nonsense.

To simplify notation, let us drop the subscript $i$ from $f$ and $p$, and understand that the $f$ and $p$ in an equation all have the same $i$ subscript. Equation (18–9) may then be written $0 = -Mp + f$, and by adding $p$ to both sides we obtain

$$p = (I - M)p + f. \tag{18–11}$$

The column sums of the absolute values of $(I - M)$ will be less than 1.0 save in the unlikely case in which less than half of a product is produced by establishments primarily engaged in its production. The iterative process for the solution of (18–11) will therefore converge. In this process, we take as a first approximation of $p$, $p^{(0)} = f$ and then define successive approximations by

$$p^{(k+1)} = (I - M)p^{(k)} + f \tag{18–12}$$

To see the economic interpretation of Eq. (18–12), let us write out the equation for the use of a product, say steel, in making another product, say $j$:

$$p_j^{(k+1)} = f_J - \sum_{\substack{l=1 \\ l \neq j}}^{n} m_{jl} p_l^{(k)} + (1 - m_{jj}) p_j^{(k)} \tag{18–13}$$

The first term on the right of Eq. (18–13) tells us to begin with the steel purchases by the establishments in industry $j$. The second term directs us to remove the amounts of steel needed for making the secondary products of those establishments, using our present estimate $(p^{(k)})$ of the technology of those products. Finally, the last term causes us to add back the steel used in making product $j$ in other industries. The amount of steel added by the third term is exactly equal to the amount stolen, via second terms, from other industries on account of their production of product $j$:

$$(1 - m_{jj})p_j = \sum_{\substack{l=1 \\ l \neq j}}^{n} m_{lj} p_j \quad \text{since} \quad \sum_{l=1}^{n} m_{lj} = 1$$

It is now clear how to keep the negative elements out of $p$. When the "removal" term, the second on the right of Eq. (18–12), is larger than the "primary use" term, the $f_j$, we simply scale down all components of the removal term to leave a zero

Table 18.1. Comparison of Row-transfer and Purified Tables in Millions of Dollars. Electric Apparatus and Motors Row

| Buyer | Row-transfer | Secondary | Primary | Purified |
|---|---|---|---|---|
| Aluminium | 4 | 0 | 4 | 2.5 |
| Wire and other nonferrous | 30 | 25 | 5 | 0 |
| Heating, plumbing products | 91 | 13 | 78 | 73 |
| Hardware, plating, wire products | 23 | 3 | 20 | 17 |
| Engines and turbines | 38 | 19 | 19 | 11 |
| Metal working machinery | 99 | 14 | 85 | 97 |
| Pumps, blowers, and compressors | 193 | 59 | 134 | 146 |
| Household appliances | 150 | 20 | 130 | 119 |

balance. Then instead of adding back the "total-stolen-from-other-industries" term, $(1 - m_{jj})p_j$ all at once, we added it back bit-by-bit, as it is captured. If a plundered industry runs out of steel with only a third of the total amount of plundering claims satisfied, we simply add only a third of each plundering product's claim into the product's cell in $p$.

This procedure has been programmed and applied to the U.S. 1958 matrix. As should be expected, the results do not often differ greatly from the primary flows. The resulting table appears fairly reliable and considerably neater conceptually than any other method of handling the secondary product. Table 18.1 shows a portion of the electric motors and apparatus row of the row-transfer and the purified tables.

## Constancy of Capital Coefficients

In the first section, we were concerned only with total equipment expenditures; in this section, we study its division among the various types of machinery. More specifically, we ask: What changes have there been in the share of equipment spending going to each of six types of metal-cutting machine tools: boring machines, drilling machines, gear-cutting machines, grinding machines, lathes, and milling machines?

The *American Machinist/Metalworking Manufacturing* magazine periodically conducts a Census of Machine-Tool ownership. In the 1963 census, the most recent one published, the major machine-tool using industries were asked how many of each of these metal-cutting tools they owned and how many of these had been bought within the last ten years, that is from 1953 to 1962. After a little grouping necessary to match the AM/MM purchasing industries with our model's investment sectors, we had 27 purchasing industries. For each of these 27 industries, we summed the equipment investment (in 1965 constant dollars) from 1953 through 1962 and divided the sum into the stock of each type of machine tool that, in 1963, had been bought in the last ten years. For each tool, then, we had a ten-year average of the number of tools bought per million dollars of equipment investment by each of the 27 purchasing industries.

For each of the last ten years, we applied these averages to investment in the 27

Table 18.2.  Rates of Change of Machine-Tool Coefficients 1958–1967

| Machine type | Percent change per year | Standard error |
|---|---|---|
| Boring | $-1.5$ | 1.3 |
| Drilling | 4.5 | 0.7 |
| Gear-cutting and finishing | 1.5 | 2.2 |
| Grinding | 0.7 | 1.0 |
| Lathes | 2.0 | 0.9 |
| Milling | 1.0 | 1.4 |

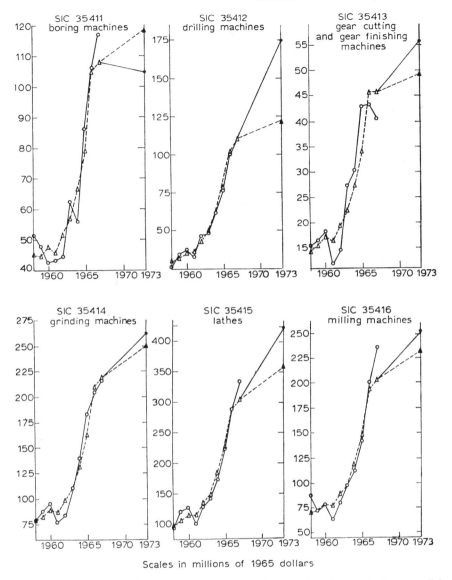

Scales in millions of 1965 dollars

Fig. 18.1. U.S. purchases of metal-cutting machine tools. ○: actual figure; △: trend coefficient indicator; ●: trend coefficient indicator for 1973; ▲: constant coefficient indicator for 1973.

purchasing industries and summed over these industries to prepare what we may call the constant-coefficient indicator for each type of machine tool. For comparison with it, we constructed the series for domestic purchases of each of these six tools, that is, manufacturers' shipments plus imports minus exports. All series were deflated to constant dollars.

To find the rate of change of the shares, we regressed on time the logarithm of the quotient of actual purchases divided by the constant-coefficient indicator. Table

18.2 shows the resulting coefficients and Figure 18.1 shows the graphs of actual and predicted series.

A quick glance at these figures leaves no doubt that the constant-coefficient indicator, coupled with a simple trend, is a valuable guide in forecasting machine-tool sales. For some types, such as drilling machines and lathes, there seems little room for improvement; for other types, such as boring and gear-cutting machines, the fit is less close. In particular, the 1967 observation was badly missed in most of the types. Table 18.2 shows that most of the trend terms were weak. Only the 4.5 percent per year increase for drilling machines and the 2.0 percent per year increase for lathes were statistically significant at customary significance levels. The other four ranged from $+1.5$ percent to $-1.5$ percent, but were not significantly different from zero. For them, the constant-coefficient indicator would have worked just about as well without the trend.

All in all, it appears that the assumptions of input-output analysis, which certainly include steady trends in the coefficients, hold up very well in the case of machine tools.

Now what of the future? The graphs show the projections of our model for 1973, a target five years in the future. The circle indicates the forecast based on the trend coefficient indicator; the triangle, the forecast by the constant-coefficient indicator. A slowing down in growth stands out. This slowing down in equipment investment characterizes all of our forecasts. It will be a good test of our present equations to see whether it occurs.

**References**

Almon, C. 1963. A modified Leontief dynamic model for consistent forecasting or indicative planning. *Econometrica 31*: 665–678.

Almon, C. 1966. *The American Economy to 1975*. New York: Harper and Row.

American Machinist/Metalworking Manufacturing. 1963. *The Ninth American Machinist Inventory of Metalworking Equipment*. New York: McGraw-Hill.

Cambridge, Department of Applied Economics. 1962. *A Programme for Growth. I. A Computable Model of Economic Growth*. Cambridge: Chapman and Hall.

Leontief, W. 1953. *Studies in the Structure of American Economy*. New York: Oxford University Press.

Manne, A. 1964. Key sectors of the Mexican economy. In: Studies in Process Analysis, Cowles Foundation Monograph 18. New York: John Wiley.

# 19

On the Theory of Dynamic Input-Output
Models with Different Time Profiles
of Capital Construction and
Finite Lifetime of Capital Equipment

LEIF JOHANSEN

## Background

There exists a well-developed theory of "classical" or standard dynamic input-output models with uniform construction or gestation periods for all sectors and types of capital equipment, and with infinite durability of capital. In practice different construction periods are very important, ranging from, say, one year up to some seven or eight years. For some sectors finite durability of capital equipment is also important, although, for many purposes, perhaps less so than the differences in construction periods. In practical planning dynamic input-output models with different construction periods seem to have been used mainly in Eastern Europe, but examples can also be found in some developing countries. Experience particularly in Eastern Europe suggests that the length of the construction periods is of considerable importance for the growth potential of an economy. In this paper we shall be concerned particularly with the growth aspect, although construction periods are obviously important also for the generation of possible cycles.

In addition to the formulation of the dynamic system in terms of quantities of inputs and outputs, we shall also study the corresponding price system. In connection with the more complicated dynamic pattern in the present context, it will be elucidating to consider the price version of the input-output model as a system of equations representing investment calculations rather than equations representing price fixation on the basis of cost calculations.

As far as possible we shall exploit results for the simplest case of a dynamic input-output model. For convenience we put down the equations for this model.

Let there be $n$ sectors, and let $x_{it}$ and $y_{it}$ represent total output and final demand for sector $i$ in period $t$. Let $a_{ij}$ be the current input coefficients and $b_{ij}$ be the coefficients for capital stock produced in sector $i$ and used in sector $j$. The ordinary

This article was published in *The Journal of Economic Theory*, Volume 19, Number 2, December, 1978, pp. 513–523. © Academic Press. Reprinted by permission.

input-output equations are then

$$x_{it} - \sum_{j=1}^{n} a_{ij}x_{jt} - \sum_{j=1}^{n} b_{ij}(x_{j,t+1} - x_{jt}) = y_{it} \qquad (i = 1, \ldots, n) \qquad (19\text{-}1)$$

In order to study the growth possibilities in such a system, it is useful to study first the corresponding homogeneous system (i.e., the system for which all $y_{it} = 0$). For this system we have a balanced growth path with a growth rate $\lambda$, i.e., a growth path at which

$$x_{it} = x_i(1 + \lambda)^t \qquad (19\text{-}2)$$

where the composition of output $x_1, \ldots, x_n$ and the growth rate $\lambda$ must satisfy the following equations:

$$x_i - \sum_{j=1}^{n} a_{ij}x_j - \lambda \sum_{j=1}^{n} b_{ij}x_j = 0 \qquad (i = 1, \ldots, n) \qquad (19\text{-}3)$$

The value of $\lambda$ is determined as the real positive root of the equation

$$|I - A - \lambda B| = 0 \qquad (19\text{-}4)$$

where $I$ is the unit matrix, $A$ is the matrix of $a_{ij}$, and $B$ is the matrix of $b_{ij}$.

The value of $\lambda$ represents the growth potential of the economy in the sense that it is the growth rate which is achievable if all outputs are used as inputs into production again (as current input and investment). If there is some nonproductive consumption, represented by $y_{it} > 0$, then the growth rate will be some proportion of $\lambda$ determined by the savings ratio.

The price equation corresponding to this system is

$$p_i = \sum_{j=1}^{n} p_j a_{ji} + r \sum_{j=1}^{n} p_i b_{ji} \qquad (i = 1, \ldots, n) \qquad (19\text{-}5)$$

where $p_1, \ldots, p_n$ are prices and $r$ is the interest rate. The value of $r$ is determined by an equation similar to (19–4), and we have $r = \lambda$. This equality represents the fact that the rate of interest is equal to the rate at which capital is able to grow if all outputs are ploughed back as productive inputs.

For some purposes it is useful to have the following formula for the change in $\lambda$ (and $r$) when the coefficients $a_{ij}$ and $b_{ij}$ are changed:

$$d\lambda = -\frac{\sum_{i,j}(p_i x_j da_{ij} + \lambda p_i x_j db_{ij})}{\sum_{i,j} p_i b_{ij} x_j} \qquad (19\text{-}6)$$

This formula can be derived very simply by differentiating the equations in (19–3), multiplying the resulting equations by $p_i$, taking the sum over $i$, and finally using Eq. (19–5). (A. Bródy (1970) uses this formula for assessing the consequences for numerical computations of $\lambda$ of errors in the data for $a_{ij}$ and $b_{ij}$. However, the formula also has important economic interpretations, giving directly the consequences for the growth rate—and the interest rate—of various sorts of combined changes in the coefficients of the input-output system.)

Equations (19–3) and (19–4) above were written down for the case of a closed model, i.e., the case in which the elements $y_{it}$ in (19–1) are zero. If we interpret the elements $y_{it}$ in (19-2) as consumption, then we could again study the growth rate under proportional growth, i.e., growth in which all elements of output and consumption grow at the same rate. Then the growth rate will be equal to the investment or savings ratio multiplied by the intrinsic growth rate $\lambda$ determined by (19-4), when the investment or savings ratio is calculated in terms of total income and total consumption evaluated by means of the equilibrium prices determined by the price equations (19-5). For an elaboration of this observation, see L. Johansen (1973).

The equations of the model as written above do not specify labour explicitly. The model could be conceived as a pure capital accumulation model, with labour abundance, or one may in a well-known way interpret one sector, say number $n$, as a labour sector, defining the coefficients on the row $a_{n1}, \ldots, a_{nn}$ as labour input coefficients, and the column coefficients $a_{1n}, \ldots, a_{nn}$ as consumption requirements per unit of work. The price $p_n$ would then be the minimum wage rate necessary to cover the corresponding consumption expenditures. (If we interpret sector $n$ in this way, and still have final demand items $y_{1t}, \ldots, y_{nt}$, then these items are final demand categories in excess of necessary consumption, for instance interpreted as consumption out of nonwage income.)

This remark on the treatment of labour applies also to the more general model set out in the following sections.

## The Model with Different Time Profiles for Capital Construction and Finite Life-time of Capital Equipment

Let $b_{ij}$ represent the amount of capital equipment produced in sector $i$ and needed per unit of output in sector $j$, as before. However, we now decompose $b_{ij}$ into elements $b_{ij1}, b_{ij2}, \ldots, b_{ijT}$ such that

$$b_{ij1} + b_{ij2} + \cdots + b_{ijT} = b_{ij}$$

Here $b_{ij\theta}$ is input in capital construction which must be delivered $\theta$ periods before the piece of production capacity, of which it is a part, is to be ready for use, for $\theta = 1, \ldots, T$. $T$ is accordingly the longest gestation period, and the distribution of $b_{ij}$ over $b_{ij1}, \ldots, b_{ijT}$ gives the time profile of this particular input into capital construction. We could of course introduce $T_{ij}$ for the longest gestation period for deliveries belonging to $b_{ij}$. However, we get a simpler notation system by using $T$ for the longest gestation period for the whole system, implying that $b_{ij\theta} = 0$ for $\theta > T_{ij}$.

On the output side we let capacity created in sector $i$ in a certain period last for $S_i$ periods. In other words, the capacity remains constant for $S_i$ periods, and then suddenly drops to zero. It is possible to extend the model so as to cover also more general cases (and this has recently been done by M. Åberg and H. Persson [1981]).

In setting out the quantity equations for this model it is convenient to write down separately equations for output and equations for changes in capacity. As above we let $x_{it}$ represent output from sector $i$ in period $t$. We let production capacity in the same sector in the same period be $K_{it}$, so that $x_{it} \leqslant K_{it}$. Furthermore, production capacity to be completed so as to be ready for use in period $t$ will be denoted by $k_{it}$. This is a gross concept, representing net addition to capacity as well as replacement investment.

Instead of (19–1) we now have the following equation for output:

$$x_{it} - \sum_{j=1}^{n} a_{ij}x_{jt} - \sum_{\theta=1}^{T} \sum_{j=1}^{n} b_{ij\theta}k_{j,t+\theta} = y_{it} \qquad (i = 1, \dots, n) \qquad (19\text{–}7)$$

The double sum in this formulation represents deliveries to capacity which is to be ready for use 1, 2, ..., $T$ periods ahead.

Total capacity in period $t$ is determined as total capacity in the preceding period, $t - 1$, plus gross addition to capacity, $k_{it}$, minus capacity which is becoming obsolete in period $t$, which is the same as capacity created $S_i$ periods earlier, i.e., $k_{i,t-s_i}$. This gives the following equation

$$k_{it} = k_{i,t-1} + k_{it} - k_{i,t-s_i} \qquad (i = 1, \dots, n) \qquad (19\text{–}8)$$

We now want to study balanced growth paths with full capacity utilization, i.e., we assume

$$x_{it} = K_{it} \qquad (i = 1, \dots, n) \qquad (19\text{–}9)$$

so that

$$x_{it} = x_{i,t-1} + k_{it} - k_{i,t-s_i} \qquad (i = 1, \dots, n) \qquad (19\text{–}10)$$

Our system now consists of (19–7) and (19–10) with $x_{it}$ and $k_{it}$ as variables.

We again study the homogeneous system and consider especially the possibility of proportional growth in all sectors. This requires both output and capacity creation to grow at the same rate, i.e.

$$x_{it} = x_i(1 + \lambda)^t, \quad k_{it} = k_i(1 + \lambda)^t \qquad (i = 1, \dots, n) \qquad (19\text{–}11)$$

Inserting this in Eqs. (19–7) and (19–10) we obtain, after cancellation of the factor $(1 + \lambda)^t$:

$$x_i - \sum_{j=1}^{n} a_{ij}x_j - \sum_{\theta=1}^{T} \sum_{j=1}^{n} b_{ij\theta}k_j(1 + \lambda)^{\theta} = 0, \qquad (19\text{–}12)$$

$$x_i = x_i(1 + \lambda)^{-1} + k_i - k_i(1 + \lambda)^{-s_i} \qquad (19\text{–}13)$$

We can now use Eq. (19–13) to express $k_i$ in terms of $x_i$. (We assume throughout that $\lambda > 0$.) Inserting the resulting expression into (19–12) we have

$$x_i - \sum_{j=1}^{n} a_{ij}x_j - \lambda \sum_{j=1}^{n} \left[ \sum_{\theta=1}^{T} b_{ij\theta} \frac{(1 + \lambda)^{s_j + \theta - 1}}{(1 + \lambda)^{s_j} - 1} \right] x_j = 0 \qquad (19\text{–}14)$$

This system of equations is comparable to (19–3) for the standard dynamic input-output model. It reduces to this form if $T = 1$ and $S_j \to \infty$. Then the sum in brackets in (19–14) contains only one element which tends to $b_{ij1}$ when $S_j \to \infty$.

For the further consideration of system (19–14) it is convenient to introduce a sort of total capital coefficients defined (for $\lambda \neq 0$) by

$$b_{ij}^* = b_{ij}^*(\lambda) = \sum_{\theta=1}^{T} b_{ij\theta} \frac{(1+\lambda)^{s_j+\theta-1}}{(1+\lambda)^{s_j}-1} \qquad (19\text{--}15)$$

We write $b_{ij}^*(\lambda)$ to signify that these are functions of the growth rate $\lambda$. With these symbols introduced the system (19–14) can be written as

$$x_i - \sum_{j=1}^{n} a_{ij}x_j - \lambda \sum_{j=1}^{n} b_{ij}^*(\lambda)x_j = 0 \qquad (19\text{--}16)$$

Corresponding to the determination of $\lambda$ by (19–4) in the standard case $\lambda$ is now to be determined by

$$|I - A - \lambda B^*(\lambda)| = 0 \qquad (19\text{--}17)$$

where $B^*(\lambda)$ is the matrix of the coefficients defined by (19–15) as functions of $\lambda$.

The introduction of the "coefficients" $b_{ij}^*(\lambda)$ is useful because it makes the system similar to the system for the standard case, the "only" difference being that the constant matrix $B$ in the standard case is replaced by a matrix $B^*(\lambda)$ whose elements depend on $\lambda$. One step in the same direction is taken in papers by V. Z. Belenkii et al. (1973–1974) and by V. A. Volkonskii (1975) who introduce similar coefficients which reflect the time profiles of capital construction. They are somewhat more general in that they assume different growth rates for different sectors, but less general in that they do not allow for different durabilities. On the other hand W. F. Gossling (1972, 1975) uses a similar type of "coefficients" to reflect replacement needs in the case of finite durabilities of capital equipment, but does not allow for different time profiles of capital construction.

Some further comments on the economic interpretation of the "coefficients" $b_{ij}^*(\lambda)$ will be given in the section entitled, "Alternative Derivation of the Price Equations."

The system now described will determine a unique value of the growth rate $\lambda$. If we are able to solve (19–4) numerically for $\lambda$, then it will also be a simple matter to solve (19–17), at least approximately, by iteration or interpolation. Some further observations on these aspects are given in the appendix notes.

From the formulation just given one can see the direction of the effect of the time profiles introduced in the present formulation of the dynamic input-output model. Consider first the capital construction profile. Suppose that we have the correct solution for a special set of profiles. Then increase the production or gestation periods by shifting some parts of the total coefficients $b_{ij}$ towards longer lags, keeping the totals fixed as given by $b_{ij} = \sum_\theta b_{ij\theta}$. If we calculate $b_{ij}^*(\lambda)$ for the new profile of construction periods, but with the old value of $\lambda$, then it follows from (19–15) that the values of $b_{ij}^*(\lambda)$ will increase, i.e., the elements of the matrix $B^*(\lambda)$ in (19–17) will increase. The old value of $\lambda$ does then no longer satisfy Eq. (19–17), and the new correct value must be lower than the previous one. (The formal proof is given in the last section.) In other words, extentions of construction periods diminish the growth rate.

Consider next the durabilities of capital equipment $S_j$. If some $S_j$ is increased, then all $b_{ij}^*(\lambda)$ for this $j$ decline, for the initially given value of $\lambda$. Then this value of $\lambda$ does no longer satisfy (19–17), and the new solution for $\lambda$ will give a higher value than the previous one. In other words, increases in the durability of capital equipment increase the growth rate.

## Prices, Interest Rate, and Investment Profitability

In the standard case the price of the dynamic input-output model can be written down in the form (19–5), simply as current input cost plus an interest rate applied to the value of capital stock in each sector. In the case of different construction periods and finite lifetime of capital equipment this cannot be done so simply, because we cannot write down a simple expression for the value of capital per unit of output as done by $\sum_j p_j b_{ji}$ in (19–5). It seems better to approach the price problem via the investment profitability calculations. The equilibrium requirement now is that the present value of investment outlays and the present value of future revenue achieved by establishing a piece of production capacity in a sector should be in balance.

Now consider sector $i$. For the comparison of present value of outlays with the present value of revenues it is immaterial to which point of time we refer the various items by discounting. We consider the point of time at which a piece of capacity is being completed and ready for use. For establishing a capacity of one unit in sector $i$ we must invest an amount $\sum_j p_j b_{ji1}$ one period earlier, an amount $\sum_j p_j b_{ji2}$ two periods earlier, etc. The value of this stream of investment outlays, calculated at the point of time when the capacity is completed, will be

$$\sum_{\theta=1}^{T} \sum_{j=1}^{n} p_j b_{ji\theta}(1 + r)^\theta \tag{19–18}$$

When the capacity has been installed, it will produce one unit of output per period for $S_i$ periods, beginning with the immediately commencing period. The revenue earned per period will be $(p_i - \sum_j p_j a_{ji})$, so that the present value of the stream of revenues will be

$$\left( p_i - \sum_{j=1}^{n} p_j a_{ji} \right)\left[ 1 + \frac{1}{1+r} + \left( \frac{1}{1+r} \right)^2 + \cdots + \left( \frac{1}{1+r} \right)^{S_i - 1} \right]$$

$$= \frac{(1 + r)^{S_i} - 1}{r(1 + r)^{S_i - 1}} \left( p_i - \sum_{j=1}^{n} p_j a_{ji} \right) \tag{19–19}$$

Equilibrium now requires that (19–18) and (19–19) be equal:

$$\sum_{\theta=1}^{T} \sum_{j=1}^{n} p_j b_{ji\theta}(1 + r)^\theta = \frac{(1 + r)^{S_i} - 1}{r(1 + r)^{S_i - 1}} \left( p_i - \sum_{j=1}^{n} p_j a_{ji} \right) \tag{19–20}$$

This formula can be reorganized so as to make it comparable with the formulas for the volume system. We obtain

$$p_i - \sum_{j=1}^{n} p_j a_{ji} - r \sum_{j=1}^{n} p_j \left[ \sum_{\theta=1}^{T} b_{ji\theta} \frac{(1 + r)^{S_i + \theta - 1}}{(1 + r)^{S_i} - 1} \right] = 0 \tag{19–21}$$

This is comparable with (19–14). Introducing the notations from (19–15) we can write the equation as

$$p_i - \sum_{j=1}^{n} p_j a_{ji} - r \sum_{j=1}^{n} p_j b_{ji}^*(r) = 0 \qquad (19\text{–}22)$$

corresponding to (19–16). For the determination of the rate of interest $r$ this yields the same equation as for the determination of the growth rate $\lambda$ in the volume system, i.e.,

$$|I - A - rB^*(r)| = 0 \qquad (19\text{–}23)$$

This means that we have the rate of interest equal to the growth rate, i.e., $r = \lambda$, just as for the standard system set out in the introductory section.

We may conclude as follows: By requiring present value calculations for investments to be in balance, we have obtained a system of price equations which stand in relation to the volume equations in exactly the same way as for the price equations in the standard dynamic input-output model. We have also obtained a way of representing the generalized system, allowing for different time profiles of capital construction and finite lifetime of capital equipment, which is exactly the same as for the standard system, except for the fact that we have to replace the constant capital input coefficients $b_{ij}$ by "coefficients" $b_{ij}^*(\lambda)$ or $b_{ij}^*(r)$ which are not constant, but depend on the growth rate or the rate of interest.

## Alternative Derivation of the Price Equations

As already pointed out we do not have a simple measure of capital value, independent of the rate of interst, in the present model. Furthermore, if we consider a piece of production capacity through its history, it will begin to increase in value from the first investment; it will reach the maximal value at the point of time when it is ready for use, and from then on decline in value towards the expiration of its lifetime. Under these circumstances we cannot build up a price equation from the cost side by taking costs into account simply by multiplying capital value by the rate of interest. This is why we preferred the investment calculation point of view used in the preceding section.

However, as an alternative it is possible to build up the price equations from cost considerations by using an annuity method of calculating interest charges and amortization. We then write the price equations in the form

$$p_i = \sum_{j=1}^{n} p_j a_{ji} + D_i \qquad (19\text{–}24)$$

where $D_i$ represents capital costs.

We now interpret $D_i$ as a constant amount every year over the lifetime of the capacity such that its present value is equal to the investment made. The value of capital invested per unit of capacity in sector $i$, calculated at the point of time when the capacity is ready for use, is

$$J_i = \sum_{\theta=1}^{T} \sum_{j=1}^{n} p_j b_{ji\theta}(1 + r)^\theta \qquad (19\text{--}25)$$

Accordingly, we get the following relation between $D_i$ and $J_i$:

$$D_i\left[1 + \frac{1}{1+r} + \cdots + \left(\frac{1}{1+r}\right)^{s_i-1}\right] = J_i \qquad (19\text{--}26)$$

which gives

$$D_i = \frac{r(1+r)^{s_i-1}}{(1+r)^{s_i} - 1} J_i \qquad (19\text{--}27)$$

If we combine this with (19–25) and insert in (19–24), then we get precisely the price equations (19–21).

If we add the amounts $D_i$ over the lifetime of the equipment, then we get more than $J_i$. We can consider $D_i$ as covering both amortization and interest charges. The essential thing is that the total annuity is worth $J_i$, but we might consider $D_i$ as made up of a declining interest charge and an increasing amortization amount, according to the "constant total annuity" principle of amortization (see e.g., J. Lesourne [1963, pp. 248–249]).

We may conclude this section by giving yet another interpretation which is perhaps less convincing as a representation of cost accounting, but which may nevertheless be of interest because of the light it sheds on the interpretation of the capital coefficients $b_{ij}^*(\lambda)$ or $b_{ij}^*(r)$.

Consider first a breakdown of the investment $J_i$ defined by (19–25) according to sectors of origin $j$, i.e., consider

$$p_j \sum_{\theta=1}^{T} b_{ji\theta}(1 + r)^{\theta-1} \qquad (19\text{--}28)$$

This is the value of capital produced in sector $j$ which must be invested in sector $i$ per unit of output from sector $i$. (As a minor deviation from (19–25) we have here calculated the value of the investment at the point of time of the last input into the project, i.e., one period before it is actually used. This is the reason for the exponent $\theta - 1$ instead of $\theta$. In the present context this provides a better connection with previous formulas.) However, the capacity created lasts only for $S_i$ periods. Thus, if we want to sustain the production of one unit of output from sector $i$ for an indefinite time, then this investment must be repeated every $S_i$ period. In this sense we may say that the capital requirement for capital produced by sector $j$ and used in sector $i$, per unit of output from sector $i$, will be

$$p_j \sum_{\theta=1}^{T} b_{ji\theta}(1 + r)^{\theta-1}\left[1 + \left(\frac{1}{1+r}\right)^{s_i} + \left(\frac{1}{1+r}\right)^{2s_i} + \cdots\right]$$

$$= p_j\left[\sum_{\theta=1}^{T} b_{ji\theta}(1 + r)^{\theta-1}\right]\frac{1}{1 - (1+r)^{-s_i}} = p_j b_{ji}^*(r) \qquad (19\text{--}29)$$

The last equality follows by a simple reorganization of formula (19–15). According to this, total capital value per unit of output in sector $i$ in the sense now given will be

$$\sum_{j=1}^{n} p_j b_{ji}^*(r) \tag{19-30}$$

and the last term in the price equation (19–22) can be interpreted as interest on this extended capital concept.

Comparing the two approaches presented in this section, we may say that there is a choice between two viewpoints: We may consider capital costs as consisting of interest and amortization on existing capital equipment, or, we may consider capital costs as consisting of interest on the extended capital concept which is the present value of the stream of investments necessary to sustain a constant flow of output when an investment has to be repeated at certain intervals corresponding to the finite durability of capital equipment.

## Evaluation of Alternative Techniques

For the standard dynamic input-output model we have formula (19–6) for the effects of changes in the technological coefficients on the growth rate. We now seek a similar expression for the present case with different time profiles of capital construction and finite lifetime of capital equipment. In connection with (19–6) we described how this formula could be derived by simple differentiation. If we apply the same procedure to formula (19–16) and use the price equations (19–22), then we get

$$d\lambda = -\frac{\sum_{i,j}(p_i x_j \, da_{ij} + \lambda p_i x_j \, db_{ij}^*(\lambda))}{\sum_{i,j} p_i b_{ij}^*(\lambda)x_j} \tag{19-31}$$

In this formula the differential $db_{ij}^*(\lambda)$ must be evaluated so as to be expressed in terms of changes in the basic coefficients $b_{ij\theta}$. We must then also take into account that $db_{ij}^*(\lambda)$ depends on $d\lambda$. We have

$$db_{ij}^*(\lambda) = \sum_{\theta=1}^{T} \frac{\partial b_{ij}^*(\lambda)}{\partial b_{ij\theta}} db_{ij\theta} + \frac{\partial b_{ij}^*(\lambda)}{\partial \lambda} d\lambda \tag{19-32}$$

where $b_{ij}^*(\lambda)$ is defined by (19–15). (We use partial differentiation of $b_{ij}^*(\lambda)$ since $b_{ij}^*(\lambda)$ is a function of all $b_{ij\theta}$ as well as of $\lambda$, although we have for simplicity not indicated this explicitly in the notations.)

The partial derivatives of $b_{ij}^*(\lambda)$ with respect to $b_{ij\theta}$ are seen immediately from (19–15). The derivative with respect to $\lambda$ can also be written down on the basis of (19–15), but does not simplify in any very convenient way. We, therefore, retain it as $\partial b_{ij}^*(\lambda)/\partial \lambda$. Then we get

$$d\lambda = -\frac{\sum_{i,j}(p_i x_j \, da_{ij} + \lambda p_i x_j \sum_{\theta=1}^{T}[(1+\lambda)^{s_j+\theta-1}/((1+\lambda)^{s_j}-1)]db_{ij\theta})}{\sum_{i,j}(p_i b_{ij}^*(\lambda)x_j + \lambda p_i(\partial b_{ij}^*(\lambda)/\partial \lambda)x_j)}$$

$$\tag{19-33}$$

The denominator in this expression can be shown to be positive. For this purpose introduce

$$\tilde{b}_{ij}(\lambda) = \lambda b_{ij}^*(\lambda) \tag{19-34}$$

Using the definition (19–15) this can be written as

$$\tilde{b}_{ij}(\lambda) = \sum_{\theta=1}^{T} b_{ij\theta}(1+\lambda)^\theta \bigg/ \sum_{\tau=0}^{s_j-1} (1+\lambda)^{-\tau} \tag{19-35}$$

from which it is easily seen that

$$\frac{\partial \tilde{b}_{ij}(\lambda)}{\partial\lambda} > 0 \tag{19-36}$$

On the other hand, we have

$$\frac{\partial \tilde{b}_{ij}(\lambda)}{\partial\lambda} = b_{ij}^*(\lambda) + \lambda \frac{\partial b_{ij}^*(\lambda)}{\partial\lambda} \tag{19-37}$$

Then the positivity of the denominator in (19–33) follows.

The direction of the change in the growth rate by changes in the coefficients, accordingly, will be determined by the sign of the numerator. Thus, if we have a technological change affecting the input coefficients in a certain sector $j$, then the effect on $\lambda$ is determined by the sign of

$$\sum_{i=1}^{n} p_i\, da_{ij} + \lambda \sum_{i=1}^{n} p_i \sum_{\theta=1}^{T} \frac{(1+\lambda)^{s_j+\theta-1}}{(1+\lambda)^{s_j}-1}\, db_{ij\theta} \tag{19-38}$$

or, if we prefer, by the sign of

$$\frac{(1+r)^{s_j}-1}{r(1+r)^{s_j-1}} \sum_{i=1}^{n} p_i\, da_{ij} + \sum_{i=1}^{n} p_i \sum_{\theta=1}^{T} (1+r)^\theta\, db_{ij\theta} \tag{19-39}$$

In this expression the first term is the present value (referring to the point of time when a piece of capacity is ready for use) of the changes in current input costs over the coming $S_j$ periods, and the second term is the present value of the changes in investment costs. If the expression is positive, the value of the growth rate $\lambda$ will decline, and vice versa. Thus, the effect of a change in technology on the growth rate can be indicated by a present value calculation of the implied cost changes, using the equilibrium prices and the equilibrium rate of interest. From this criterion it is clear, as pointed out before, that a change in the time profile of capital construction which shifts some capital input from a longer to a shorter gestation lag, while keeping the total capital input as defined by $\sum_\theta b_{ij\theta}$ constant, will increase the growth rate. However, for the exact numerical evaluation of the effect one must use the full expression (19–33).

## The Model with Positive Consumption Demand, Savings and Growth

We now return to the model as given by (19–7) to (19–10) with positive final, nonproductive demand, i.e., $y_{it} \geq 0$ for $i = 1, \ldots, n$, and $y_{it} > 0$ for at least some $i$.

We shall refer to $y_{it}$ as consumption. (As already suggested, if there is a labour sector among $1, \ldots, n$, then $y_{it}$ could be interpreted as consumption out of nonwage income.)

When there is positive consumption, then the growth rate will no longer be the $\lambda = r$ previously considered. We now consider proportional growth of production and consumption at a rate $\gamma$. Inserting this and arranging the equations in the same way as we did in the second section, we now get

$$x_i - \sum_{j=1}^{n} a_{ij}x_j - \gamma \sum_{j=1}^{n} b_{ij}^*(\gamma)x_j = y_i \qquad (i = 1, \ldots, n) \qquad (19\text{--}40)$$

where $b_{ij}^*(\gamma)$ corresponds to $b_{ij}^*(\lambda)$ as defined by (19–15), only with the growth rate $\lambda$ replaced by $\gamma$.

The system (19–40) is no longer homogeneous in $x_1, \ldots, x_n$, and $\gamma$ cannot be found by an equation like (19–17). Something must also be added in order to make the problem of finding $\gamma$ determinate. This missing element must clearly be related to the saving or investment proportion in the economy. By inspection of (19–40) it seems plausible that, if we keep $x_1, \ldots, x_n$ normalized to a certain level, and then increase consumption $y_1, \ldots, y_n$, then the proportions between $x_1, \ldots, x_n$ will have to be readjusted, and $\gamma$ will have to be reduced in order to make the equations hold. (From (19–34) to (19–36) we know that $\gamma b_{ij}^*(\gamma)$ decreases if $\gamma$ decreases.) The increase in consumption as compared with the level of production, means a reduced investment proportion. Accordingly, we should expect a higher consumption proportion to be associated with a lower rate of growth $\gamma$. To make these considerations more precise we have to bring the prices into the picture. We then use the equilibrium prices and the rate of interest as determined in the third section.

For simplicity we introduce the following notations:

$$v_i = p_i - \sum_{j=1}^{n} p_j a_{ji} = \text{gross value added per unit of output in sector } i$$

$$\text{(profits if there is a labour sector among } 1, \ldots, n);$$

$$V = \sum_{i=1}^{n} v_i x_i = \text{total gross value added};$$

$$C = \sum_{i=1}^{n} p_i y_i = \text{value of total consumption};$$

$$J = V - C = \text{value of gross investment}.$$

Now multiply Eq. (19–40) by price $p_i$ and take the sum over $i$. This gives

$$J = V - C = \gamma \sum_{i=1}^{n} \sum_{j=1}^{n} p_i b_{ij}^*(\gamma)x_j \qquad (19\text{--}41)$$

On the other hand, multiply through the equations in (19–22)—the equations determining the equilibrium prices and the interest rate—by $x_i$ and take the sum over $i$. This gives

$$V = r \sum_{i=1}^{n} \sum_{j=1}^{n} p_i b_{ij}^*(r)x_j \qquad (19\text{--}42)$$

Combining (19–41) and (19–42) we get for the growth rate $\gamma$:

$$\gamma = \frac{J}{V} rR \quad \text{where} \quad R = R(r, \gamma) = \frac{\sum_{i,j} p_i b_{ij}^*(r) x_j}{\sum_{i,j} p_i b_{ij}^*(\gamma) x_j} \tag{19–43}$$

This expression consists of the gross investment quota $J/V$ (investment in proportion to profits if there is a labour sector among $1, \ldots, n$) multiplied by the rate of interest $r$ which reflects the productivity of capital, corrected by a term $R = R(r, \gamma)$ which in general depends on both $r$ and $\gamma$ and on prices and output composition.

For the standard case $S_j = \infty$ and $T = 1$, $b_{ij}^*(r)$ and $b_{ij}^*(\gamma)$ both reduce to a constant $b_{ij1} = b_{ij}$. Then the proportion to the right in (19–43) is identically equal to unity and we have

$$\gamma = \frac{J}{V} r \qquad \text{(for $S_j = \infty$ and $T = 1$)} \tag{19–44}$$

In this case there is no scrapping or depreciation. $J/V$ is the investment proportion in total income, and the growth rate is the product of this investment proportion and the rate of interest $r$, which reflects capital productivity. Three points are worth observing in this connection.

1. The formula just given corresponds to the classical growth formula for one-sector models.

2. In deriving this growth formula we have used the equilibrium prices in calculating the value of gross investment $J$ and total value added $V$. Accordingly, it appears that these equilibrium prices serve excellently the aggregation, since (19–44) is an exact formula under the conditions stated when $J$ and $V$ are calculated by means of the equilibrium prices, i.e., there is no aggregation error.

3. An interesting corollary of this result is a *dynamic non-substitution theorem*. If there is a choice between different techniques in the various sectors and we want to maximize the growth rate $\gamma$ under a fixed investment proportion $J/V$, then we should use the production techniques which maximize the intrinsic growth rate $\lambda = r$. Criteria for this maximization follow from the methods of evaluation of alternative techniques given in the previous section. The nonsubstitution aspect of this is that the technologies thus determined should be used regardless of the composition of consumption. [This nonsubstitution interpretation has been elaborated in L. Johansen (1973)].

When the conditions stipulated for (19–44) do not hold, then the proportion to the right in (19–43) is variable. We may trace the connection between the gross investment quota $J/V$ and the growth rate $\gamma$ in the following way. Consider the system (19–40) and keep $y_1, \ldots, y_n$ fixed. For a given value of $\gamma$ we can then solve the system for $x_1, \ldots, x_n$. (If we like we could require a normalization of $x_1, \ldots, x_n$ and readjust the level of $y_1, \ldots, y_n$.) By inserting the given value of $\gamma$ and the resulting $x_1, \ldots, x_n$ into (19–43) we can calculate the implied gross investment quota $J/V$. By repeating this calculation for different values of $\gamma$ we can trace a curve relating $J/V$ and $\gamma$. For $\gamma = r$ it is clear that there is no room for $y_1, \ldots, y_n > 0$, so at this end we have $\gamma = r$ and $J/V = 1$. At the other end we have $\gamma = 0$. This case needs a special treatment since the definition (19–15) of $b_{ij}^*(\gamma)$ is not valid for this case. Clearly $b_{ij}^*(\gamma) \to \infty$ as $\gamma \to +0$. On the other hand, for the case $\gamma = 0$ we could replace $\gamma b_{ij}^*(\gamma)$ by

$$\tilde{b}_{ij}(0) = \frac{1}{S_j} \sum_{\theta=1}^{T} b_{ij\theta} \qquad (19\text{-}45)$$

which follows from (19–35). We then get from (19–41) and (19–42)

$$\frac{J}{V} = \sum_{i,j} p_i \frac{\sum_\theta b_{ij\theta}}{S_j} x_j \bigg/ r \sum_{i,j} p_i b_{ij}^*(r) x_j \qquad \text{for } \gamma = 0. \qquad (19\text{-}46)$$

In other words, we have a positive gross investment quota for a zero growth rate, which is of course as it should be.

The fact that $\gamma b_{ij}^*(\gamma)$ increases with $\gamma$ (see (19–36)) suggests that there is a positive association between $\gamma$ and $J/V$. We have

$$\sum_{i,j} p_i [\gamma b_{ij}^*(\gamma)] x_j = \frac{J}{V} r \sum_{i,j} p_i b_{ij}^*(r) x_j$$

and for the increasing $\gamma$ the left-hand side would tend to increase while prices, $r$ and $b_{ij}^*(r)$ remain the same. However, this consideration is not quite sufficient to establish a monotonous positive association since the composition of output $x_1, \ldots, x_n$ in general changes with the growth rate. It only establishes a presumption for normal cases.

The natural idea now is, of course, to introduce the concepts of *net* investment and *net* income and try to establish an equation like

$$\gamma = \frac{\text{Net investment}}{\text{Net income}} \cdot r = sr = s\lambda \qquad (19\text{-}47)$$

where net investment is gross investment $J$ minus depreciation, net income is gross value added $V$ minus depreciation, and $s$ is introduced for the rate of saving out of net income. In fact, when $r$ and $\gamma$ are determined as explained before, then Eq. (19–47) can be shown to hold as a consequence of reasonable definitions of the concepts involved.

In order to see this it is necessary to introduce the notion of total value of existing capital. We write this as $K_t(r)$ for period $t$, and $K(r)$ for the base point of time. Since each piece of capital now lives through a history with a construction period of, in general, more than one time unit, and a productive life with gradually declining remaining lifetime, the value of capital cannot be calculated independently of the rate of interest. This is signified by the notations just given. The value of capital will, of course, also depend upon prices, but this is true also for the standard case.

We may now define net income as

$$\text{Net income} = rK(r) \qquad (19\text{-}48)$$

i.e., as the product of the interest rate and the value of capital stock. The rationale of this is that the rate of interest is equal to the intrinsic growth rate $\lambda$ as shown before. This means that, if nothing is extracted for consumption, then the value of total capital stock would increase by $\lambda K_t(r) = rK_t(r)$ from one period to the next, and income is defined as the value which can be extracted from the system without reducing total wealth.

If we now introduce

$$I = \text{value of total net investment}$$

and use $C$ for consumption as introduced above, then the concepts should satisfy the following equation

$$C + I = rK(r) \tag{19-49}$$

Furthermore, since we are studying proportional growth at a rate $\gamma$, we must also have

$$I = \gamma K(r) \tag{19-50}$$

According to the definition of the rate of saving we have

$$I = rK(r) - C = srK(r) \tag{19-51}$$

From (19–50) and (19–51) follows (19–47) as sought for.

It follows from this development that the comments made in connection with (19–44) for the standard case also hold for the general case, including the "dynamic non-substitution theorem." In deriving these results for the general case, we have only used some reasonable requirements which the definitions of the aggregated concepts ought to satisfy, in addition to the fact that $r = \lambda$ is the rate at which wealth will increase if nothing is extracted for consumption. We have not yet shown in detail how the various concepts could be calculated. However, they are implied by the formulas we have already given.

Let $D$ be total depreciation (at the base point of time). Then the net investment quota involved in (19–47) is $s = (J - D)/(V - D)$. Then (19–43) and (19–47) together imply

$$\frac{J}{V} R = \frac{J - D}{V - D} \tag{19-52}$$

where $R = R(r, \gamma)$ is defined in (19–43). This gives the following expression for depreciation, when we also use (19–43) to eliminate $V$:

$$D = J \frac{1 - R}{1 - (\gamma/r)} \tag{19-53}$$

Net income will be

$$V - D = V \frac{1 - (\gamma/rR)}{1 - (\gamma/r)} \tag{19-54}$$

From the definitional requirement (19–48) we have for the value of capital stock

$$K(r) = \frac{V}{r} \frac{1 - (\gamma/rR)}{1 - (\gamma/r)} \tag{19-55}$$

from which net investment according to (19–50) will be

$$I = V \frac{\gamma}{r} \frac{1 - (\gamma/rR)}{1 - (\gamma/r)} = J \frac{R - (\gamma/r)}{1 - (\gamma/r)} \qquad (19\text{–}56)$$

where the last equality uses (19–43). It is easily seen that this is the same as $J - D$ as calculated from (19–53).

In the formulas above $\gamma/rR$ should be replaced by $J/V$ as given by (19–46) for the case of $\gamma = 0$.

The use of the formulas given above should perhaps be recapitulated. Suppose we want to calculate the growth rate corresponding to a given net investment quota. We must then first solve the price system and get the interest rate $r$ (equal to the intrinsic growth rate $\lambda$). Then the growth rate $\gamma$ follows simply from (19–47). Having established the growth rate $\gamma$ we can use (19–40) to find output composition $x_1, \ldots, x_n$ for a given consumption composition $y_1, \ldots, y_n$. (We can normalize either the level of output or the level of consumption.) Having now both prices and outputs, $r$ and $\gamma$, we can calculate $R = R(r, \gamma)$ according to the definition in (19–43). Value added $V$ and gross investment $J$ follow easily from the definitions, and then the other aggregated magnitudes, i.e., total depreciation $D$, net income ($V - D$), the value of capital stock $K(r)$, and net investment follow from (19–53) to (19–56).

## Appendix

### *Notes on the Solution of the Characteristic Equation*

The characteristic equation determining the intrinsic growth rate of the dynamic input-output model was given in the second section in the following form

$$|I - A - \lambda B^*(\lambda)| = 0 \qquad (19\text{–}A1)$$

where the elements of the matrix $B^*(\lambda)$ are given by (19–10) in the text.

In this appendix we shall present some observations on the solution of (19–A1). We are only interested in real and positive solutions. Furthermore, we shall exploit as far as possible what is known about the standard case in which we have a constant matrix $B$ instead of $B^*(\lambda)$.

In order to study the existence and possible uniqueness of the solution of equation (19–A1), it is convenient to introduce an additional variable $z$ and replace (19–A1) by the following system in two equations and two variables, $\lambda$ and $z$:

$$|I - A - zB^*(\lambda)| = 0 \qquad (19\text{–}A2)$$

$$z = \lambda \qquad (19\text{–}A3)$$

From (19–A2) we may consider $z$ as a function of $\lambda$, $z = g(\lambda)$. (It is known that there will exist one real positive solution $z$ to (19–A2) for a given $\lambda$ if $A$ is productive and irreducible, which we shall assume.) The function will be continuous for $\lambda > 0$ since $z$ will be a continuous function of $B^*$ and $B^*$ is a continuous function of $\lambda$.

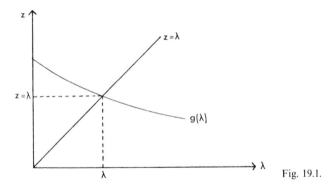

Fig. 19.1.

Let us first consider the case in which all $S_j = \infty$. Then we have

$$b_{ij}^*(\lambda) = \sum_{\theta=1}^{T} b_{ij}{}^{\theta}(1 + \lambda)^{\theta-1} \qquad (19\text{–}A4)$$

Then $b_{ij}^*(\lambda)$ is an increasing function of $\lambda$ for all $i, j$ for which some $b_{ij\theta} > 0$ for some $\theta > 1$; otherwise $b_{ij}^*(\lambda)$ is constant.

Now consider first $\lambda = 0$ in (19–A2). The value $\lambda = 0$ raises no special problems in this case since $b_{ij}^*(\lambda)$ is well defined by (19–A4) also for $\lambda = 0$. Then the system is the same as

$$|I - A - zB| = 0 \qquad (19\text{–}A5)$$

in which $B$ is the matrix of

$$b_{ij} = \sum_{\theta=1}^{T} b_{ij\theta} \qquad (19\text{–}A6)$$

This is the same as the characteristic equation of a dynamic input-output model of the standard type for which all investments in a project have been compressed into one period. For this system there exists one real positive solution $z$, i.e., we have $g(0) > 0$. If we now increase $\lambda$, then some or all elements of $B^*(\lambda)$ will increase while none will decline. It is then known that the value of the solution $z$ to (19–A2) will decline. (This also follows from (19–6).) Accordingly $g(\lambda)$ is a declining function of $\lambda$. It follows from this that there will exist one unique solution for $\lambda$ in (19–A1) corresponding to the unique solution for $z$ and $\lambda$ in (19–A2) to (19–A3) as illustrated in Figure 19.1.

If we now go back to the general case in which we have $b_{ij}^*(\lambda)$ given by (19–15), then the reasoning will be more complicated. For a very small value of $\lambda$ the value of $b_{ij}^*(\lambda)$ will be very large, and it will approach infinity as $\lambda \to +0$. It follows from this that $g(\lambda)$ will now tend to zero as $\lambda \to 0$. In other words, for very small values of $\lambda$, $g(\lambda)$ is an increasing function of $\lambda$. On the other hand, if we go on increasing $\lambda$, then sooner or later all $b_{ij}^*(\lambda)$ for which $b_{ij\theta} > 0$ for some $\theta > 1$ will start increasing, and then $g(\lambda)$ will be a declining function of $\lambda$. (This effect will dominate the effect of those $b_{ij}^*(\lambda)$ for which $b_{ij\theta} = 0$ for $\theta > 1$, since these $b_{ij}^*(\lambda)$ will tend to constants as $\lambda$ increases.)

In this case the solution may be as illustrated in Figure 19.2. In order to secure

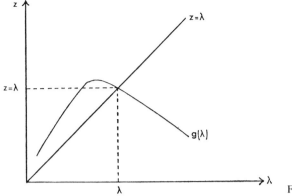

Fig. 19.2.

the existence of a solution $\lambda > 0$ in this case, the curve for $g(\lambda)$ must not lie below the line $z = \lambda$ throughout. In order to see when this will hold it is useful to rewrite Eq. (19–A2) into the form

$$\left| I - A - \left( \frac{z}{\lambda} \right) [\lambda B^*(\lambda)] \right| = 0 \qquad (19\text{–}A7)$$

or

$$|I - A - \rho \tilde{B}(\lambda)| = 0 \text{ where } \rho = \frac{z}{\lambda} \qquad (19\text{–}A8)$$

and where $\tilde{B}(\lambda) = \lambda B^*(\lambda)$ has positive elements defined by (19–34) to (19–35). It was also observed by (19–36) that the elements of $\tilde{B}(\lambda)$ increase with the value of $\lambda$, and at least some of them increase strictly. From these facts follows that $\rho$ is a strictly declining function of $\lambda$.

From (19–35) we see that, when $\lambda \to 0$ the elements $\tilde{b}_{ij}(\lambda)$ tend towards

$$\tilde{b}_{ij}(0) = \frac{1}{S_j} \sum_{\theta=1}^{T} b_{ij\theta} \qquad (19\text{–}A9)$$

which are purely physical characteristics: investments of various sorts regardless of gestation periods, divided by the lifetime of capital in the sector. $\rho$ is a continuous function of the elements $\tilde{b}_{ij}$ and will tend towards a value $\rho_0$ determined by

$$|I - A - \rho_0 \tilde{B}(0)| = 0 \qquad (19\text{–}A10)$$

when $\lambda$ tends to zero.

Now let us see what this means in connection with Figure 19.2. If $\rho_0$ determined by (19–A10) is $\rho_0 < 1$, then we will never get a solution $\lambda > 0$ since the function $g(\lambda) = z = \rho\lambda$ then starts from the origin by being below the ray $z = \lambda$, and will not intersect it for any $\lambda > 0$ since $\rho$ is a declining function of $\lambda$. On the other hand if $\rho_0 > 1$, then $g(\lambda)$ will be above $z = \lambda$ in the figure immediately to the right of the origin, and since we know that $g(\lambda)$ will eventually decline, there will exist a solution $\lambda > 0$.

As a by-product of this line of reasoning we also see that the solution for $\lambda$ is unique. Since $\rho = z/\lambda$ is the slope of the ray from the origin to a point on the curve $z = g(\lambda)$ in Figure 19.2, if there is an intersection between $z = \lambda$ and $z = g(\lambda)$ for some $\lambda > 0$, there cannot be another intersection to the right of the first one.

The treatment of the existence and uniqueness problem given above is partly influenced by the mathematical treatment of a generalized model by M. Åberg and H. Persson (1981). They approach the problems in a somewhat different way, but their approach suggested the crucial role of the fact that the elements $\lambda b_{ij}^*(\lambda)$ increase with $\lambda$.

We saw above that the existence of a solution $\lambda > 0$ depends on the existence of a solution $\rho_0 > 1$ to Eq. (19–A10). Introducing $\sigma = \rho_0 - 1$ we can write the equation as

$$|I - (A + \tilde{B}(0)) - \sigma\tilde{B}(0)| = 0 \qquad (19\text{–}A11)$$

We now have a solution $\lambda > 0$ if there is a positive solution for $\sigma$ in (19–A11). Such a solution exists if $A + \tilde{B}(0)$ is a productive matrix in the usual input-output sense. The interpretation of $A + \tilde{B}(0)$ is straightforward: it consists of the current input-output coefficients in $A$ plus capital coefficients summed over gestation lags and averaged out over the lifetime of capital equipment in the sector concerned, see (19–A9).

From the formulation (19–A2) to (19–A3) it is clear that there would be no difficulty in solving the present problem numerically if we are able to solve the characteristic equation for the standard case of a constant matrix $B$. A natural idea is to solve the equations iteratively, i.e., to start with an initial guess about the value of $\lambda$, say $\lambda_0$, insert this in (19–A2) and solve for $z$, next insert this value for $\lambda$ and solve again for $z$, and so on. This would generate $\lambda_0, \lambda_1, \lambda_2, \ldots, \lambda_q, \ldots$ according to

$$|I - A - \lambda_{q+1}B^*(\lambda_q)| = 0 \qquad (q = 0, 1, 2, \ldots) \qquad (19\text{–}A12)$$

This process would probably converge in most cases, but there seems to be nothing in what has been demonstrated above which shows that it will always converge from any initial guess $\lambda_0$. However, even if the process (19–A12) should not work well, it would be easy to find the solution with any desired degree of accuracy by calculating the value of $z$ for selected values of $\lambda$ and then proceed by interpolations.

### References

Belenkii, V. Z., V. A. Volkonskii, and N. V. Pavlov. 1973–1974. Dynamic input-output models in planning and price calculations and economic analysis. *Matekon* **10**: pp. 74–101.

Bródy, A. 1970. *Proportions, Prices and Planning*. Budapest: Akadémiai Kiadó, and Amsterdam: North Holland.

Gossling, W. F. 1972. *Productivity Trends in a Sectoral Macroeconomic Model*. London: Input-Output Publishing Company.

Gossling, W. F. 1975. Relative prices and wages-bills, under steady growth-rates. In *Capital Coefficients and Dynamic Input-Output Models*, ed. W. F. Gossling. London: Input-Output Publishing Company.

Lesourne, J. 1963. *Economic Analysis and Industrial Management*. Englewood Cliffs, N.J.: Prentice-Hall.

Johansen, L. 1973. The rate of growth in dynamic input-output models: Some observations along lines suggested by O. Lange and A. Bródy. In *Jahrbuch der Wirtschaft Osteuropas*, Band 4, pp. 69–88. Munich: Günter Olzog Verlag.

Volkonskii, V. A. 1975. Industry price levels and the norm of effectiveness in the system of optimal planning. *Matekon* **12**: 28–46.

Åberg, M. and H. Persson. 1981. A note on a closed input-output model with finite lifetimes and gestation lags. *Journal of Economic Theory* 24(3): pp. 446–452.

# 20

## Developing Ex Ante Input-Output Flow and Capital Coefficients

W. HALDER FISHER and CECIL H. CHILTON

The input-output (I/O) model provides the economist and the business planner a modeling framework within which a wide variety of estimates or forecasts can be reconciled and brought into mutual consistency.

Most past I/O applications have involved the generation of transactions tables from collected statistics, with subsequently derived direct coefficients. While this provides statistical descriptions of the specific real time period, its usefulness is a function of the accuracy of the survey and the representativeness of the period. The worst shortcoming of this statistical (ex post) approach to coefficients is the inevitable time lapse between survey and table. This lag would not be serious if the technical relationships expressed by the coefficients were highly stable. This, however, is not the case. In addition to technological change (conceptually, the main cause of changes in the coefficients), coefficients are affected by changes in relative prices, in output product-mixes, and/or in capacity utilization rates. Thus, the six-year gestation times of recent I/O tables for the U.S. economy have greatly limited their usefulness.

In addition, there is no such thing as a truly *normal* year. Therefore, although *realistic*, the results of a given statistical survey need not (indeed, probably cannot) be *typical*. Abnormalities introduced by the business cycle, exogenous shocks, or step functions seriously impair their generality.

These shortcomings of traditional statistical (ex post) approaches—along with the fact that statistics for such forecasts were not collected often enough in the past—led Battelle-Columbus to experiment with a new method of I/O table construction, which involves the direct generation of technical coefficients by means of judgmental technological forecasts (or back-casts). These ex ante coefficients are then combined with other estimates to derive transaction tables for either future or past years. Comparisons of these two approaches follow.

### The Statistical (Ex Post) Approach

Most of the I/O tables now being produced for government or business use involve the traditional statistical (ex post) approach exemplified by the Office of Business

This article was published in *Input-Output Techniques*, edited by Andrew Bródy and Anne P. Carter, © North-Holland Publishing Co., 1972, pp. 398–405. Reprinted by permission.

Economics of the U.S. Department of Commerce (OBE). This approach constitutes an attempt to measure, as precisely as possible, the actual business situation of a specified period. Sellers are asked to distribute the value of their total output over all buyers; and buyers are asked to report by sources the value of all purchases. An attempt is then made to reconcile these two often quite different sets of numbers into a single set of interindustry sales/purchases and final demands that balance meaningfully in both directions.

The first step in generating either a statistical or an ex ante I/O table is to define the sectors into which the economy will be divided. Thereafter, however, the two methods diverge.

The second step in the statistical approach involves surveys of sellers and buyers (classified by sectors) and the construction of a working table in which each cell contains two entries: (a) the total value of sales which sector $I$ reported making to sector $J$; and (b) the total value of purchases which sector $J$ reported making from sector $I$. These two numbers can (and usually do) differ significantly because of: (1) imperfections in each firm's knowledge of its markets; (2) imperfect coverage *by the survey* of representative groups of buyer and seller firms in given markets; (3) the fact that buyers tend to think in terms of prices paid, while sellers tend to think in terms of prices received; and (4) errors in classifying particular firms or establishments.

These sources of cell-by-cell error are obvious, and much time and money are required to examine each cell in the matrix and to substitute a single entry for the reported two. Also during this portion of the exercise, there is a strong temptation to improve and refine the individual cells—at great expense generally unmatched by improvements in the table.

### The Judgmental (Ex Ante) Approach

In contrast to the above, the judgmental (ex ante) approach to an I/O table is made via the direct coefficients. Direct coefficients indicate the proportions in which *purchased inputs* and *values added* are combined to create *output*. If a given sector achieves its output by means of a single pure technology, its direct coefficients consist of a single, clearly defined set of proportions. Most sectors, however, are made up of many establishments utilizing many different technologies. Such a sector's coefficients are weighted composites of several "pure" coefficients. Only if the matrix were finely disaggregated could each component technology be shown in its "pure" form.

After the sectors have been defined, the second step in the ex ante approach consists of expressing its current or projected technology in coefficient terms. In some instances we may have access to statistics which throw light on these proportions. Nevertheless, especially if we are projecting a future (or estimating a hypothetical) technology, we must usually turn to the knowledge and judgment of industry experts. This is a crucial part of the Battelle method. A great deal of preparatory time must be combined with meticulous field interviews to assure that a valid and meaningful set of expert judgments is obtained and converted into coefficients. We return to this point later; for now, let us assume that the entire input column has been expressed as coefficients, with purchased inputs and values added summing to unity.

Before the I/O table can be completed, *every* sector's coefficients must be established to the satisfaction of the experts involved. We need not, however, specifically consider all the interindustry relationships between this and other sectors. (This is one of the greatest advantages of the ex ante over the ex post methodology.) It is also necessary that we estimate the total dollar values of the final demands that each productive sector must supply. This must be done, and in essentially the same terms, in either approach.

After the direct input coefficient matrix has been established, it is inverted by means of the Leontief procedure and multiplied by the final demand vector in order to generate the dollar values of total outputs. These dollar values are then entered into the total input vector and distributed vertically in proportion to the direct coefficients, thus producing a dollar-flow matrix.

In summarizing this ex ante procedure, we can say that major intellectual efforts are expended in two activities: (a) establishing a column of direct coefficients for each sector, and (b) estimating the final demands. The remaining operations are carried out by the computer. Moreover, the mathematics of computation assure that the table of dollar flows (transactions) is always precisely balanced and internally consistent.

## Significance of the Differences Between the Ex Ante and Ex Post Approaches

The differences between these two methods of the I/O table construction are fundamental. The traditional approach involves cell-by-cell collection of sales/purchase statistics and their tedious reconciliation into a single balanced table. The ex ante approach involves the generation for each sector of a set of direct coefficients measuring the relevant state-of-the-art. The first requires much tedious statistical and accounting work; the second requires access to specialized, often rare, expertise.

For tables of similar size describing the same economic situation, the ex ante approach generally is both less expensive and less difficult than the ex post approach, assuming availability of the necessary expertise. Moreover, the ex post approach can be applied only to situations which can be or have been surveyed, and the table usually is far out of date by the time all data are assembled. To define a future situation by this method is methodologically ambiguous.

It is philosophically difficult to compare these two approaches qualitatively. To the extent that its potentialities for error have been overcome (at a considerable cost in time and money), the ex post table may be termed more "realistic" than the ex ante table. But if the surveyed period were quite abnormal this realism would be neither typical nor meaningful. On the other hand, if the experts whose knowledge and judgment are utilized in constructing ex ante coefficients fail for any reason to take account of an important factor, the ex ante table might be thoroughly unrealistic. The ex post table, at best, is a factually realistic reflection of the period for which statistics have been collected; the ex ante table, while not necessarily factually descriptive of a specific period, can be a functionally and conceptually correct delineation of a given past or future stage of technology.

We often need descriptions, in I/O terms, of specific past or future times. If this need relates to a past situation, the ex post table can be constructed only if the relevant information happened to have been collected. In the absence of collected data, this approach becomes impossible and we must fall back on an ex ante approach. In the same vein, if we need an I/O description of the future, only the ex ante approach can give full effect to newly emerging technologies.

There are many situations for which the ex ante approach provides a good description and few for which the ex post approach is clearly better. When we add to this the fact that the former generally is more flexible, is easier to apply, and is much less expensive, it becomes obvious why Battelle chose to follow this route. There is also another aspect of the ex ante approach that especially recommended it to Battelle: by definition, Battelle's staff is made up of technological experts, familiar with technologies of the past and present, and working to create the technologies of the future. Thus, when provided with the technical guidance that channels their expertise in the proper direction, this staff provides exactly the kinds of knowledge and judgement for which ex ante I/O methods call.

## Research Applications

Battelle researchers, as part of the "Aids to Corporate Thinking" (ACT) program, first developed and applied the ex ante method within an I/O context during 1966–1967. Since that time, it has been twice reapplied to I/O flow coefficients and once to capital stock coefficients. These applications will be reviewed briefly before taking up the details of the method itself.

### *Act II*

The initial application of this approach took place as part of the 1966–1967 generation of Battelle's 82-sector forecasts of the U.S. economy for 1975. The base data from which this exercise took its departure consisted of:

1. The 70-order table for the U.S. in 1947—modified by the Harvard Project into general comparability with the 1958 table—and further disaggregated to 82 sectors by Battelle.

2. The OBE's 82-order table for the U.S. in 1958.

3. A limited number of forecasts of major 1970 coefficients, made also by the Harvard Project.

4. A table of intermediate flow coefficients (at 82-sector detail) mathematically extrapolated through 1947 and 1958 (and occasionally 1970) to 1975; these extrapolations also were made by Battelle.

The overall methods for selecting and interviewing experts were developed at this time. They have been further refined, but not substantially altered, in subsequent experiments.

### Act IV

In conjunction with the 1969–1970 continuation of research in the ACT program, all secondary transfers were removed from the 1958 and 1975 coefficients. The resulting so-called "pure technology" coefficients were then submitted to experts for review in the same manner used by the original 1966–1967 activity. Although three to four years had elapsed between the original research and the reviews, and despite the fact that major adjustments had been made in the projections by the corrections for secondary transfers, the reviewers displayed a high degree of confidence in the projections. By and large, their revisions consisted of a large number of "fine tunings", with relatively few substantial adjustments of earlier results.

In this connection two significant relationships became apparent: First, the selection of the interviewer is just as important as the selection of experts to be interviewed. And, second, the experts, if properly chosen, are generally capable of improving the usefulness of statistical (ex post) coefficients. The first of these two findings merely underlines a wellknown rule of survey statistics. The second, however, increased our confidence in the method itself and therefore should be further elaborated.

During the original (ACT II) interviews, many experts expressed puzzlement over and disagreement with particular coefficients in the U.S. tables for 1958. Almost all these points of disagreement have been traced back to the convention adopted by the OBE for dealing with secondary output. When the effects of the OBE's transfers of secondary output were removed from the tables, the vast majority of the disagreements were resolved.

### Disaggregation of Nonferrous Metals

Also during 1969–1970, as a separate exercise from ACT IV, the single Primary Nonferrous Metals Manufactures sector was disaggregated into six new subsectors. Insofar as the disaggregation of the nonferrous metals row was concerned, the task was not particularly difficult and was carried out in conjunction with the above-mentioned ACT IV review. Disaggregation of the column proved more difficult and resulted in further refinements of the method.

The only base data available for the columnwise disaggregation consisted of the aggregated sector coefficients. The Battelle methodology (see below) therefore had to be applied as a two-step approach: first, the researcher worked with a single expert in nonferrous metals process-economics to establish preliminary 1958 and 1975 coefficients for each subsector; and, second, each of these was reviewed with other (subsector) experts in order to establish final input structures for each subsector. Results have been highly satisfactory. In fact, specific weaknesses have been uncovered in the U.S. tables for 1958 that were derived in the traditional manner from survey statistics. These weaknesses do not seem to arise from current methods of ex post table construction as much as they do from the use of "establishment" conventions in conducting the U.S. Census of Manufactures.[1]

### Application to Capital Coefficients

A research project recently completed for the "Scientific American" magazine involved the systematic application of Battelle's ex ante method of the task of constructing a complete matrix of capital coefficients for the U.S. In general, the selection of experts, the conduct of interviews, and the approach "by the column" were carried over from the I/O applications. The *criteria* for selecting the experts were changed somewhat, because of the nature of the problem; and the statistical base used in preparation for the dialogues was quite different from any used in the earlier exercises.

The only capital data available for U.S. government sources were capital flow statistics. Stock-concept data were obtained from the National Planning Association, but were confined to the manufacturing industries (Waddell, et al., 1966). For nonmanufacturing sectors, data were made available by the Harvard Project and by the NPA which were based on the 1958 capital flow matrix of the Bureau of Labor Statistics. The Harvard data had been adjusted toward a stock concept for both manufacturing and nonmanufacturing industries at 82-sector detail. The NPA data were at 4-digit SIC detail, but treated manufacturing and the nonmanufacturing sectors differently. Most of these complications affected only the statistical preparations for fieldwork or the post-field refinement procedures.

### The Battelle Technique

Unlike other methods of technology forecasting that usually try to date the likely future occurrence of a specific technological event. Battelle's researchers must forecast the kind of technology a given sector would be using in a given year. This is a significant distinction, since the forecasts cannot be easily played back and forth between a panel of experts and a secretariat (e.g., the Delphi method).

A second important consideration is introduced by the sheer immensity of forecasting an 82-sector I/O matrix containing over 6800 cells. Every cell in the matrix—even value added and historically empty cells—must be considered, in order both to anticipate the effects of technological change and to provide adequate statistical control.

In order to take account of these two aspects of its forecasting problem, the Battelle research team decided to take the following steps away from present-day forecasting techniques:

1. To use only one or two experts for each sector, but to be extremely selective in choosing them.
2. To provide each expert with one set of coefficients based on a recent past situation and, where possible, with one or more sets of coefficients representing econometric projections to the target year.

3. To let the interviewer provide for continuing interaction between the expert and (a) his earlier statements, (b) the benchmark data, (c) the supplied projections, (d) a second expert, or (e) background knowledge possessed by the interviewer.
4. To reduce uncontrollable (open-ended) freedom for error by forecasting every cell in a sector's input structure.
5. To have the interviewer act as a constant monitor, reminding the expert of relevant concepts and definitions and probing for full explanations.

## Field Interviews

Field work was carried out by a small number of individual interviewers. We felt it important to use a minimal number of interviewers in order to minimize the degree to which differences in personal "style" might introduce inadvertent biases into the results.

### *Advance Preparation*

In order to facilitate communication in the field, we prepared worksheets for each sector in which were displayed the detailed benchmark data relevant to the particular investigation.

A key element in the input-output coefficient exercise turned out to be the set of extrapolated 1975 coefficients which gave the several experts "something to shoot at". After briefing, they were asked to consider the validity of the individual extrapolations: "Assuming that the 1958 coefficients adequately describe the sector technology in that year, is the given extrapolation compatible with the future you envisage for this sector?"

Having such benchmark data available made it feasible to select sector experts without regard for access to private operating and engineering records. In fact, the approach to the interview emphasized that Battelle was seeking general technical expertise, not confidential company data; and this approach opened many doors that otherwise might have remained closed.

### *Selecting the Experts*

Probably the most crucial steps in this method of estimating coefficients involve the selection of experts and the conduct of the dialogues. Both I/O and capital coefficients projects were carried out by *columns*, rather than by *rows*. It is our strong conviction that the complexity of the U.S. economy assures that few know who ultimately purchases and uses a given sector's output; while many experts know what their sectors purchase as inputs.

In selecting experts for sector forecasts, care was taken to obtain both technical and business understanding. We felt that, although essential, technical knowledge would lead to "science fiction" unless tempered by business understanding.

Therefore, each expert was chosen to provide the following mix of expertise:

Knowledge of the industry's technical research and innovations—in the laboratory or pilot-plant, and planned for broader use.

Understanding of the past, current, and future technical trends in the industry, especially as determinants of input-mix.

Acquaintance with the firms and persons in the industry and clear understanding of their habits and personalities as decision-makers.

Historical familiarity with the industry's pace of technological innovations, and an understanding of the business factors affecting them.

Some of the experts interviewed were engineers, technologists, and executives in representative companies (e.g., a tobacco company, an automobile manufacturer, a major broad-based insurance company). Others worked in closely related activities qualifying them as expert observers of a sector (e.g., trade magazine editors and trade association executives in leather tanning, highway construction, hotel management). Finally, we chose a number of Battelle engineers, technical economists, and technologists who, by experience and research involvement, were experts on particular sectors (e.g., steel, electronic components, railroad transportation, livestock).

If a single expert could not be found with both business and technical knowledge, we tried to find two persons, one for each, and to interview them together. Where necessary for full coverage, more than one expert might be interviewed in sequence.

### Conduct of Interview

The typical interview lasted one to one-and-a-half hours, although lengths ranged from thirty minutes to three hours. In a few cases, more than one sector was covered in an interview. Although appointments were normally made to see one expert per interview, on several occasions the interviewee invited one or more of his associates to participate; and these sessions understandably tended to be longer than the one-to-one meetings.

The interview customarily opened with the Battelle investigator describing the overall objective of the project, its sponsorship, the expected public availability of results, and its relationship to input/output analysis.

In the studies involving input/output coefficients, the objective of the interview was stated as forecasting the average technological profile of the sector in 1975 (then 8 or 9 years away). This required expert judgments as to the commercialization of yet-proved technologies and the rates of diffusion in technological advances. In the capital coefficient study, the objective of the interview was stated as updating the 1958 coefficients to represent the most advanced technology available (in 1969–1970) that could be employed if the subject sector's facilities were rebuilt within the next five years. Thus, this latter project did not involve technological forecasting; rather, it assumed universal use of today's best technology.

The interviewer next briefed the expert on sector definitions, I/O conventions, all relevant constraints, and the statistical basis for the entire projection. The expert was then asked to discuss, *in his own context*, the trends affecting the input structure of the target sector. Generally useful as thought-starters (if necessary) were questions about investment in facilities for labor savings, for quality control, and for waste disposal; questions about kinds of raw materials and mix of products; and questions about the dynamics of technological innovation. This usually nonquant-itative discussion of technological change was recorded by the interviewer and became the basis for further dialogue.

Once the summary of technological trends was completed, the interviewer carried the expert through a cell-by-cell scrutiny of their impacts upon each input coefficient for that industry, including its value added. Each empty cell was examined separately to determine whether or not new interindustry markets might emerge as a result of anticipated changes in technology.

The interviewee was also asked to inspect the 1958 coefficients to detect gross errors or abnormalities. In the capital coefficients study, few significant questions were raised as to the validity of the 1958 figures, and these questions were more likely to arise during the discussion of sector trends than during the inspection process itself. It will be recalled that this was not the case in the I/O studies.

Finally, the interviewee was asked to suggest quantitative values that represen-ted the effects of the trends on the elements of investment or operating inputs. Responses could be in either relative or absolute terms: e.g., a 1958 coefficient might be said to rise by 20 percent, by 31 percentage points, or it might change from 0.0157 to 0.0175. The interviewee expressed the changes in the manner most comfortable and convenient to him, but was asked to justify each change for the project files.

Most experts thought more in terms of pluses than in minuses, i.e., the more easily identified trends involved additions to investment or operating inputs. No special attempt was made to achieve a balance of pluses and minuses during the interview, but merely to achieve an acceptable internal set of numerical relation-ships, and the experts were quite willing to let Battelle undertake the balancing (normalization) process.

Generally speaking, the most difficult part of the interviews was estimating changes in value added or the capital/output ratio. These numbers are affected by changes in labor inputs, capital investments, and values of output, with trends often in opposite directions. For example, a sector might use capital more intensively to save labor, while concurrent engineering improvements were making the new capital more productive than the old.

### Interview Follow-up

Computation of the 1975 coefficients involved two steps: numerical expression of all changes suggested by the expert, and normalization of all numbers to total 1.0000. There were few deviations from this procedure: for example, if the expert specified absolute numerical values for one or more coefficients, these values would be excluded from the normalization.

In cases where the expert disagreed significantly with 1958 base data, these data were changed and normalized to make their relationship with the 1975 numbers comparable. Finally, all rough notes taken during the interview were rewritten for the permanent file.

### The Modular "Peel-Back" for Capital Coefficients

In order to simplify its generation of manufacturing and nonmanufacturing capital expansion factors, the NPA resorted to modular treatment of certain common groups of capital inputs. This meant that a single module, assumed to have a fixed composition involving many separate capital input items, would enter as a unit into many different industry tables. Although the module's own composition would be fixed internally, its expansion factor (coefficient) value could vary from one industry to another.

Certain modules (especially those affected by computer technology, concern for internal working environments, or concern for pollution control) were expected to change composition between 1958 and 1975. These changes were made by means of a special set of field interviews similar to those already described.

### Selected Examples

Two examples illustrate some of the problems met and results obtained. The first is taken from the input-output coefficients and the second from the capital coefficients.

*Purchasing sector*: ordnance
*Supplying sector*: communication equipment
The 1975 trial coefficient was 0.03435, identical with the 1958 coefficient. Our first expert predicted that the ordnance sector would significantly increase its purchase of communication equipment by 1975 because of increased output of guided missiles; he recommended a coefficient of 0.05000 (subsequently normalized to 0.04531). The exclusion of secondary transfers into this sector resulted in a revised coefficient of 0.04614.

Some three years later, a second expert reviewed the sector and recommended that this coefficient be reduced to 0.04000 (normalized to 0.04009) on the basis that shifting military priorities would reduce the emphasis on guided missiles.

*Purchasing sector*: railroad transportation
*Supplying sector*: new construction
The 1958 base data on capital flows showed a coefficient of 0.274476. The expert considered this as too low to represent an adequate capital stock of roadbed, stations, signal towers, etc., and suggested an adjusted figure of 0.400000. For 1975, the coefficient would be lower because of fewer stations, the consolidation of small yards, and the fact that some functions are being taken over by shippers and forwarders. The suggested new coefficient was 0.350000.

**NOTE**

1. For instance, the U.S. tables for 1958 and 1963 failed to show any sales by the forestry industry to nonferrous metals refining. However, "poling" is still standard practice in the copper industry—that is, adding green softwood logs to the molten metal to stir the metal and reduce its exposure to oxygen.

**Reference**

Waddell, R. M., P. M. Ritz, J. D. Norton, and M. K. Wood, 1966. Capacity expansion planning factors. Washington: National Planning Association.

# IV
# Economic Analysis

# 21

# Factor Endowment Change and Comparative Advantage: The Case of Japan, 1956–1969

PETER S. HELLER

## Introduction

It has long been recognized that a country's relative factor endowment will strongly influence the focus of its comparative advantage in international trade. Within the context of the postwar Japanese economy, this paper will consider the impact of a rapid change in factor endowment on the structure of comparative advantage. Several questions will be addressed. In 1954 Japan was a "labor-abundant" economy,[1] midway in the factor endowment spectrum between the developed (MDCs) and the less developed (LDCs) economies. A dualism in its economy and trade structure prevailed. Has the rapid erosion of its labor abundance affected the technological character of its trade? As Japan's factor endowment has converged toward that of the MDCs, has the dualism in its trade structure been similarly eroded?

In the first section we shall briefly describe the structure of Japan's economy during the period and the change that occurred in its factor endowment. The second section will discuss the Leontief methodology used in evaluating the technology of a trade bundle, and some of the index number problems that arise when this methodology is applied in a dynamic context. The third section will discuss the change in the technology of exports and imports over the period is examined. The last section will examine the allocative efficiency of Japan's export structure.

## Setting

In many respects Japan provides an interesting case study for the analysis of these issues. Within the last two decades it has evolved from a battle-torn, moderately-developed economy to the third largest economy in the world and its economic growth rate has far outpaced that of other MDCs and LDCs. This structural transformation in its production structure has been accompanied by a striking shift

This article was published in *The Review of Economics and Statistics*, Volume 58, Number 3, August, 1976, pp. 283–292. © By the President and Fellows of Harvard College. Reprinted by permission.

in its factor endowment position. Rapid rates of capital formation and demographic restraint led to the disappearance of "easy labor market conditions" in the factor markets by the early 1960s, and Japan's factor endowment position has rapidly converged toward a capital-abundant position characteristic of the MDCs. Has Japan's comparative advantage in trade similarly shifted toward the more capital intensive sectors of its economy?

In the mid 1950s, Japan's factor endowment position was midway between the MDCs and LDCs. Its low per capita income and labor-abundant factor market placed its capital-labor endowment below the former; large inflows of American aid and high domestic savings rates supported a capital infrastructure surpassing that of most LDCs. Japan's historical emphasis on education and its receptiveness to foreign technology suggest a labor force rich in human capital endowment relative to most LDCs and some MDCs.[2]

Japan's labor-abundant conditions were a critical support for its dualistic production structure in the 1950s. They allowed a dichotomy between the factor price ratios facing the large and small firm sectors of the economy. Although the small firms found capital relatively inaccessible, reflecting government priorities, the wage rates of their employees were also considerably *below* the rates paid in the large-scale enterprises of any given sector. This dualism in factor-price ratios was mirrored by a technological dualism within any sector.[3] Labor intensive and capital intensive firms coexisted within sectors. Government support enabled the large capital intensive sectors to survive and develop. At the same time, the labor intensive sectors—the traditional source of Japan's international competitiveness—were constrained by the unavailability of capital to maintain this labor intensive technology focus. This dualism in industrial structure extended to Japan's trade structure. Tatemoto and Ichimura's study of Japan's 1951 trade structure[4] reveals that it clearly exported capital intensively to the LDCs and labor intensively to the MDCs.

As the labor market tightened in the early 1960s, the viability of this industrial dualism was eroded.[5] Wage differentials began to narrow as wage rate increases in the small firm sector outpaced those of the larger firms, without, however, any substantial lessening in the differential accessibility to the capital markets. The ability of the small firms to remain competitive in world markets in the export of labor-intensive commodities was threatened. Do we find a shift in Japan's comparative advantage toward the relatively capital intensive sectors of its economy, or have the labor intensive sectors continued to dominate exports, with only a shift in the capital intensity of the latter's production processes?

Moreover, since 1955 Japan's factor endowment has converged toward that of the MDCs. Between 1955 and 1960 the aggregate capital-labor ratio in its economy rose at an annual rate of 6.72 percent, and this increased to 8.34 percent between 1960 and 1967. Its per capita income also converged to MDC levels. Most of its LDC trading partners have not kept pace with this rate of growth, leading to an increasing divergence in their relative factor endowments. Although some of Japan's partners have grown rapidly (Taiwan and Philippines, in particular), the continued high rates of labor force growth in the LDCs have more than kept pace with the rate of industrial growth.[6] Would this shift in Japan's relative factor endowment position bode a shift in both the commodity composition and the regional destination of its exports?

Any shift in Japan's *import* structure due to these trends would be influenced also by the gradual freeing of its restrictive policies toward imports. Tariff and nontariff barriers were used to shield established, inefficient sectors (particularly in agriculture and the traditional consumption goods[7] sector) and to subsidize the development of many new industrial sectors. The effect was to bias the mix of imports toward goods for which import substitution was not yet viable, goods embodying new technological innovations, and raw materials necessary for industrial production. Consumption imports were restrained.[8] Substantial liberalization of these import barriers did not begin until 1964. As late as 1969 residual import restriction policies were still maintained on agricultural products, petroleum, coal and sulphur, bovine cattle leather, and digital computers.[9]

## Methodology

A well-defined procedure, initially developed by Leontief,[10] exists for measuring the technological characteristics associated with the production of a bundle of commodities. Applying a set of industry-specific factor coefficients to an input-output table of an economy, a measure of the direct and indirect requirements for labor, capital and skilled labor associated with the production of a unit of any industry's output can be obtained. To calculate the capital or skill intensity associated with a unit bundle of exports or imports, one simply weights these factor requirements by the relative shares of each industry in the given trade bundle.

For our analysis, we shall use the following notation. Let $k_x(i, t)$ be the ratio of the capital to labor requirements, direct and indirect, associated with the production of a bundle of exports $(x)$ in period $t$, using the technology of year $i$ (similarly $k_m(i, t)$ connotes the same measure for imports). Let $l_x(i, t)$ indicate the share of skilled workers in the labor force required to produce a bundle of exports in period $t$, using the technology of year $i$.

Our analysis relies on technology data for the Japanese economy in three years: 1955, 1960, and 1970. The sixty-sector input-output tables[11] and capital coefficients[12] corresponding to each period are deflated to 1960 prices. Estimates of output per employed laborer and per skilled worker, respectively, in each of the sixty sectors were obtained from Economic Planning Agency (E.P.A.) and Census Department statistics, respectively. Our skill coefficient is the proportion of professional and technical workers employed in each sector's labor force.[13]

Statistics on the commodity composition and destination of trade flows are derived from the commodity trade statistics (series D) of the United Nations. Each SITC category is imputed to a sector in the input-output table. All Japanese exports are assumed "competitive"; for imports, we exclude primary commodity imports (such as agricultural commodities and industrial nonprocessed raw materials) since these are industries for which Japan has only limited production capacity or potential due to its meager natural resource base. This effectively excludes a significant fraction of Japan's total imports, particularly from the less developed world.[14,15] All trade flows are deflated to 1960 prices.[16]

The interpretation of the estimated factor intensity of Japan's export and import structure over the period is subject to obvious index number difficulties. Specifically, any change in the computed technology characteristics may reflect changes in one or more of the following: (1) the relative factor and commodity price structure, (2) the interindustrial production structure, (3) the industry-specific factor coefficients, and (4) the commodity mix of trade. We have removed one source of variation by our deflation of all data to 1960 prices.

The relative influence of each of the remaining factors may be examined by holding the others constant. For example, shifts in the factor-intensity of Japan's exports over time that arise strictly from changes in the export bundle's *composition* can be estimated by holding the technology constant, that is, by using the input-output table and factor coefficients of a given year in the analysis of each year's exports. Alternatively, we could evaluate a *given* year's trade bundle by the technologies of different years.

There are several problems with this partial approach. First, in the constant technology case, an examination of the technology of export bundles of alternative years could yield theoretically ambiguous results according to the particular *base year* technology chosen. For example, applying the 1960 technology, we might find that Japan's economy in this period had been such as to reverse the relative factor intensity of the industries exporting to Korea and the United States. More likely, if the growth in capital intensity of the rapidly growing export sectors exceeds that of other sectors, use of a later year's technology would suggest a more extreme change in the capital intensity of exports arising from compositional changes in the export bundle.

Second, if one were to compare the technology of Japan's export bundles to an LDC and to an MDC over time, one would expect that by holding the technology constant, any differences would be biased downward. This problem arises because the structure of exports is influenced by shifts in the factor endowment structure. For example, the increase in the capital and skill intensity of production in Japan should be reflected by a capital intensive shift of the composition of its export bundles. Those export bundles embodying the largest share of capital intensive goods would be more capital intensive, relative to other export bundles using a 1968 technology, than with a 1960 technology.

Third, all our analysis is carried out in constant 1960 prices. This may distort our analysis in the opposite direction, but the degree of distortion (or even its direction) is not fully apparent without further analysis.[17]

### Factor Endowment Change and the Trade Structure: 1956–1969

In this section we shall examine whether the disappearance of the labor-abundant economy is reflected in the technological character of Japan's trade from 1955 to 1968. In Table 21.1 we present the capital and skill intensity associated with the

Table 21.1.   The Capital and Skill Intensity of Japan's Exports and Imports: 1956–1968 (*in 1960 Prices*)

Part A: In current year technology: capital and skill intensity $(k_x (i, t))^a$, $(l_x (i, t))^b$

| Commodity bundle | $k_x$ (55, 56) | $k_x$ (60, 60) | $k_x$ (60, 63) | $k_x$ (70, 68) | $l_x$ (55, 56) | $l_x$ (60, 60) | $l_x$ (70, 68) |
|---|---|---|---|---|---|---|---|
| Total exports | 1734 | 2334 | 2594 | 4826 | 1.46% | 2.57% | 4.38% |
| Total exports to MDCs[d] | 1522 | 2173 | 2.441 | 4735 | 0.96 | 2.38 | 4.35 |
| Total exports to LDCs[d] | 1867 | 2496 | 2697 | 4784 | 1.86 | 2.78 | 4.33 |
| Total Exports to Communist bloc | n.a. | 3202 | 3264 | 5463 | n.a. | 3.73 | 5.00 |

Part B: In constant 1960 technology: capital intensity and skill intensity of exports and imports

| Commodity bundle | (*in U.S. dollars*) | | | |
|---|---|---|---|---|
| | 1956 | 1960 | 1963 | 1968 |
| | Capital intensity | | | |
| Total exports | 2291 | 2334 | 2594 | 2820 |
| Exports to MDCs | 2024 | 2173 | 2441 | 2771 |
| Exports to LDCs | 2444 | 2496 | 2697 | 2810 |
| Exports to Communist bloc | n.a. | 3202 | 3264 | 3146 |
| Total imports | 2169 | 2425 | 2367 | 2624 |
| Imports from MDCs | 2290[c] | 2561 | 2401 | 2663 |
| Imports from LDCs | 1759[c] | 1967 | 2014 | 2106 |
| | Skill intensity | | | |
| Total exports | 2.56% | 2.57% | 3.00% | 3.51% |
| Exports to MDCs | 2.21 | 2.38 | 2.80 | 3.50 |
| Exports to LDCs | 2.70 | 2.63 | 3.11 | 3.46 |
| Exports to Communist bloc | n.a. | 3.73 | 3.96 | 3.81 |
| Total imports | 2.77 | 2.36 | 2.47 | 2.93 |
| Imports from MDCs | 3.12[c] | 2.66 | 2.74 | 3.16 |
| Imports from LDCs | 1.72[c] | 1.48 | 1.71 | 2.33 |

[a] $k_x (i, t)$ is the ratio of the direct and indirect capital requirements (in U.S. dollars) to the indirect and direct total labor requirements (in numbers of laborers) required to produce a unit bundle (in U.S. dollars) of exports of period $t$, using the technology of year $i$. Note, $k_m (i, t)$ is the equivalent measure for imports.

[b] $l_x(i, t)$ is the ratio of the direct and indirect requirements of professional and technical workers to the direct and indirect total labor requirements (both in numbers of laborers) required to produce a unit bundle of exports of period $i$, using the technology of period $i$. Note that $l_m (60, t)$ is the equivalent measure for imports.

[c] Where we assume that $k_m$ (60, 60), ratios of total MDC and LDC imports to total imports, can be applied to 1956 imports.

[d] For MDCs, this includes all countries falling within the U.N. definition of a developed country for the purposes of the E1 category of the SITC statistics. For LDCs, this includes all countries falling within the E2 category. The Communist bloc countries include all countries falling within the E3 category of the SITC statistics. See U.N., *Commodity Trade Statistics*.

production of representative *export* bundles of Japan from 1955 to 1968. In part A, the capital and skill intensity measures are calculated using the technology associated with each year's production process. The increase in the capital intensity of Japan's exports over the period is immediately apparent, rising from $1734 per man in 1955 to $4826 per man in 1968. The skill intensity rises from 1.46 percent to 4.38 percent. This reflects both changes in the composition of the export bundles as well as the general increase in the capital and skill intensity of all production processes over the period. The effect of the former alone is shown in part B of Table 21.1, where we have held the technology constant to that prevailing in 1960.[18] The increasing values of $k_x$ and $l_x$ indicate a rising share of capital intensive sectors in the export bundles.[19]

The technological dualism associated with exports to the developed and less developed world, initially noted by Tatemoto and Ichimura for trade in 1951, is also apparent throughout the period. Japan's exports to the LDCs are more capital and skill intensive than its exports to the MDCs. However, there is also a clear convergence in the technologies associated with Japan's exports to the two blocs. Between 1956 and 1963, the gap between the $k_x$ of each fell from $420 to $250; by 1968 it was less than $75. By 1968 the gap in skill intensities was eliminated and the export bundles to the MDCs had become more skill intensive.[20]

Two aspects of this convergence are worth noting. First, there is a strong difference in the rate of change of the technologies associated with the export bundles to the LDCs and MDCs. Between 1956 and 1969, $k_x$ rose by 34 percent for the MDCs as compared with only 16 percent for the LDCs; similarly, the skill intensity $l_x$ rose 59 percent for exports to the MDCs, but only 28 percent of the LDCs. Second, by 1968 the technology of exports to the MDCs had converged to the high *initial* (1956) level of the capital and skill intensity of exports to the LDCs. In fact, the latter were *at least* as capital intensive as Japan's *own* imports from the MDCs.[21] Despite the import-substitution inspired impetus to the pre-1960 development of this capital intensive sector, Japan was nevertheless able to compete with other MDCs in the sale of these goods to LDC markets.[22] The convergence process after 1960 indicates that this competitiveness was extended to MDC markets as well.

From Table 21.1, the growth in the skill intensity of these export bundles markedly exceeds the growth in their capital intensity.[23] The compositional change in the trade bundle implied—toward industries with high skill requirements relative to capital inputs—is surprising. The stock of capital has grown more rapidly than the skilled labor force[24]—9.94 percent annually relative to 4.38 percent—between 1960 and 1967—1970.[25] Hence, one might have expected an increase in the relative price of skilled manpower relative to capital, which would increase the competitive advantage of firms that are capital rather than skill intensive. In this case, the compositional change would be the reverse. Our conflicting results may be an artifact that arises from holding the technology used in Table 21.1 constant. With changes in technology included, one would hypothesize that industries that in 1960 intensively used skilled labor relative to capital made the appropriate technological substitutions in subsequent periods. This is partially verified in Table 21.2. Application of the 1970 technology reveals a

Table 21.2.   The Relative Capital Factor Intensity of Trade Under Alternative Technologies

| Technology year $i$ | $k_x(i, 60)/k_x(i, 56)$ | | | $k_x(i, 68)/k_x(i, 56)$ | | | $k_x(i, 68)/k_x(i, 60)$ | | |
|---|---|---|---|---|---|---|---|---|---|
| Trade bundle   $i =$ | 1955 | 1960 | 1970 | 1955 | 1960 | 1970 | 1955 | 1960 | 1970 |
| Total exports | $1.020^a$ | 1.018 | 1.015 | 1.255 | 1.230 | 1.242 | 1.231 | 1.208 | 1.224 |
| Exports to MDCs | 1.066 | 1.074 | 1.076 | 1.412 | 1.369 | 1.378 | 1.325 | 1.275 | 1.281 |
| Exports to LDCs | 1.022 | 1.021 | 1.012 | 1.159 | 1.149 | 1.149 | 1.134 | 1.125 | 1.136 |

| Technology year $i$ | $l_x(i, 68)/l_x(i, 60)$ | | | $l_x(i, 60)/l_x(i, 56)$ | | |
|---|---|---|---|---|---|---|
| Trade bundle   $i =$ | 1955 | 1960 | 1970 | 1955 | 1960 | 1970 |
| Total exports | 1.610 | 1.365 | 1.373 | 1.076 | 1.003 | 1.015 |
| Exports to MDCs | 1.890 | 1.470 | 1.440 | 1.296 | 1.076 | 1.077 |
| Exports to LDCs | 1.420 | 1.315 | 1.260 | 1.065 | 0.974 | 1.024 |

$^a$ This equals $[k(55, 60)]/[k(55, 56)]$ where $k_x(55, 60)$ is the ratio of the direct and indirect capital requirements (in U.S. dollars) to the direct and indirect total labor requirements, using the 1955 technology for the export bundle of 1960.

lower (greater) growth in the skill (capital) intensity of the export bundles between 1960 and 1968.

We earlier noted that the rate of technological change or of adjustment to changes in the factor price ratio may differ across industries over time. Any resulting change in the relative factor intensity rankings of industries would then conceivably reverse the results obtained from comparing export bundles over time with a fixed period of technology. Were such reversals significant for the Japanese economy over the period, and do they undermine the basic thrust of our results?

Although changes in the ranking of sectors by capital intensity are not unusual for the Japanese economy between 1956 and 1969, one does not observe extreme shifts in any particular sector's interindustry ranking. For example, between 1960 and 1968 the capital intensity of the mining sectors increased particularly fast, and their sectoral ranking rose. The capital intensity of basic chemicals and steel in 1968 was 2.04 and 2.14 times higher, respectively, than their levels in 1960. The capital intensity of other products grew less rapidly, with lower such multiples. (The multiples for leather products, miscellaneous fabrics and spinning, and chemical fibres were only 1.76, 1.66, and 1.46, respectively.)

To test whether these reversals substantively modify our results, we compare (Table 21.2) the growth in the capital and skill intensity of the export bundles of 1956, 1960, and 1968. For example, in Table 21.2, columns 1 to 3 measure the relative capital intensity of exports in 1960 and 1956, using each of the three different technologies. If the effect of technological change had been neutral across sectors the ratios would be the same, regardless of technology period chosen.

For total exports between 1956 and 1960, the choice of technology year makes little difference to the revealed rate of change in their capital intensity. The convergence process of exports to the MDCs is more strongly indicated by the later period technologies, since the increase in their capital intensity between 1956 and 1960 rises from 6.6 percent to 7.4 percent. For the LDCs, the technological change over the period has virtually no impact on the relative factor intensity of exports in the two earlier periods. Similarly, between 1960 and 1968 application of the later year technology leads to only a slight increase in the capital intensity of the 1968 trade bundle.

For skill intensity, the choice of technology year does significantly influence the rate of technological change observed in the export bundles over time, particularly for the MDCs. Application of the 1955 technology suggests more rapid change in skill intensity between 1955 and 1968 than would the later period technologies, as the sharp differentials in the skill intensities across sectors in 1955 were significantly eroded. For example, by the technology of 1955 the industries in which Japan would specialize with the MDCs in 1968 appear significantly more skill intensive than the industries in which it specialized in 1956 and 1960. Use of the later period technologies substantially erodes this differential.

Finally, the erosion of the dualism in Japan's domestic economy and its convergence to the factor endowment structure of other industrialized countries is also reflected in the factor intensity of its imports and overall trade structure. The technological dualism in Japan's competitive import structure is the opposite of that found for exports. Not surprisingly, Japan's *imports* from the MDCs are considerably more capital and skill intensive than its imports from LDCs (Table 21.1) throughout the period.[26] Consequently, the factor intensity of its overall trade (exports relative to imports) with the LDCs is consistently more capital and skill intensive than with the MDCs. Nevertheless, Japan's *overall* trade with the latter has become increasingly capital intensive. The capital and skill intensities of Japan's exports to the MDCs were only 88 percent and 71 percent of its imports in 1956; by 1967–1969 these ratios had risen to 104 percent and 111 percent, respectively.[27] Japan's exports to the LDCs were 35 percent to 50 percent more capital and skill intensive than its imports from them over the *entire* period.

## The Allocative Efficiency of Japan's Export Structure

The above results implicitly suggest that Japan's trade structure was responsive to changes in its internal factor endowment mix and external market conditions. Was Japan allocatively efficient in its mix of exports over the period, in the sense that it was realizing the maximum gains from trade?

Let us define a period $t$s export bundle as "allocatively efficient" in an "absolute" sense if given the technology and factor prices of period $t$, the unit export bundle of period $t$ required the *lowest* amount of factors of production to produce, for all factors, of the unit export bundles of any other period $t - n$ or $t + n$. The earlier and later period bundles should reflect the technology and prices prevailing in their

Table 21.3.  Absolute Direct and Indirect Factor Requirements for $1000 of Exports Using Alternative
Technologies: 1956–1968 (*in 1960 Prices*)

| Technology year / Year: trade bundle | 1955 | | | 1960 | | | 1970 | | |
|---|---|---|---|---|---|---|---|---|---|
| | $L^a$ | $S^b$ | $K^c$ | $L$ | $S$ | $K$ | $L$ | $S$ | $K$ |
| *Total exports* 1956 | 1.587 | 0.0232 | 2752 | 1.099 | 0.0281 | 2517 | 0.516 | 0.0162 | 2006 |
| *to MDCs* 1956 | 1.736 | 0.0166 | 2643 | 1.242 | 0.0274 | 2514 | 0.629 | 0.0177 | 2162 |
| *to LDCs* 1956 | 1.519 | 0.0282 | 2837 | 1.031 | 0.0278 | 2520 | 0.463 | 0.0155 | 1925 |
| *Total exports* 1960 | 1.531 | 0.0241 | 2707 | 1.058 | 0.0272 | 2469 | 0.487 | 0.0155 | 1922 |
| *to MDCs* 1960 | 1.607 | 0.0125 | 2609 | 1.132 | 0.0269 | 2460 | 0.535 | 0.0162 | 1977 |
| *to LDCs* 1960 | 1.460 | 0.0289 | 2780 | 0.991 | 0.0261 | 2474 | 0.441 | 0.0151 | 1861 |
| *Total exports* 1963 | 1.375 | 0.0285 | 2725 | 0.938 | 0.0281 | 2434 | 0.416 | 0.0154 | 1840 |
| *to MDCs* 1963 | 1.427 | 0.0248 | 2654 | 0.996 | 0.0278 | 2431 | 0.452 | 0.0157 | 1885 |
| *to LDCs* 1963 | 1.346 | 0.0318 | 2767 | 0.909 | 0.0283 | 2452 | 0.394 | 0.0152 | 1818 |
| *Total exports* 1968 | 1.247 | 0.0315 | 2715 | 0.830 | 0.0291 | 2341 | 0.351 | 0.0154 | 1696 |
| *to MDCs* 1968 | 1.238 | 0.0290 | 2662 | 0.831 | 0.0291 | 2302 | 0.357 | 0.0155 | 1690 |
| *to LDCs* 1968 | 1.269 | 0.0356 | 2746 | 0.839 | 0.0290 | 2358 | 0.352 | 0.0152 | 1682 |

[a] Total units of labor required per thousand dollar units of output.
[b] Total units of skilled labor per thousand dollar units of output.
[c] In U.S. dollars.

own periods and thus would require a greater absolute amount of factors if produced in a different period.

In Table 21.3 we calculated the absolute labor, capital, and skilled labor requirements associated with the production of exports in 1956, 1960, and 1968, using each period's technology. The results do not satisfy our criterion of absolute economic efficiency. Regardless of the technology year chosen, the lowest total labor requirements are associated with the later export bundles. The capital requirements similarly decline for all bundles evaluated with the 1960 and 1968 technologies; with the 1955 technology, only the exports to the MDCs have higher capital requirements in the later years. Only the skilled labor requirements of an export bundle are lowest in their associated technology year.

In itself, this does *not* imply that Japan was allocatively inefficient in its production process, but it does suggest the apparent opportunity cost of the relative shortage of skilled labour in the earlier periods. In order to produce export bundles with lower skilled labor requirements, other export bundles were foregone with substantially lower absolute amounts of capital and labor. For example, a $1 million bundle of Japan's 1960 exports to MDCs required 300 workers and $158,000 of capital stock more than would have been required to produce the 1968 export bundle, with the same 1960 technology. The saving was 2.2 units of skilled labor. Given the wage differential that prevailed during the period, it is likely that factors other than production technology and relative factor prices explain the particular choice of each period's export bundles. Although it is possible that the later periods' export bundles could have been produced for export in the earlier periods at lower resource cost, Japan's comparative advantage at that time in world markets may have dictated against the later export bundles. The alternative allocational options may have been developed in the later periods only with (1) improved perception of the relative returns to exporting alternative commodities, (2) increased production capacity in the more profitable commodities, and (3) the development of export marketing channels.


## Conclusion

The principal conclusion emerging from this analysis is that the change in Japan's factor endowment strongly altered its comparative advantage in trade. The composition of its export bundle shifted toward the capital intensive sectors, and this shift was reinforced by a relatively faster deepening in the capital intensity of these sectors. Whatever viability had previously been attached to a dualism in Japan's export structure—exporting capital intensively to the MDCs, labor intensively to the LDCs—was deeply eroded by the disappearance of labor-abundant conditions in Japan's factor markets. The convergence in the technological characteristics of Japan's exports to the LDCs and MDCs suggests that the range of technology for which Japan is internationally competitive has narrowed. It is primarily focused on capital and skill intensive sectors. Our analysis of the "gains from trade" also suggests that the change in the composition of the export bundle has probably been an efficient change, with a lower absolute resource cost associated with its production relative to alternative past bundles.

These results suggest that in countries where the relative factor endowment position is rapidly changing, industrialization policies predicted on a country's "dynamic comparative advantage" may not be totally unreasonable. However, the Japanese case also highlights the importance, during this import-substitution industrialization process, of continued promotion of exports from the established, internationally competitive, labor intensive sectors. Equally relevant is the ability to render competitive in international markets the new industries themselves. Further research on the export promotion strategies of the Japanese is required.


**NOTES**

1. We will use the term "labor-abundant" to signify an economy characterized by easy labor market, tight credit conditions, with a high proportion of workers in low marginal productivity occupations. The term "labor-surplus" is often associated with many theoretical constructs of questionable applicability to Japan at this time, and has been avoided.

2. These assertions can be easily substantiated. For example, Japan's capital-labor endowment (fixed capital per employee) *and* skill endowment (skilled employees as a percentage of the labor force) in 1971 were dominated by Canada, United States, Austria, Belgium, France, Germany, Netherlands, Norway, Australia, and Israel, while Japan dominated Italy, Portugal, Spain, Mexico, Hong Kong, Korea, Pakistan, and Taiwan. A similar correspondence exists for these variables and GDP per capita. See Hufbauer (1971).

3. Dualism in the economy was primarily reflected by the large differences in the value-added productivity of labor, the capital-labor ratio and the average wage rate of labor across firms of different size, both within and across industries. Most studies suggest the dualism to be a consequence of distortions in labor, product, and capital markets. However, other recent studies suggest that the differentials in the wage and interest rates faced by large and small firms may

merely reflect differences in the quality of labor and in the composition of loans, respectively. See Broadbridge (1966), Fukuchi and Oguchi (1969, 1972), Minami (1973), and Engwall (1974).

4. This was observed by Tatemoto and Ichimura (1959).

5. For further discussion of the character and impact of the factor endowment shift, see Economic Planning Agency, *Economic Survey of Japan* (1960/61–1968/69) and Shinohara (1968).

6. For data on relative sectoral growth rates, see United Nations (1972) and Turnham (1970).

7. Prior to 1964, nontariff and tariff barriers existed on a wide range of commodities, such as meat, spices, carbon black, chemicals (such as sulphuric and nitric acid, crude benzoil, resins), cement, leather, perfumes, coffee, cocoa, butter, glass instruments, footwear, clothing, radios, sewing machines, motorcycles, vegetables, electric bulbs, binoculars, fish, office equipment, stoves jewels, and musical instruments.

8. On the other hand, some Japan specialists argue that at least prior to the late 1960s, Japanese consumers were biased toward traditional Japanese consumer goods, and thus that its market was shielded by these behavioral factors. See Ohkawa and Rosovsky (1961).

9. Kojima (1971).

10. Leontief (1954).

11. The 1955 and 1960 tables were provided by the E.P.A. in 1960 prices. Since the 1970 input-output table exists only in current market prices, we applied price deflators from independent sources to render this to 1960 prices. See *Japan Statistical Yearbook* (1966, 1973), and Bank of Japan (1973).

12. The capital coefficients were available only on a twenty sector basis and although these have been matched with the corresponding industries in the sixty sector table, there is obviously a loss in accuracy. Deflated capital coefficients were available only until 1967. Consequently, the 1970 technology data reflect the 1970 input-output table and the *1967* capital and labor coefficients. (See E.P.A., *Input-Output Data Series for Planning Use*, 1970). The data also have not been corrected for the level of capacity utilization in any given year. We did not attempt to derive capital coefficients from the annual investment and depreciation statistics of the *Kogyo Tokei Hyo* (Census of Manufactures) because (1) the capital coefficients for 1955 and 1960 would have been based upon extremely poor and noncomparable statistics since the quality of the data has improved considerably over the last 15 years; (2) an accurate linkage between the sectors of the *Kogyo Tokei Hyo* and of the input-output tables for the manufacturing sectors would be quite difficult since the *Kogyo Tokei Hyo* statistics are on *establishment* rather than *product* basis, and (3) we still would have lacked data for the nonmanufacturing sectors.

13. These coefficients were available for 1955, 1960, and 1970. See *1960 Population Census of Japan* and *1970 Population Census*.

14. For example, in 1963 Japan's imports of SITC commodity classes 0–4 were 75 percent of its total imports, 63.5 percent of its imports from the MDCs, 94 percent of its imports from the LDCs. In 1969 these statistics had fallen slightly to 69.8 percent, 57.5 percent, and 86 percent, respectively.

15. It could be argued that our inclusion of "indirect" factor requirements biases the results in that the high proportion of raw materials imported imputes a factor bias which is unrepresentative of its actual production structure. In fact, the results obtained do not significantly differ when only the direct factor requirements of a given trade bundle are calculated.

16. The deflators were supplied by the E.P.A. and the Bank of Japan. Export and import deflators are available for each sector. See Bank of Japan, *Export and Import Price Index Annual*, 1969, 1965.

17. For example, changes in the factor price ratio $(P_L/P_K)$ would yield shifts in the technology utilized in any production process, in the relative prices observed for capital and labor intensive commodities, and in the resulting mix of production as between these classes of commodities. The bias arising from not incorporating these changes in prices would depend upon the elasticity of substitution as between factors of production in each sector, and the price elasticity of demand for these sectors in both domestic and world markets. For example, suppose an increase in $(P_L/P_K)$ between 1955 and 1968. If the elasticities of substitution in production are less than unity, the relative commodity prices of capital and labor intensive commodities would probably decline between 1955 and 1968 (assuming the demand elasticity for each commodity was not high). Deflation to 1960 dollars would lead to an overvaluation (undervaluation) of capital (labor) intensive goods in Japan's exports in the later periods.

18. In what follows, $k_x$, $k_m$, $l_x$, and $l_m$ will correspond to the 1960 technology unless indicated. Similarly, we shall omit the time subscript $t$ unless indicated.

19. This does not reflect a decline in the absolute volume of labor intensive commodities, but rather a rising share for the more capital intensive sectors.
20. In 1967 the skill intensity of exports to the MDCs and LDCs was 3.29 percent and 3.39 percent, respectively. By 1969 these had become 3.56 percent and 3.38 percent, respectively.
21. In 1956 the capital and skill intensities of these imports were $2290 per laborer and 3.12 percent, respectively. See Table 21.1.
22. The competitiveness of Japanese producers may be explained by (1) their willingness to accept low profit margins on what was, at that time, a residual element in their total sales; (2) the export promotional incentives offered by the Japanese government, in addition to basic capital subsidies; (3) the newness of the capital plant structures in many LDCs, yielding great latitude as to the type of capital equipment purchase; (4) the newness of LDC markets.
23. The growth in both the capital and skill intensity of the later period export bundles in Table 21.1 reflects the positive correlation between the capital and skill intensity of Japanese industry. The rank order correlation coefficient between the capital and skill intensity rankings of the nonservice industry sectors (agriculture, mining and manufacturing) was 0.846. Although this complementarity relative to the use of unskilled labor is significant, it does not preclude substitutability between capital and skilled labor, as the results of Table 21.2 suggest.
24. E.P.A. *Input-Output Data Series for Planning Use*, 1970; Office of Prime Minister, *1960 Population Census of Japan*, and *1970 Population Census*.
25. The growth rate for capital was calculated over 1960–1967; for skilled labor, 1960–1970.
26. The size of this differential is probably understated. If the structure of Japan's preliberalization trade regime was, in fact, biased against capital intensive goods, particularly those for which Japan was developing a domestic production capacity, there is an underrepresentation of these goods in the observed import bundle. This would be particularly true for the imports from the MDCs.
27. These results are derived from Table 21.1, part B and measure (i) $k_x$ (60, $t$)/$k_m$ (60, $t$) and (ii) $l_x$ (60, $t$)/$l_m$ (60, $t$).

### References

Bank of Japan. 1973. *Price Indexes Annual*.
————. 1972. *Economic Statistics Annual*.
————. 1965, 1969. *Export and Import Price Index Annual*.
Blumenthal, T. 1972. Exports and economic growth: The case of postwar Japan. *Quarterly Journal of Economics* 86 (November): 617–731.
Broadbridge, S. 1966. *Industrial Dualism in Japan*. Chicago: Aldine Publishing Co.
Engwall, L. 1974. Industrial structure in a dual economy. *Hitotsubashi Journal of Economics* 14 (February).
Fukuchi, T., and N. Oguchi. 1969. Niju kozo no keiryo keizaigakuteki mokei. *The Economics Studies Quarterly* 20 (April): 61–72.
————. 1972. The trend of the dual structure and the backwash effect in Japan. *The Economic Studies Quarterly* 23 (August): 60–69.
Heller, P. S. 1975. Factor endowment change and the structure of comparative advantage: The case of Japan, 1956–1969. Center for Research on Economic Development Discussion Paper No. 42 University of Michigan.
Hufbauer, G. 1971. The impact of national characteristics and technology on the commodity composition of trade in manufactured goods. In *The Technology Factor in International Trade*, ed. R. Vernon. New York: National Bureau of Economic Research.
Japan, Government of, Economic Planning Agency. 1960/61–1969/70. *Economic Survey of Japan*. Tokyo: *The Japan Times*.
————. 1960, 1963, 1965, and 1970. *Input-Output Tables*.
————. 1970. *Input-Output Data Series for Planning Use*. (May).
Japan, Government of Ministry of International Trade and Industry. 1958–1967. *Census of Manufactures*.
Japan, Government of, Office of Prime Minister. 1960. *Population Census of Japan*. Vol. 2, part 4.
————. 1966, 1971, 1973. *Japan Statistical Yearbook*.

————. 1970. *1970 Population Census*. Vol. 5, part 1, division 1: p. 504.

Kojima, K. 1970. Structure of comparative advantage in industrial countries: A verification of the factor proportions theorem. *Hitotsubashi Journal of Economics* 11 (June): 2–29.

————. 1971. Nontariff barriers to Japan's trade. Japan Economic Research Center Discussion Paper No. 16 (December).

Leontief, W. 1954. Domestic production and foreign trade: The american capital position re-examined. *Economia Internationale* 7 (February): 3–32.

Lowinger, T. 1972. The neo-factor proportions theory of international trade: An Empirical Investigation. *American Economic Review* 62: 675–681.

Minami, R. 1973. *The Turning Point in Economic Development*. Tokyo: Kinokinya.

Naya, S. 1967. Natural resources, factor mix and factor reversal in international trade. *American Economic Review* 57 (May): 561–570.

Ohkawa, K., and H. Rosovsky. 1961. The indigenous components in the modern Japanese economy. *Economic Development and Cultural Change* 9 (April): 476–501.

Rybczynski, T. M. 1955. Factor endowment and relative commodity prices. *Economia* 22 (November): 336–341.

Shinohara, M. 1968. Patterns and some structural changes in Japan's postwar industrial growth. In *Economic Growth*, ed. L. Klein and K. Ohkawa. Homewood, Illinois: Irwin.

Tatemoto, M., and S. Ichimura. 1959. Factor proportions and foreign trade: The case of Japan. *Review of Economics and Statistics* 41 (November): 442–446.

Turnham, D. 1970. *The Employment Problem in Less Developed Countries: A Review of Evidence*. Paris: O.E.C.D.

United Nations. 1956–1969. *Commodity Trade Statistics*. series D. New York: United Nations.

————. 1972. *Industrial Development Survey*. Vol. 4. New York: United Nations.

# 22

## Factor Proportions, Linkages, and the Open Developing Economy

JAMES RIEDEL

The purpose of this paper is to examine the theoretical rationale underlying the growth of "footloose," import-dependent industry observed in many of the most successful developing countries (Hong Kong, Taiwan, S. Korea, for example). A second objective is to develop empirical formulations appropriate for analyzing the resource allocation consequences of a "footloose" industrial structure in a developing country. It is argued that previous applications of input-output techniques to factor-intensity measurement have in general ignored the implications of trade in intermediate inputs. The Leontief test of the Heckscher-Ohlin trade theory is perhaps the first and certainly the most widely adopted application of input-output techniques to the measurement of the factor intensity of production (Leontief, 1953, 1956). The first section of this paper will attempt to demonstrate that the procedure developed by Leontief is not strictly appropriate in an open economy which utilizes imported as well as domestically supplied inputs. An alternative formulation is developed in this paper, which when compared to Leontief's formulation yields a measure of the domestic resource cost or saving resulting from the use of imported rather than domestically produced inputs. In the second section, the formulations developed in the first section are applied to the Taiwan economy in an effort to demonstrate the resource allocation consequences of an import-dependent, "footloose" industrial structure typical of the island economies in East Asia which dominate the exclusive group of superlative economic development performers.

### Factor Intensity Measurement and Import Inputs

Perhaps the most important contribution of the Leontief test aside from its interesting revelations about neoclassical trade theory, was its recognition that the factor intensity of production of any given commodity is determined not only by the factor requirements at the last stage of production, but also by the factor requirements at each intermediate stage. Applying input-output techniques,

This article was published in *The Review of Economics and Statistics*, Volume 57, Number 4, November, 1975, pp. 487–494. © By the President and Fellows of Harvard College. Reprinted by permission.

Leontief measured the total (direct and indirect) labor and capital required to produce 1 unit of commodity $j$ by

$$L_j = \sum_i l_i r_{ij}$$

$$K_j = \sum_i k_i r_{ij}$$

(22-1)

where $l_i$ and $k_i$ are the direct labor-output and capital-output ratios, respectively, at the $i^{\text{th}}$ stage of production, and $r_{ij}$ are elements of the inverted Leontief matrix $[I - A]^{-1}$. What is ignored in applying this measure of total factor-intensity is that it implicitly assumes that all intermediate goods are produced domestically. The elements $r_{ij}$ of $[I - A]^{-1}$ indicate the total input (output) required per unit, which may of course be supplied domestically or imported. If intermediate goods are imported, what is relevant in calculation of total factor intensity is the factor requirements in producing the goods which exchange (exported) for imported inputs, not the factor requirements of producing the imported inputs directly. In an open economy, therefore, total factor intensity is appropriately measured as the weighted sum of (1) the per unit labor and capital required at the last stage of production and in producing domestic inputs, and (2) the labor and capital cost implicit in earning the foreign exchange with which to purchase the imported inputs required per unit.

The labor and capital cost of earning one unit of foreign exchange (in equilibrium) is the labor and capital required to produce one unit of exports which in turn can be defined as the average labor and capital requirements per unit output in each sector of the economy weighted by the distribution of exports from each sector.[1] The labor and capital required at the last stage of production and in the production of domestically supplied inputs which go into the production of exports is

$$L_{Fx} = \sum_j \left[ \sum_i l_i s_{ij} \right] e_j = \sum_j L_{jx} e_j$$

and

(22-2)

$$K_{Fx} = \sum_j \left[ \sum_i k_i s_{ij} \right] e_j = \sum_j K_{jx} e_j$$

where $s_{ij}$ are elements of the inverted domestic Leontief matrix $[I - D]^{-1}$, and $e_j$ is the proportionate share of the $j^{\text{th}}$ commodity in total exports.[2] Of course the production of exports itself requires imported inputs. If $M_{jx}$, defined as[3]

$$M_{jx} = \sum_i M_i s_{ij}$$

is the total (direct and indirect) import requirement per unit output of commodity $j$, then

$$M_{Fx} = \sum_j \left[ \sum_i M_i \cdot s_{ij} \right] e_j = \sum_j M_{jx} \cdot e_j$$

is the direct and indirect import requirement per unit export. Thus to produce one unit of exports we need $L_{Fx}$ units of labor and $K_{Fx}$ units of capital at the last stage of production and for domestically produced inputs; and we need $L_{Fx}M_{Fx}$ and $K_{Fx}M_{Fx}$ of labor and capital, respectively, to produce additional exports to finance the imports which were employed in the original production of one unit of exports. In addition, we recognize that the additional exports (required to finance the imports used in the first round) also require imported inputs. In the first round $M_{Fx}$ units of imports (foreign exchange = exports) are required; thus in the second round $M_{Fx} \cdot M_{Fx}$ units of imported inputs are required, which in turn will entail the employment of $L_{Fx} (M_{Fx2})$ labor and $K_{Fx} (M_{Fx2})$ capital in the production of exports with which to finance these additional imports. The second round of additional exports, likewise, requires imported inputs ($M_{Fx2} \cdot M_{Fx}$) and consequently more exports and hence the employment of still more labor and capital, and so on. The sum of all labor and capital required in the production of one unit of exports (i.e., foreign exchange) is thus

$$L_{Fx} + L_{Fx}M_{Fx} + L_{Fx}M_{Fx2} + L_{Fx}M_{Fx3} + \cdots + L_{Fx}M_{Fxn} = \frac{L_{Fx}}{1 - M_{Fx}}$$

and                                                                                         (22–3)

$$K_{Fx} + K_{Fx}M_{Fx} + K_{Fx}M_{Fx2} + K_{Fx}M_{Fx3} + \cdots + K_{Fx}M_{Fxn} = \frac{K_{Fx}}{1 - M_{Fx}}$$

since $0 \leqslant M_{Fx} \leqslant 1$. For any given commodity $j$, therefore, the factor intensity of production as measured by total factor requirement per unit outputs is given by

$$L'_j = L_{jx} + M_{jx}\left[\frac{L_{Fx}}{1 - M_{Fx}}\right]$$

and                                                                                         (22–4)

$$K'_j = K_{jx} + M_{jx}\left[\frac{K_{Fx}}{1 - M_{Fx}}\right]$$

which expresses the two components of total factor cost in an open economy: (1) employment and capital cost at the last stage of production and producing domestic inputs; and (2) employment and capital cost implicit in earning the foreign exchange (fraction of one unit of foreign exchange) with which to purchase imported inputs required directly and indirectly in the production of commodity $j$.

Recognizing the interrrelatedness of the economy, not only explicit but also implicit relationships, points to the fact that the factor intensity of a given production activity is dependent not only upon the technology in the final stage of production and in each and every sector of the economy from which it is supplied, but also in an open economy upon the technology which underlies the structure of foreign trade. In other words when an economy, e.g., Taiwan, imports steel, machinery, synthetic fibre and other relatively capital intensive intermediate goods with foreign exchange earned by exporting transistor radios, plastic toys, garments, and the like, it is implicitly substituting the latter labor-intensive goods for the former capital-intensive goods in the production process. One can easily see that to evaluate the factor intensity of the structure of production on the assumption that all inputs are supplied domestically, can easily produce misleading conclusions in such an economy.

The orthodox measure of total factor intensity, as developed by Leontief, indicates the factor intensity at the last stage of production and of the goods which go into the production process. The measure formulated above differs in that it indicates the factor intensity of the goods which exchange (exported) for intermediate inputs not produced domestically. Therefore, a comparison of the two measures reveals the net factor cost or saving derived from the utilization of imported rather than domestically supplied intermediate goods.[4] For example, $L_j < L'_j$ and $K_j > K'_j$ indicate that the importation of intermediate inputs (i.e., the implicit substitution of exports for otherwise domestically supplied inputs) reduces the overall capital requirement in the economy, but entails a greater demand for labor than would be the case if all intermediate inputs were supplied domestically. In a labor-abundant, capital-scarce, less-developed country (LDC) presumably such a trade-off indicates that the importation of intermediate inputs is in accordance with neoclassical principles of comparative advantage, though of course the resource allocation consequences of any such trade-off can be precisely weighed and evaluated only if one has knowledge of the shadow prices of labor and capital, from which the net resource cost (saving) can be derived.

It should be recognized that the total factor intensity of production in an open economy is highly sensitive to the structure of exports. In such cases where the export pattern is drastically out of line with comparative advantage considerations, the comparison of factor intensities under the existing structure of production with those under a hypothetical structure which assumes all intermediate inputs are supplied domestically will yield little in terms of "explaining" the existing structure, or as a guide for planning and policy. It may be, for example, that relatively capital-intensive intermediate inputs are imported in a given LDC with foreign exchange earned by exporting equally, or even relatively more capital-intensive commodities. In such a case, even though we may find $K < K'$ and $L > L'$, the optimal solution would not be the substitution of domestic inputs for imports (though it might be an improvement), but rather a restructuring of exports toward more labor-intensive commodities. If the relevant problem is one of deriving an ex ante measure of comparative advantage then certainly a more general concept of "domestic resource cost" than the one discussed above is required.[5]

## Factor Intensity in the Taiwan Economy

Among the countries struggling to industrialize none has been more successful than several East Asian island economies (Hong Kong, Singapore, Taiwan, S. Korea); among this exclusive group perhaps Taiwan has witnessed the most remarkable gains. The manufacturing sector in Taiwan has grown at a compound growth rate of 18 percent per annum over the last decade, 1961–1971, providing the engine of growth in real per capita income of 7 percent per annum over the same period.[6] The industrialization process of the East Asian island economies exhibits 2

Table 22.1.   Total Labor and Capital Requirement per Million NT$ Output in the Taiwan Economy: 1969/1970

| Sector | | Existing production structure | | Assumed production structure | | Resource cost (saving) from utilization of imported inputs per million NT$ output | |
|---|---|---|---|---|---|---|---|
| | | Total labor requirement per million NT$ output $L'_j$ (man-years) | Total capital requirement per million NT$ output $K'_j$ (NT$) | Total labor requirement per million NT$ output $L_j$ (man-years) | Total capital requirement per million NT$ output $K_j$ (NT$) | $L_j - L'_j$ (man-years) | $K_j - K'_j$ (NT$) |
| Agriculture | 1 | 80.037 | 603,896 | 80.813 | 601,590 | 0.776 | -2,306 |
| Forestry | 2 | 53.039 | 269,754 | 52.867 | 269,570 | -0.172 | -184 |
| Fishing | 3 | 47.377 | 961,436 | 46.018 | 955,390 | -1.359 | -6,046 |
| Mining | 4 | 29.526 | 963,949 | 28.149 | 924,120 | -1.377 | -39,829 |
| Sugar | 5 | 50.725 | 1,823,587 | 50.694 | 1,840,263 | -0.031 | 16,676 |
| Canned Food | 6 | 51.305 | 856,071 | 47.867 | 854,255 | -3.438 | -1,816 |
| Tobacco | 7 | 13.713 | 532,005 | 12.438 | 505,772 | -1.275 | -26,233 |
| Alcoholic Beverage | 8 | 13.582 | 591,609 | 13.655 | 591,080 | 0.073 | -529 |
| M.S.G. (flavoring) | 9 | 21.098 | 958,907 | 20.522 | 959,403 | -0.576 | 496 |
| Wheat Flour | 10 | 29.730 | 1,356,392 | 60.212 | 1,015,072 | 30.482 | -341,320 |
| Edible Vegetable Oil | 11 | 34.929 | 1,030,485 | 63.621 | 707,109 | 28.692 | -323,376 |
| Nonalcoholic Beverage | 12 | 19.751 | 902,586 | 18.874 | 903,496 | -0.877 | 910 |
| Tea | 13 | 63.492 | 928,672 | 63.213 | 924,052 | -0.279 | -4,620 |
| Miscellaneous Food | 14 | 40.990 | 1,023,690 | 51.356 | 909,457 | 10.366 | -114,233 |
| Artificial Fibre | 15 | 22.113 | 2,015,453 | 16.915 | 2,351,993 | -5.198 | 336,540 |
| Artificial Fabric | 16 | 27.896 | 1,771,792 | 23.098 | 2,209,460 | -4.798 | 437,668 |
| Cotton Fabrics | 17 | 34.591 | 1,558,800 | 50.735 | 1,405,319 | 16.144 | -153,481 |
| Wool and Worst Fabric | 18 | 26.706 | 1,164,921 | 46.492 | 1,131,493 | 19.786 | -33,428 |
| Apparel | 19 | 37.000 | 1,127,671 | 42.058 | 1,114,892 | 5.058 | -12,779 |
| Lumber | 20 | 41.887 | 689,096 | 41.830 | 687,947 | -0.057 | -1,149 |
| Plywood | 21 | 28.977 | 1,142,292 | 40.513 | 753,849 | 11.536 | -388,443 |
| Bamboo, Rattan Products | 22 | 38.812 | 500,977 | 38.804 | 485,407 | -0.008 | -15,570 |
| Paper/Pulp | 23 | 30.019 | 1,287,610 | 29.177 | 1,339,397 | -0.842 | 51,787 |
| Printing/Publications | 24 | 24.034 | 950,372 | 23.435 | 976,681 | -0.599 | 26,309 |

| | | | | | | |
|---|---|---|---|---|---|---|
| 25 | Leather and Products | 29.159 | 1,064,862 | 41.405 | 1,043,152 | 12.246 | −21,710 |
| 26 | Rubber and Products | 28.123 | 927,432 | 26.852 | 953,379 | −1.271 | −25,947 |
| 27 | Chemical Fertilizer | 21.978 | 2,260,389 | 20.334 | 2,288,390 | −1.644 | 28,001 |
| 28 | Medicines | 21.712 | 1,011,302 | 19.712 | 1,026,313 | −2.000 | 15,011 |
| 29 | Plastic and Products | 21.446 | 1,128,948 | 18.581 | 1,176,112 | −2.865 | 47,164 |
| 30 | Petroleum | 14.381 | 1,140,883 | 11.948 | 1,112,679 | −2.433 | −28,204 |
| 31 | Nonedible Vegetable Oil | 44.952 | 691,017 | 47.147 | 644,227 | 2.195 | −46,790 |
| 32 | Miscellaneous Industrial Chemicals | 22.510 | 1,469,730 | 20.451 | 1,489,438 | −2.059 | 19,708 |
| 33 | Miscellaneous Chemical Manufactures | 23.656 | 838,167 | 21.566 | 929,033 | −2.090 | 90,866 |
| 34 | Cement | 11.111 | 1,119,058 | 10.549 | 1,123,138 | −0.562 | 4,080 |
| 35 | Cement Products | 28.823 | 1,345,015 | 27.437 | 1,365,772 | −1.386 | 20,757 |
| 36 | Glass Products | 20.430 | 1,338,766 | 19.273 | 1,336,531 | −1.157 | −2,235 |
| 37 | Miscellaneous Nonmetal Mineral Products | 27.673 | 876,703 | 27.016 | 874,754 | −0.657 | −1,949 |
| 38 | Steel and Iron | 23.679 | 1,168,287 | 17.351 | 1,264,854 | −6.328 | 96,567 |
| 39 | Steel and Iron Products | 30.747 | 1,134,838 | 23.934 | 1,239,167 | −6.813 | 104,329 |
| 40 | Aluminium | 25.407 | 3,066,601 | 22.719 | 3,224,579 | −2.688 | 157,978 |
| 41 | Aluminium Products | 31.188 | 2,392,612 | 29.267 | 2,482,649 | −1.921 | 90,037 |
| 42 | Miscellaneous Metal Products | 23.397 | 901,889 | 19.054 | 931,106 | −4.343 | 29,217 |
| 43 | Machinery | 29.797 | 1,054,242 | 25.366 | 1,115,222 | −4.431 | 60,980 |
| 44 | H. H. Electrical Appliances | 20.753 | 950,926 | 17.975 | 968,512 | −2.778 | 17,586 |
| 45 | Communication Equipment | 27.231 | 921,659 | 22.796 | 891,637 | −4.435 | −30,022 |
| 46 | Other Electrical Appliances | 29.734 | 1,056,710 | 25.746 | 1,139,976 | −3.988 | 83,266 |
| 47 | Shipbuilding | 26.304 | 1,118,904 | 22.358 | 1,168,932 | −3.946 | 50,028 |
| 48 | Motor Vehicles | 21.275 | 978,290 | 16.276 | 990,103 | −4.999 | 11,813 |
| 49 | Other Transport Equipment | 26.389 | 826,433 | 23.512 | 826,073 | −2.877 | −360 |
| 50 | Miscellaneous Manufactures | 33.010 | 1,080,684 | 30.475 | 1,202,021 | −2.535 | 121,337 |
| 51 | Construction | 33.460 | 684,924 | 32.283 | 677,890 | −1.177 | −7,034 |
| 52 | Services | 13.840 | 437,690 | 13.248 | 425,780 | −0.592 | −11,910 |

*Source:* See Appendix Table 22.A.1.

Table 22.2.   Total Labor and Capital Requirement per Million NT$ Output in the Taiwan Economy: 1969/1970

| Sector | | Existing structure of production | | Assumed structure of production | | Resource cost (saving) from utilization of imported manufacturing inputs per million NT$ output | |
|---|---|---|---|---|---|---|---|
| | | Total labor requirement per million NT$ output $L'_j$ (man-years) | Total capital requirement per million NT$ output $K'_j$ (NT$) | Total labor requirement per million NT$ output $L_j$ (man-years) | Total capital requirement per million NT$ output $K_j$ (NT$) | $L_j - L'_j$ (man-years) | $K_j - K'_j$ (NT$) |
| Sugar | 5 | 11.340 | 1,575,431 | 11.272 | 1,594,894 | -0.068 | 19,463 |
| Canned Food | 6 | 12.608 | 583,634 | 12.292 | 592,628 | -0.316 | 8,994 |
| Tobacco | 7 | 4.610 | 457,675 | 4.109 | 436,934 | -0.501 | -20,741 |
| Alcoholic Beverage | 8 | 5.463 | 510,620 | 5.429 | 515,501 | -0.034 | 4,881 |
| M.S.C. (flavoring) | 9 | 9.775 | 772,756 | 9.723 | 783,896 | -0.052 | 11,140 |
| Wheat Flour | 10 | 4.077 | 580,187 | 4.858 | 696,343 | 0.781 | 116,156 |
| Edible Vegetable Oil | 11 | 2.605 | 255,147 | 3.311 | 364,316 | 0.706 | 109,169 |
| Nonalcoholic Beverage | 12 | 9.524 | 762,261 | 9.286 | 716,609 | -0.238 | -45,652 |
| Tea | 13 | 8.892 | 601,640 | 8.739 | 605,319 | -0.153 | 3,679 |
| Miscellaneous Food | 14 | 11.345 | 564,058 | 11.536 | 607,916 | 0.191 | 43,858 |
| Artificial Fibre | 15 | 11.158 | 1,884,467 | 9.433 | 2,156,745 | -1.725 | 272,278 |
| Artificial Fabric | 16 | 15.725 | 1,609,235 | 14.416 | 2,011,998 | -1.309 | 402,763 |
| Cotton Fabric | 17 | 17.123 | 1,049,778 | 17.489 | 1,164,170 | 0.366 | 114,392 |
| Wool and Worst Fabric | 18 | 10.969 | 763,698 | 11.544 | 854,601 | 0.575 | 90,903 |
| Apparel | 19 | 23.981 | 881,855 | 24.135 | 941,360 | 0.154 | 59,505 |
| Lumber | 20 | 7.310 | 461,446 | 7.336 | 466,571 | 0.026 | 5,125 |
| Plywood | 21 | 7.130 | 495,025 | 7.716 | 552,250 | 0.586 | 57,225 |
| Bamboo, Rattan Products | 22 | 18.929 | 314,439 | 18.806 | 326,774 | -0.123 | 12,335 |
| Paper/Pulp | 23 | 13.341 | 1,042,730 | 13.378 | 1,073,105 | 0.037 | 30,375 |
| Printing/Publications | 24 | 15.917 | 814,141 | 15.876 | 829,807 | -0.041 | 15,666 |
| Leather and Products | 25 | 15.227 | 774,344 | 15.321 | 786,539 | 0.094 | 12,195 |
| Rubber and Products | 26 | 15.113 | 730,870 | 15.494 | 727,172 | 0.381 | -3,698 |
| Chemical Fertilizer | 27 | 8.664 | 1,929,177 | 8.379 | 1,960,787 | -0.285 | 31,610 |

| | | | | | | | |
|---|---|---|---|---|---|---|---|
| 28 | Medicines | 13.069 | 860,065 | 12.771 | 861,803 | −0.298 | 1,738 |
| 29 | Plastic and Products | 7.653 | 735,924 | 9.492 | 964,896 | 1.839 | 228,972 |
| 30 | Petroleum | 2.732 | 796,314 | 2.885 | 830,662 | 0.153 | 34,348 |
| 31 | Nonedible Vegetable Oil | 5.071 | 418,681 | 3.922 | 360,239 | −1.149 | −58,442 |
| 32 | Miscellaneous Industrial Chemicals | 6.549 | 1,055,364 | 6.257 | 1,088,114 | −0.292 | 32,750 |
| 33 | Miscellaneous Chemical Manufactures | 10.725 | 673,913 | 8.290 | 671,965 | −2.435 | −1,948 |
| 34 | Cement | 3.329 | 917,293 | 3.408 | 926,679 | 0.079 | 9,386 |
| 35 | Cement Products | 19.185 | 1,117,752 | 19.127 | 1,131,574 | −0.058 | 13,822 |
| 36 | Glass Products | 10.675 | 1,087,004 | 10.597 | 1,098,810 | −0.078 | 11,806 |
| 37 | Miscellaneous Nonmetal Mineral Products | 16.395 | 568,455 | 16.432 | 577,662 | 0.037 | 9,207 |
| 38 | Steel and Iron | 10.859 | 1,001,907 | 9.398 | 1,036,102 | −1.461 | 34,195 |
| 39 | Steel and Iron Products | 18.788 | 1,013,539 | 17.289 | 1,051,502 | −1.499 | 37,963 |
| 40 | Aluminium | 13.297 | 2,804,758 | 12.561 | 2,942,837 | −0.736 | 138,079 |
| 41 | Aluminium Products | 21.891 | 2,185,159 | 21.378 | 2,263,730 | −0.513 | 78,571 |
| 42 | Miscellaneous Metal Products | 13.041 | 770,477 | 12.427 | 764,900 | −0.614 | −5,577 |
| 43 | Machinery | 19.076 | 918,375 | 18.795 | 939,369 | −0.281 | 20,994 |
| 44 | H. H. Electrical Appliances | 11.512 | 803,418 | 11.602 | 813,154 | 0.090 | 9,736 |
| 45 | Communication Equipment | 14.055 | 783,021 | 14.595 | 712,088 | 0.540 | −70,933 |
| 46 | Other Electrical Appliances | 18.903 | 919,718 | 18.968 | 961,866 | 0.065 | 42,148 |
| 47 | Shipbuilding | 15.959 | 1,029,078 | 16.458 | 1,054,337 | 0.499 | 25,259 |
| 48 | Motor Vehicles | 10.874 | 850,916 | 9.961 | 834,634 | 0.913 | −16,282 |
| 49 | Other Transport Equipment | 18.458 | 725,262 | 18.889 | 707,567 | 0.431 | −17,695 |
| 50 | Miscellaneous Manufactures | 18.932 | 883,516 | 18.280 | 975,565 | −0.652 | 92,049 |

*Source:* See Appendix Table 22.A.1.

outstanding characteristics: (1) strong orientation toward exports; (2) heavy reliance on imported inputs. In Taiwan, for example, 30 percent of manufactured output was exported in 1969, while 14 percent of the value of manufactured output was contributed by imported inputs.[7] Applying the formulations developed above we attempt to demonstrate the rationale of these key features of industrialization in Taiwan in terms of neoclassical, factor-proportions theory.

Table 22.1 presents measures of total labor and capital requirements per million NT$ (New Taiwan Dollar) in each of 52 sectors of the Taiwan economy given the existing structure of production and alternatively assuming all intermediate inputs were domestically supplied. The difference between factor requirements under the two alternative production structures has been calculated to reveal the capital and labor cost or saving attributable to the importation of intermediate inputs. It is of course inappropriate to sum the factor requirements over all sectors since the factor requirements of any one sector are measured in terms of total requirements throughout the entire economy. However, taking the average factor cost (saving), weighted by the distribution of output over all sectors, reveals that the utilization of imported inputs in the Taiwan economy on the average entailed a saving of 0.189 man-years of employment and an additional capital cost of NT$ 224 per million NT$ output. The apparent paradox (of the Leontief type) which this result poses is resolved upon closer inspection of Table 22.1. It is precisely those sectors which rely heavily on imported natural raw material inputs that exhibit the paradoxical $L > L'$, $K < K'$: wheat for flour; cotton for textiles; timber for plywood; hides for leather products, etc. In sectors more dependent upon imports of processed intermediate inputs the expected result ($L < L'$, $K > K'$) is found. It is of course well known that neoclassical factor-proportions theory is unable to explain trade in natural resources; and this fact has been used to resolve the paradox which Leontief discovered in the United States trade pattern as well (Seija Naya, 1967).

Total factor requirements per million NT$ output in each of 46 manufacturing sectors—excluding the indirect requirements in the primary sector (1–4), construction (51) and services (52)—under the two alternative production structures are presented in Table 22.2. Calculation of the implicit factor costs of imported inputs (i.e., the export equivalent) is based on average factor requirements of manufactured exports.[8] Comparisons of $(L, K)$ and $(L', K')$ in Table 22.2 reveal the resource (labor and capital) cost or saving resulting from the importation of manufactured intermediate inputs, assuming these inputs were imported with foreign exchange earned by exporting strictly manufactured commodities.

Abstracting from trade in nonmanufactured goods clearly resolves the paradox we found in our previous results. On the average (weighted by the distribution of output in the manufacturing sector) the trade-off of manufactured exports for imports of manufactured intermediate inputs saved the economy NT$ 59,893 in capital and entailed an additional labor cost of 0.083 man-years per million NT$ output—as compared with the alternative of supplying all manufactured inputs domestically. Although we would need to know the shadow prices of capital and labor to precisely calculate the net resource cost of this trade off, it is quite apparent that resource savings in terms of capital well outweigh resource costs in terms of labor.[9] In other words, import-dependent, "footloose" industry in a developing country such as Taiwan would appear quite justifiable in terms of resource allocation as judged by strictly factor-proportions considerations.

## NOTES

1. The analysis throughout abstracts from international capital flows. The assumption is that the balance on goods and services is in equilibrium; imports can be traded only for exports—an assumption neither uncommon nor extremely restrictive.
2. According to standard input-output notation,

$$[D] = \{d_{ij}\}$$

where,

$$d_{ij} = a_{ij} - m_{ij},$$

is the *domestic* input required per unit, equal to the total input required per unit less imported inputs.
3. $M_i = \sum_j m_{ji}$ is the total direct per unit imported intermediate input requirement.
4. An implicit assumption is that all imported intermediates are substitutable.
5. Such as developed by Bruno in 1963 and elaborated in 1972.
6. *Taiwan Statistical Data Book*, CIECD, Executive Yuan, Taipei, Republic of China, 1972.
7. According to *Input-Output Table for Taiwan, 1969*, CIECD, Executive Yuan, Taipei, Republic of China. The percent of value contributed by imported inputs is calculated by

$$\sum_j \left[ \sum_i M_i s_{ji} \right] q_j$$

where $q_j$ is the proportion of the $j^{\text{th}}$ sector in total value of manufacturing output.

8.
$$\frac{L_{Fx}}{1 - M_{Fx}} = \frac{\sum_{j=5}^{50} \left[ \sum_{i=5}^{50} l_i s_{ij} \right] e_j}{1 - \sum_{j=5}^{50} \left[ \sum_{i=5}^{50} M'_s s_{ij} \right] e_j}$$

where $e_j$ is the proportion share of the $j^{\text{th}}$ sector in total export of manufactures.
9. The yearly average wage in the manufacturing sector in Taiwan in 1969 was NT$ 16,000.

### References

Bruno, M. 1963. *Interdependence, Resource Use and Structural Change in Israel*. Jerusalem: Bank of Israel.
———. 1972. Domestic resource costs and effective protection: Clarification and synthesis. *Journal of Political Economy* 80 (January/February): pp. 16–33.
Leontief, W. W. 1953. Domestic productions and foreign trade: The American capital position re-examined. *Proceedings of the American Philosophical Society* 97(4): pp. 332–349.
———. 1956. Factor proportions and the structure of American trade: Further theoretical and empirical analysis. *Review of Economics and Statistics* 38 (November): 386–407.
Naya, S. 1967. Natural resources, factor mix and factor reversals in international trade. *American Economic Review*, 57 (May): 561–570.

Appendix

Table 22.A.1.

| Sector | Per unit imported input requirement $M_i$ | Direct labor requirement per million NT$ output $l_i$ (man-years) | Direct capital requirement per million NT$ output $k_i$ (NT$) | $M_j^x = \sum_{i=1}^{52} M_i s_{ij}$ | $L_j^x = \sum_{i=1}^{52} l_i s_{ij}$ (man-years) | $K_j^x = \sum_{i=1}^{52} k_i s_{ij}$ (NT$) | Sectoral share of total exports $e_j$ | Sectoral share of total output $q_j$ |
|---|---|---|---|---|---|---|---|---|
| Agriculture | 0.0188 | 56.970 | 246,000 | 0.0895 | 77.187 | 513,060 | 0.053 | 0.130 |
| Forestry | 0.0029 | 50.000 | 184,500 | 0.0197 | 52.412 | 249,760 | 0.006 | 0.008 |
| Fishing | 0.1245 | 36.256 | 585,200 | 0.1920 | 41.263 | 766,570 | 0.043 | 0.017 |
| Mining | 0.0118 | 22.847 | 738,100 | 0.0589 | 27.650 | 904,170 | 0.004 | 0.016 |
| Sugar | 0.0021 | 8.550 | 1,298,692 | 0.0651 | 48.652 | 1,757,515 | 0.034 | 0.010 |
| Canned Food | 0.0280 | 7.787 | 242,365 | 0.1554 | 46.356 | 698,351 | 0.068 | 0.011 |
| Tobacco | 0.0627 | 3.081 | 321,641 | 0.0839 | 11.041 | 446,853 | 0.002 | 0.014 |
| Alcoholic Beverage | 0.0039 | 4.000 | 400,000 | 0.0395 | 12.324 | 551,520 | 0.001 | 0.010 |
| M.S.G. (flavoring) | 0.0108 | 5.939 | 280,677 | 0.0844 | 18.410 | 873,247 | 0.004 | 0.002 |
| Wheat Flour | 0.6964 | 3.111 | 533,878 | 0.7147 | 6.970 | 631,021 | 0.000 | 0.006 |
| Edible Vegetable Oil | 0.6614 | 2.148 | 210,167 | 0.6805 | 13.258 | 339,825 | 0.000 | 0.008 |
| Nonalcoholic Beverage | 0.0359 | 5.568 | 367,042 | 0.0982 | 16.624 | 802,920 | 0.001 | 0.002 |
| Tea | 0.0024 | 7.508 | 448,014 | 0.0613 | 61.540 | 866,457 | 0.009 | 0.002 |
| Miscellaneous Food | 0.1491 | 8.123 | 260,209 | 0.3066 | 31.226 | 712,513 | 0.014 | 0.024 |
| Artificial Fibre | 0.3367 | 4.489 | 1,280,546 | 0.4128 | 8.967 | 1,596,490 | 0.017 | 0.009 |
| Artificial Fabric | 0.2320 | 5.764 | 597,855 | 0.4416 | 13.833 | 1,323,599 | 0.088 | 0.021 |
| Cotton Fabrics | 0.2796 | 9.095 | 557,415 | 0.4866 | 19.095 | 1,064,935 | 0.041 | 0.021 |
| Wool and Worst Fabric | 0.3508 | 6.156 | 427,474 | 0.4685 | 11.786 | 689,426 | 0.010 | 0.006 |
| Apparel | 0.0490 | 16.623 | 404,585 | 0.2355 | 29.500 | 888,655 | 0.029 | 0.016 |
| Lumber | 0.0084 | 6.111 | 371,877 | 0.0463 | 40.412 | 642,105 | 0.014 | 0.008 |
| Plywood | 0.5505 | 5.518 | 370,496 | 0.6074 | 9.634 | 525,873 | 0.044 | 0.008 |
| Bamboo, Rattan Products | 0.0276 | 15.856 | 115,481 | 0.0923 | 35.873 | 407,299 | 0.014 | 0.006 |
| Paper/Pulp | 0.1309 | 7.614 | 520,819 | 0.2369 | 22.475 | 1,047,173 | 0.000 | 0.012 |
| Printing/Publications | 0.0469 | 10.685 | 422,377 | 0.1418 | 19.518 | 806,455 | 0.004 | 0.009 |
| Leather and Products | 0.3169 | 8.689 | 354,884 | 0.4346 | 15.319 | 623,773 | 0.005 | 0.002 |
| Rubber and Products | 0.3247 | 8.147 | 241,660 | 0.4020 | 15.321 | 519,430 | 0.012 | 0.005 |

| | | | | | | | | | |
|---|---|---|---|---|---|---|---|---|---|
| Chemical Fertilizer | 27 | 0.0703 | 3.390 | 1,241,972 | 0.2437 | 14.217 | 2,013,051 | 0.005 | 0.008 |
| Medicines | 28 | 0.1408 | 8.524 | 515,814 | 0.2036 | 15.228 | 804,663 | 0.002 | 0.004 |
| Plastic and Products | 29 | 0.1970 | 3.887 | 392,530 | 0.3353 | 10.768 | 788,642 | 0.051 | 0.017 |
| Petroleum | 30 | 0.3053 | 1.781 | 676,605 | 0.3432 | 3.451 | 792,559 | 0.017 | 0.023 |
| Nonedible Vegetable Oil | 31 | 0.1259 | 2.500 | 200,000 | 0.1724 | 39.462 | 516,043 | 0.002 | 0.001 |
| Misc. Industrial Chemicals | 32 | 0.1902 | 3.088 | 665,978 | 0.2922 | 13.205 | 1,173,168 | 0.003 | 0.014 |
| Misc. Chemical Manufactures | 33 | 0.3528 | 4.262 | 191,486 | 0.4089 | 10.634 | 423,162 | 0.004 | 0.009 |
| Cement | 34 | 0.0547 | 1.559 | 783,263 | 0.0965 | 8.038 | 1,021,118 | 0.009 | 0.011 |
| Cement Products | 35 | 0.0414 | 15.377 | 588,430 | 0.1503 | 24.036 | 1,192,471 | 0.000 | 0.002 |
| Glass Products | 36 | 0.0381 | 6.921 | 735,390 | 0.1224 | 16.532 | 1,214,539 | 0.006 | 0.003 |
| Miscellaneous Nonmetallic Mineral Products | 37 | 0.0451 | 14.249 | 392,638 | 0.1036 | 24.374 | 771,557 | 0.004 | 0.005 |
| Steel and Iron | 38 | 0.3173 | 3.218 | 384,765 | 0.4704 | 8.699 | 690,864 | 0.011 | 0.015 |
| Steel and Iron Products | 39 | 0.3701 | 10.507 | 395,387 | 0.4970 | 14.919 | 630,418 | 0.016 | 0.008 |
| Aluminium | 40 | 0.1455 | 5.922 | 1,545,269 | 0.2824 | 16.414 | 2,779,985 | 0.002 | 0.003 |
| Aluminium Products | 41 | 0.0198 | 12.830 | 597,857 | 0.1922 | 25.067 | 2,197,543 | 0.003 | 0.001 |
| Miscellaneous Metal Products | 42 | 0.3010 | 7.026 | 297,688 | 0.3589 | 11.967 | 537,631 | 0.005 | 0.003 |
| Machinery | 43 | 0.2268 | 11.505 | 348,764 | 0.3889 | 17.412 | 659,536 | 0.024 | 0.013 |
| H.H. Electrical Appliances | 44 | 0.1699 | 5.291 | 356,492 | 0.2756 | 11.976 | 671,212 | 0.005 | 0.007 |
| Communication Equipment | 45 | 0.4184 | 5.366 | 219,267 | 0.5028 | 11.219 | 411,352 | 0.076 | 0.020 |
| Other Electrical Appliances | 46 | 0.2975 | 10.901 | 343,635 | 0.3947 | 17.164 | 656,117 | 0.007 | 0.011 |
| Shipbuilding | 47 | 0.3320 | 9.722 | 548,862 | 0.3842 | 14.069 | 728,967 | 0.004 | 0.003 |
| Motor Vehicles | 48 | 0.2732 | 4.002 | 333,663 | 0.3725 | 9.412 | 600,229 | 0.002 | 0.013 |
| Other Transport Equipment | 49 | 0.1862 | 13.036 | 336,504 | 0.2828 | 17.383 | 539,411 | 0.006 | 0.003 |
| Miscellaneous Manufactures | 50 | 0.3031 | 12.876 | 445,818 | 0.3749 | 21.071 | 700,187 | 0.026 | 0.005 |
| Construction | 51 | 0.0375 | 20.392 | 125,700 | 0.1369 | 29.100 | 545,980 | 0.003 | 0.060 |
| Services | 52 | 0.0243 | 9.787 | 293,700 | 0.0516 | 12.197 | 385,320 | 0.188 | 0.353 |

Sources: *Input-Output Tables Taiwan 1969*, CIECD, Executive Yuan, Taipei, Taiwan, Republic of China.
*Report on Industrial and Commercial Surveys*, No. 2 (1969), Ministry of Economic Affairs, Taipei, Taiwan, Republic of China.
*Taiwan Agricultural Yearbook 1972*, Department of Agriculture, Provincial Government of Taiwan, Republic of China.

Note: $M_{f}^{x} = \sum_{j=1}^{52} M_{j}^{x} = 0.2662$; $L_{f}^{x} = \sum_{j=1}^{52} L_{j}^{x} e_{j} = 23.368$ (man-years); $K_{f}^{x} = \sum_{j=1}^{52} K_{j}^{x} e_{j} = 744758$

$\frac{L_{f}^{x}}{1 - m_{f}^{x}} = 31.846$ (man-years); $\frac{K_{f}^{x}}{1 - M_{f}^{x}} = 1,014,931$ (NT$).

(NT$).

Appendix
Table 22.A.2.

| Sector | | Per unit imported manufactured input requirement $M_i$ | $M_j^x = \sum_{i=1}^{52} M_i s_{ij}$ | $L_j^x = \sum_{i=5}^{50} l_i s_{ij}$ (man-years) | $K_j^x = \sum_{i=5}^{50} k_i s_{ij}$ (NT$) | Sectoral share of total manufactured exports $e_j$ | Sectoral share of total manufactured output $q_j$ |
|---|---|---|---|---|---|---|---|
| Sugar | 5 | 0.0013 | 0.0400 | 10.807 | 1,537,698 | 0.049 | 0.025 |
| Canned Food | 6 | 0.0135 | 0.1181 | 11.034 | 472,225 | 0.096 | 0.026 |
| Tobacco | 7 | 0.0620 | 0.0702 | 3.674 | 391,453 | 0.003 | 0.035 |
| Alcoholic Beverage | 8 | 0.0020 | 0.0195 | 5.203 | 492,225 | 0.002 | 0.024 |
| M.S.G. (flavoring) | 9 | 0.0005 | 0.0372 | 9.279 | 737,664 | 0.005 | 0.006 |
| Wheat Flour | 10 | 0.0000 | 0.0087 | 3.961 | 571,980 | 0.001 | 0.013 |
| Edible Vegetable Oil | 11 | 0.0007 | 0.0113 | 2.454 | 244,488 | 0.000 | 0.019 |
| Nonalcoholic Beverage | 12 | 0.0344 | 0.0733 | 8.547 | 693,114 | 0.001 | 0.006 |
| Tea | 13 | 0.0001 | 0.0336 | 8.444 | 569,944 | 0.013 | 0.005 |
| Miscellaneous Food | 14 | 0.0136 | 0.0440 | 10.759 | 522,551 | 0.020 | 0.057 |
| Artificial Fibre | 15 | 0.3293 | 0.3893 | 5.969 | 1,517,223 | 0.024 | 0.022 |
| Artificial Fabric | 16 | 0.2234 | 0.4032 | 10.350 | 1,228,784 | 0.125 | 0.051 |
| Cotton Fabric | 17 | 0.0263 | 0.0642 | 16.268 | 989,216 | 0.058 | 0.050 |
| Wood and Worst Fabric | 18 | 0.1200 | 0.1699 | 8.705 | 603,424 | 0.014 | 0.015 |
| Apparel | 19 | 0.0347 | 0.0952 | 22.712 | 792,049 | 0.041 | 0.039 |
| Lumber | 20 | 0.0031 | 0.0259 | 6.965 | 437,014 | 0.020 | 0.019 |
| Plywood | 21 | 0.0105 | 0.0261 | 6.779 | 474,649 | 0.062 | 0.019 |
| Bamboo, Rattan Products | 22 | 0.0181 | 0.0495 | 18.270 | 276,744 | 0.019 | 0.016 |
| Paper/Pulp | 23 | 0.1270 | 0.1895 | 10.815 | 863,966 | 0.008 | 0.030 |
| Printing/Publications | 24 | 0.0417 | 0.1149 | 14.386 | 705,751 | 0.006 | 0.022 |
| Leather and Products | 25 | 0.1867 | 0.2550 | 11.828 | 533,791 | 0.007 | 0.005 |
| Rubber and Products | 26 | 0.3176 | 0.3503 | 10.443 | 400,416 | 0.018 | 0.012 |
| Chemical Fertilizer | 27 | 0.0363 | 0.1108 | 7.187 | 1,824,655 | 0.007 | 0.020 |
| Medicines | 28 | 0.1384 | 0.1759 | 10.724 | 694,131 | 0.003 | 0.010 |
| Plastics and Products | 29 | 0.0072 | 0.0486 | 7.005 | 690,078 | 0.072 | 0.042 |

| | | | | | | | |
|---|---|---|---|---|---|---|---|
| Petroleum | 30 | 0.0321 | 0.0428 | 755,939 | 2.162 | 0.024 | 0.055 |
| Nonedible Vegetable Oil | 31 | 0.1143 | 0.1402 | 286,424 | 3.202 | 0.003 | 0.002 |
| Misc. Industrial Chemicals | 32 | 0.0643 | 0.1089 | 952,634 | 5.098 | 0.004 | 0.034 |
| Misc. Chemical Manufactures | 33 | 0.3468 | 0.3785 | 316,857 | 5.679 | 0.005 | 0.021 |
| Cement | 34 | 0.0138 | 0.0437 | 876,069 | 2.746 | 0.013 | 0.026 |
| Cement Products | 35 | 0.0396 | 0.1147 | 1,009,551 | 17.656 | 0.000 | 0.004 |
| Glass Products | 36 | 0.0199 | 0.0582 | 1,032,102 | 9.899 | 0.009 | 0.007 |
| Miscellaneous Nonmetal Mineral Products | 37 | 0.0083 | 0.0395 | 531,193 | 15.868 | 0.005 | 0.012 |
| Steel and Iron | 38 | 0.3106 | 0.4469 | 580,326 | 4.901 | 0.015 | 0.039 |
| Steel and Iron Products | 39 | 0.3668 | 0.4814 | 559,412 | 12.370 | 0.022 | 0.020 |
| Aluminium | 40 | 0.1053 | 0.1931 | 2,622,598 | 10.723 | 0.003 | 0.007 |
| Aluminium Products | 41 | 0.0180 | 0.1336 | 2,059,128 | 20.110 | 0.004 | 0.003 |
| Miscellaneous Metal Products | 42 | 0.2902 | 0.3326 | 456,720 | 8.607 | 0.007 | 0.008 |
| Machinery | 43 | 0.2198 | 0.3649 | 574,148 | 14.212 | 0.035 | 0.032 |
| H.H. Electrical Appl. | 44 | 0.1561 | 0.2362 | 580,600 | 8.364 | 0.007 | 0.016 |
| Communication Equipment | 45 | 0.4121 | 0.4786 | 331,536 | 7.675 | 0.108 | 0.047 |
| Other Electrical Appliances | 46 | 0.2957 | 0.3733 | 567,567 | 13.927 | 0.009 | 0.026 |
| Shipbuilding | 47 | 0.3287 | 0.3704 | 679,663 | 11.022 | 0.005 | 0.006 |
| Motor Vehicles | 48 | 0.2710 | 0.3539 | 517,066 | 6.157 | 0.003 | 0.031 |
| Other Transport Equip. | 49 | 0.1825 | 0.2586 | 481,313 | 15.011 | 0.009 | 0.006 |
| Miscellaneous Manufactures | 50 | 0.2776 | 0.3243 | 577,589 | 14.609 | 0.037 | 0.011 |

Source: See Appendix table 22.A.1.

Note: $M_{f}^{x} = \sum_{j=5}^{50} M_{j}^{x}e_{j} = 0.2156$; $L_{f}^{x} = \sum_{j=5}^{50} L_{j}^{x}e_{j} = 10,475$ (man-years); $K_{f}^{x} = \sum_{j=5}^{50} K_{j}^{x}e_{j} = 739,962$ (NT\$); $\dfrac{L_{f}^{x}}{1-M_{f}^{x}} = 13,331$ (man-years); $\dfrac{K_{f}^{x}}{1-M_{f}^{x}} = 943,347$ (NT\$).

# 23

## Interindustry Relations of a Metropolitan Area

WERNER Z. HIRSCH

A great deal of interest in metropolitan area analysis centers around the economic impact of autonomous forces upon any of the area's sectors and upon its totality. Researchers have often felt that even approximate knowledge of the structural relationship of the local economy would greatly facilitate their work.

An interindustry flow table has been prepared for the St. Louis area, one of the ten largest metropolitan areas in the United States. From such a table it is but a short step to technical coefficients and the inverse matrix. Furthermore, from these tables, income and employment multipliers can be estimated. Both are powerful tools in assessing the impact of final demand changes upon the economic activity of a metropolitan area.

Much serious criticism has been leveled in the past against area technical coefficients which are deduced from national ones pertaining to an outdated period. This study undertakes to estimate 1955 coefficients with the help of data obtained directly from a sample of private and public bodies which together constitute the St. Louis metropolitan area economy. Based upon this information some activity and impact projections are attempted.

### Theoretical Scheme of Model

A metropolitan area interindustry relations model divides the local economy into identifiable sectors along two lines—product (or industry) and geography. Thus, transactions are identified as taking place between specific industries at specific locations. Final demand determines the activity level of the metropolitan area's processing sectors.

The interindustry relations model of St. Louis conforms to Leontief's balanced regional model as modified by Moore and Petersen.[1] It uses Moore and Petersen's Utah model as a point of departure. However, three modifications have been introduced: First, the model of the St. Louis metropolitan area relies for its

This article was published in *The Review of Economics and Statistics*, Volume 41, Number 4, November, 1959, pp. 360–369. © By the President and Fellows of Harvard College. Reprinted by permission.

implementation upon company records, from which structural relationships are estimated; secondly, a detailed export matrix has been prepared; thirdly, the household and local government sectors are treated as part of the endogenous segment of the economy.[2] The latter modification was made since in metropolitan areas the activities of households and local governments are closely related to the general level of economic activity in the area. Furthermore, this study is interested in the local multiplier effect resulting from new household income generation in the area.

The St. Louis interindustry flow table is composed of three subtables:

1. The local matrix, in the upper left, represents local sales to local sectors;
2. the export matrix, in the upper right, represents local sales to nonlocal sectors;
3. the import matrix, in the lower left, represents nonlocal sales to local sectors.

Nonlocal activities could be broken down further into those that take place in the rest of the nation and those in foreign countries. Because relatively little trade was found to take place between St. Louis and foreign countries and because the study is not especially concerned with foreign trade, no such breakdown has been made. Furthermore, since the main concern is with output, income, and employment impacts on the St. Louis metropolitan area, and not the impact upon nonlocal supplying industries, the import matrix is compressed into a single import row.

The interindustry relations model of St. Louis is an open, nondynamic equilibrium model. Goods and service flows are gross rather than net and include intrasector transactions.

**Implementation of the Model**

The study area is coterminous with the Standard Metropolitan Area of St. Louis, which is composed of St. Louis City and the counties of St. Louis and St. Charles in Missouri and of Madison and St. Clair in Illinois. The area is clearly delineated, since it is far removed from other metropolitan areas and is surrounded by a sparsely populated and mainly agricultural hinterland. Virtually the entire labor force both lives and works inside this area.

The economy is distinguished by its diversification, with ten industries each accounting for more than four percent of manufacturing employment. In terms of 1955 manufacturing employment, the food and kindred products industry takes first place, accounting for about 14 percent of manufacturing employment. The transportation equipment industry is second, with about 11 percent. The area's growth over the last fifty years or so has been rather slow, the average metropolitan area having grown about one and one half times as fast as St. Louis.

The population of the St. Louis metropolitan area for 1955 was estimated at about 1.8 million. The year 1955 was selected for the implementation of the model since it was both recent and relatively normal. Furthermore, it made possible output comparisons with the 1954 Census of Manufactures and Business.

## Classification System

The industrial classification system divides the economy into sixteen manufacturing sectors, eleven nonmanufacturing sectors, the household sector, three government sectors, and gross private capital formation. Making explicit allowance for the recording of inventory changes, the industrial classification system breaks the St. Louis economy into 33 industrial sectors.

The 50-industry classification of the Interindustry Relations Study for 1947 is used as a point of departure. Some sectors are combined, and all those industries which were relatively unimportant in St. Louis were lumped together as either "miscellaneous manufacturing" or "undistributed."

## Data Sources

Input and output data were obtained for most large and medium-sized companies operating in the St. Louis area. In response to personal pleas by the heads of Washington University and St. Louis University, each of these companies assigned one of its key officials to work with the research staff of this study for a three-month period. Each company prepared its own input-output table for 1955. To assure uniformity in reporting, detailed oral and written instructions were given to each participant as well as a booklet into which the input and output data were placed.

Aside from collecting primary financial data from local companies, sector output totals were estimated on the basis of Census and trade association data. These estimates helped check the primary output data and gauge the adequacy of the sample data for those sectors in which only part of the companies cooperated with the study. Control totals were then compared with aggregate output data obtained by aggregating the outputs of local companies. In some instances, these company data covered the entire sector. When this was not the case, the output of cooperating firms was "blown up" on the basis of employment data, to produce an estimate of the sector's total output. The two sets of output totals were compared and, if any discrepancies appeared, their origin was traced and adjustments were made to produce a reasonable total.

## Construction of Transaction Table

The St. Louis interindustry flow table is by and large a composite of the area's firms' input-output tables, where independently derived control totals are used to check and, in case of need, to adjust output totals. Thus, the input-output tables of area firms were aggregated into industrial sector tables. In those sectors in which only partial firm coverage had been obtained, it was necessary to infer from the sample data to the entire sector. Input and output distributions of participating firms were assumed to be representative of those of the entire sector. While it is difficult to assess the adequacy of this assumption, it appears to be met more closely by some industries than by others. In a few subsectors for which no firm input-output tables were available, technical coefficients from the 200 × 200 Interindustry Flow Table for 1947 were used as a point of departure. Adjustments were made for recognizable dissimilarities in the area's production processes, marketing practices, and product mix. The geographical origin of inputs and destination of outputs were obtained from local companies or on the basis of the same inputs to similar industries.

The government sectors pose some special problems. Their inputs represent goods and service purchases, including transfer payments and subsidies. New public construction and maintenance is considered to have been purchased from the construction industry. In cooperation with this study, local governments prepared input-output tables. Current expenditures of state government in the St. Louis metropolitan area primarily cover office operations, maintenance activities, and welfare payments, all of which were estimated and inputs allocated with the help of state officials. Detailed studies were made of the operations and purchases of the federal government in the area. On their basis inputs were estimated and allocated.

Government outputs are represented by tax receipts. On the basis of an analysis of the tax system and structure of each of the three government sectors, taxes were separated into those received from households and from business and industry. The issue of tax shifts was altogether neglected. Taxes paid by commerce and industry were allocated according to the sector's employment. Since the counterpart of this entry is the industrial sector's tax payment, which is obtained from company records, government output allocations were readily checked.

The household sector also calls for some special consideration. Inputs include personal consumption expenditures of industrial sectors, direct personal taxes, and personal savings, which are allocated to gross private capital formation. Household output figures comprise wages and salaries including bonuses and retirement benefits, depreciation, interest, dividends, noncorporate profits, retained earnings, and various subsidy payments. Since no appropriate consumer purchase data for the St. Louis metropolitan area exist, household inputs and outputs were estimated in the following manner:

The Interindustry Relations Study for 1947 was taken as a point of departure. The input percentages of households by industrial sectors were calculated. To estimate total household inputs in the St. Louis metropolitan area, it was assumed first that about 70 percent of 1947 United States household demand had originated in urban households. With the help of an estimate of the percent of United States urban families residing in the St. Louis area, 1947 St. Louis household inputs were estimated. After adjusting for population and income increases a first estimate of 1955 St. Louis household inputs was obtained. It was divided tentatively among the various industrial sectors in line with 1947 national relationships. The *Study of Consumer Expenditures, Incomes and Savings* indicates that St. Louis families spent in 1950 about $100 more than the average urban family.[3] This amount was adjusted for income changes to 1955 and distributed among those industrial sectors in which local demand exceeded national demand. In this fashion a refined input estimate was obtained. The geographic allocation of inputs was made on the basis of detailed analyses as to which goods and services are or are not locally produced.

Household outputs were taken from company information which took the form of sector payments to households. Checks were made with the help of various secondary sources of data.

An interindustry transaction table is a double-entry table, where every cell stands for an input as well as an output. Except for a few cases, these two figures are independently derived in the St. Louis model. If there were differences, they were reconciled.

## St. Louis Area Transactions Structure

The St. Louis metropolitan area produced an output totalling $15.6 billion in 1955. Of this output, $11.6 billion was sold to the local economy, while $4.0 billion was exported. Thus, about 74 percent of local output was produced for direct local use. Both exports and imports varied greatly from sector to sector. For instance, sector 7 (products of petroleum and coal) imported almost 72 percent of its inputs, whereas sector 24 (medical, educational, and nonprofit organizations) and sector 29 (local government) each imported less than 2 percent.

Table 23.A1 of the Appendix is a 29 × 29 technical coefficient matrix, computed from the gross flows after adjusting for inventory changes. Each element in a given column is divided by the column total adjusted for inventory changes. Each coefficient represents the percentage of direct input required from the different local sectors that are not part of final demand.

Finally, in Table 23.A2 of the Appendix direct and indirect requirements per dollar of final demand are shown. This table is derived from Table 23.A1. It is the transposed inverse of the difference between an identity matrix and the technical coefficient matrix $(I - A)^{-1}$, often simply referred to as the inverse matrix.

## Activity and Impact Projections

Within the framework of this interindustry flow model, economic activities of industrial sectors in a metropolitan area are linked together in a maze of interdependencies with one another and with their counterparts outside the area. In the short run, these relationships, once quantified, can be assumed to remain unchanged, so that the impact of final demand changes on the local economy can be assessed and compared. In the long run, the structural relations are likely to change, but it might be possible to obtain tentative projections of future technical coefficients by studying potential technical improvements and location shifts.

### Input Projections

The general approach employed by this study, i.e., to seek the active cooperation of experts and decisionmakers in the metropolitan area, also proved its usefulness in the making of input projections. Final demand facing a local industrial sector is mainly determined by national and international demands as well as the local industry's location advantages in relation to actual and potential markets. These two points were considered when final demand—i.e., local sales to gross private capital formation, federal and state governments, and inventories, as well as to all 33 nonlocal sectors—was projected. In all cases much interest centered around the joint effort of industrialists and researchers in estimating the final demand matrix.

In addition to short-term projections, a ten-year projection to 1965 was attempted. Between 1955 and 1965 total final demand facing the St. Louis metropolitan area was estimated to increase in 1955 prices from $5.8 billion to $9.0 billion, i.e., 38 percent. On the assumption that technical coefficients would remain unchanged, this demand increase would go hand in hand with an increase in local purchases—again in 1955 prices—from $11.6 billion in 1955 to $15.1 billion in 1965, i.e., 30 percent. The input changes of the various sectors differed greatly. For instance, the input of sector 8—leather and leather products—was projected to decline by about 19 percent, while that of sector 15—other transportation equipment—would increase by about 119 percent.

### Income Multiplier

To evaluate and compare the income impact of final demand changes, income multipliers were calculated. To the extent that 1955 conditions prevail and final demand changes are met by the industrial sector, the direct income change is measured by the sectors' household coefficients given in Table 23.A.1 of the Appendix. They have been reproduced in column 1 of Table 23.1.

For example, a $1 increase (or decrease) in the final demand for St. Louis food and kindred products increased (or decreased) income directly originating in that sector by $0.14. This figure is quite small compared to that of sector 24 (medical, educational, and nonprofit organizations), i.e., $0.77. Among the manufacturing sectors the direct income change is smallest in sector 7 (products of petroleum and coal)—($0.08), and highest in sector 5 (printing and publishing)—($0.45).

There are at least four main reasons for differences in the direct income effect. While the relative wage level, labor intensity, and labor productivity have some bearing, perhaps the most decisive factor appears to be the relative importance of imported inputs. Generally, the more highly integrated an area and sector, the fewer its imports, the larger the share of local inputs and incomes, and the larger the direct income change.

But clearly this is not the entire income effect. The increase (or decrease) in a local industrial sector's output, produced by a change in its final demand, unleashes forces which affect the output of all those local industrial sectors which directly or indirectly supply this sector. The interindustry flow model makes possible the tracing and evaluation of this chain reaction.

A rather simple and somewhat unrealistic Model I visualizes changes in final demand to lead to production adjustments not only in the specific local sector but also in all those which are directly or indirectly linked to it; however, it assumes that neither consumer expenditures nor investment expenditures for new plant and equipment are affected.

Model II makes allowance also for consumer expenditure adjustments, which come about because of output changes and which lead to a chain reaction of

Table 23.1 Income Interactions in the St Louis Metropolitan Area, 1955

| Industrial sector | Direct income change (1) | Direct and indirect income change (2) | Indirect income change (3) | Multiplier (Model I) (4) | Direct indirect and induced income change (5) | Induced income change (6) | Indirect and induced income change (7) | Multiplier (Model II) (8) |
|---|---|---|---|---|---|---|---|---|
| 1. Food and kindred products | 0.14 | 0.23 | 0.09 | 1.77 | 0.36 | 0.13 | 0.22 | 2.57 |
| 2. Textiles and apparel | 0.32 | 0.41 | 0.09 | 1.28 | 0.64 | 0.23 | 0.32 | 2.00 |
| 3. Lumber and furniture | 0.35 | 0.49 | 0.14 | 1.41 | 0.77 | 0.28 | 0.42 | 2.20 |
| 4. Paper and allied products | 0.26 | 0.39 | 0.13 | 1.50 | 0.62 | 0.23 | 0.36 | 2.38 |
| 5. Printing and publishing | 0.45 | 0.56 | 0.11 | 1.24 | 0.87 | 0.31 | 0.42 | 1.93 |
| 6. Chemicals | 0.25 | 0.32 | 0.07 | 1.28 | 0.51 | 0.19 | 0.26 | 2.04 |
| 7. Products of petroleum and coal | 0.08 | 0.13 | 0.05 | 1.72 | 0.22 | 0.09 | 0.14 | 2.75 |
| 8. Leather and leather products | 0.38 | 0.47 | 0.09 | 1.25 | 0.75 | 0.28 | 0.37 | 1.97 |
| 9. Iron and steel | 0.35 | 0.46 | 0.11 | 1.30 | 0.73 | 0.27 | 0.38 | 2.08 |
| 10. Nonferrous metals | 0.27 | 0.40 | 0.13 | 1.51 | 0.64 | 0.24 | 0.37 | 2.37 |
| 11. Plumbing and heating supplies; fabricated structural metal products | 0.36 | 0.45 | 0.09 | 1.27 | 0.71 | 0.26 | 0.35 | 1.97 |
| 12. Machinery (except electrical) | 0.31 | 0.44 | 0.13 | 1.44 | 0.70 | 0.26 | 0.39 | 2.26 |
| 13. Motors and generators; radios; and other electric machinery | 0.44 | 0.53 | 0.09 | 1.22 | 0.84 | 0.31 | 0.40 | 1.91 |
| 14. Motor vehicles | 0.17 | 0.28 | 0.11 | 1.72 | 0.45 | 0.17 | 0.28 | 2.65 |
| 15. Other transportation equipment | 0.33 | 0.37 | 0.04 | 1.13 | 0.59 | 0.22 | 0.26 | 1.79 |
| 16. Miscellaneous manufacturing | 0.37 | 0.53 | 0.16 | 1.43 | 0.85 | 0.32 | 0.48 | 2.30 |
| 17. Coal; gas; electric power; water | 0.26 | 0.35 | 0.09 | 1.35 | 0.58 | 0.23 | 0.32 | 2.23 |
| 18. Railroad transportation | 0.39 | 0.51 | 0.12 | 1.29 | 0.81 | 0.30 | 0.42 | 2.08 |
| 19. Other transportation | 0.43 | 0.54 | 0.11 | 1.25 | 0.86 | 0.32 | 0.43 | 2.00 |
| 20. Trade | 0.61 | 0.73 | 0.12 | 1.19 | 1.16 | 0.43 | 0.55 | 1.90 |
| 21. Communications | 0.44 | 0.49 | 0.05 | 1.10 | 0.79 | 0.30 | 0.35 | 1.80 |
| 22. Finance and insurance; rentals | 0.34 | 0.50 | 0.16 | 1.48 | 0.84 | 0.34 | 0.50 | 2.47 |
| 23. Business and personal services | 0.57 | 0.74 | 0.17 | 1.29 | 1.16 | 0.42 | 0.59 | 2.03 |
| 24. Medical, educational and nonprofit organ. | 0.77 | 0.86 | 0.09 | 1.11 | 1.34 | 0.48 | 0.57 | 1.74 |
| 25. Undistributed | 0.36 | 0.49 | 0.13 | 1.36 | 0.82 | 0.33 | 0.46 | 2.28 |
| 26. Eating and drinking places | 0.35 | 0.51 | 0.16 | 1.48 | 0.82 | 0.31 | 0.47 | 2.34 |
| 27. Capital construction and maintenance | 0.40 | 0.59 | 0.19 | 1.47 | 0.93 | 0.34 | 0.53 | 2.32 |

interindustry reactions in income, output, and once more on consumer expenditures. There can be different versions of Model II, depending on the assumptions about the income-consumption function. By and large, consumption expenditures tend to vary less than income. Although under such conditions the income-consumption function would tend to be curvilinear, for simplicity's sake this study assumes a linear relationship. For this reason the income multiplier of Model II will tend somewhat to overstate the income effect.

Model I can be implemented by computing the inverse matrix $(I - A)^{-1}$, after moving the household sector into final demand. In this study, also, local government was moved into final demand, and the corresponding $27 \times 27$ inverted matrix was calculated. (This matrix has not been reproduced here.) The direct and indirect income change per dollar change of final demand was calculated, for instance, for St. Louis food products in 1955, by multiplying each figure in the food products row of the $27 \times 27$ inverted matrix by the appropriate household coefficients and then summing the products. The figure in the food column and row of the inverted $27 \times 27$ matrix is 1.043425 and the household coefficient of the food sector is 0.144564 (see Table 23.A.1 of the Appendix). Their product, 0.150842, is added to the product of 0.004746 in the food row and textile column and 0.320126 (the household coefficient of the textile sector), i.e., 0.001519. The sum of the 27 products of column 1 is 0.225609 or 0.23.

These direct and indirect income changes are presented in column 2 of Table 23.1, and in column 3 the direct income changes are given. In column 4 the interindustry income multiplier of Model I is presented, which shows the direct and indirect income change in relation to the direct one.

Model II also is readily implemented. The direct, indirect, and induced income changes per dollar change in final demand, assuming a linear and homogeneous income-consumption function, are given in the household column of the $29 \times 29$ inverted matrix of Table 23.A2 of the Appendix. They are reproduced in column 5 of Table 23.1. In column 6 the induced income changes are represented and in column 7 the indirect and induced ones. Finally, in column 8 the interindustry income multiplier of Model II, reflecting both indirect and induced income effects, is given. For instance, the largest indirect and induced multiplier was 2.75 for sector 7—products of petroleum and coal—i.e., final demand changes of sector 7 will lead to total income changes 2.75 times as large as those in the sector itself.

A perusal of Table 23.1 leads to some interesting insights into the different income changes and multipliers and their relationships. First of all, the interindustry income multipliers for St. Louis in 1955 are perhaps smaller than one might have expected. The highest multiplier is 2.75, and even this figure is probably somewhat on the high side, since it assumes a linear income-consumption relationship.

While a general comparison is difficult with the Utah income multipliers for 1947, both studies have two comparable sectors: iron and steel, and nonferrous metals.[4] Both Utah sectors produced substantially lower direct income changes than did their St. Louis counterparts. In the case of the nonferrous metals sector the difference was especially noticeable—0.09 in Utah versus 0.27 in St. Louis. For both sectors the income multipliers of Models I and II were much larger in Utah than in St. Louis. Again nonferrous metals showed the most extreme dissimilarities, the income multiplier of Model II for Utah being 14.67 and that for St. Louis 2.37. In addition to the four factors discussed earlier the dissimilarities may be due to product-mix differences in the two areas. High direct income effects go hand in hand with relatively low multipliers not only when Utah and St. Louis are compared; this inverse relationship holds also for many of the 27 St. Louis sectors.

The 1955 St. Louis interindustry multipliers of Model I of the 16 manufacturing sectors varied between 1.13 and 1.77 and those of Model II between 1.79 and 2.75. In both cases the highest multiplier is only about one and one half times as large as the lowest one.

St. Louis interindustry multipliers of Model I are highly correlated with those of Model II. The following regression equation well describes the 1955 relationship:

$$Y = 0.35 + 1.29X$$

where $Y$ is the 1955 St. Louis interindustry income multiplier of Model II, and $X$ is the 1955 St. Louis interindustry income multiplier of Model I.

The correlation coefficient is 0.97, which is highly significant at an $a$ of 0.05, as is also the regression coefficient. In St. Louis a 1 point increase in the interindustry income multiplier of Model I was on the average associated in 1955 with a 1.29 point increase in the income multiplier of Model II.

Knowledge of the multiplier is sometimes less important than that of the absolute income changes that can be expected to accompany an exogenous force of a stated magnitude. For instance, in planning an industrial development program for a metropolitan area it is often useful to know which industrial sector, if increased, will produce the greatest possible overall income increase per $1 million final demand change. In 1955 sector 5—printing and publishing—had the greatest overall income impact on the St. Louis economy. A $1 million final demand increase raised the area's income by about $870,000. At the other end of the spectrum was sector 7—products of coal and petroleum—with an anticipated increase of merely $220,000.

In 1955 the relative importance of the direct, indirect, and induced income changes varied greatly from one St. Louis sector to the next (see Table 23.2). The relative importance of direct income changes is smallest for sectors 7 (products of petroleum and coal), 14 (motor vehicles), and 1 (food and kindred products); and largest for sectors 24 ( medical, educational, and nonprofit organizations), 15 (other transportation equipment), and 21 (communications). When the indirect income change is considered, these sectors reverse their order. Sectors 21, 15, and 24 have the smallest relative indirect income effects; and sectors 1, 14, and 7 the largest. While the other transportation equipment, communications, and medical, educational, and nonprofit organization sectors jointly occupy similar positions, they apparently do so for different reasons. The relatively unimportant indirect local income changes of the other transportation equipment sector can be traced to its heavy reliance on imported inputs which generate income changes outside but not inside the area. On the other hand, medical, educational, and nonprofit organizations use few inputs other than those of local households. Thus, very little

Table 23.2   Direct, Indirect, and Induced Income Change as Percent of Total Income Change, St Louis Metropolitan Area, 1955 (*in Percents*)

| Industrial sector | Direct income change | Indirect income change | Induced income change |
|---|---|---|---|
| 1.  Food and kindred products | 0.39 | 0.25 | 0.36 |
| 2.  Textiles and apparel | 0.50 | 0.14 | 0.36 |
| 3.  Lumber and furniture | 0.46 | 0.18 | 0.36 |
| 4.  Paper and allied products | 0.42 | 0.21 | 0.37 |
| 5.  Printing and publishing | 0.52 | 0.13 | 0.35 |
| 6.  Chemicals | 0.49 | 0.14 | 0.37 |
| 7.  Products of petroleum and coal | 0.36 | 0.23 | 0.41 |
| 8.  Leather and leather products | 0.51 | 0.12 | 0.47 |
| 9.  Iron and steel | 0.48 | 0.15 | 0.37 |
| 10. Nonferrous metals | 0.42 | 0.20 | 0.38 |
| 11. Plumbing and heating supplies; fabricated structural metal products | 0.51 | 0.13 | 0.36 |
| 12. Machinery (except electrical) | 0.44 | 0.18 | 0.38 |
| 13. Motors and generators; radios; other electrical machinery | 0.52 | 0.11 | 0.37 |
| 14. Motor vehicles | 0.38 | 0.24 | 0.38 |
| 15. Other transportation equipment | 0.56 | 0.07 | 0.37 |
| 16. Miscellaneous manufacturing | 0.44 | 0.19 | 0.37 |
| 17. Coal; gas; electric power; water | 0.45 | 0.15 | 0.40 |
| 18. Railroad transportation | 0.48 | 0.15 | 0.37 |
| 19. Other transportation | 0.50 | 0.13 | 0.37 |
| 20. Trade | 0.52 | 0.10 | 0.38 |
| 21. Communications | 0.56 | 0.06 | 0.38 |
| 22. Finance and insurance; rentals | 0.40 | 0.20 | 0.40 |
| 23. Business and personal services | 0.49 | 0.15 | 0.36 |
| 24. Medical, educational and nonprofit organ. | 0.57 | 0.07 | 0.36 |
| 25. Undistributed | 0.44 | 0.16 | 0.40 |
| 26. Eating and drinking places | 0.43 | 0.20 | 0.37 |
| 27. Capital construction and maintenance | 0.43 | 0.20 | 0.37 |

indirect income can be generated by sector 24. Apparently in connection with sector 21 both factors play a role.

The relative importance of induced income changes varies relatively little. Probably part of the responsibility rests with the linear income-consumption function. The relative importance of induced income changes is largest in sector 8 (leather and leather products) and smallest in sector 5 (printing and publishing).

### Employment Multiplier

What about the employment impact of final demand changes? The direct employment effect of a change in final demand of a sector is the slope of its employment-output function. Whenever possible this function was approximated by aggregating annual company employment and output data, respectively, to

Table 23.3    Employment Interaction in the St Louis Metropolitan Area, 1955

| | Employment change (in man-years) per one million dollar final demand change | | Interindustry employment increase | |
|---|---|---|---|---|
| Industrial sector | Direct (1) | Direct and indirect (2) | Indirect change (3) | Multiplier (4) |
| 1. Food and kindred products | 29 | 36 | 7 | 1.24 |
| 2. Textiles and apparel | 105 | 130 | 25 | 1.24 |
| 3. Lumber and furniture | 97 | 134 | 37 | 1.38 |
| 4. Paper and allied products | 66 | 92 | 26 | 1.39 |
| 5. Printing and publishing | 90 | 115 | 25 | 1.27 |
| 6. Chemicals | 44 | 53 | 9 | 1.20 |
| 7. Products of petroleum and coal | 12 | 14 | 2 | 1.18 |
| 8. Leather and leather products | 91 | 113 | 22 | 1.25 |
| 9. Iron and steel | 63 | 82 | 19 | 1.30 |
| 10. Nonferrous metals | 46 | 63 | 17 | 1.37 |
| 11. Plumbing and heating supplies; fabricated structural metal products | 109 | 138 | 29 | 1.27 |
| 12. Machinery (except electrical) | 88 | 119 | 31 | 1.35 |
| 13. Motors and generators; radios; other electrical machinery | 103 | 133 | 30 | 1.29 |
| 14. Motor vehicles | 20 | 29 | 9 | 1.45 |
| 15. Other transportation equipment | 64 | 72 | 7 | 1.12 |
| 16. Miscellaneous manufacturing | 68 | 98 | 30 | 1.45 |
| 17. Coal; gas; electric power; water | 67 | 88 | 21 | 1.31 |
| 18. Railroad transportation | 67 | 88 | 21 | 1.32 |
| 19. Other transportation | 113 | 148 | 35 | 1.31 |
| 20. Trade | 199 | 258 | 59 | 1.29 |
| 21. Communications | 100 | 112 | 12 | 1.12 |
| 22. Finance and insurance; rentals | 51 | 75 | 24 | 1.47 |
| 23. Business and personal services | 104 | 147 | 43 | 1.41 |
| 24. Medical, educational and nonprofit organizations | 223 | 283 | 60 | 1.27 |
| 25. Undistributed | 52 | 73 | 21 | 1.40 |
| 26. Eating and drinking places | 104 | 155 | 51 | 1.49 |
| 27. Capital construction and maintenance | 55 | 81 | 26 | 1.47 |

represent the local industrial sector for a number of years. The output data were deflated and brought to a common base, 1955. A regression line was fitted to these time series data. In the absence of such data, sector employment and output data were used to compute an employment-output ratio.

The direct employment changes (in man-years) per one million dollar final demand change are presented in column 1 of Table 23.3. The employment impact is lowest for final demand changes of sector 7 (products of petroleum and coal), i.e., merely 12 man-years per million dollar final demand change. For sector 24 (medical, educational, and nonprofit organizations) the impact is greatest: 223 man-years.

To estimate the direct and indirect employment change per one million dollar final demand change, the household and local government sectors once more have been moved into the final demand sector. Each cell value in the inverted 27 × 27 matrix indicates the direct and indirect production requirement changes that result from a one dollar change in the final demand of a given local industrial sector. As in Model 1 of the income multiplier, no allowance is made for consumption and investment changes. Production changes can be converted into employment changes with the help of each sector's employment-output function. By adding each column of the data so derived, the direct and indirect employment change of the sector indicated at the top of the column has been obtained. These figures are given in column 2 of Table 23.3. The indirect employment change is presented in column 3 of Table 23.3.

By and large, sectors with high direct employment changes have also high indirect employment changes, and vice versa. However, there are some interesting exceptions. For example, the St. Louis communications sector had the ninth largest direct but only the 22nd largest indirect employment effect. Many of the nonhousehold inputs of this sector are either labor extensive or are imported, or both. Sectors 15 (other transportation equipment), 2 (textiles and apparel), 8 (leather and leather products), and 11 (plumbing and heating supplies and fabricated structural metal) fall into this same category. The opposite appears to hold for sectors 3 (lumber and furniture), 4 (paper), 12 (machinery—except electrical), 16 (miscellaneous manufacturing), 22 (finance, insurance, and rentals), and 27 (construction).

The interindustry employment multiplier relating direct plus indirect employment changes to direct ones indicates how much total St. Louis employment is affected by changes in man-year employment in a given local sector. It varies from a low of 1.12 in sectors 15 (other transportation equipment) and 21 (communications) to a high of 1.49 in sector 26 (eating and drinking places). There appears to exist no significant relation between the direct employment change and the employment multiplier.

### Exports and area stability

In addition to its employment and income multiplier, the economic stability of a local industrial sector depends upon the nature of its local and export demand, particularly on the latter. In this connection it is of interest to know the relative importance of exports and how diffused these exports are.

Exports accounted for only about 26 percent of the St. Louis area's 1955 output. Generally, a smaller and less diversified area would tend to have a higher export figure. However, this relatively small overall export percentage can be misleading. For instance, virtually the entire output of sectors 15 (other transportation equipment)—97 percent—and 8 (leather and leather products)—90 percent—is exported. And yet some sectors, such as 26 (eating and drinking places), 28 (households), and 29 (local government) export none of their output. In Table 23.4 the relative importance of exports of the sectors is given.

The diffusion of sector exports, i.e., to how many different outside sectors exports

Table 23.4   Relative Importance of Exports from the St Louis Metropolitan Area, 1955

| Industrial sector | Total sales ($000) (1) | Local sales ($000) (2) | Exports ($000) (3) | Exports as % of total sales (4) |
|---|---|---|---|---|
| 1. Food and kindred products | $1,126,209 | $396,075 | $730,134 | 64.83% |
| 2. Textiles and apparel | 181,522 | 74,428 | 107,094 | 59.00 |
| 3. Lumber and furniture | 79,218 | 56,035 | 23,183 | 29.26 |
| 4. Paper and allied products | 130,668 | 90,082 | 40,586 | 31.06 |
| 5. Printing and publishing | 140,823 | 96,380 | 44,443 | 31.56 |
| 6. Chemicals | 494,752 | 74,389 | 420,363 | 84.96 |
| 7. Products of petroleum and coal | 657,631 | 176,240 | 481,391 | 73.20 |
| 8. Leather and leather products | 148,687 | 14,327 | 134,360 | 90.36 |
| 9. Iron and steel | 275,226 | 102,243 | 172,983 | 62.85 |
| 10. Nonferrous metals | 151,163 | 28,709 | 122,454 | 81.01 |
| 11. Plumbing and heating supplies; fabricated structural metal products | 223,490 | 142,298 | 81,192 | 36.33 |
| 12. Machinery (except electrical) | 233,802 | 120,129 | 113,673 | 48.62 |
| 13. Motors and generators; radios; other electrical machinery | 202,970 | 35,908 | 167,062 | 82.31 |
| 14. Motor vehicles | 585,402 | 338,951 | 246,451 | 42.10 |
| 15. Other transportation equipment | 293,121 | 9,625 | 283,496 | 96.72 |
| 16. Miscellaneous manufacturing | 440,937 | 266,585 | 174,352 | 39.54 |
| 17. Coal; gas; electric power; water | 198,476 | 169,721 | 28,755 | 14.49 |
| 18. Railroad transportation | 370,904 | 159,198 | 211,706 | 57.08 |
| 19. Other transportation | 222,192 | 165,781 | 56,411 | 25.39 |
| 20. Trade | 788,594 | 712,908 | 75,686 | 9.60 |
| 21. Communications | 89,867 | 75,559 | 14,308 | 15.92 |
| 22. Finance and insurance; rentals | 668,002 | 655,040 | 12,962 | 1.94 |
| 23. Business and personal services | 383,365 | 372,914 | 10,451 | 2.73 |
| 24. Medical, educational and nonprofit organizations | 152,575 | 147,570 | 5,005 | 3.28 |
| 25. Undistributed | 606,638 | 369,679 | 236,959 | 39.06 |
| 26. Eating and drinking places | 176,552 | 176,552 | — | — |
| 27. Capital construction and maintenance | 723,890 | 723,890 | — | — |
| 28. Households | 3,868,595 | 3,868,595 | — | — |
| 29. Local government | 208,413 | 208,413 | — | — |

are shipped, varies greatly. For instance, while about 97 percent of the St. Louis output of other transportation equipment was exported, these exports were destined to but two industrial sectors: 64 percent of the output went to the federal government and 33 percent to other transportation. Quite a different picture existed in relation to sector 4 (paper and allied products), which exported to virtually each and every industrial sector. But no single sector accounted for more than 4 percent of total output.

The extremely heavy dependence of the other transport equipment sector— accounting in 1955 for about 11 percent of St. Louis manufacturing employment— upon two export markets, both of which often encounter substantial demand changes, makes it a very unstable sector. The secondary and tertiary effects, however, are kept to a minimum by very low interindustry income and employment multiplier (see Tables 23.1 and 23.3). Consequently, the overall employment and income changes are somewhat less than they would be otherwise. Should more suppliers of this sector locate in the St. Louis metropolitan area the multipliers will tend to go up. Economic instability will increase unless the new local firms supply many additional sectors.

## Conclusions

This interindustry relations study of the St. Louis metropolitan area is an attempt to develop more adequate methods for a better understanding of the economic fabric of a metropolitan area. It endeavors to shape tools that can help assess and anticipate income, output, and employment impacts. Much work remains to be done, both to improve input and output estimates, and to perfect the theoretical framework within which they are employed. More information on income-consumption functions is needed to enhance the quality of the income and employment multipliers. In the meantime, considering impact estimates merely as rough approximations, both the private and public sectors of the area's economy can be, and in part already have been, guided in their decisions by such estimates.

### NOTES

1. Frederick T. Moore and James W. Petersen, "Regional Analysis: An Interindustry Model of Utah," *Review of Economics and Statistics*, Volume 37 (November 1955).
2. Moore and Petersen place households in the processing segment when they estimate induced income effects.
3. Wharton School of Finance and Commerce (Philadelphia, 1957).
4. Op. cit., 375.

## Appendix

Table 23.A.1.   Technical Coefficients: Direct Purchases per Dollar of Output, St. Louis Metropolitan Area, 1955[a]

| Industrial sector | Food and kindred products 1 | Textiles and apparel 2 | Lumber and furniture 3 | Paper and allied products 4 | Printing and publishing 5 |
|---|---|---|---|---|---|
| 1. Food and kindred products | 0.041125 | 0.000821 | — | 0.004485 | — |
| 2. Textiles and apparel | 0.003869 | 0.083026 | 0.034970 | 0.002801 | — |
| 3. Lumber and furniture | 0.000192 | 0.000556 | 0.023997 | 0.002478 | 0.000014 |
| 4. Paper and allied products | 0.017158 | 0.007977 | 0.004181 | 0.114795 | 0.005048 |
| 5. Printing and publishing | 0.002042 | 0.000242 | 0.000380 | 0.003161 | 0.089594 |
| 6. Chemicals | 0.001478 | 0.003234 | 0.018144 | 0.00447 | 0.003779 |
| 7. Products of petroleum and coal | 0.002076 | 0.000914 | 0.004029 | 0.008747 | 0.000471 |
| 8. Leather and leather products | 0.000003 | 0.000474 | 0.002319 | 0.000077 | — |
| 9. Iron and steel | 0.000206 | — | 0.032626 | — | — |
| 10. Nonferrous metals | 0.000053 | 0.000017 | — | 0.000329 | 0.002182 |
| 11. Plumbing and heating supplies, fabricated structural metal products | 0.032212 | 0.000259 | 0.017168 | 0.002112 | 0.000157 |
| 12. Machinery (except electrical) | 0.00292 | 0.003983 | 0.004029 | 0.002717 | 0.006317 |
| 13. Motors and generators, radios, other electrical machinery | 0.000423 | — | — | — | — |
| 14. Motor vehicles | 0.000144 | — | — | 0.000130 | — |
| 15. Other transportation equipment | — | — | — | — | — |
| 16. Miscellaneous | 0.019220 | 0.018830 | 0.008971 | 0.006161 | — |
| 17. Coal, gas, electric power, and water | 0.004873 | 0.003686 | 0.004701 | 0.011189 | 0.004513 |
| 18. Railroad transportation | 0.013618 | — | 0.020133 | 0.014502 | 0.005297 |
| 19. Other transportation | 0.010621 | 0.008880 | 0.016142 | 0.015176 | 0.005262 |
| 20. Trade | 0.010374 | 0.027556 | 0.018815 | 0.022553 | 0.006624 |
| 21. Communications | 0.001779 | 0.001504 | 0.001989 | 0.001179 | 0.006075 |
| 22. Finance, insurance, rentals | 0.002937 | 0.009459 | 0.013038 | 0.005357 | 0.009711 |
| 23. Business and personal services, etc. | 0.012299 | 0.007691 | 0.019385 | 0.006321 | 0.012149 |
| 24. Medical, educational, nonprofit | 0.000005 | 0.000006 | — | 0.001209 | 0.000014 |
| 25. Undistributed | 0.001673 | 0.003652 | 0.042458 | 0.050494 | 0.046259 |
| 26. Eating and drinking places | — | — | — | — | — |
| 27. Capitalized construction and maintenance | 0.002999 | 0.001278 | 0.002192 | 0.003911 | 0.002339 |
| 28. Households | 0.144564 | 0.320126 | 0.349205 | 0.262520 | 0.447300 |
| 29. Local government | 0.003245 | 0.001113 | 0.001571 | 0.001944 | 0.001662 |

Table 23.A.1   Continued

| Industrial sector | Misc. 16 | Coal, gas, etc. 17 | Railroad trans. 18 | Other trans. 19 |
|---|---|---|---|---|
| 1. Food and kindred products | 0.007612 | 0.000141 | 0.002435 | 0.004069 |
| 2. Textiles and apparel | 0.006677 | 0.000272 | 0.000084 | 0.000135 |
| 3. Lumber and furniture | 0.002303 | 0.000922 | 0.000423 | 0.000356 |
| 4. Paper and allied products | 0.021959 | 0.000378 | 0.000262 | 0.000459 |
| 5. Printing and publishing | 0.000102 | 0.000957 | 0.002842 | 0.002291 |
| 6. Chemicals | 0.001027 | 0.001512 | 0.009064 | 0.001818 |
| 7. Products of petroleum and coal | 0.005105 | 0.077188 | 0.033003 | 0.046181 |
| 8. Leather and leather products | 0.000196 | 0.000186 | 0.000027 | 0.000009 |
| 9. Iron and steel | 0.039255 | 0.001930 | 0.002041 | 0.000005 |
| 10. Nonferrous metals | 0.005770 | 0.000574 | 0.000175 | 0.000045 |
| 11. Plumbing and heating supplies, fabricated structural metal products | 0.006525 | 0.002342 | 0.000693 | 0.001683 |
| 12. Machinery (except electrical) | 0.001932 | 0.002675 | 0.003289 | 0.000284 |
| 13. Motors and generators, radios, other electrical machinery | — | 0.016758 | — | — |
| 14. Motor vehicles | 0.000344 | 0.002504 | 0.000113 | 0.022030 |
| 15. Other transportation equipment | 0.000002 | 0.000171 | 0.001316 | 0.007737 |
| 16. Miscellaneous | 0.053577 | 0.002806 | 0.001394 | 0.002345 |
| 17. Coal, gas, electric power and water | 0.014949 | 0.032649 | 0.002993 | 0.005500 |
| 18. Railroad transportation | 0.014370 | 0.017488 | 0.046570 | 0.002327 |
| 19. Other transportation | 0.010151 | 0.004172 | 0.010795 | 0.012309 |
| 20. Trade | 0.013223 | 0.003497 | 0.002373 | 0.029686 |
| 21. Communications | 0.001777 | 0.001270 | 0.002952 | 0.005711 |
| 22. Finance, insurance, rentals | 0.042131 | 0.007245 | 0.004117 | 0.035915 |
| 23. Business and personal services, etc. | 0.014686 | 0.005970 | 0.019784 | 0.031621 |
| 24. Medical, educational, nonprofit organizations | 0.000014 | 0.000015 | — | — |
| 25. Undistributed | 0.060176 | 0.001013 | — | 0.004186 |
| 26. Eating and drinking places | — | — | — | 0.000689 |
| 27. Capitalized construction and maintenance | 0.004708 | 0.049890 | 0.093679 | 0.013353 |
| 28. Households | 0.373596 | 0.255492 | 0.392808 | 0.432621 |
| 29. Local government | 0.002529 | 0.028824 | 0.008867 | 0.011962 |

[a] Each entry shows direct purchases from St Louis area industry named at left by industry named at top per dollar of output

| Chemicals 6 | Products of petroleum and coal 7 | Leather and leather products 8 | Iron and steel 9 | Non-ferrous metals 10 | Plumbing heating, etc. 11 | Machinery 12 | Motors, generators, etc. 13 | Motor vehicles 14 | Other trans. equip. 15 |
|---|---|---|---|---|---|---|---|---|---|
| 0.002622 | 0.001718 | 0.001920 | 0.000007 | — | 0.000013 | — | — | — | — |
| 0.000513 | 0.000009 | — | — | 0.001078 | 0.000197 | 0.000120 | 0.000133 | 0.012985 | 0.000535 |
| 0.000318 | 0.000030 | 0.020363 | 0.001046 | 0.000714 | 0.000971 | 0.010522 | 0.001586 | 0.001207 | 0.001794 |
| 0.007472 | 0.002800 | 0.012327 | 0.000258 | 0.004472 | 0.006579 | 0.003127 | 0.006494 | 0.002314 | 0.000529 |
| 0.001985 | — | 0.005652 | 0.000283 | 0.000628 | 0.001674 | 0.003456 | 0.007119 | — | 0.000331 |
| 0.057643 | 0.009617 | 0.004527 | 0.002958 | 0.017418 | 0.001374 | 0.000509 | 0.011475 | — | 0.001842 |
| 0.000124 | 0.024347 | 0.000602 | 0.008179 | 0.006840 | 0.001522 | 0.000466 | 0.001163 | 0.002174 | 0.003074 |
| 0.000036 | 0.000052 | 0.056340 | 0.000036 | 0.000198 | 0.000157 | 0.001347 | 0.000301 | 0.000195 | 0.000310 |
| 0.002741 | 0.000271 | — | 0.018054 | 0.003955 | 0.055223 | 0.029833 | 0.014830 | 0.021932 | 0.024952 |
| 0.006756 | 0.000329 | 0.000020 | 0.001217 | 0.016267 | 0.028473 | 0.001091 | 0.053126 | — | 0.001006 |
| 0.007987 | 0.007044 | 0.008103 | 0.008404 | 0.007145 | 0.031615 | 0.036830 | 0.034143 | 0.026793 | 0.003855 |
| 0.000051 | 0.000193 | — | 0.002162 | 0.003698 | 0.016103 | 0.055885 | 0.017574 | 0.019494 | 0.001600 |
| | — | — | 0.000956 | 0.004326 | 0.004847 | 0.024191 | 0.012958 | 0.000678 | 0.004322 |
| 0.000041 | — | — | — | — | 0.002323 | 0.000719 | 0.000251 | 0.154335 | 0.000089 |
| | — | — | 0.000062 | — | — | — | — | — | 0.020053 |
| 0.001373 | 0.001949 | 0.000694 | 0.020481 | 0.007151 | 0.005167 | 0.013926 | 0.003927 | 0.012314 | 0.001846 |
| 0.007622 | 0.008347 | 0.006622 | 0.011500 | 0.011842 | 0.008275 | 0.006099 | 0.011839 | 0.004368 | 0.004312 |
| 0.018784 | 0.024381 | 0.005355 | 0.033002 | 0.017339 | 0.008942 | 0.012096 | 0.006632 | 0.015992 | 0.003098 |
| 0.016161 | 0.011032 | 0.003557 | 0.005341 | 0.006483 | 0.002117 | 0.005851 | 0.006774 | 0.004315 | 0.002538 |
| 0.005915 | 0.000892 | 0.013263 | 0.011420 | 0.021910 | 0.011327 | 0.025586 | 0.002050 | 0.003857 | 0.003845 |
| 0.003151 | 0.001354 | 0.002681 | 0.002961 | 0.000968 | 0.002949 | 0.009045 | 0.002621 | 0.001123 | 0.002016 |
| 0.003814 | 0.002998 | 0.010879 | 0.008059 | 0.003580 | 0.005071 | 0.006523 | 0.004459 | 0.001106 | 0.004520 |
| 0.003680 | 0.001904 | 0.011216 | 0.001922 | 0.002620 | 0.006946 | 0.008695 | 0.008765 | 0.005349 | 0.000931 |
| 0.000043 | — | — | 0.000018 | 0.000046 | 0.000004 | — | 0.000025 | — | — |
| 0.000345 | 0.034185 | 0.015574 | 0.076010 | 0.064943 | — | 0.005470 | 0.000818 | 0.017691 | — |
| 0.010931 | 0.002348 | 0.006345 | 0.005479 | 0.007720 | 0.002206 | 0.010376 | 0.002853 | 0.003087 | 0.007802 |
| 0.253126 | 0.077457 | 0.379563 | 0.353244 | 0.267169 | 0.356461 | 0.308201 | 0.438178 | 0.165318 | 0.331181 |
| 0.003595 | 0.003760 | 0.003665 | 0.003092 | 0.001250 | 0.001540 | 0.003024 | 0.001084 | 0.001219 | 0.003132 |

| Trade 20 | Communications 21 | Finance, insurance, etc. 22 | Business and personal services 23 | Medical, educational, etc. 24 | Undistributed 25 | Eating and drinking 26 | Capitalized construction 27 | Households 28 | Local government 29 |
|---|---|---|---|---|---|---|---|---|---|
| 0.003131 | 0.001391 | — | 0.003193 | 0.056287 | 0.001182 | 0.150788 | 0.000054 | 0.074979 | 0.014879 |
| 0.000268 | — | 0.000253 | 0.000702 | 0.001193 | 0.002637 | 0.001393 | 0.0006637 | 0.008851 | 0.000619 |
| 0.000919 | 0.003282 | 0.004382 | 0.000047 | 0.003048 | — | 0.000085 | 0.015761 | 0.004021 | 0.002804 |
| 0.016755 | 0.000423 | 0.002897 | 0.003665 | 0.005094 | 0.001512 | 0.004599 | 0.004014 | 0.001610 | 0.002865 |
| 0.004669 | 0.006565 | 0.006966 | 0.061607 | 0.017871 | 0.022907 | 0.002226 | — | 0.005236 | 0.006516 |
| 0.001212 | 0.000356 | 0.000599 | 0.001630 | 0.024099 | 0.000279 | 0.001677 | — | 0.000896 | 0.000364 |
| 0.005036 | 0.001168 | 0.019166 | 0.006125 | 0.003475 | 0.040886 | 0.001246 | 0.021495 | 0.011036 | 0.007471 |
| 0.000010 | 0.000189 | — | 0.000175 | — | — | — | 0.000011 | 0.000626 | — |
| 0.001849 | — | — | 0.000759 | — | 0.008948 | — | 0.021219 | 0.000053 | 0.000749 |
| 0.000024 | — | — | — | 0.000577 | — | 0.000164 | 0.001852 | 0.000016 | 0.000005 |
| 0.001214 | 0.000178 | — | 0.003253 | 0.003291 | — | 0.002135 | 0.039693 | 0.002100 | 0.001651 |
| 0.000257 | 0.007723 | 0.000606 | 0.010674 | 0.001993 | 0.062062 | 0.000181 | 0.012694 | 0.001444 | 0.001070 |
| | — | — | 0.000063 | 0.002032 | — | — | — | 0.001660 | 0.000725 |
| 0.001041 | 0.006065 | 0.000067 | 0.021992 | — | 0.058775 | 0.000119 | 0.001254 | 0.014133 | 0.003148 |
| 0.000236 | — | — | — | — | — | — | — | 0.000332 | 0.000038 |
| 0.001874 | 0.001647 | 0.004286 | 0.015562 | 0.013892 | 0.036275 | 0.018952 | 0.067691 | 0.016858 | 0.009486 |
| 0.014522 | 0.004329 | 0.116069 | 0.017678 | 0.020991 | — | 0.032936 | 0.001073 | 0.000593 | 0.010585 |
| 0.002487 | 0.000334 | 0.000325 | 0.001748 | 0.001055 | — | 0.009555 | 0.027361 | 0.002807 | 0.003296 |
| 0.007043 | 0.001124 | 0.003584 | 0.002309 | 0.004799 | — | 0.007318 | 0.019927 | 0.018890 | 0.011501 |
| 0.005874 | 0.002092 | 0.20684 | 0.016535 | 0.010011 | 0.036643 | 0.080022 | 0.069506 | 0.125547 | 0.004597 |
| 0.004640 | 0.020797 | 0.004451 | 0.011352 | 0.003016 | — | 0.000895 | 0.001533 | 0.011356 | 0.003733 |
| 0.059377 | 0.008679 | 0.043582 | 0.053226 | 0.001659 | — | 0.034539 | 0.009482 | 0.121615 | 0.029461 |
| 0.071918 | 0.003260 | 0.011837 | 0.050174 | 0.006031 | — | 0.022645 | 0.033201 | 0.050570 | 0.008776 |
| | | 0.003630 | 0.010322 | 0.002471 | — | — | 0.00010 | 0.029391 | 0.006108 |
| | | | 0.005679 | | 0.017096 | — | — | 0.028119 | — |
| 0.003650 | — | 0.000906 | — | 0.006097 | — | — | — | 0.044457 | — |
| 0.003918 | 0.023268 | 0.096949 | 0.004894 | 0.018808 | 0.010115 | 0.005511 | 0.000246 | 0.000687 | 0.143206 |
| 0.611204 | 0.443967 | 0.337039 | 0.571774 | 0.768748 | 0.358362 | 0.348583 | 0.400362 | 0.009433 | 0.706894 |
| 0.009998 | 0.016513 | 0.041765 | 0.007729 | 0.000426 | 0.040103 | 0.008842 | 0.001807 | 0.029356 | — |

by latter.

Table 23.A.2   Direct and Indirect Requirements per Dollar of Final Demand, St Louis Metropolitan Area, 1955[a]

| Industrial sector | Food and kindred products 1 | Textiles and apparel 2 | Lumber and furniture 3 | Paper and allied products 4 | Printing and publishing 5 |
|---|---|---|---|---|---|
| 1. Food and kindred products | 1.075363 | 0.008768 | 0.003046 | 0.024828 | 0.009378 |
| 2. Textiles and apparel | 0.058831 | 1.098091 | 0.004812 | 0.017065 | 0.011353 |
| 3. Lumber and furniture | 0.069734 | 0.048205 | 1.029818 | 0.013834 | 0.015639 |
| 4. Paper and allied products | 0.061255 | 0.010974 | 0.007467 | 1.136783 | 0.016044 |
| 5. Printing and publishing | 0.078225 | 0.010128 | 0.005609 | 0.014807 | 1.114245 |
| 6. Chemicals | 0.048709 | 0.006485 | 0.003811 | 0.014187 | 0.010815 |
| 7. Products of petroleum and coal | 0.021509 | 0.002688 | 0.001599 | 0.005792 | 0.004748 |
| 8. Leather and leather products | 0.068896 | 0.009435 | 0.026944 | 0.022447 | 0.019683 |
| 9. Iron and steel | 0.065537 | 0.008807 | 0.005970 | 0.008404 | 0.014130 |
| 10. Nonferrous metals | 0.057244 | 0.008936 | 0.006028 | 0.012856 | 0.013161 |
| 11. Plumbing and heating supplies, fabricated structural metal products | 0.063556 | 0.008491 | 0.005835 | 0.015079 | 0.013849 |
| 12. Machinery (except electrical) | 0.062743 | 0.008731 | 0.016229 | 0.012020 | 0.016497 |
| 13. Motors and generators, radios, other electrical machinery | 0.074836 | 0.009863 | 0.007250 | 0.016256 | 0.021875 |
| 14. Motor vehicles | 0.040351 | 0.022156 | 0.004771 | 0.008513 | 0.008417 |
| 15. Other transportation equipment | 0.052623 | 0.007361 | 0.005694 | 0.006348 | 0.009612 |
| 16. Miscellaneous | 0.084371 | 0.017853 | 0.008380 | 0.035396 | 0.016869 |
| 17. Coal, gas, electric power and water | 0.052780 | 0.007153 | 0.005721 | 0.007104 | 0.011319 |
| 18. Railroad transportation | 0.075616 | 0.009553 | 0.007168 | 0.009129 | 0.017717 |
| 19. Other transportation | 0.081748 | 0.010437 | 0.006267 | 0.009982 | 0.018955 |
| 20. Trade | 0.107511 | 0.013630 | 0.008583 | 0.030608 | 0.028714 |
| 21. Communications | 0.072002 | 0.009202 | 0.008817 | 0.008199 | 0.019839 |
| 22. Finance, insurance, rentals | 0.076855 | 0.010405 | 0.012093 | 0.013233 | 0.023153 |
| 23. Business and personal services | 0.108223 | 0.014599 | 0.007953 | 0.016932 | 0.090422 |
| 24. Medical, educational, nonprofit organizations | 0.179149 | 0.016937 | 0.011669 | 0.020340 | 0.040880 |
| 25. Undistributed | 0.075316 | 0.013616 | 0.006380 | 0.012180 | 0.039427 |
| 26. Eating and drinking places | 0.231365 | 0.011800 | 0.005844 | 0.018653 | 0.018640 |
| 27. Capitalized construction and maintenance | 0.083900 | 0.012499 | 0.022492 | 0.017172 | 0.018112 |
| 28. Households | 0.138595 | 0.017469 | 0.009235 | 0.013920 | 0.023165 |
| 29. Local government | 0.135137 | 0.016350 | 0.013698 | 0.017601 | 0.029093 |

Table 23.A.2   *Continued*

| Industrial sector | Misc. 16 | Coal, gas, etc. 17 | Railroad trans. 18 | Other Trans. 19 |
|---|---|---|---|---|
| 1. Food and kindred products | 0.032954 | 0.017356 | 0.019983 | 0.021542 |
| 2. Textiles and apparel | 0.040917 | 0.024393 | 0.007021 | 0.026852 |
| 3. Lumber and furniture | 0.036656 | 0.030328 | 0.031902 | 0.038266 |
| 4. Paper and allied products | 0.02874 | 0.032289 | 0.024710 | 0.034302 |
| 5. Printing and publishing | 0.027716 | 0.030733 | 0.015020 | 0.028396 |
| 6. Chemicals | 0.017762 | 0.023597 | 0.026991 | 0.031201 |
| 7. Products of petroleum and coal | 0.010655 | 0.015913 | 0.029361 | 0.017880 |
| 8. Leather and leather products | 0.024146 | 0.030007 | 0.014494 | 0.023840 |
| 9. Iron and steel | 0.047290 | 0.034254 | 0.043936 | 0.025307 |
| 10. Nonferrous metals | 0.034354 | 0.032177 | 0.028513 | 0.025569 |
| 11. Plumbing and heating supplies, fabricated structural metal products | 0.028700 | 0.030729 | 0.020026 | 0.021441 |
| 12. Machinery (except electrical) | 0.038621 | 0.029216 | 0.023303 | 0.025643 |
| 13. Motors and generators, radios, other electrical machinery | 0.030429 | 0.037581 | 0.018204 | 0.029470 |
| 14. Motor vehicles | 0.013492 | 0.019544 | 0.026666 | 0.017842 |
| 15. Other transportation equipment | 0.020154 | 0.021610 | 0.010370 | 0.018025 |
| 16. Miscellaneous | 1.086829 | 0.047169 | 0.027667 | 0.034275 |
| 17. Coal, gas, electric power and water | 0.025046 | 1.052703 | 0.029114 | 0.022171 |
| 18. Railroad transportation | 0.032820 | 0.027623 | 1.061179 | 0.035000 |
| 19. Other transportation | 0.030268 | 0.036053 | 0.013522 | 1.036043 |
| 20. Trade | 0.037823 | 0.056295 | 0.015175 | 0.037644 |
| 21. Communications | 0.026537 | 0.027367 | 0.009031 | 0.021926 |
| 22. Finance, insurance, rentals | 0.039728 | 0.151325 | 0.015717 | 0.029572 |
| 23. Business and personal services, etc. | 0.053107 | 0.059538 | 0.015544 | 0.033578 |
| 24. Medical, educational, nonprofit organizations | 0.055700 | 0.059488 | 0.016586 | 0.040498 |
| 25. Undistributed | 0.066844 | 0.026066 | 0.013058 | 0.023689 |
| 26. Eating and drinking places | 0.049082 | 0.065089 | 0.022374 | 0.031868 |
| 27. Capitalized construction and maintenance | 0.101068 | 0.032989 | 0.041857 | 0.046442 |
| 28. Households | 0.043335 | 0.040845 | 0.013776 | 0.038518 |
| 29. Local government | 0.059584 | 0.052389 | 0.021754 | 0.049189 |

[a] Each entry shows, per dollar of deliveries to final demand by industry named at left, the total dollar production directly

| | Chemicals 6 | Products of petroleum and coal 7 | Leather and leather products 8 | Iron and steel 9 | Non-ferrous metals 10 | Plumbing, heating etc. 11 | Machinery 12 | Motors, generators, etc. 13 | Motor vehicles 14 | Other trans. equip. 15 |
|---|---|---|---|---|---|---|---|---|---|---|
| | 0.003377 | 0.013713 | 0.000331 | 0.005002 | 0.001595 | 0.038671 | 0.004193 | 0.001711 | 0.009471 | 0.000338 |
| | 0.006087 | 0.018717 | 0.001110 | 0.004014 | 0.000740 | 0.006686 | 0.009575 | 0.001878 | 0.015513 | 0.000475 |
| | 0.023129 | 0.028801 | 0.003221 | 0.040109 | 0.001516 | 0.026649 | 0.013969 | 0.002468 | 0.022036 | 0.000649 |
| | 0.008080 | 0.031146 | 0.000648 | 0.004578 | 0.001162 | 0.009394 | 0.011935 | 0.002052 | 0.019151 | 0.000550 |
| | 0.007552 | 0.025519 | 0.000760 | 0.004847 | 0.003244 | 0.008790 | 0.017013 | 0.002616 | 0.023861 | 0.000580 |
| | 1.063406 | 0.015694 | 0.000483 | 0.006268 | 0.008055 | 0.014287 | 0.004235 | 0.006151 | 0.012378 | 0.000481 |
| | 0.011615 | 1.034456 | 0.000252 | 0.002605 | 0.000902 | 0.010204 | 0.004567 | 0.000845 | 0.007915 | 0.000266 |
| | 0.008291 | 0.021886 | 1.060398 | 0.005039 | 0.001008 | 0.016617 | 0.006933 | 0.002148 | 0.018643 | 0.000497 |
| | 0.006337 | 0.033105 | 0.000683 | 1.024627 | 0.002443 | 0.016748 | 0.013294 | 0.003342 | 0.022421 | 0.000607 |
| | 0.012584 | 0.029999 | 0.000785 | 0.010945 | 1.018046 | 0.017331 | 0.014445 | 0.006635 | 0.019800 | 0.000492 |
| | 0.004979 | 0.021879 | 0.000819 | 0.062657 | 0.031012 | 1.041139 | 0.023616 | 0.007720 | 0.019706 | 0.000475 |
| | 0.004037 | 0.021031 | 0.002162 | 0.039681 | 0.004601 | 0.048987 | 1.066498 | 0.028224 | 0.018134 | 0.000511 |
| | 0.016570 | 0.024758 | 0.001089 | 0.022425 | 0.056742 | 0.045154 | 0.026205 | 1.016442 | 0.020228 | 0.000580 |
| | 0.002149 | 0.017019 | 0.000691 | 0.032398 | 0.001658 | 0.038689 | 0.030231 | 0.002943 | 1.194896 | 0.000352 |
| | 0.004258 | 0.018908 | 0.000846 | 0.029236 | 0.001987 | 0.010223 | 0.006197 | 0.006160 | 0.013649 | 1.020847 |
| | 0.004962 | 0.034432 | 0.000973 | 0.048101 | 0.007338 | 0.016796 | 0.013574 | 0.002865 | 0.025330 | 0.000643 |
| | 0.005084 | 0.099443 | 0.000724 | 0.007170 | 0.002439 | 0.011883 | 0.008739 | 0.019259 | 0.017142 | 0.000619 |
| | 0.013421 | 0.059276 | 0.000742 | 0.008833 | 0.001357 | 0.013028 | 0.011094 | 0.002332 | 0.019826 | 0.002014 |
| | 0.005612 | 0.072210 | 0.000764 | 0.005576 | 0.000991 | 0.011858 | 0.007997 | 0.002528 | 0.046367 | 0.008536 |
| | 0.005666 | 0.038112 | 0.001013 | 0.007575 | 0.001131 | 0.012767 | 0.009488 | 0.003410 | 0.039671 | 0.001005 |
| | 0.003170 | 0.021894 | 0.000895 | 0.004679 | 0.000790 | 0.008954 | 0.014276 | 0.002334 | 0.025302 | 0.000487 |
| | 0.004575 | 0.055410 | 0.000772 | 0.007692 | 0.001387 | 0.013482 | 0.009047 | 0.004471 | 0.020792 | 0.000608 |
| | 0.006592 | 0.039758 | 0.001217 | 0.008322 | 0.001487 | 0.016224 | 0.022032 | 0.003902 | 0.054686 | 0.000738 |
| | 0.030539 | 0.039115 | 0.001151 | 0.007398 | 0.002341 | 0.018655 | 0.011876 | 0.005956 | 0.030608 | 0.000857 |
| | 0.003940 | 0.065018 | 0.000823 | 0.019104 | 0.001387 | 0.013601 | 0.075133 | 0.003958 | 0.090121 | 0.000528 |
| | 0.005396 | 0.028178 | 0.000721 | 0.005705 | 0.001359 | 0.016070 | 0.006992 | 0.002957 | 0.020258 | 0.000622 |
| | 0.004565 | 0.049703 | 0.000894 | 0.032485 | 0.004445 | 0.051825 | 0.022092 | 0.003120 | 0.024888 | 0.000796 |
| | 0.004982 | 0.037056 | 0.001305 | 0.006121 | 0.001155 | 0.013298 | 0.009934 | 0.003873 | 0.034609 | 0.000913 |
| | 0.006037 | 0.046647 | 0.001158 | 0.011297 | 0.001817 | 0.020800 | 0.012761 | 0.004535 | 0.034911 | 0.000970 |

| | Trade 20 | Communications 21 | Finance, insurance, etc. 22 | Business and personal services 23 | Medical, educational, etc. 24 | Undistributed 25 | Eating and drinking 26 | Capitalized construction 27 | Households 28 | Local government 29 |
|---|---|---|---|---|---|---|---|---|---|---|
| | 0.064574 | 0.007896 | 0.059849 | 0.042498 | 0.011362 | 0.017495 | 0.016287 | 0.016916 | 0.357935 | 0.019748 |
| | 0.123512 | 0.011791 | 0.109826 | 0.059090 | 0.020207 | 0.028583 | 0.029338 | 0.021692 | 0.644351 | 0.029676 |
| | 0.135424 | 0.014714 | 0.133599 | 0.081661 | 0.024409 | 0.076039 | 0.035229 | 0.030387 | 0.774638 | 0.038076 |
| | 0.118245 | 0.011299 | 0.102050 | 0.056894 | 0.020829 | 0.081593 | 0.028274 | 0.027014 | 0.620820 | 0.032209 |
| | 0.134736 | 0.020262 | 0.141960 | 0.079081 | 0.027386 | 0.082951 | 0.039677 | 0.030308 | 0.874335 | 0.041460 |
| | 0.081290 | 0.011465 | 0.081669 | 0.043777 | 0.016013 | 0.019799 | 0.023179 | 0.030266 | 0.510358 | 0.026315 |
| | 0.034563 | 0.005030 | 0.036793 | 0.019716 | 0.006819 | 0.044105 | 0.009838 | 0.013569 | 0.216499 | 0.015393 |
| | 0.122917 | 0.014577 | 0.124740 | 0.069473 | 0.023389 | 0.045128 | 0.033856 | 0.031001 | 0.745325 | 0.037052 |
| | 0.120221 | 0.014480 | 0.119217 | 0.058413 | 0.033756 | 0.106746 | 0.033071 | 0.033201 | 0.728076 | 0.038379 |
| | 0.121732 | 0.011256 | 0.102206 | 0.055368 | 0.019924 | 0.092019 | 0.028915 | 0.089281 | 0.635200 | 0.032290 |
| | 0.115786 | 0.014317 | 0.112950 | 0.061542 | 0.022191 | 0.032525 | 0.032234 | 0.027385 | 0.709879 | 0.032788 |
| | 0.130217 | 0.014619 | 0.115130 | 0.065262 | 0.021904 | 0.035522 | 0.031784 | 0.035126 | 0.698488 | 0.034467 |
| | 0.124936 | 0.015867 | 0.130520 | 0.072303 | 0.026160 | 0.036450 | 0.037957 | 0.032772 | 0.836826 | 0.037536 |
| | 0.072376 | 0.008694 | 0.070561 | 0.041903 | 0.014045 | 0.041175 | 0.020352 | 0.020347 | 0.448201 | 0.022244 |
| | 0.088939 | 0.011129 | 0.092689 | 0.044904 | 0.018344 | 0.022938 | 0.026724 | 0.026129 | 0.589133 | 0.029432 |
| | 0.141374 | 0.015579 | 0.175489 | 0.082113 | 0.026720 | 0.100519 | 0.038429 | 0.039164 | 0.845026 | 0.044473 |
| | 0.091682 | 0.010779 | 0.097186 | 0.053115 | 0.018381 | 0.025454 | 0.026367 | 0.076046 | 0.580716 | 0.055670 |
| | 0.126464 | 0.016135 | 0.128829 | 0.086272 | 0.025595 | 0.030983 | 0.036888 | 0.124161 | 0.812688 | 0.044651 |
| | 0.156040 | 0.019815 | 0.170183 | 0.100686 | 0.027341 | 0.037273 | 0.039855 | 0.045193 | 0.860392 | 0.051380 |
| | 1.174615 | 0.023475 | 0.238555 | 0.162529 | 0.037110 | 0.042643 | 0.056111 | 0.046251 | 1.155186 | 0.062640 |
| | 0.116499 | 1.033307 | 0.126878 | 0.062521 | 0.024642 | 0.028266 | 0.035730 | 0.049202 | 0.787818 | 0.050395 |
| | 0.151867 | 0.018551 | 1.177921 | 0.083075 | 0.030565 | 0.032811 | 0.039374 | 0.139899 | 0.844534 | 0.084015 |
| | 0.186759 | 0.030696 | 0.233810 | 1.142031 | 0.047305 | 0.052261 | 0.052866 | 0.047497 | 1.162033 | 0.061071 |
| | 0.203218 | 0.023638 | 0.201648 | 0.106856 | 1.044050 | 0.049047 | 0.066692 | 0.058091 | 1.335543 | 0.057593 |
| | 0.158605 | 0.013353 | 0.127739 | 0.065920 | 0.025693 | 1.054178 | 0.037168 | 0.041592 | 0.816533 | 0.076676 |
| | 0.202446 | 0.014795 | 0.167138 | 0.094538 | 0.026074 | 0.031670 | 1.037535 | 0.039083 | 0.820197 | 0.048226 |
| | 0.206699 | 0.017070 | 0.158609 | 0.111828 | 0.029306 | 0.041560 | 0.042252 | 1.033682 | 0.925460 | 0.043939 |
| | 0.219835 | 0.023029 | 0.228290 | 0.112057 | 0.048285 | 0.053102 | 0.070558 | 0.041321 | 1.557196 | 0.064242 |
| | 0.206429 | 0.024661 | 0.232349 | 0.114283 | 0.047827 | 0.049217 | 0.060574 | 0.185586 | 1.333529 | 1.058433 |

and indirectly required from St Louis area industry named at top.

# 24

# The Fundamental Structure
of Input-Output Tables:
An International Comparison

DAVID SIMPSON AND JINKICHI TSUKUI

## Introduction

The structure of production of an economic system, represented by the matrix of input-output coefficients, has traditionally been held to be determined by technology. Thus, the coefficients are sometimes called "technical" coefficients.

If this is, in fact, the case one should expect to discover a productive structure which is common to all economic systems having a like technology.[1] For example, one should expect to find certain characteristics of the input-output matrices of all industrialized countries, which have a technological origin.

Previous studies have compared the productive structure of different economic systems in purely economic, or even arithmetic terms, so that they may fairly be described as taxonomic. On the other hand, this paper suggests that there are certain fundamental elements which may be found in the productive structure of modern economic systems which are purely technical in character. It is demonstrated that the economic systems of Japan and the United States, although superficially dissimilar, contain almost identical patterns of industries which are strongly interrelated. This pattern, or framework, of productive relations has several interesting properties, which are found to be shared by the pattern of interindustry relations in other economies. The theoretical implications of this discovery are briefly discussed.

## Method of Analysis

Beginning with the hypothesis that there should be a common technological element in the structure of production of different countries we prepared and analysed comparable input-output tables of the United States and Japan (Figures 24.1 and 24.2). These two countries were chosen because they offered the possibility

This article was published in *The Review of Economics and Statistics*, Volume 47, Number 4, November, 1965, pp. 434–446. © By the President and Fellows of Harvard College. Reprinted by permission.

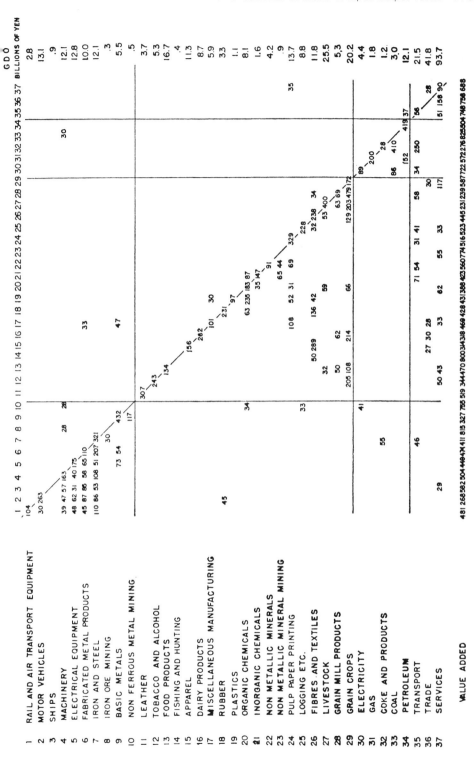

Fig. 24.1. United States, 1947.

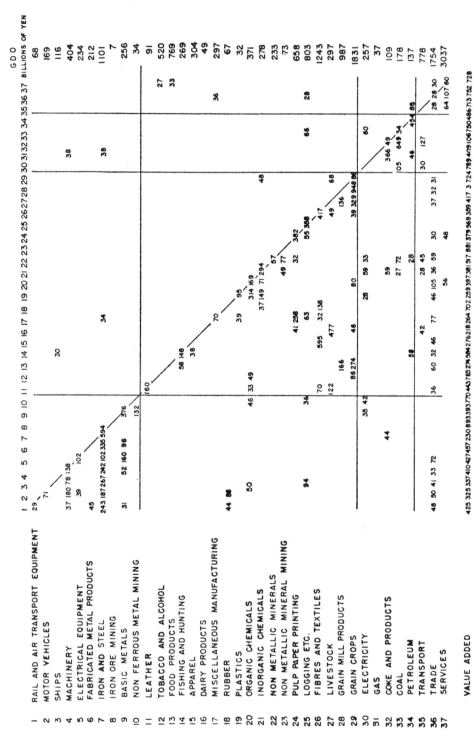

Fig. 24.2. Japan, 1955.

of preparing tables of reasonably homogeneous yet detailed sectors. We were not interested in an international comparison per se.

In addition to technology, input-ouput tables of different countries may be expected to reflect a number of other real factors, such as the structure of final demand and quantitative and qualitative variations in factor endowments.

Such "real" differences between countries are likely to be augmented in the construction of input-output tables by differences in accounting procedures (valuation of transactions at producers' or purchasers' prices, distinction between capital and current transactions, the establishment or the activity as the basic unit of account, the treatment of secondary flows, etc.), and differences in methods of aggregation.

By choosing the input coefficient matrix of countries for comparison, rather than the flow table, we diminish the effect of differences in final demand. We cannot eliminate, however, the influence of differences in relative prices.[2] We can only hope that these will not vary enough to obscure an underlying common structure. We tried to minimize the differences in the composition of industries in the two countries by carefully compiling comparable sectors from more detailed tables. The 192-order table for the United States, 1947, and the 124-order table for Japan, 1955, were each aggregated to 38 sectors—the most detailed classification consistent with reasonable homogeneity and sector size.

The process of refinement can then be carried one stage further by discarding the smaller coefficients from both tables. The simplest rule, which we have used, is to discard all those

$$a_{ij} \leqslant \left(\frac{100}{n}\right)\%$$

where $n$ is the number of sectors. In so doing, we may have reduced the influence of accounting distortions in the structure of production, for the larger coefficients represent flows which are likely to have been more accurately observed in the basic industrial accounts. Furthermore, while small differences may exist in similar processes between countries, the raw material or intermediate good undergoing processing is likely to be the same. It should be emphasized that we have not imposed a peculiar standardizing procedure on the table. Any other method of distinguishing the significant relations in the coefficient matrix would be bound to produce a similar result.

While there are many small coefficients and few large coefficients in the typical matrix, the relative importance of the coefficients eliminated by the application of our rule can be gauged from Tables 24.1 and 24.2, which show the larger coefficients retained as a percent of the sum of all intermediate coefficients in each column.

We are left then with a matrix of larger coefficients, which reveals the pattern of industries which are strongly related one to another. This pattern might be described as the skeleton of the productive system, or the framework, or, to give it a third name, the fundamental structure of production.

When we compare the skeleton or framework matrix which we have derived for

Table 24.1   United States, 1947—Coverage of the Framework Matrix

| Industry column number | Framework coefficients as percent of sum of coefficients rows 1–37 | Framework coefficients as percent of sum of coefficients rows 1–34 | Intrabloc coefficients as percent of sum of coefficients rows 1–37 | Intrabloc coefficients as percent of sum of coefficients rows 1–34 |
|---|---|---|---|---|
| 1 | 72.4 | 80.5 | 78.4 | 87.2 |
| 2 | 80.6 | 85.9 | 77.9 | 83.0 |
| 3 | 61.0 | 64.2 | 64.1 | 76.1 |
| 4 | 74.4 | 82.2 | 80.0 | 88.4 |
| 5 | 66.1 | 74.0 | 71.3 | 79.9 |
| 6 | 70.5 | 98.9 | 81.7 | 89.5 |
| 7 | 76.7 | 79.3 | 69.4 | 79.9 |
| 8 | 15.1 | 18.7 | 36.2 | 44.7 |
| 9 | 81.6 | 88.5 | 89.0 | 96.6 |
| 10 | 55.5 | 67.7 | 22.9 | 27.9 |
| 11 | 63.8 | 72.2 | 86.3 | 97.6 |
| 12 | 75.9 | 78.5 | 82.9 | 95.2 |
| 13 | 69.2 | 73.0 | 76.2 | 91.0 |
| 14 | 25.0 | 34.2 | 46.0 | 63.0 |
| 15 | 84.0 | 87.9 | 88.3 | 98.0 |
| 16 | 83.3 | 89.2 | 84.9 | 96.0 |
| 17 | 65.9 | 63.2 | 62.3 | 72.4 |
| 18 | 75.2 | 82.9 | 80.2 | 88.4 |
| 19 | 74.9 | 81.8 | 89.8 | 98.1 |
| 20 | 71.2 | 72.3 | 70.6 | 83.6 |
| 21 | 64.1 | 62.4 | 74.1 | 91.1 |
| 22 | 58.6 | 58.5 | 56.6 | 71.5 |
| 23 | 24.3 | — | 24.8 | 36.6 |
| 24 | 74.4 | 70.0 | 81.4 | 95.4 |
| 25 | 70.0 | 67.0 | 65.6 | 80.7 |
| 26 | 75.7 | 84.5 | 86.1 | 96.2 |
| 27 | 86.6 | 95.0 | 89.5 | 98.4 |
| 28 | 84.1 | 86.9 | 85.3 | 96.9 |
| 29 | 77.2 | 69.1 | 53.3 | 88.4 |
| 30 | 75.2 | 65.8 | 61.0 | 83.2 |
| 31 | 82.2 | 89.6 | 91.1 | 99.2 |
| 32 | 88.1 | 85.2 | 63.5 | 89.5 |
| 33 | 17.1 | 20.0 | 16.0 | 18.7 |
| 34 | 84.5 | 90.3 | 85.5 | 91.4 |
| 35 | 57.1 | n.a. | 48.4 | n.a. |
| 36 | 65.3 | n.a. | 72.3 | n.a. |
| 37 | 49.0 | n.a. | 42.3 | n.a. |

Japan and the United States in the foregoing manner, both have five distinct properties.

### *Decomposability*

When industries are grouped according to their physical qualities, and blocs are arranged in the following order:

Table 24.2   Japan, 1955—Coverage of the Framework Matrix

| Industry column number | Framework coefficients as percent of sum of coefficients rows 1–37 | Framework coefficients as percent of sum of coefficients rows 1–34 | Intrabloc coefficients as percent of sum of coefficients rows 1–37 | Intrabloc coefficients as percent of sum of coefficients rows 1–34 |
|---|---|---|---|---|
| 1 | 83.0 | 83.8 | 69.0 | 77.5 |
| 2 | 83.7 | 84.6 | 68.6 | 76.0 |
| 3 | 79.9 | 81.4 | 57.9 | 63.9 |
| 4 | 78.8 | 81.1 | 79.2 | 87.6 |
| 5 | 76.1 | 75.5 | 66.5 | 79.0 |
| 6 | 79.4 | 85.5 | 84.2 | 90.7 |
| 7 | 82.9 | 86.4 | 83.5 | 87.1 |
| 8 | — | — | 18.7 | 23.8 |
| 9 | 79.6 | 85.5 | 79.6 | 85.5 |
| 10 | 53.5 | 67.2 | 14.3 | 18.0 |
| 11 | 75.6 | 77.8 | 85.1 | 95.8 |
| 12 | 62.8 | 73.8 | 78.6 | 92.4 |
| 13 | 76.9 | 79.6 | 80.4 | 93.3 |
| 14 | 73.6 | 77.8 | 58.4 | 69.0 |
| 15 | 88.5 | 91.7 | 88.0 | 98.1 |
| 16 | 77.5 | 89.0 | 75.7 | 93.4 |
| 17 | 77.9 | 78.4 | 73.5 | 85.5 |
| 18 | 58.7 | 65.1 | 68.1 | 75.5 |
| 19 | 85.3 | 89.2 | 80.8 | 91.2 |
| 20 | 79.8 | 75.3 | 62.4 | 88.3 |
| 21 | 89.2 | 89.9 | 59.9 | 68.3 |
| 22 | 83.4 | 82.6 | 71.6 | 59.1 |
| 23 | — | — | 40.3 | 50.5 |
| 24 | 82.9 | 82.1 | 77.9 | 91.0 |
| 25 | 83.1 | 88.6 | 87.7 | 93.6 |
| 26 | 78.8 | 84.0 | 89.4 | 95.3 |
| 27 | 86.1 | 91.0 | 85.9 | 98.0 |
| 28 | 98.3 | 99.1 | 95.6 | 99.6 |
| 29 | 85.1 | 87.9 | 78.6 | 93.5 |
| 30 | 64.0 | 66.0 | 59.7 | 79.2 |
| 31 | 82.6 | 89.7 | 75.0 | 81.4 |
| 32 | 92.4 | 92.9 | 7.8 | 9.3 |
| 33 | 64.0 | 73.1 | 39.6 | 45.2 |
| 34 | 88.3 | 93.6 | 89.3 | 94.6 |
| 35 | 62.7 | n.a. | 36.2 | n.a. |
| 36 | 80.2 | n.a. | 59.3 | n.a. |
| 37 | 55.1 | n.a. | 34.9 | n.a. |

i. Metal
ii. Nonmetal
iii. Energy
iv. Services

then the matrix is decomposable in such a way that the blocs follow a triangular order with respect to one another. This property is sometimes described as "bloc-triangularity."

### Bloc Independence

But the matrix of industries grouped in the foregoing manner shows the still stronger property of independence of the blocs. That is to say that no industry is related to another outside its own bloc. This property is sometimes described as "bloc-diagonality." With the exception of the services bloc, the blocs in Figures 24.1 and 24.2 are almost independent of one another.

### Triangularity

The framework matrix shows not only these two properties, but, in addition, it has the property of being almost triangular in all its elements.[3]

### Physical Homogeneity of Blocs

The industries in each bloc have a common physical characteristic.[4]

In other words, what we may call the fundamental structure of production is described by a matrix having generally bloc-independent and triangular properties, where the blocs are composed of industries with easily recognizable technical qualities. These properties appear to hold for the fundamental structure, not only of the United States and Japanese economies, but also of other industrialized countries (see Figures 24.3, 24.4, and 24.5).

## Interpretation of Results

Theoretical studies of the properties of matrices have discussed the consequences of decomposability, i.e., of the existence of separable submatrices. Empirical studies of the structure of input-output tables of different countries, on the other hand, have generally devoted themselves to an arrangement of industries in a triangular order, the criteria for the arrangement being quite mechanical.[5] The most common approach has been to compare the rankings of industries which result when the triangularity of each country's table is, in some sense, maximized. Two broad conclusions may be drawn from these studies. First, in the absence of an absolutely triangular and/or completely decomposable table, no unique criterion for the ordering of industries can be derived. Secondly, although a particular ordering of industries in an input-output table may facilitate a recursive approximation to a computational solution, there is nothing theoretically instructive about the juxtaposition, on purely mechanical grounds, of rather heterogeneous industries.

The structure of Figures 24.1 and 24.2 may perhaps be illustrated by a brief discussion of the relations between the different forms of decomposable matrices.

1. If a matrix can be divided into six parts, $A'$, $A''$, $B'$, $B''$, $C$, and 0 as shown in Figure 24.6, where 0 is a zero submatrix (all elements zero), and $A$ and $B$ are square

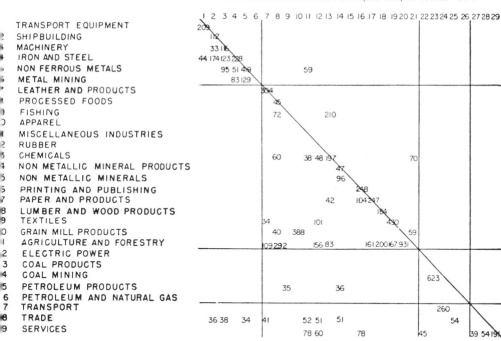

Fig. 24.3. Norway, 1950.

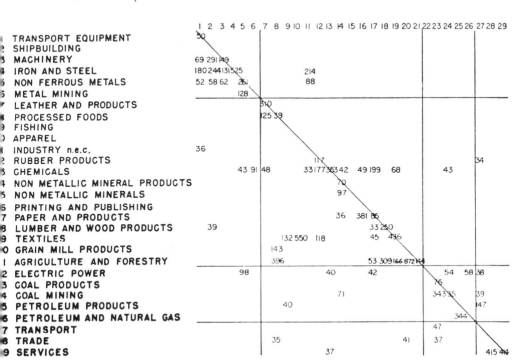

Fig. 24.4. Italy, 1950.

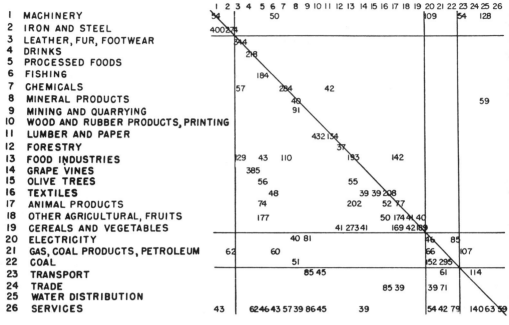

Fig. 24.5. Spain, 1957.

matrices, then the matrix is said to be decomposable. The form of the matrix (a) may also be described as bloc-triangular.

2. If either $A'$ or $A''$ has all zero elements, then, by proper rearrangement of industries, we can always obtain the matrix in the form (b). A similar statement holds for submatrix $B$.

3. If either $A'$ or $A''$ and either $B'$ or $B''$ have all zero elements, then the matrix can be made perfectly triangular (c) by proper rearrangement of industries.

4. If $C$ has all zero elements, then the matrix satisfies the property of bloc independence, sometimes described as bloc-diagonality, (d).

5. If both conditions (c) and (d) are satisfied, we obtain a matrix which is *both* triangular *and* bloc-independent, (e).

With the evident exception of the services bloc of industries, the fundamental structure of production revealed in Figure 24.1 and 24.2 is very close to the structure of the matrix illustrated in Figure 24.6(e).

A closer inspection of Figure 24.1 and 24.2 shows the following points:

a. The order of industries in both figures is identical.

b. Of the sectors frequently treated as endogenous in input-output tables, maintenance construction and waste products are excluded.

c. The industries of the metals bloc are more closely related to each other than are the industries of the nonmetals bloc, a fact which explains the prevalence of industrial complexes of metal industries.

d. The three principal cores of the nonmetals bloc are the chemical, agricultural, and textile industries.

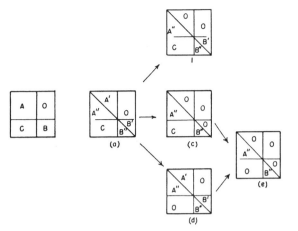

Fig. 24.6.

e. Tables 24.1 and 24.2 shows the extent to which the framework coefficients in each column cover the total intermediate inputs of that industry. Generally, the coverage is better than 65 percent in the United States table, and better than 75 percent in the Japanese table, and improves by about 5 percent when the services bloc is disregarded.

f. Within each bloc, industries have been arranged so that they form a triangular hierarchy. (Those industries which have only a diagonal coefficient are ranked within the bloc according to their proportion of intermediate and total output.) As Figures 24.1 and 24.2 show, the triangular ordering is not flawless, but it can be improved without destroying the bloc concept by rearranging the blocs in the manner illustrated in Figure 24.7.

Industries were arranged within the blocs in the following way: Whenever the order could be made perfectly triangular in one table, that order was maintained in the corresponding bloc in the other table. Thus, the order of metal industries was determined by the Japanese table and the order of the energy industries by the perfectly triangular order in the United States table. In the Japanese table, all of the industries in the nonmetals bloc down to 28 are in a perfectly triangular order, and the United States table indicates that industry 29 must be placed below 28. In the services bloc, the position of industry 35 is clear, while the order of 36 and 37 is indeterminate. Our arrangement of these two industries is therefore tentative. It would, of course, be possible to improve triangularity[6] by allowing the order of industries to differ between the two countries. Likewise, triangularity might possibly be further improved by changing the order of industries between blocs.

But we prefer to maintain our bloc arrangement as well as the identical order of industries as a basis for the comparison of the productive structure of other countries. We should like to point out that the degree of triangularity within the blocs obtained with our common industrial order is quite satisfactory.

Bearing in mind that the minimum value of coefficients admitted to the framework in the two tables is 0.027, and that several coefficients are ten times that order of magnitude, the importance of the remaining above-diagonal coefficients is

| | | 11 | 12 | 13 | 14 | 15 | 16 | 17 | 1 | 2 | 3 | 4 | 5 | 6 | 7 | 8 | 9 | 10 | 18 | 19 | 20 | 21 | 22 | 23 | 24 | 25 | 26 | 27 | 28 | 29 | 30 | 31 | 32 | 33 | 34 | 35 | 36 | 37 | GDO BILLIONS OF YEN |
|---|---|---|---|---|---|---|---|---|---|---|---|---|---|---|---|---|---|---|---|---|---|---|---|---|---|---|---|---|---|---|---|---|---|---|---|---|---|---|---|
| 11 | LEATHER | 160 | | | | | | | | | | | | | | | | | | | | | | | | | | | | | | | | | | | | | 91 |
| 12 | TOBACCO AND ALCOHOL | | | | | | | | | | | | | | | | | | | | | | | | | | | | | | | | | | | | | 27 | 520 |
| 13 | FOOD PRODUCTS | | | | | | | | | | | | | | | | | | | | | | | | | | | | | | | | | | | | | 33 | 679 |
| 14 | FISHING AND HUNTING | | | 58 | 148 | | | | | | | | | | | | | | | | | | | | | | | | | | | | | | | | | | 269 |
| 15 | APPAREL | | | | 38 | | | | | | | | | | | | | | | | | | | | | | | | | | | | | | | | | | 304 |
| 16 | DAIRY PRODUCTS | | | | | | | | | | | | | | | | | | | | | | | | | | | | | | | | | | | | | | 49 |
| 17 | MISCELLANEOUS MANUFACTURING | | | | | | | 70 | | | | | | | | | | | | | | | | | | | | | | | | | | | | | 36 | | 297 |
| 1 | RAIL AND AIR TRANSPORT EQUIPMENT | | | | | | | | 29 | | | | | | | | | | | | | | | | | | | | | | | | | | | | | | 68 |
| 2 | MOTOR VEHICLES | | | | | | | | | 71 | | | | | | | | | | | | | | | | | | | | | | | | | | | | | 169 |
| 3 | SHIPS | | | | 30 | | | | | | | | | | | | | | | | | | | | | | | | | | | | | | | | | | 116 |
| 4 | MACHINERY | | | | | | | | 37 | 180 | 78 | 138 | | | | | | | | | | | | | | | | | | | | 38 | | | | | | | 404 |
| 5 | ELECTRICAL EQUIPMENT | | | | | | | | | 39 | | | 102 | | | | | | | | | | | | | | | | | | | | | | | | | | 234 |
| 6 | FABRICATED METAL PRODUCTS | | | | | | | | 45 | | | | | | | | | | | | | | | | | | | | | | | | | | | | | | 212 |
| 7 | IRON AND STEEL | | | | | | | 34 | 243 | 137 | 267 | 242 | 102 | 335 | 594 | | | | | | | | | | | | | | | | | 38 | | | | | | | 1101 |
| 8 | IRON ORE MINING | | | | | | | | | | | | | | | | | | | | | | | | | | | | | | | | | | | | | | 7 |
| 9 | BASIC METALS | | | | | | | | 31 | | | 92 | 160 | 96 | | | 316 | | | | | | | | | | | | | | | | | | | | | | 256 |
| 10 | NON FERROUS METAL MINING | | | | | | | | | | | | | | | | 132 | | | | | | | | | | | | | | | | | | | | | | 34 |
| 18 | RUBBER | | | | | | | | 44 | 88 | | | | | | | | | | | | | | | | | | | | | | | | | | | | | 67 |
| 19 | PLASTICS | | | | | | | 39 | | | | | | | | | | | | 95 | | | | | | | | | | | | | | | | | | | 32 |
| 20 | ORGANIC CHEMICALS | 33 | 49 | | | | | | | | 50 | | | | | | | 46 | | 314 | 169 | | | | | | | | | | | | | | | | | | 371 |
| 21 | INORGANIC CHEMICALS | | | | | | | | | | | | | | | | | | 37 | 149 | 71 | 294 | | | | | | | | 48 | | | | | | | | | 278 |
| 22 | NON METALLIC MINERALS | | | | | | | | | | | | | | | | | | | | | | 57 | | | | | | | | | | | | | | | | 233 |
| 23 | NON METALLIC MINERAL MINING | | | | | | | | | | | | | | | | | | | | | 49 | 77 | | | | | | | | | | | | | | | | 73 |
| 24 | PULP PAPER PRINTING | | | | | | 41 | 258 | | | | | | | | | | | | | | | 32 | | 382 | | | | | | | | | | | | | | 658 |
| 25 | LOGGING ETC. | | | | | | | 63 | | | 94 | | | | | | | 36 | | | | | | | 55 | 358 | | | | | | | | 66 | | | 28 | | 803 |
| 26 | FIBRES AND TEXTILES | 70 | | | | 595 | | 32 | | | | | | | | | | | 138 | | | | | | | | 417 | | | | | | | | | | | | 1243 |
| 27 | LIVESTOCK | 122 | | | | | 477 | | | | | | | | | | | | | | | | | | | | 49 | | | 68 | | | | | | | | | 297 |
| 28 | GRAIN MILL PRODUCTS | | | 166 | | | | | | | | | | | | | | | | | | | | | | | | 136 | | | | | | | | | | | 987 |
| 29 | GRAIN CROPS | | 86 | 274 | | | 46 | | | | | | | | | | | | | | 80 | | | | | | 39 | 329 | 948 | 88 | | | | | | | | | 1831 |
| 30 | ELECTRICITY | | | | | | | | | | | | | | | | 35 | 42 | | 28 | | 59 | 33 | | | | | | | | | | | 60 | | | | | 257 |
| 31 | GAS | | | | | | | | | | | | | | | | | | | | | | | | | | | | | | | | | | | | | | 37 |
| 32 | COKE AND PRODUCTS | | | | | | | | | | | | | | 44 | | | | | | | 59 | | | | | | | | | | 366 | 48 | | | | | | 109 |
| 33 | COAL | | | | | | | | | | | | | | | | | | | | | 27 | 72 | | | | | | | | 105 | | 649 | 34 | | | | | 178 |
| 34 | PETROLEUM | | | | 58 | | | | | | | | | | | | | | | | | | 28 | | | | | | | | | 46 | | | 454 | 88 | | | 137 |
| 35 | TRANSPORT | | | | | | 42 | | | | | | | | | | | | | | | 28 | 45 | | | | | | | | 30 | | 127 | | | | | | 778 |
| 36 | TRADE | 36 | | 60 | 32 | 46 | | 77 | 48 | 50 | 41 | 33 | 72 | | | | | | | 46 | 105 | 36 | 59 | | 30 | | | 37 | 32 | 31 | | | | | | 28 | 28 | 30 | 1754 |
| 37 | SERVICES | | | | | | | | | | | | | | | | | | | | 56 | | | | 48 | | | | | | | | | | | 64 | 107 | 80 | 3037 |

VALUE ADDED 443785 274584276218 264425325337410427467230893393770 702259397381517 881379 569359417 3 724789409106750486713752728

Fig. 24.7 Japan 1955

evidently slight. In Figure 24.2, ten out of 14 are less than 0.05, while in the United States Figure 24.1, ten out of 11 coefficients are less than 0.05. The remaining coefficients are:

---

Organic chemicals →
    Inorganic chemicals      (United States 0.087)
Livestock         →
    Grain crops          (Japan      0.068)
Logging        →
    Coal mining[a]        (Japan      0.066)
Electricity      →
    Coal mining         (Japan      0.060)
Petroleum     →
    Transport           (Japan      0.088)

---

[a] This is probably pit-props: American coal mines use more capital-intensive methods.

g. Although comparisons of individual coefficients in the two tables are vitiated by the aforementioned shortcomings in comparability, one might expect to find some similarity in the order of magnitude of coefficients in each bloc.

In the metals bloc, although there are fewer nonzero coefficients in Japan, the order of magnitude of the coefficients in the two tables is remarkably alike. Most distinct is the fact that each industry preceding respectively iron and steel and machinery receives an input from these industries.

In the upper, or final, part of the nonmetals bloc the off-diagonal coefficients are all zero in the United States and almost zero in the Japanese table. In the lower, or basic, part of the bloc, the differences in the sizes of the coefficients are more pronounced, possibly reflecting greater differences in factor productivity between the two countries in the nonmetal industries (particularly agriculture). However, the pattern of nonzero coefficients in both countries is clearly centered on the agricultural, chemical, and textiles industries. In particular, the following industries are similarly related in both countries:

---

Fibres and textiles →
    Rubber            (0.136, U.S.; 0.138, Japan)
Organic chemicals →
    Plastics           (0.314, U.S.; 0.235, Japan)
Grain crops        →
    Organic chemicals (0.065, U.S.; 0.079, Japan)
Livestock          →
    Textiles & fibres    (0.053, U.S.; 0.048, Japan)

---

In the energy bloc the inputs from coal to electricity and from coal to coke are of comparable magnitude in both tables.

Because of the heterogeneity of such industries as trade and services, it would be unwise to draw any conclusions from the similarity of coefficients in the services bloc.

h. It is clear from the coefficients outside the blocs in Figures 24.1 and 24.2 that the blocs are not completely independent of one another. Sources of incompleteness can be described as (*i*) services bloc industries, (*ii*) energy industries, and (*iii*) other industries.

*i.* In view of the universal role of services, trade, and transport as inputs, the fact that these three industries are the principal source of bloc interdependence should not be unexpected. The term "services" covers an unusually heterogeneous collection of activities, and the size of the industry is enormous in both countries; in the United States it forms about 13 percent, and in Japan about 15 percent of total gross domestic output. The trade sector, which is responsible for much of the interdependence in the Japanese table, is also large. However, most of the input coefficients to other industries from industries in the services bloc are very small. In the United States table, 12 out of 20 nonzero coefficients are less than 0.05 (all of the input coefficients of trade are less than 0.03), in Japan, 17 out of 25 are less than 0.05. However, we are not anxious to extend the concept of decomposability to the services industries. We would prefer to exclude the bloc from further analysis of the fundamental structure of production.

*ii.* The energy inputs to other industries which occur in the Japanese table are all small. The three inputs which occur in both tables are the inputs of coke to iron and steel (Japan 0.044; U.S. 0.054), electricity to nonferrous metal mining (0.041 for both countries), and petroleum to transport (Japan 0.088; U.S. 0.037). The first two reflect in each case a distinct technological process, respectively, the blast furnace, and the extraction of copper (and zinc, in the United States) from its ore by an electrolytic process. The recurrence of inputs of electric energy in the Japanese table may possibly be explained by the high price of electricity.

*iii.* Once more we can observe that the coefficients connecting industries of different blocs are nearly all small.

Miscellaneous industries has been assigned to the nonmetals bloc. Ideally it would be possible to split it into two parts and thus eliminate an obvious source of bloc interdependence.

One relation between the metals and nonmetals industries to appear in both Figures 24.1 and 24.2 is the input of rubber to motor vehicles (0.045, U.S.; 0.088, Japan). Whereas we intuitively think that modern production frequently mixes physically different goods like rubber and steel, the framework of large coefficients demonstrates that this is, in fact, an *atypical* process. Also, in both tables, nonferrous metal mining uses organic chemicals (0.034, U.S.; 0.046, Japan) and logging (0.033, U.S.; and 0.036, Japan). The former coefficient represents the use of explosives, and the latter (presumably) the use of timber for shoring. Thus, there is no physical combination of metals with nonmetals in these two processes of production.

An example of physical combination of metal with nonmetal is the use of wood in the building of ships (Figure 24.2). Clearly, this is a peculiar feature of the Japanese productive structure, which can be attributed to the number of wooden fishing boats produced by the shipbuilding industry.

We have described the important exceptions to a fundamental structure of production, represented by the large coefficients of the input-output matrix, which is clearly both bloc and triangular in form. The identification of this particular structure is not dependent on a particular method of aggregation. We have applied our procedure of discarding the smaller coefficients directly to tables of other countries with different industrial classifications,[7] and have found the same fundamental structure [see Figures 24.3 (Norway), 24.4 (Italy), and 24.5 (Spain)]. To perform this test, it was necessary only to arrange the industries in an order as close to that obtaining in Figures 24.1 and 24.2 as the classifications permitted, and the eliminate coefficients $\leqslant (100/n)$ percent in value (where $n =$ the number of industries). We are confident that this test, when applied to the table of any industrialized country will reveal the same fundamental structure of production, i.e., a matrix which is generally bloc *and* triangular in form, and in which the blocs are composed of metal industries, nonmetals, energy, and services.

**Theoretical Implications**

It would be rash to argue from the examples of Japan and the United States alone, but the recurrence of the same pattern of interindustry relations in the input-output tables for different countries in different years, despite all the well-known statistical difficulties, compels us to conclude that there does exist a common fundamental structure of production in modern economic systems.

The stability of this structure, its scope in time and place, must be the subject of further research. All we wish to do here is report its existence, and comment briefly upon its implications.

Undoubtedly, the most striking feature of the structure is its bloc property. By itself, the existence of separable blocs would be merely a curiosum. What gives it importance is the composition of the blocs, namely that one bloc is composed of *all* metal industries and *only* metal industries, another of nonmetal industries, another of energy, and another of services. The independence of metals of all other industries, and their strong mutual relations implies that, despite the enormous technological progress since the industrial revolution, combination between metals and nonmetals in production processes has been limited. As was noted in (h *iii*) above, the only significant physical combination between metals and nonmetals occurring in both the United States and Japanese tables was the input of rubber into motor vehicles. While there are evidently other examples which spring to mind, the tables show that their *empirical* significance is overshadowed by the inputs of physically similar materials. Whether the tendency for substitution or combination of metals and nonmetals is increasing or dimishing with time remains a debatable point. On the one hand, it may be argued that at an early stage of their development metals necessarily had to be combined with other materials, for example ships were built of wood and iron, but the further development of the metal industries has led to a preference for all-metal ships. On the other hand, it may be argued that the progress of technology is constantly opening up possibilities of physical combination which did not exist previously.

The precise nature of technical constraints upon substitution of materials is not our concern here. However, we might point to the obvious physical properties of metals—in comparison to nonmetals, they are generally durable, malleable, strong, and homogeneous in quality.[8]

The triangular property of the matrix, on the other hand, seems to be less directly influenced by technology. However, chronological order suggests a reason why an output at one stage of the process of production might not be used as an input at an earlier stage.

Clearly, the economic system is circular in the long run, i.e., produced capital goods themselves become inputs in production. Indeed, it is through the use of capital goods that metal industries and nonmetal industries are connected. But, limiting ourselves only to the current flow of material goods and services, and leaving aside problems of aggregation, it seems that, while the original matrix shows circular relations such that it can seldom be satisfactorily triangulated, the fundamental structure is triangular in form.

Although the framework matrices we have derived for the various countries approach more closely a triangular pattern than they do a bloc pattern, it is the bloc character of the framework which we wish to emphasize. In order to preserve this meaningful pattern we have avoided improvements in the triangularity of the table which would require either changing the order of industries between blocs (i.e., changing the composition of the blocs), or changing the order of industries in one country. We have changed the order of industries in both countries only where it improved the triangularity in both the United States and Japan. The industries of the other countries, Norway, Italy, and Spain, were then arranged in an order resembling, as closely as the different classifications would permit, the hierarchy established for the United States and Japan. This hierarchy within the blocs is intuitively reasonable as it stands, but it could be modified if the matrices of other countries, prepared in comparable detail, indicated that a general improvement would result.

Given a large number of sufficiently comparable tables it should be possible to identify which parts of the framework matrix of each country represent its peculiarities[9] and which parts represent the fundamental structure of production, which, in turn, represents the common technology. The limited comparison which we have been able to make suggests that the difference in the technology between countries is not so great as is frequently supposed, and is overshadowed by the similarity. So far as imports of competing goods are concerned, we can be sure that the prices of domestically produced and imported goods will be identical, and that the quantities of inputs used in their production will be similar. In this case, the coefficient matrix will not be different from that obtaining in a state of self-sufficiency.

On the other hand, the input coefficients of goods not produced domestically (i.e., whose supply is entirely imported) will be missing from the coefficient matrix. But even though the existence of noncompeting imports means that one or more column vectors of coefficients are missing from the original matrix, the bloc and triangularity properties of the framework matrix remain unchanged.

Granted the stability among countries of the fundamental structure in its present form, a range of possible empirical applications arises. It offers a rational basis for grouping industries, either for aggregation, or for detailed partial analysis. It offers a frame of reference for the analysis of the structure of production of individual countries. In distinguishing the common properties of the structure of production in various economic systems, it offers a method of organizing data in historical as well as interregional studies.

In this short paper, the authors have reported what appears to them to be an important empirical regularity, the existence of a fundamental structure of production. The refinement and development of this concept requires much further research.

**NOTES**

1. We mean by technology all feasible transformations of goods and factors. In this sense, the word should be carefully distinguished from the actual combinations of factors of production and intermediate goods used in production, which are determined by the prevailing system of prices. In this context, it may be appropriate to advance the following suggestion: While the primary factors of production may possibly be substituted for each other as their relative prices change, inputs of intermediate goods are inelastic with respect to price changes for reasons dictated by technology.

2. The structure of relative prices in a country is a function of technology and value added:

$$p = v[I - A]^{-1}$$

where $p$ is the row vector of prices, and $v$ the row vector of value added, and $[I - A]$ is the input-output matrix. Given the same technology in different countries, differences in relative prices depend upon differences in value added among countries. However, there is reason to think that the structure of value added may not vary much among countries, since differences in factor inputs are likely to be offset by differences in factor prices.

3. When the industries are arranged in the following categories, then the triangularity of the matrix is improved:

   iia. Nonmetal, final
   ia. Metal, final
   ib. Metal, basic
   iib. Nonmetal, basic
   iii. Energy
   iv. Services

   For an illustration, see Figure 24.7.

4. At first, we were inclined to think that industries having strong input-output relations with one another could be classified as organic or inorganic until the results obtained forced us to modify our hypothesis.

5. The comparative analysis of input-output tables of different countries was first undertaken by Chenery and Watanabe in their important paper, "International Comparisons of the Structure of Production," *Econometrica*, 20 (Oct. 1958), which showed that such tables were nearly triangular. While the results presented here can be regarded, in a sense, as an improvement upon those obtained by Chenery and Watanabe, the authors wish to emphasize that their grouping of industries in blocs is an a priori one, based upon the physical characteristics of the industries and not upon the tables.

6. Triangularity may be measured by some function of the above-diagonal coefficients. For a description of the procedure, see the apendix. See also Chenery and Watanabe, *op. cit.*, 496.

7. The Norwegian and Italian tables were taken from H. B. Chenery and P. G. Clark, *Interindustry Economics* (New York: Wiley, 1959) 218–221. The Spanish table is the "Tablas Input-Output de la economía Española, año 1957," prepared by M. de Torres Martínez, et al.

8. Construction is perhaps the most important example of the physical combination of metal and nonmetal in production. Typically, metals are combined with mineral products (both inorganic

materials). Buildings are required to fulfill a particular mechanical function characterized by lack of motion.

9. For example, the large input, in Figure 24.3, of fishing to chemicals represents the outputs of the whaling industry, in which Norway has specialized. The familiar problems of comparability of input-output tables make the exact identification of the common technology perhaps impossible, but, as a rule of thumb, we can neglect small coefficients which do not recur in the different tables.

## Appendix

### *Methods of Triangulation*

If we wish to minimize the simple sum of the elements above the principal diagonal of any matrix, we can adopt the following procedure.

#### PRELIMINARY

In any ordering of industries, the elements on the principal diagonal always remain on the diagonal, hence they may be disregarded.

In any ordering, elements which occupy a symmetric position with respect to the principal diagonal always remain symmetric, i.e., the element symmetric to the $ij^{th}$ element is the $ji^{th}$, and if industry $i$ is moved to the position of $i'$, and $j$ to $j'$, then $a_{ij}$ becomes $a_{i'j'}$, and $a_{ji}$ becomes $a_{j'i'}$. Accordingly, whenever any element moves across the diagonal as a result of reordering, its symmetric element will cross the diagonal in the opposite direction, and we need consider only *differences* in the values of symmetric elements, i.e., if $a_{ij} > a_{ji}$ we may write 0 for $a_{ji}$ and write $(a_{ij} - a_{ji})$ for $a_{ij}$. Conversely, if $a_{ij} < a_{ji}$ we may write 0 for $a_{ij}$ and $(a_{ji} - a_{ij})$ for $a_{ji}$.

Thus, we can greatly reduce the number of nonzero elements and make our task much easier.

#### CHANGING THE ORDER OF INDUSTRIES

Suppose that we are considering moving industry $i$ downwards in the hierarchy of industries:

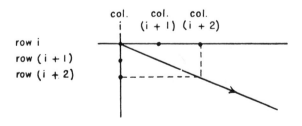

In that case, we should first compare the elements $a_{i(i+1)}$ and $a_{(i+1)i}$.

If $a_{i(i+1)} > a_{(i+1)i}$, i.e., if, in the terms of our preliminary procedure of treating only differences, $a_{i(i+1)} > 0$, then industry $i$ should be moved down to position $(i+1)$.

If $a_{i(i+1)} \leqslant a_{(i+1)i}$ we next compare the sum of the elements $a_{i(i+1)}$ and $a_{i(i+2)}$ with the sum of $a_{(i+1)i}$ and $a_{(i+2)i}$.

If $a_{i(i+1)} + a_{i(i+2)} > a_{(i+1)i} + a_{(i+2)i}$, then industry $i$ should be moved down to position $(i+2)$. On the other hand, if $a_{i(i+1)} + a_{i(i+2)} \leqslant a_{(i+1)i} + a_{(i+2)i}$, we next compare the sum of the three elements of row $i$ following $a_{ii}$ with the sum of the three elements of column $i$ which follow $a_{ii}$.

This procedure is repeated so that whenever

$$\sum_{k=1}^{r} a_{i(i+k)} > \sum_{k=1}^{r} a_{(i+k)i}$$

industry $i$ is moved down to position $(i+r)$. If this inequality is not satisfied, industry $i$ should not be moved downwards in the hierarchy of industries.

The criterion for moving any industry $i$ *upwards* in the hierarchy is analogous to that just described:

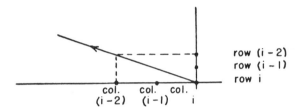

First, compare the elements $a_{i(i-1)}$ and $a_{(i-1)i}$.

If $a_{(i-1)i} > a_{i(i-1)}$, then industry $i$ should be moved up to position $(i-1)$.

If $a_{(i-1)i} \leqslant a_{i(i-1)}$, we next compare the sum of elements $a_{(i-1)i}$ and $a_{(i-2)i}$ with the sum of elements $a_{i(i-1)}$ and $a_{i(i-2)}$.

The procedure is repeated so that whenever

$$\sum_{k=1}^{r} a_{(i-k)i} > \sum_{k=1}^{r} a_{i(i-k)}$$

industry $i$ is moved upwards in the hierarchy to position $(i-r)$.


**AN ILLUSTRATION**

If the foregoing procedure is followed for each industry in the matrix, no further improvement can be made toward the objective of minimizing the simple sum of the above-diagonal elements by changing the position of any *one* industry. However, even if the criteria we have formulated are satisfied for each industry independently, improvement may still be possible by moving groups of industries, as the following example indicates.

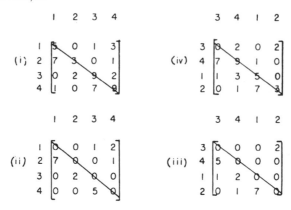

Applying our criteria to each industry in matrix (*i*), it is clear that triangularity cannot be improved by changing the position of any single industry. The same conclusion can be reached more quickly by considering matrix (*ii*), the matrix of differences in the values of symmetric elements of matrix (*i*). For reasons also discussed in the first section of this appendix, the diagonal elements have been disregarded.

But the example shows that when the positions of two industries are simultaneously changed, the sum of the above-diagonal elements may still be reduced. When industries 1 and 2 change places with 3 and 4, the sum of the above-diagonal elements is reduced from 4, in (*ii*) to 2 in (*iii*). Matrix (*iv*) likewise shows an improvement over (*i*).

### AN ALTERNATIVE OBJECTIVE

Another reason commonly advanced for triangulating a matrix is that the result can be used for approximations to numerical solutions. Let us define two matrices, $A_N$ and $A_0$, which are perfectly triangular, so that $A_N$ includes only zero terms above its principal diagonal, while $A_0$ includes only zero terms below its principal diagonal. Then their sum will be

$$A = A_N + A_0$$

If we are given a matrix, $A$, and an output vector, $X$, we can derive the vector $Y$ from the familiar relation

$$X = AX + Y$$

or

$$X = A_N X + A_0 X + Y$$

therefore,

$$Y = (I - A)X = (I - A_N)X - A_0 X$$

On the other hand, we may obtain an approximate solution for $Y$ by triangulating matrix $A$ and neglecting the above-diagonal elements

$$Y^* = (I - A_N)X$$

The error of this estimate is

$$\Delta Y = Y^* - Y = A_0 X$$

so that minimizing the error is equivalent to minimizing the value of the sum of the above-diagonal elements of the flow matrix. But this is a trivial case since, knowing $X$, we can compute the true value of $Y$ without resorting to matrix inversion. A more interesting case is that in which $X$ is unknown. Then the output associated with a given final demand, $Y$, will be

$$X = (I - A)^{-1}Y$$

which may be written

$$X = (I - A_N)^{-1}(A_0 X + Y)$$

Should we decide to estimate $X$ by triangulating matrix $A$ so that we neglect the elements of $A_0$, the estimate will be

$$X^* = (I - A_N)^{-1}Y$$

whereas the true value of $X$ is

$$X = (I + A_N + A_N^2 + A_N^3 + \cdots)(A_0 X + Y)$$

Thus the computational error will be

$$\Delta X = X - X^* = A_0 X + (A_N A_0 + A_N^2 A_0 + A_N^3 A_0 + \cdots)X$$

where $\Delta X$ is nonnegative, since $A_N$, $A_0$, and $X$ are all nonnegative. The sum of the elements of each succeeding vector grows progressively smaller. But since all the vectors are positive, the total error will be further diminished as each subsequent vector is taken into account. Minimizing only the sum of the elements of the first term, the vector $A_0 X$, implies minimizing the simple sum of the above-diagonal elements of the flow matrix. But when the second term is taken into account it is clear that the location of the above-diagonal elements is important. Hence, a reordering of industries which reduces the value of the sum of the elements of $A_0 X$ may be more than offset by an increase in the value of the sum of the elements of the second term. Of course, if $A_0$ is zero there will be no error of computation. But if a perfect triangulation is impossible, which is generally the case, the size of the error is a function not only of $A_0$ but also of $Y$, through $X$. This means that the formula for the "best" triangular order, in the above sense, depends in each case upon the particular bill-of-goods. In other words, there is no unique method of triangulation.

# 25

# Economic Interrelatedness

CHÍOU-SHUANG YAN and EDWARD AMES

This paper associates with every input-output matrix an *order matrix*, and defines for every submatrix of that order matrix an *interrelatedness function*. For certain classes of submatrices, the interrelatedness function has an economic interpretation. On closed input-output matrices, where irreducibility is of theoretical importance, irreducibility is associated with a condition on the interrelatedness function. On open input-output matrices, the function is a way of interpreting certain propositions found in economic history. Interrelatedness is also associated with diversification (and hence specialization) in a simple way. It is empirically possible (using Leontief's data for the U.S. economy in 1919 and 1929) to examine the way in which interrelatedness has changed over time, and to make predictions about hypotheses such as technological convergence, which have so far been difficult to subject to empirical analysis. Thus, the interrelatedness function is a useful measure in a number of economic contexts.

It is assumed that readers are familiar with Leontief's closed system, which consists of two matrix equations (actually identities):

$$\theta = (A - E)Q$$

$$\theta = (A_T - E)P$$

Here $\theta$ is a vector of zeros, $E$ is the identity matrix, $Q$ and $P$ are (equilibrium) vectors of quantities, and prices, respectively. The matrix $A$ has elements $a_{ij}$, referring to the quantity of commodity $i$ that is used to produce one unit of commodity $j$. $A_T$ is the transpose of $A$. These equations are said to be identities, in that all output is allocated to some industry (including producers' savings). Leontief's hypothesis is that the matrix $A$ is independent of $P$ and $Q$, and (for short periods of time) it is constant. The elements of $A$ are, in the nature of things, nonnegative.

1. If $A$ has nonnegative elements, then $A^2$, $A^3$, ... also have nonnegative elements. In particular, if we consider an arbitrary power $A^K$ of $A$, then the elements $a_{ij}^{(K)}$ of $A^K$ will be sums of products of elements of $A$. That is

$$a_{ij}^{(K)} = \sum a_{ir_1} a_{r_1 r_2} \cdots a_{r_K j}$$

where $\sum$ represents $(K)$-fold summation.

This article was published in *The Review of Economic Studies*, Volume 32, Number 4, October, 1965, pp. 290–310. © The Society for Economic Analysis Ltd. Reprinted by permission.

We shall define *an order matrix B* for any matrix $A$ having nonnegative elements. $B$ has the same number of rows and columns as $A$. We examine the sequence $a_{ij}$, $a_{ij}^{(2)}$, $a_{ij}^{(3)}$, ..., and find the first nonzero term in the sequence. If this term is the $K$th $(K = 1, 2, ...)$ then we set $b_{ij} = K$. If there is no finite $K$ for which $a_{ij}^{(K)} \neq 0$, then we set $b_{ij} = \infty$.

A simple example of an order matrix is obtained from the matrix

$$A_1 = \begin{bmatrix} 0 & 0 & 0 & 0 & a_{15} \\ 0 & 0 & 0 & 0 & a_{25} \\ 0 & 0 & 0 & 0 & a_{35} \\ 0 & 0 & 0 & 0 & a_{45} \\ a_{51} & a_{52} & a_{53} & a_{54} & a_{55} \end{bmatrix}$$

In this case, $A_1^2$ has strictly positive elements:

$$A_1^2 = \begin{bmatrix} a_{15}a_{51} \cdots a_{15}a_{55} \\ a_{25}a_{51} \cdots a_{25}a_{55} \\ \hline a_{55}a_{51} \cdots \sum_1^5 a_{5i}a_{i5} \end{bmatrix}$$

Consequently, the order matrix corresponding to $A_1$ is $B_1$:

$$B_1 = \begin{bmatrix} 2 & 2 & 2 & 2 & 1 \\ 2 & 2 & 2 & 2 & 1 \\ 2 & 2 & 2 & 2 & 1 \\ 2 & 2 & 2 & 2 & 1 \\ 1 & 1 & 1 & 1 & 1 \end{bmatrix}$$

Next, we define an *interrelatedness function*. Let $B\begin{pmatrix} i_1 \cdots i_r \\ j_1 \cdots j_s \end{pmatrix}$ be an arbitrary submatrix of an order matrix $B$, consisting of the elements of rows $i_1 \ldots i_r$ and columns $j_1 \ldots j_s$ of $B$. Then the interrelatedness function $R$ corresponding to this submatrix is defined by

$$R\begin{pmatrix} i_1 \cdots i_r \\ j_1 \cdots j_s \end{pmatrix} = \frac{1}{rs} \sum_{v=1}^{s} \sum_{w=1}^{s} \frac{1}{b_{i_v j_w}}$$

In particular, if $r = s = 1$, then $R\begin{pmatrix} i \\ j \end{pmatrix} = \frac{1}{b_{ij}}$. (It will be noted that the interrelatedness function is the reciprocal of the harmonic mean of the elements of the appropriate submatrix of $B$.) $R$ is thus uniquely defined for any submatrix of $B$, and hence for any matrix $A$ with nonnegative elements.

Observe that $0 \leqslant R \leqslant 1$; $R = 1$ if the corresponding elements of $A$ are strictly positive; and $R = 0$ if all the corresponding elements of $B$ are infinite. In the foregoing example, $R\begin{pmatrix} 12345 \\ 12345 \end{pmatrix} = 0.64$; for $i = 1, 2, 3, 4$, $R\begin{pmatrix} i \\ 12345 \end{pmatrix} = 0.6$ and $R\begin{pmatrix} 5 \\ 12345 \end{pmatrix} = 1.0$.

One further observation about the interrelatedness function is needed. If $n_1$, $n_2, \ldots$ denote the number of elements of a submatrix of the order matrix whose values are equal to $1, 2, \ldots$, then the interrelatedness function may be rewritten

$$R\begin{pmatrix} i_1 \ldots i_r \\ j_1 \ldots j_s \end{pmatrix} = \frac{1}{rs} \sum_{K=1}^{\infty} \frac{n_K}{K}$$

In particular, we may isolate the first term in this series, so that

$$R\begin{pmatrix} i_1 \ldots i_r \\ j_1 \ldots j_s \end{pmatrix} = \frac{n_1}{rs} + \sum_{K \neq 1} \frac{n_K}{Krs}$$

The first right-hand term measures the proportion of elements of the order submatrix which are equal to 1; that is, the proportion of the elements of the submatrix of $A$ which are nonzero. This term will be referred to as the *index of diversification*; the second term will be referred to as the *index of indirect relatedness.*[1]

The concept of *diversification* has economic content if the submatrix in question is $(1 \times n)$, or $(n \times 1)$, or $(n \times n)$. In the first case, the term $n_1/n$ represents the proportion of industries to which a given industry sells; in the second case, it represents the proportion of industries from which a given industry buys; in the third case, $n_1/n^2$ represents the proportion of the mathematically possible kinds of direct relations which are actually realized. Diversification is taken to be the opposite of specialization, in which an industry has few customers, or suppliers, and an economy is characterized mainly by indirect relations.

The matrix $B$ and the function $R$ depend not only on the number, but also the location of the nonzero elements of $A$. In the following example, $A_2$ differs from $A_1$ only in that the nonzero element in the first row is in the second, rather than the fifth column:

$$A_2 = \begin{bmatrix} 0 & a_{12} & 0 & 0 & 0 \\ 0 & 0 & 0 & 0 & a_{25} \\ 0 & 0 & 0 & 0 & a_{35} \\ 0 & 0 & 0 & 0 & a_{45} \\ a_{51} & a_{52} & a_{53} & a_{54} & a_{55} \end{bmatrix}$$

From this, we obtain

$$B_2 = \begin{bmatrix} 3 & 1 & 3 & 3 & 2 \\ 2 & 2 & 2 & 2 & 1 \\ 2 & 2 & 2 & 2 & 1 \\ 2 & 2 & 2 & 2 & 1 \\ 1 & 1 & 1 & 1 & 1 \end{bmatrix}$$

The interrelatedness of rows 2–5 is unchanged. That of row 1 is 0.5 rather than 0.6; that of the entire matrix is 0.66 rather than 0.68.

A matrix is said to be *reducible*, if it is possible to permute its rows and columns in such a way that $A$ has block triangular form:

$$A = \begin{pmatrix} A_{11} & 0 \\ A_{21} & A_{22} \end{pmatrix}$$

Clearly if such a form exists, $A$ contains a block of zeros which occupies some set of rows $i_1, \ldots, i_r$, and some set of columns $j_1 \ldots j_s$; these sets are a partition of the integers $1 \ldots n$; the numbers of the rows and columns have no elements in common; and $r + s = n$, the order of the matrix. The powers of $A$, moreover, are block-diagonal with the same form, so that the elements of the order matrix corresponding to $A_{12}$ are infinite. Hence the interrelatedness function corresponding to the zero block of $A$ equals zero. Conversely, if there exists an interrelatedness function equal to zero for a submatrix of $r$ rows and $s$ columns, where $r + s = n$, then it is possible to rearrange the rows and columns of $A$ into block-triangular form. Thus, $A$ is irreducible (cannot be put in block-triangular form by permuting rows and columns) if and only if $R$ is strictly positive for every submatrix of $A$ consisting of $r$ rows and $(n - r)$ columns (where $r = 1, \ldots, n - 1$).[2]

2. Irreducibility is an important theoretical property of closed Leontief systems, for it ensures that the price and output structure corresponding to a given input-output matrix is uniquely determined.

If $A$ is itself an irreducible matrix with nonnegative elements, a well-known theorem[3] states that it has a positive characteristic value, of modulus greater than any other such value; corresponding to this "maximal" characteristic value there is a unique (to within a scalar factor) characteristic vector with positive components. Moreover,[4] if both $A$ and its transpose have characteristic vectors with strictly positive components corresponding to this maximal characteristic value, then $A$ is either irreducible, or else may be written in the block diagonal form:

$$A = \begin{bmatrix} A_{11} & 0 & \cdots & 0 \\ 0 & A_{22} & \cdots & 0 \\ 0 & 0 & & \\ & & \ddots & \\ 0 & 0 & \cdots & A_{gg} \end{bmatrix}$$

where all the $A_{ii}$ are irreducible. The case where $g$ exceeds 1 is of no economic interest. It would mean that the input-output matrix $A$ referred to $g$ isolated economies, and this possibility may therefore be ignored. Now we consider the vectors $P$ and $Q$. If $P$ has zero components, the input-output matrix includes some free goods; if $Q$ has zero components, it includes goods which are not produced. In either case, the corresponding industries have no income.

In fact, this possibility is operationally meaningless, since data cannot be collected on industries with zero income. Effectively, then, the matrix $A$ is irreducible, provided that it can be shown that $P$ and $Q$ and the vectors corresponding to the maximal characteristic value (1 in this case) of $A$. Suppose that this were not the case. Then there would be some number $r > 1$ which was the maximal characteristic value of $A$, and hence of $A_T$. Assume that for some vector $x$, $A_Tx = rx$. Then the inner product of $Q$ and $x$ would be:

$$(Q, x) = (AQ, x) = (Q, A_Tx) = (Q, rx) = r(Q, x)$$

since $r > 1$, $(Q, x) = 0$. But $Q$ has strictly positive and $x$ nonnegative components. Therefore $(Q, x)$ cannot equal zero, and $r$ can only equal one. The closed Leontief matrix is irreducible, and there can be no submatrix of $A$, of order $r \times s$, where $r + s = n$, and where the rows and columns have no number in common, such that the corresponding interrelatedness function equals zero.

3. The interrelatedness function corresponding to $1 \times 1$ submatrices (i.e., single elements) of $A$ has an economic meaning. Suppose that $R\binom{i}{j} = K^{-1}$. If $K = 1$, then $a_{ij} \neq 0$, so that commodity $i$ is used directly in the production of commodity $j$. If $K = 2$, then there exists at least one $r$ such that $a_{ir}a_{rj} \neq 0$, even though $a_{ij} = 0$. That is, $i$ is used to produce $r$, and $r$ is used to produce $j$, but $i$ is not used directly in the production of $j$. That is, $i$ is *indirectly* used in the production of $j$. Thus, the smaller is $R\binom{i}{j}$ the less directly is $i$ used in the production of $j$—the less closely related are $i$ and $j$.

If $A$ is irreducible, there is nonzero interrelatedness between every pair of industries. That is, every element of the order matrix $B$ is finite. To see that this statement is true, we consider a vector $V$ with nonnegative components, and arrange these components in such a way that the first $r$ are zeros, and the last $(n - r)$ are strictly positive. We simultaneously rearrange the rows and columns of the irreducible matrix $A$, and rewrite $A$ in block form: Then we consider the vector:

$$\begin{pmatrix} A_{11} & A_{12} \\ A_{21} & A_{22} \end{pmatrix}\begin{pmatrix} 0 \\ V_2 \end{pmatrix} = \begin{pmatrix} w_1 \\ w_2 \end{pmatrix}$$

The elements of $w_2$ are positive, since $V_2$ consists of strictly positive and nonnegative elements. At least one element of $w_1$ must also be strictly positive, since otherwise $A_{12}$ would equal zero, and $A$ would be reducible, contrary to hypothesis. Consequently, the vector $w$ contains strictly fewer zero elements than the vector $V$. As a special application of this result[5] we observe that each column of $A^{K+1}$ contains strictly few zero elements than the corresponding column of $A^K$. Since $A$ is of finite order, all elements of its order matrix $B$ must be finite, as was to be shown.

If $A$ is irreducible, at least one nondiagonal element of each column of $A$ must be nonzero. By the preceding paragraph, therefore, the largest number in the order matrix of an irreducible matrix is $(n)$. The interrelatedness functions $R\binom{i}{j}$ corresponding to individual elements of an irreducible $A$, then, are the reciprocals of integers less than or equal to $(n)$; and any row or column of $B$ having $m_1$ as its maximum element must also contain every integer less than $m_1$.

In a closed input-output matrix of even moderate order—say the $42 \times 42$ matrices used by Leontief in his pioneer work, *The Structure of the American Economy*, it would be a major task to calculate the order matrix, if in fact it were necessary to compute 40 powers of the original input-output matrix. However, it turned out that the 1919 and 1929 matrices in his book were essentially like the matrix $A_1$ shown above. That is, every industry used the services of households, and every industry sold to households. In addition, of course, these were interindustrial sales, so that there were nonzero elements scattered throughout $A$. The study of interrelatedness in closed input-output matrices, thus, does not seem very interesting, but we shall find a more interesting application of the concepts of interrelatedness.

4. We shall now discuss *technological* interrelatedness, rather than interrelatedness in general. The discussion relates to "open" matrices rather than the "closed" matrices so far discussed. The open input-output matrix is a submatrix of the closed matrix. Certain of the "industries" (in particular, the household and government sectors) are deleted on the grounds that they are "final demand"; the relations remaining are described in the literature as "inter-industrial." The matrix of the open system is described in this paper as the "technological matrix." In contrast to the matrix of the closed system, the technological matrix need not be irreducible in theory; in fact the matrices under study are reducible.[6] For any given technological matrix, we may construct an order matrix. This order matrix is not a submatrix of the order matrix of the closed input-output matrix, but must be separately computed.

The 1919 technological matrix developed by Leontief turned out (after some permutations of rows and columns) to be of the block form:

$$A_3 = \begin{pmatrix} 0 & 0 & 0 \\ 0 & A_{22} & 0 \\ A_{31} & A_{32} & A_{33} \end{pmatrix}$$

The industries in the top row of blocks:

canning and preserving,
bread and bakery products,
tobacco manufactures,
other paper products, and
construction,

sold only to final consumers. Those in the second row of blocks,

liquors and beverages,
automobiles,
clothing, leather shoes,
butter, cheese, etc., and
other food industries,

in 1919 sold only to themselves and to final consumers ($A_{22}$ had only diagonal elements); in 1929 $A_{22}$ had a few nonzero nondiagonal elements. It may easily be verified that every finite power of a matrix of the form of $A_3$ is a matrix of the same form (as regards zero and nonzero blocks) as $A_3$. Moreover, $A_3$ itself is clearly reducible. Technological interrelatedness does not exist between certain pairs of industries, and hence certain industries show zero interrelatedness. However, the industries in the third block exhibited nonzero technological interrelatedness. The corresponding part of the order matrix contained no term of order exceeding three. Thus, either a pair of industries turned out to be technologically unrelated or they turned out to have interrelatedness of $1/3$, $1/2$, or $1$. Interrelatedness of order three between industries $i$ and $j$ corresponds to the following situation: industry $i$ has a customer, industry $r$, whose customer industry $s$ sells directly to industry $j$, so that $a_{ir}a_{rs}a_{sj}$ is positive. None of the industries $i, r, s, j$, of course is a final output or final demand industry.

These results rule out certain hypotheses about vertical integration in the economy. A common form of these hypotheses is that industries may be grouped into "stages" representing well-defined production and distribution processes. For instance, (1) extractive industries sell to (2) basic processing industries, which sell to (3) fabricating industries, which sell to (4) distributors, which sell to (5) ultimate consumers. If this hypothesis were valid, it would be possible to arrange the closed input-output matrix in the form:

$$A_4 = \begin{bmatrix} 0 & A_{12} & 0 & 0 & 0 \\ 0 & 0 & A_{23} & 0 & 0 \\ 0 & 0 & 0 & A_{34} & 0 \\ 0 & 0 & 0 & 0 & A_{45} \\ A_{51} & A_{52} & A_{53} & A_{54} & A_{55} \end{bmatrix}$$

The order matrix corresponding to this closed system is given by $B_4$:

$$B_4 = \begin{bmatrix} 5 & 1 & 2 & 3 & 4 \\ 4 & 4 & 1 & 2 & 3 \\ 3 & 3 & 3 & 1 & 2 \\ 2 & 2 & 2 & 2 & 1 \\ 1 & 1 & 1 & 1 & 1 \end{bmatrix}$$

The technological matrix corresponding to $A_4$ is obtained by deleting the last row and the last column. Corresponding to this matrix is the order matrix

$$B_4' = \begin{bmatrix} \infty & 1 & 2 & 3 \\ \infty & \infty & 1 & 2 \\ \infty & \infty & \infty & 1 \\ \infty & \infty & \infty & \infty \end{bmatrix}$$

The fact that empirical technological matrices are block triangular is roughly consistent with the existence of an order matrix such as $B_4'$. But $B_4$, the corresponding closed order matrix, contains terms of order 3, 4, and 5, whereas

Table 25.1    Technological Interrelatedness in the U.S. Economy, 1919 and 1929

| Order of Interrelatedness | Number of elements of technological order matrix 1919 | 1929 |
|:---:|:---:|:---:|
| 1 | 358 | 521 |
| 2 | 563 | 600 |
| 3 | 317 | 241 |
| $\infty$ | 443 | 319 |
| Total | 1681 | 1681 |
| $R$ | 0.444 | 0.536 |

empirical order matrices contain only terms of order 1 and 2. This circumstance is enough to permit rejection of the hypothesis.

This result permits a comment on the "vertical disintegration hypothesis." Young and others[7] have said that as industries have grown, production processes formerly carried on within a single firm tend to be divided among several firms, so that several subindustries tend to take the place of one industry. Our data, of course, use a fixed set of industrial classifications, so that they would necessarily be incapable of showing this differentiation process. The disintegration process would have the effect of increasing the importance, over time, of the diagonal elements of the input-output matrix—the sales by each industry to itself. The fact that there is little evidence of vertical chains of the type exemplified by the matrix $A_4$, however, does show that vertical disintegration, if it occurs, leads to the formation of subindustries which are sufficiently similar to be aggregated together, when the economy is presented as an input-output model of moderate order (Leontief's data refer to 41 industries other than final output and consumption). It is not clear, of course, what would be true if the number of industries considered separately was substantially increased.[8]

5. We may now consider the empirical connections between diversification and interrelatedness. This connection may be viewed either in terms of the economy as a whole or of particular industries. Table 25.1 gives a frequency distribution of the elements of the 1919 and 1929 technological order matrices, and the values of the interrelatedness function for these matrices. Interrelatedness increased from 0.444 to 0.536 in the decade. The increase in diversification was equal to 0.097. That is, it exceeded the increase in interrelatedness, because there was a small decrease (0.005) in indirect relatedness in the economy.

If individual industries are grouped according to their interrelatedness characteristics (Table 25.2), it may be observed that on balance, indirect relatedness appears to have been distributed among industries in 1929 in much the same way as it had been in 1919, although diversification had clearly increased. This circumstance is partly the result of the nature of indirect relatedness: an industry may have very little indirect relatedness either because it has no relatedness of any kind with other industries; or because it has direct relatedness with almost all industries (is almost completely diversified). On the average, the index of indirect

Table 25.2  Number of Industries Given Interrelatedness, Diversification, and Indirect Relatedness, 1919 and 1929

| Index value | 1919 | | | 1929 | | |
|---|---|---|---|---|---|---|
| | Inter-relatedness | Diversi-fication | Indirect relatedness | Inter-relatedness | Diversi-fication | Indirect relatedness |
| 0.00–0.09 | 11 | 18 | 14 | 8 | 15 | 12 |
| 0.10–0.19 | | 7 | 2 | | 4 | 5 |
| 0.20–0.29 | | 4 | 2 | | 7 | 3 |
| 0.30–0.39 | 1 | 6 | 17 | 1 | 4 | 16 |
| 0.40–0.49 | 8 | 2 | 6 | 7 | 1 | 5 |
| 0.50–0.59 | 9 | | | 8 | 1 | |
| 0.60–0.69 | 6 | | | 6 | 1 | |
| 0.70–0.79 | 2 | 2 | | 2 | 2 | |
| 0.80–0.89 | 2 | 2 | | 3 | 3 | |
| 0.90–0.99 | 2 | | | 6 | 3 | |

relatedness is small in industries where diversification is either very small (under 0.1) or very large (above 0.7).

Only one industry—iron mining—showed decreased interrelatedness from 1919 to 1929. In 1919 iron mining had sold to "blast furnaces" and to "steel works and rolling mills," in 1929 it sold only to the former. In half the industries (20) changes in both diversification and indirect relatedness were small (less than 0.05 in either direction). But in one-fifth of the industries (eight in number) increases in diversification were greater than 0.10 and were accompanied by declines in indirect relatedness. Thus, the changes observed in Table 25.1 appear mainly to be associated with increased diversification in a relatively small number of industries.

An increase in diversification in any one industry necessarily increases interrelatedness in other industries as well. To illustrate the consequences of such an increase, we present the matrix $A_5$ with order matrix $B_5$:

$$A_5 = \begin{bmatrix} 0 & a_{12} & 0 & 0 & 0 \\ 0 & 0 & 0 & a_{24} & 0 \\ 0 & 0 & 0 & 0 & a_{35} \\ a_{41} & a_{42} & a_{43} & a_{44} & a_{45} \\ a_{51} & 0 & 0 & 0 & 0 \end{bmatrix}$$

$$B_5 = \begin{bmatrix} 3 & 1 & 3 & 2 & 3 \\ 2 & 2 & 2 & 1 & 2 \\ 2 & 3 & 5 & 4 & 1 \\ 1 & 1 & 1 & 1 & 1 \\ 1 & 2 & 4 & 3 & 4 \end{bmatrix}$$

Suppose, however, that a single element (say $a_{41}$) is set equal to zero in $A_5$. Then we

Table 25.3    Output Interrelatedness in Individual U.S. Industries, 1919

| 0–0.25 | | 0.26–0.50 | | 0.51–0.75 | | 0.76–1.00 | |
|---|---|---|---|---|---|---|---|
| Canning and preserving | 0 | Flour and grist mill products | 0.41 | Agriculture | 0.63 | Other iron and steel and electrical mfg. | 0.89 |
| Bread and bakery products | 0 | Iron mining | 0.41 | Sugar, glucose and starch | 0.52 | Refined petroleum | 0.88 |
| Tobacco mfg. | 0 | Blast furnaces | 0.45 | Slaughtering and meat packing | 0.54 | Coal | 0.95 |
| Other paper products | 0 | Nonferrous metal mining | 0.39 | Steel works and rolling mills | 0.66 | Electric utilities | 0.95 |
| Liquors and beverages | 0.024 | Coke | 0.45 | Smelting and refining | 0.57 | | |
| Butter, cheese, etc. | 0.049 | Paper and wood pulp | 0.40 | Brass, bronze, copper, etc., manufacturing | 0.57 | | |
| Automobile | 0.024 | Printing and publishing | 0.43 | Nonmetal minerals | 0.71 | | |
| Clothing | 0.024 | Other textile products | 0.48 | Petroleum and natural gas | 0.60 | | |
| Leather shoes | 0.024 | Other leather products | 0.40 | Manuf. gas | 0.67 | | |
| Other food industries | 0.049 | Rubber mfg. | 0.50 | Chemical | 0.69 | | |
| | | | | Lumber and timber products | 0.68 | | |
| | | | | Other wood products | 0.52 | | |
| | | | | Yarn and cloth | 0.55 | | |
| | | | | Leather tanning | 0.545 | | |
| | | | | Industries, n.e.s. | 0.55 | | |
| | | | | Transportation | 0.72 | | |

obtain a new matrix $\bar{A}_5$. The corresponding order matrix is

$$\bar{B}_5 = \begin{bmatrix} 4 & 1 & 3 & 2 & 3 \\ \bar{3} & 2 & 2 & 1 & 2 \\ \bar{2} & 3 & 5 & 4 & 1 \\ 2 & 1 & 1 & 1 & 1 \\ \bar{1} & 2 & 4 & 3 & 4 \end{bmatrix}$$

In this representation the three underscored numbers are one greater than in the corresponding matrix for $A_5$. The interrelatedness functions of industries 1, 2, and 4 is, thus, less for these industries if $\bar{A}_5$ holds than if $A_5$ holds. Finally, the average interrelatedness of the economy is 0.57 if $\bar{A}_5$ holds and 0.61 if $A_5$ holds.

Table 25.4   Output Interrelatedness in Individual U.S. Industries, 1929

| 0–0.25 | | 0.26–0.50 | | 0.51–0.75 | | 0.76–1.00 | |
|---|---|---|---|---|---|---|---|
| Canning and preserving | 0 | Flour and grist mill products | 0.43 | Agriculture | 0.68 | Other iron and steel and electric mfg | 0.95 |
| Bread and bakery products | 0 | Butter, cheese, etc. | 0.41 | Sugar, glucose, and starch | 0.54 | Petroleum and natural gas | 0.93 |
| Tobacco mfg. | 0 | Iron mining | 0.37 | Slaughtering and meat packing | 0.57 | Refined petroleum | 0.99 |
| Clothing | 0.024 | Nonferrous metal mining | 0.48 | Blast furnaces | 0.54 | Coal | 0.96 |
| Liquors and beverages | 0.06 | Other paper products | 0.42 | Steel works and rolling mills | 0.67 | Coke | 0.90 |
| Automobiles | 0.024 | Printing and publishing | 0.49 | Smelting and refining | 0.63 | Mfg. gas | 0.86 |
| Leather shoes | 0.024 | Other textile products | 0.48 | Brass, bronze, copper, etc., mfg. | 0.63 | Electric utilities | 0.96 |
| Other food industries | 0.07 | Other leather products | 0.43 | Nonmetal minerals | 0.73 | Chemical | 0.78 |
| | | | | Lumber and timber products | 0.69 | Construction | 0.84 |
| | | | | Other wood products | 0.54 | Transportation | 0.88 |
| | | | | Paper and wood pulp | 0.56 | | |
| | | | | Yarn and cloth | 0.68 | | |
| | | | | Leather tanning | 0.56 | | |
| | | | | Rubber mfg. | 0.59 | | |
| | | | | Industries, n.e.s. | 0.59 | | |

It is natural, therefore, to look at those industries which exhibit the greatest interrelatedness, for these are the industries which produce indirect relatedness in other industries. Tables 25.3 and 25.4 classify industries on the basis of their interrelatedness in 1919 and 1929. In both years, fuel, electric utilities, and "other iron and steel" are the most interrelated (and also the most highly diversified).

This last observation places in perspective Rosenberg's theory of technological convergence.[9] Rosenberg has urged that in the nineteenth century, the U.S. machine tool industry was a particularly important instrument in the diffusion of technological change throughout the entire economy; metal-working plants underwent a clearly marked sequence of changes in product, and the new processes required at each step in the sequence turned out to have wide applications in the manufacture of other products. Rosenberg, therefore, argues that where technology is "convergent" upon many industries, changes tend to be diffused rapidly and efficiently throughout the economy. On the other hand, where technology is "not convergent," a technological change tends to have only one or a few applications.

We should be inclined to predict that the greater is the interrelatedness of an industry, the greater would its technological convergence be. In the 1920s, the fuel and power industries showed higher interrelatedness than "other iron and steel and electrical manufactures" (which includes machine tools). This observation does not invalidate Rosenberg's assertion, which relates to an earlier period (1840–1910). It does give a more precise statement of the conditions which may be associated with technological convergence.

In conclusion, then, we may state that the concepts of diversification (and its converse, specialization) and interrelatedness may be given a numerical formulation; and that input-output data provide the basis for testing empirically a variety of assertions made about certain tendencies in the economic history of industrializing economies. In presenting this measure, we do not claim to have a theory in the sense of a unique prediction of the amount of interrelatedness (or diversification) to be associated with a particular level of development. It is nevertheless useful to have made the step of reducing to a more precise form or number of hitherto regrettably imprecise statements about economic development processes.

**NOTES**

1. This terminology is suggested by a reading of Harvey Leibenstein, *Economic Backwardness and Economic Growth* (New York: Wiley, 1957), Chapter 7; *Economic Theory and Organizational Analysis* (New York: Harpers, 1960), Chapters 7–8; and Edward Ames and Nathan Rosenberg, *The Progressive Division and Specialization of Industries*, Institute for Quantitative Research in Economics and Management, Purdue University, Lafayette, Indiana, Institute Paper No. 39, April, 1963. But these papers do not point out the numerical connection between interrelatedness and diversification.
2. It is also easy to show that if $R$ is strictly positive for a given submatrix, it is also positive for all submatrices which include this submatrix.
3. F. R. Gantmacher, *The Theory of Matrices*, New York: Chelsea, 1959, Vol. 2, p. 53.
4. *Ibid.*, Vol. 2, pp. 78–79.
5. Gantmacher, *op. cit.*, Vol. 2, pp. 51–52.
6. Morishima, M. and Kaneko, Y., "On the Speed of Establishing Multi-Sectoral Equilibrium," *Econometrica*, Vol. 30, 1962, p. 819, assume that the technological matrix is irreducible. This is not the case.
7. Allyn Young, "Increasing Returns and Economic Progress," *Economic Journal*, December, 1928; George Stigler, "The Division of Labor is Limited by the Extent of the Market," *Journal of Political Economy*, June, 1951; Edward Ames and Nathan Rosenberg, *loc. cit.*
8. Our findings are not independent of the amount and kind of aggregation used by Leontief. When industries are combined (as they must always be in empirical work) to form an input-output matrix of manageable size, a matrix $A$ of order $n$ is replaced by a matrix $\bar{A}$ of smaller order $s$. Each element of $\bar{A}$ corresponds to a block of $A$, and it will be nonzero if any element of the block differs from zero. A power of $\bar{A}$, say $\bar{A}^q$, has elements $a_{ij}^{(q)}$. These will be nonzero if it is possible to find any product $a_{ik}a_{kj}^{(q-1)}$ not equal to zero ($k = 1, \ldots, s$). The corresponding portion of the matrix $A^q$ is a sum of products of blocks $A_{ik}^{(q-1)}A_{kj}$. Even if both blocks contain a nonzero element, however, it is possible for the product to be equal to zero. Hence the $q$th power of $\bar{A}$ may have fewer zero elements than the $q$th power of $A$ has zero blocks; in any event it cannot have more. Consequently, the more aggregation is carried out upon $A$ (the smaller $s$ is, relative to $n$), the lower one should expect the calculated order of interrelatedness of the matrix to be; that is, the more closely interrelated the economy will appear to be. For lack of information about the actual importance of this fact, we cannot discuss it further.
9. Nathan Rosenberg, "Technological Change in the Machine Tool Industry, 1840–1910," *Journal of Economic History*, XXIII, 1963, and "Capital Good Technology and Economic Growth," Oxford Economics Papers, 1963.

APPENDIX—Table 25.A.1 The Technological Order Matrix for the U.S. Economy, 1919

| Selling industries \ Buying industries | 1. Agriculture | 2. Flour and grist mill products | 3. Canning and preserving | 4. Bread and bakery products | 5. Sugar, glucose, and starch | 6. Liquors and beverages | 7. Tobacco manufactures | 8. Slaughtering and meat packing | 9. Butter, cheese, etc. | 10. Other food industries | 11. Iron mining | 12. Blast furnaces | 13. Steel works and rolling mills | 14. Other iron and steel and electric manufactures | 15. Automobiles | 16. Nonferrous metal mining | 17. Smelting and refining | 18. Brass, bronze, copper, etc., manufactures | 19. Nonmetal minerals | 20. Petroleum and natural gas | 21. Refined petroleum | 22. Coal | 23. Coke | 24. Manufactured gas | 25. Electric utilities | 26. Chemical | 27. Lumber and timber products | 28. Other wood products | 29. Paper and wood pulp | 30. Other paper products | 31. Printing and publishing | 32. Yarn and cloth | 33. Clothing | 34. Other textile products | 35. Leather tanning | 36. Leather shoes | 37. Other leather products | 38. Rubber manufactures | 39. Industries n.e.s. | 40. Construction | 41. Transportation (steam railroads) |
|---|---|---|---|---|---|---|---|---|---|---|---|---|---|---|---|---|---|---|---|---|---|---|---|---|---|---|---|---|---|---|---|---|---|---|---|---|---|---|---|---|---|
| 1. Agriculture | 1 | 1 | 1 | 1 | 1 | 1 | 1 | 1 | 1 | 1 | 2 | 2 | 3 | 3 | 2 | 2 | 3 | 3 | 2 | 3 | 2 | 2 | 3 | 3 | 3 | 1 | 3 | 2 | 1 | 2 | 2 | 1 | 2 | 2 | 1 | 2 | 2 | 2 | 2 | 2 | 3 |
| 2. Flour and grist mill products | 1 | 2 | 2 | 1 | 2 | 2 | 2 | 2 | 2 | 2 | 3 | 3 | 3 | 3 | 3 | 3 | 3 | 3 | 3 | 3 | 3 | 3 | 3 | 3 | 3 | 2 | 3 | 3 | 2 | 3 | 3 | 2 | 3 | 3 | 2 | 3 | 3 | 3 | 3 | 3 | 3 |
| 3. Canning and preserving | ∞ | ∞ | ∞ | ∞ | ∞ | ∞ | ∞ | ∞ | ∞ | ∞ | ∞ | ∞ | ∞ | ∞ | ∞ | ∞ | ∞ | ∞ | ∞ | ∞ | ∞ | ∞ | ∞ | ∞ | ∞ | ∞ | ∞ | ∞ | ∞ | ∞ | ∞ | ∞ | ∞ | ∞ | ∞ | ∞ | ∞ | ∞ | ∞ | ∞ | ∞ |
| 4. Bread and bakery products | ∞ | ∞ | ∞ | ∞ | ∞ | ∞ | ∞ | ∞ | ∞ | ∞ | ∞ | ∞ | ∞ | ∞ | ∞ | ∞ | ∞ | ∞ | ∞ | ∞ | ∞ | ∞ | ∞ | ∞ | ∞ | ∞ | ∞ | ∞ | ∞ | ∞ | ∞ | ∞ | ∞ | ∞ | ∞ | ∞ | ∞ | ∞ | ∞ | ∞ | ∞ |
| 5. Sugar, glucose, and starch | 1 | 2 | 1 | 1 | 1 | 1 | 2 | 2 | 1 | 1 | 3 | 3 | 3 | 3 | 2 | 3 | 3 | 3 | 3 | 3 | 3 | 3 | 3 | 3 | 3 | 2 | 3 | 2 | 2 | 3 | 3 | 1 | 2 | 2 | 2 | 2 | 2 | 2 | 2 | 3 | 3 |
| 6. Liquors and beverages | ∞ | ∞ | ∞ | ∞ | ∞ | 1 | ∞ | ∞ | ∞ | ∞ | ∞ | ∞ | ∞ | ∞ | ∞ | ∞ | ∞ | ∞ | ∞ | ∞ | ∞ | ∞ | ∞ | ∞ | ∞ | ∞ | ∞ | ∞ | ∞ | ∞ | ∞ | ∞ | ∞ | ∞ | ∞ | ∞ | ∞ | ∞ | ∞ | ∞ | ∞ |
| 7. Tobacco manufactures | ∞ | ∞ | ∞ | ∞ | ∞ | ∞ | ∞ | ∞ | ∞ | ∞ | ∞ | ∞ | ∞ | ∞ | ∞ | ∞ | ∞ | ∞ | ∞ | ∞ | ∞ | ∞ | ∞ | ∞ | ∞ | ∞ | ∞ | ∞ | ∞ | ∞ | ∞ | ∞ | ∞ | ∞ | ∞ | ∞ | ∞ | ∞ | ∞ | ∞ | ∞ |
| 8. Slaughtering and meat packing | 1 | 2 | 2 | 1 | 2 | 2 | 2 | 1 | 1 | 2 | 2 | 2 | 3 | 3 | 2 | 2 | 3 | 3 | 2 | 3 | 2 | 2 | 3 | 3 | 3 | 1 | 3 | 2 | 2 | 3 | 2 | 1 | 2 | 2 | 1 | 2 | 2 | 2 | 2 | 2 | 3 |
| 9. Butter, cheese, etc. | ∞ | ∞ | ∞ | 1 | ∞ | ∞ | ∞ | ∞ | ∞ | 1 | ∞ | ∞ | ∞ | ∞ | ∞ | ∞ | ∞ | ∞ | ∞ | ∞ | ∞ | ∞ | ∞ | ∞ | ∞ | ∞ | ∞ | ∞ | ∞ | ∞ | ∞ | ∞ | ∞ | ∞ | ∞ | ∞ | ∞ | ∞ | ∞ | ∞ | ∞ |
| 10. Other food industries | ∞ | ∞ | ∞ | 1 | ∞ | ∞ | ∞ | ∞ | ∞ | 1 | ∞ | ∞ | ∞ | ∞ | ∞ | ∞ | ∞ | ∞ | ∞ | ∞ | ∞ | ∞ | ∞ | ∞ | ∞ | ∞ | ∞ | ∞ | ∞ | ∞ | ∞ | ∞ | ∞ | ∞ | ∞ | ∞ | ∞ | ∞ | ∞ | ∞ | ∞ |
| 11. Iron mining | 2 | 3 | 3 | 3 | 3 | 3 | 3 | 3 | 3 | 3 | 2 | 1 | 1 | 2 | 2 | 2 | 3 | 3 | 2 | 2 | 3 | 2 | 3 | 3 | 3 | 3 | 3 | 3 | 3 | 3 | 3 | 3 | 3 | 3 | 3 | 3 | 3 | 2 | 3 | 2 | 2 |
| 12. Blast furnaces | 2 | 2 | 2 | 2 | 2 | 2 | 3 | 3 | 2 | 2 | 2 | 2 | 1 | 2 | 1 | 2 | 3 | 3 | 2 | 1 | 3 | 2 | 3 | 3 | 3 | 3 | 3 | 3 | 3 | 3 | 3 | 3 | 3 | 3 | 3 | 3 | 3 | 2 | 3 | 1 | 2 |
| 13. Steel works and rolling mills | 1 | 2 | 2 | 2 | 2 | 2 | 2 | 2 | 2 | 2 | 1 | 1 | 1 | 1 | 1 | 1 | 2 | 2 | 1 | 1 | 2 | 1 | 2 | 2 | 2 | 2 | 2 | 2 | 2 | 2 | 2 | 2 | 2 | 2 | 2 | 2 | 2 | 1 | 2 | 1 | 1 |
| 14. Other iron and steel and electric manufactures | 1 | 1 | 1 | 1 | 1 | 1 | 2 | 2 | 1 | 1 | 1 | 2 | 1 | 1 | 1 | 1 | 2 | 2 | 1 | 1 | 1 | 1 | 2 | 2 | 1 | 1 | 1 | 1 | 1 | 2 | 1 | 1 | 1 | 1 | 1 | 1 | 1 | 1 | 2 | 1 | 1 |
| 15. Automobiles | ∞ | ∞ | ∞ | ∞ | ∞ | ∞ | ∞ | ∞ | ∞ | ∞ | ∞ | ∞ | ∞ | ∞ | 1 | ∞ | ∞ | ∞ | ∞ | ∞ | ∞ | ∞ | ∞ | ∞ | ∞ | ∞ | ∞ | ∞ | ∞ | ∞ | ∞ | ∞ | ∞ | ∞ | ∞ | ∞ | ∞ | ∞ | ∞ | ∞ | ∞ |
| 16. Nonferrous metal mining | 3 | 3 | 3 | 3 | 3 | 3 | 3 | 3 | 3 | 3 | 3 | 3 | 2 | 2 | 2 | 1 | 1 | 2 | 3 | 3 | 3 | 3 | 3 | 3 | 3 | 2 | 3 | 3 | 3 | 3 | 3 | 3 | 3 | 3 | 3 | 3 | 3 | 3 | 3 | 2 | 3 |
| 17. Smelting and refining | 2 | 2 | 2 | 2 | 2 | 2 | 2 | 2 | 2 | 2 | 2 | 2 | 1 | 1 | 1 | 2 | 1 | 1 | 2 | 2 | 2 | 2 | 3 | 3 | 2 | 1 | 2 | 2 | 2 | 3 | 2 | 2 | 2 | 2 | 2 | 2 | 2 | 2 | 3 | 1 | 2 |
| 18. Brass, bronze, copper, etc., manufactures | 2 | 2 | 2 | 2 | 2 | 2 | 1 | 2 | 2 | 1 | 2 | 2 | 2 | 1 | 1 | 2 | 2 | 2 | 2 | 2 | 2 | 2 | 2 | 2 | 1 | 2 | 2 | 2 | 2 | 2 | 2 | 2 | 2 | 2 | 2 | 2 | 2 | 2 | 2 | 1 | 2 |
| 19. Nonmetal minerals | 1 | 2 | 1 | 2 | 1 | 1 | 2 | 2 | 2 | 2 | 2 | 1 | 1 | 1 | 1 | 2 | 1 | 2 | 1 | 2 | 1 | 2 | 2 | 2 | 1 | 1 | 2 | 2 | 1 | 2 | 2 | 2 | 2 | 2 | 1 | 2 | 2 | 2 | 2 | 1 | 1 |
| 20. Petroleum and natural gas | 2 | 2 | 2 | 2 | 2 | 2 | 3 | 2 | 2 | 2 | 3 | 2 | 1 | 2 | 1 | 1 | 2 | 2 | 1 | 1 | 1 | 2 | 2 | 1 | 2 | 1 | 2 | 2 | 2 | 3 | 2 | 2 | 2 | 3 | 2 | 3 | 3 | 2 | 1 | 2 | 1 |
| 21. Refined petroleum | 1 | 1 | 1 | 1 | 1 | 1 | 2 | 1 | 1 | 1 | 2 | 2 | 1 | 1 | 1 | 1 | 1 | 1 | 1 | 1 | 1 | 2 | 1 | 1 | 1 | 1 | 1 | 1 | 1 | 2 | 1 | 1 | 2 | 2 | 2 | 2 | 2 | 1 | 1 | 1 | 1 |
| 22. Coal | 2 | 1 | 1 | 1 | 1 | 1 | 1 | 1 | 1 | 1 | 1 | 1 | 1 | 1 | 1 | 1 | 1 | 1 | 1 | 2 | 1 | 1 | 1 | 1 | 1 | 1 | 1 | 1 | 1 | 1 | 1 | 1 | 1 | 1 | 1 | 1 | 2 | 1 | 1 | 2 | 1 |
| 23. Coke | 2 | 2 | 2 | 1 | 1 | 2 | 2 | 2 | 2 | 2 | 2 | 1 | 1 | 1 | 1 | 2 | 1 | 1 | 1 | 2 | 1 | 2 | 1 | 1 | 2 | 1 | 2 | 2 | 1 | 2 | 2 | 1 | 2 | 2 | 2 | 2 | 2 | 2 | 1 | 2 | 2 |
| 24. Manufactured gas | 2 | 2 | 2 | 1 | 2 | 2 | 2 | 2 | 2 | 1 | 2 | 2 | 1 | 1 | 1 | 2 | 1 | 2 | 1 | 1 | 1 | 2 | 2 | 1 | 1 | 1 | 2 | 2 | 1 | 1 | 1 | 2 | 2 | 2 | 2 | 1 | 2 | 2 | 2 | 2 | 2 |
| 25. Electric utilities | 2 | 1 | 1 | 1 | 2 | 1 | 1 | 1 | 1 | 1 | 1 | 1 | 1 | 1 | 1 | 1 | 1 | 1 | 1 | 1 | 1 | 1 | 1 | 1 | 1 | 1 | 1 | 1 | 1 | 1 | 1 | 1 | 1 | 1 | 1 | 1 | 1 | 1 | 1 | 2 | 2 |
| 26. Chemical | 1 | 2 | 2 | 2 | 2 | 2 | 2 | 1 | 2 | 2 | 1 | 1 | 2 | 2 | 1 | 1 | 2 | 2 | 1 | 2 | 1 | 1 | 2 | 2 | 2 | 1 | 2 | 2 | 1 | 2 | 1 | 1 | 2 | 2 | 1 | 2 | 2 | 1 | 2 | 1 | 2 |
| 27. Lumber and timber products | 1 | 2 | 2 | 2 | 2 | 2 | 2 | 2 | 2 | 2 | 1 | 2 | 2 | 1 | 1 | 1 | 2 | 2 | 2 | 2 | 1 | 1 | 2 | 2 | 1 | 1 | 1 | 1 | 2 | 2 | 2 | 2 | 2 | 2 | 2 | 1 | 2 | 2 | 1 | 1 | 1 |
| 28. Other wood products | 2 | 2 | 2 | 2 | 2 | 2 | 1 | 3 | 2 | 2 | 2 | 3 | 2 | 1 | 2 | 2 | 3 | 3 | 2 | 2 | 2 | 2 | 3 | 2 | 2 | 2 | 2 | 1 | 2 | 3 | 2 | 2 | 2 | 2 | 2 | 1 | 2 | 2 | 3 | 2 | 2 |
| 29. Paper and wood pulp | 3 | 3 | 3 | 3 | 3 | 3 | 3 | 3 | 3 | 3 | 3 | 3 | 3 | 3 | 3 | 3 | 3 | 3 | 3 | 3 | 3 | 3 | 3 | 3 | 3 | 3 | 3 | 3 | 1 | 1 | 1 | 3 | 3 | 3 | 3 | 3 | 3 | 3 | 3 | 1 | 2 |
| 30. Other paper products | ∞ | ∞ | ∞ | ∞ | ∞ | ∞ | ∞ | ∞ | ∞ | ∞ | ∞ | ∞ | ∞ | ∞ | ∞ | ∞ | ∞ | ∞ | ∞ | ∞ | ∞ | ∞ | ∞ | ∞ | ∞ | ∞ | ∞ | ∞ | ∞ | ∞ | ∞ | ∞ | ∞ | ∞ | ∞ | ∞ | ∞ | ∞ | ∞ | ∞ | ∞ |
| 31. Printing and publishing | 2 | 2 | 3 | 3 | 2 | 2 | 3 | 2 | 3 | 3 | 2 | 2 | 2 | 2 | 3 | 2 | 2 | 3 | 2 | 3 | 2 | 2 | 2 | 3 | 3 | 2 | 2 | 3 | 3 | 3 | 1 | 3 | 3 | 3 | 3 | 3 | 3 | 3 | 3 | 3 | 1 |
| 32. Yarn and cloth | 1 | 2 | 2 | 2 | 2 | 2 | 2 | 2 | 2 | 2 | 3 | 3 | 3 | 2 | 1 | 3 | 3 | 3 | 3 | 3 | 3 | 3 | 3 | 3 | 3 | 2 | 3 | 1 | 2 | 3 | 3 | 1 | 1 | 1 | 2 | 1 | 1 | 1 | 1 | 2 | 3 |
| 33. Clothing | ∞ | ∞ | ∞ | ∞ | ∞ | ∞ | ∞ | ∞ | ∞ | ∞ | ∞ | ∞ | ∞ | ∞ | ∞ | ∞ | ∞ | ∞ | ∞ | ∞ | ∞ | ∞ | ∞ | ∞ | ∞ | ∞ | ∞ | ∞ | ∞ | ∞ | ∞ | ∞ | 1 | ∞ | ∞ | ∞ | ∞ | ∞ | ∞ | ∞ | ∞ |
| 34. Other textile products | 1 | 2 | 2 | 2 | 2 | 2 | 2 | 2 | 2 | 2 | 3 | 3 | 3 | 2 | 2 | 3 | 3 | 3 | 3 | 3 | 3 | 3 | 3 | 3 | 3 | 2 | 3 | 1 | 2 | 3 | 3 | 1 | 1 | 2 | 2 | 2 | 2 | 2 | 2 | 3 | 3 |
| 35. Leather tanning | 2 | 3 | 3 | 3 | 3 | 3 | 2 | 2 | 3 | 3 | 2 | 2 | 3 | 2 | 1 | 2 | 3 | 3 | 2 | 3 | 2 | 2 | 3 | 3 | 3 | 1 | 3 | 1 | 2 | 3 | 1 | 1 | 1 | 2 | 1 | 1 | 1 | 2 | 2 | 2 | 2 |
| 36. Leather shoes | ∞ | ∞ | ∞ | ∞ | ∞ | ∞ | ∞ | ∞ | ∞ | ∞ | ∞ | ∞ | ∞ | ∞ | ∞ | ∞ | ∞ | ∞ | ∞ | ∞ | ∞ | ∞ | ∞ | ∞ | ∞ | ∞ | ∞ | ∞ | ∞ | ∞ | ∞ | ∞ | ∞ | ∞ | ∞ | ∞ | ∞ | ∞ | ∞ | ∞ | ∞ |
| 37. Other leather products | 1 | 2 | 2 | 2 | 2 | 2 | 2 | 2 | 2 | 2 | 3 | 3 | 3 | 3 | 3 | 3 | 3 | 3 | 3 | 3 | 3 | 3 | 3 | 3 | 3 | 2 | 3 | 3 | 2 | 3 | 3 | 2 | 3 | 3 | 2 | 3 | 3 | 3 | 3 | 3 | 3 |
| 38. Rubber manufactures | 2 | 2 | 2 | 2 | 2 | 2 | 3 | 3 | 2 | 2 | 2 | 3 | 2 | 1 | 1 | 2 | 3 | 3 | 2 | 2 | 2 | 2 | 3 | 3 | 2 | 2 | 2 | 2 | 2 | 3 | 2 | 2 | 2 | 2 | 2 | 2 | 2 | 1 | 3 | 2 | 2 |
| 39. Industries n.e.s. | 1 | 2 | 2 | 2 | 2 | 2 | 2 | 2 | 2 | 2 | 2 | 3 | 2 | 1 | 2 | 2 | 3 | 3 | 2 | 2 | 2 | 2 | 3 | 3 | 2 | 2 | 2 | 1 | 2 | 3 | 2 | 1 | 1 | 2 | 2 | 2 | 2 | 2 | 2 | 1 | 2 |
| 40. Construction | ∞ | ∞ | ∞ | ∞ | ∞ | ∞ | ∞ | ∞ | ∞ | ∞ | ∞ | ∞ | ∞ | ∞ | ∞ | ∞ | ∞ | ∞ | ∞ | ∞ | ∞ | ∞ | ∞ | ∞ | ∞ | ∞ | ∞ | ∞ | ∞ | ∞ | ∞ | ∞ | ∞ | ∞ | ∞ | ∞ | ∞ | ∞ | ∞ | ∞ | ∞ |
| 41. Transportation (steam railroads) | 1 | 1 | 2 | 2 | 1 | 1 | 2 | 1 | 2 | 2 | 1 | 1 | 1 | 1 | 2 | 1 | 1 | 2 | 1 | 2 | 1 | 1 | 1 | 2 | 2 | 1 | 1 | 2 | 2 | 2 | 1 | 1 | 2 | 2 | 2 | 2 | 2 | 2 | 2 | 2 | 1 |

Buying industries

Selling industries

| Selling industries | 1. Agriculture | 2. Flour and grist mill products | 3. Canning and preserving | 4. Bread and bakery products | 5. Sugar, glucose, and starch | 6. Liquors and beverages | 7. Tobacco manufactures | 8. Slaughtering and meat packing | 9. Butter, cheese, etc. | 10. Other food industries | 11. Iron mining | 12. Blast furnaces | 13. Steel works and rolling mills | 14. Other iron and steel and electric manufactures | 15. Automobiles | 16. Nonferrous metal mining | 17. Smelting and refining | 18. Brass, bronze, copper, etc., manufactures | 19. Nonmetal minerals | 20. Petroleum and natural gas | 21. Refined petroleum | 22. Coal | 23. Coke | 24. Manufactured gas | 25. Electric utilities | 26. Chemical | 27. Lumber and timber products | 28. Other wood products | 29. Paper and wood pulp | 30. Other paper products | 31. Printing and publishing | 32. Yarn and cloth | 33. Clothing | 34. Other textile products | 35. Leather tanning | 36. Leather shoes | 37. Other leather products | 38. Rubber manufactures | 39. Industries n.e.s. | 40. Construction | 41. Transportation (steam railroads) |
|---|---|---|---|---|---|---|---|---|---|---|---|---|---|---|---|---|---|---|---|---|---|---|---|---|---|---|---|---|---|---|---|---|---|---|---|---|---|---|---|---|---|
| 1. Agriculture | 1 | 1 | 1 | 1 | 1 | 1 | 1 | 1 | 1 | 1 | 2 | 3 | 3 | 2 | 2 | 2 | 3 | 2 | 2 | 2 | 2 | 2 | 3 | 3 | 2 | 1 | 1 | 1 | 1 | 2 | 2 | 1 | 2 | 2 | 1 | 2 | 2 | 2 | 2 | 2 | 2 |
| 2. Flour and grist mill products | 1 | 1 | 1 | 2 | 1 | 2 | 2 | 2 | 2 | 2 | 3 | 3 | 3 | 3 | 3 | 3 | 3 | 3 | 3 | 3 | 3 | 3 | 3 | 3 | 3 | 3 | 3 | 3 | 3 | 3 | 3 | 3 | 3 | 3 | 3 | 3 | 3 | 3 | 3 | 3 | 3 |
| 3. Canning and preserving | ∞ | ∞ | ∞ | ∞ | ∞ | ∞ | ∞ | ∞ | ∞ | ∞ | ∞ | ∞ | ∞ | ∞ | ∞ | ∞ | ∞ | ∞ | ∞ | ∞ | ∞ | ∞ | ∞ | ∞ | ∞ | ∞ | ∞ | ∞ | ∞ | ∞ | ∞ | ∞ | ∞ | ∞ | ∞ | ∞ | ∞ | ∞ | ∞ | ∞ | ∞ |
| 4. Bread and bakery products | ∞ | ∞ | ∞ | ∞ | ∞ | 2 | ∞ | ∞ | ∞ | 1 | ∞ | ∞ | ∞ | ∞ | ∞ | ∞ | ∞ | ∞ | ∞ | ∞ | ∞ | ∞ | ∞ | ∞ | ∞ | ∞ | ∞ | ∞ | ∞ | ∞ | ∞ | ∞ | ∞ | ∞ | ∞ | ∞ | ∞ | ∞ | ∞ | ∞ | ∞ |
| 5. Sugar, glucose, and starch | 1 | 2 | 1 | 1 | 1 | 1 | 2 | 2 | 1 | 1 | 3 | 3 | 3 | 2 | 2 | 3 | 3 | 3 | 3 | 3 | 3 | 3 | 3 | 3 | 3 | 2 | 2 | 2 | 2 | 2 | 2 | 1 | 2 | 2 | 2 | 2 | 2 | 2 | 2 | 2 | 3 |
| 6. Liquors and beverages | ∞ | ∞ | ∞ | 2 | ∞ | 1 | ∞ | ∞ | ∞ | 1 | ∞ | ∞ | ∞ | ∞ | ∞ | ∞ | ∞ | ∞ | ∞ | ∞ | ∞ | ∞ | ∞ | ∞ | ∞ | ∞ | ∞ | ∞ | ∞ | ∞ | ∞ | ∞ | ∞ | ∞ | ∞ | ∞ | ∞ | ∞ | ∞ | ∞ | ∞ |
| 7. Tobacco manufactures | ∞ | ∞ | ∞ | ∞ | ∞ | ∞ | 1 | ∞ | ∞ | ∞ | ∞ | ∞ | ∞ | ∞ | ∞ | ∞ | ∞ | ∞ | ∞ | ∞ | ∞ | ∞ | ∞ | ∞ | ∞ | ∞ | ∞ | ∞ | ∞ | ∞ | ∞ | ∞ | ∞ | ∞ | ∞ | ∞ | ∞ | ∞ | ∞ | ∞ | ∞ |
| 8. Slaughtering and meat packing | 1 | 2 | 2 | 1 | 2 | 2 | 2 | 1 | 1 | 1 | 2 | 3 | 3 | 2 | 2 | 2 | 3 | 2 | 2 | 2 | 2 | 2 | 3 | 3 | 3 | 2 | 2 | 3 | 3 | 1 | 2 | 2 | 2 | 3 | 1 | 2 | 2 | 2 | 2 | 2 | 3 |
| 9. Butter, cheese, etc. | 3 | 3 | 3 | 1 | 3 | 2 | 3 | 3 | 3 | 1 | 3 | 3 | 3 | 3 | 3 | 3 | 3 | 3 | 3 | 3 | 3 | 3 | 3 | 3 | 3 | 3 | 3 | 3 | 3 | 1 | 2 | 3 | 3 | 3 | 3 | 3 | 3 | 3 | 3 | 3 | ∞ |
| 10. Other food industries | ∞ | ∞ | ∞ | 1 | ∞ | 1 | ∞ | ∞ | ∞ | 1 | ∞ | ∞ | ∞ | ∞ | ∞ | ∞ | ∞ | ∞ | ∞ | ∞ | ∞ | ∞ | ∞ | ∞ | ∞ | ∞ | ∞ | ∞ | ∞ | ∞ | ∞ | ∞ | ∞ | ∞ | ∞ | ∞ | ∞ | ∞ | ∞ | ∞ | ∞ |
| 11. Iron mining | 3 | 3 | 3 | 3 | 3 | 3 | 3 | 3 | 3 | 3 | 3 | 1 | 2 | 1 | 1 | 3 | 3 | 3 | 3 | 3 | 3 | 3 | 3 | 3 | 3 | 3 | 3 | 3 | 3 | 3 | 3 | 3 | 3 | 3 | 3 | 3 | 3 | 3 | 2 | 3 |
| 12. Blast furnaces | 2 | 2 | 2 | 2 | 2 | 2 | 2 | 2 | 2 | 2 | 2 | 2 | 1 | 1 | 1 | 2 | 2 | 2 | 2 | 2 | 2 | 2 | 3 | 2 | 2 | 2 | 2 | 2 | 2 | 2 | 2 | 2 | 2 | 2 | 2 | 2 | 2 | 2 | 1 | 2 |
| 13. Steel works and rolling mills | 1 | 2 | 2 | 2 | 2 | 2 | 2 | 2 | 2 | 2 | 1 | 1 | 1 | 1 | 1 | 1 | 2 | 1 | 1 | 2 | 1 | 1 | 2 | 1 | 1 | 1 | 1 | 2 | 1 | 1 | 2 | 2 | 2 | 2 | 2 | 1 | 1 | 1 | 1 | 1 |
| 14. Other iron and steel and electric manufactures | 1 | 1 | 1 | 1 | 1 | 1 | 1 | 1 | 1 | 1 | 1 | 2 | 1 | 1 | 1 | 1 | 2 | 1 | 1 | 1 | 1 | 2 | 1 | 1 | 1 | 1 | 1 | 1 | 1 | 1 | 1 | 1 | 1 | 2 | 1 | 1 | 1 | 1 | 1 | 1 |
| 15. Automobiles | ∞ | ∞ | ∞ | ∞ | ∞ | ∞ | ∞ | ∞ | ∞ | 1 | ∞ | ∞ | ∞ | ∞ | 1 | ∞ | ∞ | ∞ | ∞ | ∞ | ∞ | ∞ | ∞ | ∞ | ∞ | ∞ | ∞ | ∞ | ∞ | ∞ | ∞ | ∞ | ∞ | ∞ | ∞ | ∞ | ∞ | ∞ | ∞ | ∞ | 1 |
| 16. Nonferrous metal mining | 2 | 3 | 3 | 3 | 3 | 2 | 3 | 2 | 2 | 2 | 2 | 3 | 2 | 2 | 2 | 1 | 1 | 2 | 2 | 2 | 2 | 3 | 3 | 3 | 2 | 1 | 2 | 3 | 2 | 3 | 1 | 2 | 3 | 2 | 2 | 3 | 3 | 2 | 2 | 2 |
| 17. Smelting and refining | 2 | 2 | 2 | 2 | 2 | 2 | 2 | 2 | 2 | 2 | 2 | 2 | 2 | 1 | 1 | 2 | 2 | 3 | 3 | 2 | 2 | 3 | 2 | 2 | 1 | 2 | 2 | 1 | 2 | 1 | 2 | 1 | 2 | 1 | 2 | 2 | 1 | 2 | 1 | 2 |
| 18. Brass, bronze, copper, etc., manufactures | 2 | 2 | 2 | 2 | 2 | 1 | 2 | 1 | 1 | 2 | 2 | 1 | 1 | 1 | 1 | 2 | 2 | 2 | 2 | 1 | 2 | 2 | 2 | 1 | 1 | 2 | 2 | 1 | 2 | 2 | 1 | 2 | 2 | 1 | 1 | 1 | 1 | 1 | 1 | 1 |
| 19. Nonmetal minerals | 1 | 2 | 1 | 2 | 1 | 1 | 1 | 2 | 2 | 2 | 2 | 1 | 1 | 1 | 1 | 2 | 2 | 1 | 1 | 1 | 1 | 1 | 1 | 1 | 1 | 2 | 2 | 1 | 2 | 1 | 2 | 2 | 2 | 1 | 1 | 2 | 2 | 1 | 1 | 1 |
| 20. Petroleum and natural gas | 2 | 1 | 1 | 1 | 1 | 1 | 1 | 1 | 1 | 1 | 1 | 1 | 1 | 1 | 1 | 1 | 1 | 1 | 1 | 1 | 1 | 1 | 1 | 1 | 1 | 1 | 1 | 1 | 1 | 1 | 1 | 1 | 1 | 1 | 1 | 1 | 1 | 1 | 1 | 2 |
| 21. Refined petroleum | 1 | 1 | 1 | 1 | 1 | 1 | 1 | 1 | 1 | 1 | 1 | 1 | 1 | 1 | 1 | 1 | 1 | 1 | 1 | 2 | 1 | 1 | 1 | 1 | 1 | 1 | 1 | 1 | 1 | 1 | 1 | 1 | 1 | 1 | 1 | 1 | 1 | 1 | 1 | 1 |
| 22. Coal | 2 | 1 | 1 | 1 | 1 | 1 | 1 | 1 | 1 | 1 | 1 | 1 | 1 | 1 | 1 | 1 | 1 | 1 | 1 | 1 | 1 | 1 | 1 | 1 | 1 | 1 | 1 | 1 | 1 | 1 | 1 | 1 | 1 | 1 | 1 | 1 | 1 | 2 | 1 | 2 |
| 23. Coke | 2 | 1 | 1 | 1 | 1 | 1 | 1 | 1 | 1 | 1 | 2 | 1 | 2 | 1 | 1 | 2 | 1 | 1 | 1 | 1 | 1 | 1 | 1 | 1 | 1 | 1 | 1 | 1 | 1 | 1 | 1 | 1 | 1 | 2 | 1 | 1 | 1 | 2 | 1 | 1 |
| 24. Manufactured gas | 2 | 1 | 1 | 1 | 1 | 1 | 1 | 1 | 1 | 1 | 2 | 1 | 2 | 1 | 1 | 2 | 1 | 1 | 1 | 1 | 2 | 2 | 3 | 1 | 2 | 1 | 1 | 1 | 1 | 1 | 1 | 1 | 1 | 1 | 1 | 1 | 1 | 1 | 2 | 2 |
| 25. Electric utilities | 2 | 1 | 1 | 1 | 1 | 1 | 1 | 1 | 1 | 1 | 1 | 1 | 1 | 1 | 1 | 1 | 1 | 1 | 1 | 1 | 1 | 1 | 1 | 1 | 1 | 1 | 1 | 1 | 1 | 1 | 1 | 1 | 1 | 1 | 1 | 1 | 1 | 1 | 1 | 1 |
| 26. Chemical | 1 | 2 | 2 | 2 | 1 | 2 | 1 | 1 | 1 | 1 | 2 | 2 | 1 | 1 | 1 | 1 | 1 | 1 | 1 | 1 | 1 | 2 | 2 | 1 | 1 | 2 | 1 | 1 | 2 | 1 | 1 | 1 | 1 | 2 | 2 | 1 | 1 | 1 | 1 | 2 |
| 27. Lumber and timber products | 1 | 2 | 2 | 2 | 2 | 1 | 2 | 2 | 2 | 2 | 1 | 2 | 2 | 1 | 2 | 1 | 1 | 2 | 1 | 1 | 1 | 1 | 1 | 1 | 2 | 2 | 1 | 2 | 2 | 2 | 1 | 2 | 1 | 1 | 1 | 1 | 1 | 1 | 1 | 2 |
| 28. Other wood products | 2 | 2 | 2 | 2 | 2 | 2 | 1 | 2 | 2 | 2 | 2 | 3 | 2 | 1 | 2 | 2 | 3 | 2 | 2 | 2 | 2 | 3 | 3 | 2 | 1 | 2 | 1 | 1 | 2 | 2 | 2 | 1 | 2 | 2 | 2 | 2 | 2 | 1 | 2 | 2 |
| 29. Paper and wood pulp | 2 | 2 | 2 | 2 | 2 | 2 | 1 | 2 | 2 | 2 | 2 | 2 | 2 | 2 | 2 | 2 | 2 | 2 | 2 | 2 | 2 | 2 | 2 | 2 | 1 | 2 | 2 | 1 | 2 | 1 | 1 | 2 | 2 | 2 | 2 | 2 | 2 | 1 | 1 | 2 |
| 30. Other paper products | 2 | 3 | 3 | 3 | 3 | 3 | 3 | 3 | 3 | 3 | 3 | 3 | 3 | 2 | 3 | 3 | 3 | 3 | 3 | 3 | 3 | 3 | 3 | 3 | 3 | 3 | 3 | 2 | 2 | 3 | 2 | 2 | 2 | 3 | 3 | 2 | 2 | 2 | 2 | 2 |
| 31. Printing and publishing | 2 | 2 | 2 | 3 | 2 | 2 | 2 | 2 | 2 | 2 | 2 | 2 | 2 | 2 | 2 | 2 | 2 | 2 | 2 | 2 | 2 | 2 | 2 | 3 | 2 | 1 | 2 | 3 | 2 | 2 | 3 | 3 | 2 | 3 | 3 | 2 | 2 | 2 | 2 | 2 |
| 32. Yarn and cloth | 1 | 2 | 2 | 2 | 2 | 2 | 2 | 2 | 1 | 2 | 2 | 2 | 2 | 1 | 2 | 2 | 2 | 2 | 2 | 2 | 2 | 2 | 3 | 2 | 2 | 2 | 1 | 1 | 2 | 2 | 1 | 1 | 1 | 1 | 1 | 1 | 1 | 1 | 1 | 2 |
| 33. Clothing | ∞ | ∞ | ∞ | ∞ | ∞ | ∞ | ∞ | ∞ | ∞ | ∞ | ∞ | ∞ | ∞ | 1 | ∞ | ∞ | ∞ | ∞ | ∞ | ∞ | ∞ | ∞ | ∞ | ∞ | ∞ | ∞ | ∞ | ∞ | ∞ | ∞ | ∞ | 1 | ∞ | ∞ | ∞ | 1 | ∞ | ∞ | ∞ | ∞ |
| 34. Other textile products | 2 | 3 | 3 | 3 | 3 | 2 | 3 | 3 | 3 | 3 | 3 | 3 | 3 | 2 | 2 | 3 | 3 | 3 | 3 | 3 | 3 | 3 | 3 | 3 | 3 | 3 | 1 | 2 | 2 | 3 | 1 | 1 | 1 | 2 | 3 | 2 | 1 | 2 | 1 | 2 |
| 35. Leather tanning | 2 | 3 | 3 | 3 | 3 | 2 | 2 | 2 | 2 | 2 | 3 | 3 | 3 | 2 | 1 | 3 | 3 | 3 | 2 | 3 | 3 | 3 | 3 | 3 | 3 | 3 | 1 | 2 | 2 | 3 | 1 | 2 | 1 | 2 | 1 | 1 | 2 | 1 | 2 | 2 |
| 36. Leather shoes | ∞ | ∞ | ∞ | ∞ | ∞ | ∞ | ∞ | ∞ | ∞ | ∞ | ∞ | ∞ | ∞ | ∞ | ∞ | ∞ | ∞ | ∞ | ∞ | ∞ | ∞ | ∞ | ∞ | ∞ | ∞ | ∞ | ∞ | ∞ | ∞ | ∞ | ∞ | ∞ | ∞ | ∞ | ∞ | 1 | ∞ | ∞ | ∞ | ∞ |
| 37. Other leather products | 1 | 2 | 2 | 2 | 2 | 2 | 2 | 2 | 2 | 2 | 3 | 3 | 3 | 3 | 3 | 3 | 3 | 3 | 3 | 3 | 3 | 3 | 3 | 3 | 3 | 2 | 2 | 2 | 3 | 2 | 3 | 1 | 3 | 3 | 1 | 3 | 3 | 3 | 3 | 2 |
| 38. Rubber manufactures | 1 | 2 | 2 | 2 | 2 | 2 | 2 | 2 | 2 | 2 | 2 | 2 | 2 | 1 | 1 | 2 | 2 | 3 | 2 | 2 | 2 | 3 | 3 | 2 | 2 | 2 | 2 | 2 | 3 | 2 | 1 | 1 | 1 | 2 | 3 | 1 | 2 | 1 | 2 | 2 |
| 39. Industries n.e.s. | 1 | 2 | 2 | 2 | 2 | 2 | 2 | 2 | 2 | 2 | 2 | 2 | 2 | 1 | 1 | 2 | 2 | 1 | 2 | 2 | 2 | 2 | 2 | 2 | 2 | 2 | 2 | 2 | 2 | 2 | 1 | 1 | 2 | 1 | 2 | 2 | 1 | 1 | 1 | 1 |
| 40. Construction | 2 | 1 | 1 | 1 | 1 | 1 | 1 | 1 | 1 | 1 | 1 | 1 | 2 | 1 | 1 | 1 | 1 | 1 | 1 | 1 | 1 | 1 | 1 | 1 | 1 | 1 | 1 | 1 | 1 | 1 | 1 | 1 | 1 | 1 | 1 | 1 | 1 | 1 | 2 | 1 |
| 41. Transportation (steam railroads) | 1 | 1 | 1 | 2 | 1 | 1 | 1 | 1 | 1 | 1 | 1 | 1 | 1 | 1 | 1 | 1 | 1 | 1 | 1 | 1 | 1 | 1 | 2 | 2 | 1 | 1 | 1 | 1 | 1 | 1 | 2 | 1 | 2 | 1 | 1 | 2 | 2 | 1 | 2 | 2 | 2 |

# 26

## Development and Trade Dependence: The Case of Puerto Rico, 1948–1963

RICHARD WEISSKOFF and EDWARD WOLFF

The transformation of Puerto Rico from a sugar-monoculture began in the early 1950s and by 1963 created a diversified manufacturing export economy. During these years, gross domestic product more than doubled, while gross investment and exports nearly quadrupled (Table 26.1). The magnitude of the industrialization, as seen in the rise of the industrial share of GDP by 50 percent and the decline of the agricultural share by about the same extent, resulted in the doubling of per capita income while employment remained stable.

Three policies have been instrumental in bringing about this transformation. First, the government's package of "incentives" and tax exemptions all but guaranteed the "promoted" manufacturers a profitable transplantation. Second, the low nonstop airfare to New York sparked the flow of tourists to Puerto Rico and encouraged the migration of Puerto Rican laborers to the mainland. Third, the "Commonwealth" status placed Puerto Rico within the United States monetary system thereby eliminating problems of foreign exchange balances and convertibility. In achieving an apparently stable political solution, the Commonwealth has encountered little difficulty in selling its bonds on Wall Street or in attracting foreign capital.[1]

The aggregate level of trade openness remained relatively stable during the rapid industrialization.[2] The ratio of exports to GDP increased slightly, with 97 percent of all exports absorbed by the United States in both years. A 30 percent decline in the concentration ratio of the four major export commodities reflects the diversification of exports. Imports as a share of GDP fell somewhat during the period and became slightly more varied in terms both of major trading partners and of major commodities.

In many other Latin American countries, import-substituting industrialization has led to a declining dependence on imports as a share of GDP.[3] However, comparison with Puerto Rico is difficult as comparably detailed data for both the pre- and post-industrialization periods exist in few developing countries.[4]

Policy implications from the Puerto Rican experience should be drawn cautiously. Except for the former Caribbean colonies, no Latin American country

This article was published in *The Review of Economics and Statistics*, Volume 57, Number 4, November, 1975, pp. 470–477. © By the President and Fellows of Harvard College. Reprinted by permission.

Table 26.1. Growth and Trade in Puerto Rico, 1948–1963 (*Macroeconomic indicators*)

| | 1948 | 1963 |
|---|---|---|
| 1. Gross domestic product (millions of 1954 dollars) | 732 | 1762 |
| 2. Gross domestic investment (millions of 1954 dollars) | 123 | 438 |
| 3. Sales to rest of the world (millions of 1954 dollars) | 299 | 1193 |
| 4. Employment (in thousands) | 572 | 606 |
| 5. Population (in thousands) | 2175 | 2467 |
| 6. GDP per capita (1954 dollars) | 337 | 714 |
| 7. Shares of GDP (%): | | |
|     a. Personal consumption | 91.9 | 87.1 |
|     b. Government consumption | 12.4 | 14.7 |
|     c. Gross domestic investment | 16.7 | 24.8 |
|     d. Net sales to rest of world | −21.0 | −26.7 |
| 8. Shares of GDP by Industrial Origin:[a] | | |
|     a. Agriculture | 18.4 | 9.1 |
|     b. Manufacturing | 16.4 | 23.9 |
|     c. Construction and quarrying | 4.1 | 6.9 |
|     d. Services | 45.5 | 50.4 |
|     e. Government | 10.3 | 11.2 |
| 9. Measures of "openness": | | |
|     a. Exports/GDP (%) | 30.9 | 33.5 |
|     b. Imports/GDP (%) | 56.8 | 47.2 |
|     c. Exports to USA/total exports (%) | 96.8 | 96.7 |
|     d. Imports from USA/total imports (%) | 93.3 | 81.9 |
|     e. 4 commodity conc. ratio, exports to USA | 74.8 | 52.3 |
|     f. 4 commodity conc. ratio, imports from USA | 29.3 | 24.0 |

[a] Do not sum to 100% due to "Statistical Discrepancy" entry in Social Accounts.

*Sources:* line 1–6 Puerto Rico Planning Board, 1968, table 1, 8–11. 7 Ibid., from table 2, 12–13. 8 Ibid., from table 4, 16–17. 9 Calculated from Puerto Rico Planning Board, 1964.

has negotiated a comparable position vis-à-vis a metropolitan power. Even the relationships of Jamaica, Trinidad, and Barbados to England, Martinique to France, and Curaçao to Holland do not provide the same freedom with which American capital moves into Puerto Rico and Puerto Rican labor migrates to the United States. Moreover, in many Latin American economies, import restrictions were devised to encourage domestic production initially of finished consumer goods and then of more primary stages of production. By falling within the United States tariff barrier Puerto Rico enjoys no protection, save that provided by transport charges, from United States goods. Its industrialization has been propelled by lucrative profits created not from extraordinary rates of protection but from extraordinary subsidies to capital, labor, and other inputs in the form of Federal income tax exemptions, lower wages, and subsidized utilities and infrastructure.[5]

The aim of this paper is to measure the net effects of the industrialization process on the structure of import flows and to examine the resulting changes in import dependence.[6] The application of a number of dependency measures reveals a kind of stability in the level of import dependence which is hypothesized to be the result of a set of counteracting effects. Moreover, evidence of substantial import substitution is found to have occurred in the absence of tariff or trade restrictions with its major trading partner, the United States, and in the absence of any deliberate program on the part of the Puerto Rican government. As will be seen, the export-propelled industrialization, the large-scale imports of capital and intermediate goods, and the wholesale imitation of North American consumption habits have failed to create a society relatively more dependent on imports than that of the former sugar economy.

## The Model

Following a general Leontief accounting framework, let

$$A = A_d + A_m$$

$$Y = Y_d + Y_m$$

where $A$ is a 26-order interindustry coefficient matrix, $A_d$ its domestic component, and $A_m$ its import component; and $Y$ is the final demand vector, $Y_d$ its domestic component, and $Y_m$ its import component. Then

$$X = [I - A_d]^{-1} Y_d \qquad (26\text{–}1)$$

where $X$ is the vector of gross domestic output. Let $F$ be the 26-by-5 matrix of final demand coefficients, with components $C$ for household consumption, $K$ for capital formation, $N$ for net inventory change, $G$ for government expenditure, and $E$ for exports. Let $F_d$ and $F_m$ be the domestic and import components of final demand. Then $F = F_d + F_m$.

Let $D$ be the 26-by-5 matrix of the distribution of domestic final demand by component. Define

$$R = A_m [I - A_d]^{-1}$$

where $R$ is the "import inverse" matrix, used in computing the imports generated by domestic final demand.

Imports generated by each component of domestic final demand are calculated as follows:

$$M_j^{st} = R^s D_j^t \cdot 100 \qquad (26\text{–}2)$$

where
  $t = 1948, 1963$
  $s = 1948, 1963$
  $j = C, K, G, E$

Thus, the vector $M_j^{st}$ is the import bundle generated by \$100 of final demand component $j$, given import and technological requirements of years $s$ and demand mix of year $t$.

Total (direct plus induced) imports stemming from \$100 of each component of

Table 26.2.   Measures of Import Dependence

| | |
|---|---|
| 1.  Share of imports in the supply of: | |
|    a.  Intermediate sales | |
| $$\alpha_1 = \frac{[A_m X]_i}{[AX]_i}$$ | $i = 1, \ldots, 26.$ |
|    b.  Final use sales | |
| $$\alpha_2 = \frac{[Y_m]_i}{Y_i}$$ | $i = 1, \ldots, 26.$ |
| 2.  Share of imports in the purchases of: | |
|    a.  Intermediate inputs | |
| $$\beta_1 = \frac{\sum_i [A_m]_i}{\sum_i A_{ij}}$$ | $j = 1, \ldots, 26.$ |
|    b.  Total inputs (material plus value added) | |
| $\beta_2 = \sum_i [A_m]_{ij}$ | $j = 1, \ldots, 26.$ |
| 3.  Total content of final product | |
|    (column sums of import inverse) | |
| $\gamma = \sum_i R_{ij}$ | $j = 1, \ldots, 26.$ |
| 4.  Import component of final demand | |
|    a.  Direct import leakage | |
| $\delta_1 = \sum_i [F_m]_{ij}$ | $j = 1, \ldots, 5.$ |
|    b.  Induced or "generated" import leakage | |
| $\delta_2 = \sum_i [RD]_{ij}$ | $j = 1, \ldots, 5.$ |

final demand are

$$M_j^t = ([F_m]_j^t + R^t[F_a]_j^t) \cdot 100 \qquad (26\text{--}3)$$

where
$$t = 1948, 1963$$
$$j = C, K, G, E$$

## Measures of Trade Dependence

Four sets of indices were constructed to measure the change in various dimensions of import dependence (see Table 26.2). Measures $\alpha_1$ and $\alpha_2$, the share of imports in the supply of intermediate and final sales, reflect the value of imports by sector of origin to total supply (gross output plus imports).[7] The second two indices measure the ratio of imports by sector of destination to the value of material inputs (Measure $\beta_1$) and to the value of material inputs plus value added (Measure $\beta_2$). A third index, $\gamma$ measures changes in the direct and indirect import content of the final output of each sector. Changes in $\gamma$ may thus reflect changes in the import demand impact on the final goods industry or of the supplying industries or changes in domestic industrial interdependence.[8]

The last two measures assess the import leakage associated with each component of final demand. A measure of the direct import content compares the relative importance of final use imports (Measure $\delta_1$) and induced intermediate imports (Measure $\delta_2$) to total final demand.[9]

**Results**[10]

The distribution of imports in 1948 and 1963 by sector of origin is shown in columns 1 and 2 of Table 26.3. In spite of an overall difference[11] of 0.43 between the two years, the three most important imports—processed foods, textiles, and metals—accounted for 56 percent of total imports in 1948 and 57 percent of imports in 1963. Imports as a fraction of total supply (gross output plus imports) fell slightly from 28.8 percent in 1948 to 26.2 percent in 1963 (columns 3 and 4).[12] Three new sectors—leather, petroleum and coal, and mining and quarrying—were created during this period. Import substitution was observed in raw and processed foods and in manufacturing goods, except for printing, whereas imports expanded in all the service sectors except for transport.

The overall stability of intermediate imports as a share of intermediate supply, Measure $\alpha_1$ (columns 5 and 6, Table 26.3) cloaks a declining import share for manufactures offset by a rising import component of services, the latter due perhaps to continued mainland control of banking, insurance, and management services.[13] The import component of final demand (Measure $\alpha_2$) fell due to the introduction of domestic processing and canning of foods.

Movements in Measure $\beta_1$ (columns 9 and 10) pinpoint changes in shares of imports in material inputs. Upward shifts occurred in the sugarcane sector, with the increased use of fertilizer and other inputs necessitated by mechanization; in the shoe and leather goods sector, which relied on imports of American-produced leather; in petroleum and coal from the import of crude oil; in hotels and restaurants due to requirements for processed foods, printing, and metal products; and real estate. Downward shifts in imports as a share of intermediate GDO occurred in agriculture due to the substitution of domestically produced animal feeds; in furniture, due to the shift toward domestically produced upholstered furniture using imported textiles; in printing due to the substitution of domestic paper; in chemicals, trade, and transport due to the rise of domestically produced petroleum inputs in 1963. The share of imports to total inputs including value added (Measure $\beta_2$) fell from 0.17 to 0.15 (columns 11 and 12), reflecting the rise in the value-added intensity of production from 44 percent of gross output in 1948 to 52 percent in 1963.[14] This rise in value added also accounts for the decline in imports generated per dollar of final demand (Measure $\gamma$, columns 13 and 14).

Changes in the pattern of overall household consumption (Table 26.4, A, columns 1 and 2) are understandable as conventional Engel effects in the face of doubling per capita income. The shares in consumption of processed foods and leather goods fell substantially while the agricultural share was constant and textiles rose slightly.[15] The shares of nonnecessities all increased significantly: furniture, chemicals (detergents and soaps), petroleum (reflecting the increased use of automobiles), metals (consumer durables), and services.[16] With the change in consumption composition, direct leakages for all sectors held stable while indirect leakages declined from 16.9 to 11.8 percent. Nevertheless, the total import leakage of 58 percent in 1948 and 51 percent in 1963 is high compared to a 20.7 percent total import leakage for the United Kingdom and 40.0 percent for India.[17]

The stability of the direct import leakage, Measure $\delta_1$, represents the offsetting

| | Distribution of imports (by sector of origin) | | Imports/total supply (GDO + M) (by sector of origin) | | Measure $\alpha_1$ inter. imports/inter. sales | | Measure $\alpha_2$ F.D. imports/final sales | | Measure $\beta_1$ inter. imports/inter. inputs | | Measure $\beta_2$ inter. imports/total inputs | | Measure $\gamma$ column sums [R] | |
|---|---|---|---|---|---|---|---|---|---|---|---|---|---|---|
| | 1948 (1) | 1963 (2) | 1948 (3) | 1963 (4) | 1948 (5) | 1963 (6) | 1948 (7) | 1963 (8) | 1948 (9) | 1963 (10) | 1948 (11) | 1963 (12) | 1948 (13) | 1963 (14) |
| 1. Ag NEC | 9.61 | 3.12 | 0.45 | 0.22 | 0.24 | 0.15 | 0.85 | 0.28 | 0.43 | 0.26 | 0.13 | 0.07 | 0.19 | 0.11 |
| 2. Sugar cane | 0.0 | 0.0 | 0 | 0 | 0 | 0 | 0 | 0 | 0.15 | 0.31 | 0.06 | 0.08 | 0.22 | 0.13 |
| 3. Sugar milling | 0.0 | 0.01 | 0 | 0 | 0 | 0 | 0 | 0 | 0.07 | 0.09 | 0.05 | 0.07 | 0.21 | 0.18 |
| 4. Proc. foods | 25.10 | 18.79 | 0.40 | 0.34 | 0.24 | 0.40 | 0.45 | 0.32 | 0.26 | 0.23 | 0.20 | 0.16 | 0.32 | 0.24 |
| 5. Textiles | 11.18 | 12.10 | 0.55 | 0.49 | 1.00 | 0.92 | 0.39 | 0.38 | 0.84 | 0.73 | 0.51 | 0.40 | 0.53 | 0.43 |
| 6. Leather | 3.04 | 2.99 | 1.00 | 0.56 | 1.00 | 0.89 | 1.00 | 0.49 | 0 | 0.81 | 0 | 0.47 | 0 | 0.51 |
| 7. Furniture | 3.38 | 3.16 | 0.70 | 0.54 | 0.91 | 0.87 | 0.36 | 0.31 | 0.65 | 0.47 | 0.40 | 0.32 | 0.43 | 0.36 |
| 8. Paper prod. | 1.67 | 1.91 | 0.88 | 0.57 | 0.88 | 0.64 | 0.58 | -0.10 | 0.67 | 0.57 | 0.45 | 0.40 | 0.55 | 0.52 |
| 9. Printing | 0.30 | 1.08 | 0.15 | 0.45 | 0.11 | 0.61 | 0.22 | -0.62 | 0.70 | 0.39 | 0.51 | 0.17 | 0.60 | 0.22 |
| 10. Chemical | 6.96 | 7.42 | 0.67 | 0.60 | 0.54 | 0.74 | 0.90 | 0.49 | 0.82 | 0.55 | 0.61 | 0.29 | 0.66 | 0.35 |
| 11. Nonmetal | 1.39 | 1.35 | 0.36 | 0.22 | 0.31 | 0.22 | 0.58 | 0.27 | 0.49 | 0.15 | 0.27 | 0.08 | 0.37 | 0.17 |
| 12. Petro. and coal | 5.46 | 1.79 | 1.00 | 0.15 | 1.00 | 0.16 | 1.00 | 0.13 | 0 | 0.58 | 0 | 0.49 | 0 | 0.54 |
| 13. Metal ind. | 9.58 | 25.87 | 0.91 | 0.67 | 0.87 | 0.67 | 0.94 | 0.66 | 0.83 | 0.46 | 0.41 | 0.30 | 0.43 | 0.35 |
| 14. Mining | 0.78 | 4.78 | 1.00 | 0.82 | 1.00 | 0.82 | 0 | 0.83 | 0 | 0.17 | 0 | 0.11 | 0 | 0.18 |
| 15. Other manuf. | 5.07 | 4.58 | 0.71 | 0.56 | 0.84 | 0.78 | 0.64 | 0.49 | 0.77 | 0.68 | 0.57 | 0.45 | 0.62 | 0.48 |
| 16. Construction | 0.0 | 0.0 | 0 | 0 | 0 | 0 | 0 | 0 | 0.37 | 0.39 | 0.26 | 0.20 | 0.45 | 0.27 |
| 17. Hotel and rest. | 0.0 | 0.08 | 0 | 0.01 | 0 | 0.02 | 0 | 0 | 0.05 | 0.12 | 0.04 | 0.04 | 0.26 | 0.08 |
| 18. Electricity | 0.0 | 0.0 | 0 | 0 | 0 | 0 | 0 | 0 | 0.83 | 0.32 | 0.27 | 0.08 | 0.29 | 0.15 |
| 19. Water and san. | 0.0 | 0.0 | 0 | 0 | 0 | 0 | 0 | 0 | 0.42 | 0.34 | 0.12 | 0.06 | 0.17 | 0.08 |
| 20. Communication | 0.0 | 0.13 | 0 | 0.05 | 0 | 0.05 | 0 | 0.07 | 0.40 | 0.20 | 0.13 | 0.05 | 0.18 | 0.08 |
| 21. Trade | 0.0 | 0.07 | 0 | 0 | 0 | 0 | 0 | 0 | 0.24 | 0.15 | 0.06 | 0.03 | 0.11 | 0.06 |
| 22. Bus. services | 0.0 | 1.91 | 0 | 0.15 | 0 | 0.12 | 0 | 0.25 | 0.50 | 0.41 | 0.19 | 0.22 | 0.26 | 0.29 |
| 23. Personal services | 0.0 | 0.01 | 0 | 0 | 0 | 0 | 0 | 0 | 0.27 | 0.27 | 0.09 | 0.10 | 0.15 | 0.15 |
| 24. Real estate | 0.0 | 0.02 | 0 | 0 | 0 | 0 | 0 | 0 | 0 | 0.17 | 0 | 0.05 | 0.08 | 0.10 |
| 25. Transport | 6.48 | 2.14 | 0.33 | 0.14 | 0.74 | 0.14 | 0.10 | 0.14 | 0.86 | 0.21 | 0.48 | 0.09 | 0.50 | 0.18 |
| 26. Govt. services | 0.0 | 6.67 | 0 | 0.17 | 0 | 0.46 | 0 | 0.15 | 0.17 | 0.20 | 0.08 | 0.08 | 0.24 | 0.17 |
| 27. TOTAL | 100.0 | 100.00 | 0.29 | 0.26 | 0.31 | 0.32 | 0.27 | 0.23 | 0.31 | 0.32 | 0.17 | 0.15 | 0.30 | 0.25 |
| Rel. diff. | — | 0.43 | | -0.09 | | 0.04 | | -0.17 | | 0.04 | | -0.12 | | |
| Overall difference | | | | | | | | | | | | | | -0.20 |

Table 26–4.   Direct and Induced Imports per \$100 of Each Component of Final Demand $M_j = ([F_m]_j + R[F_d]_j) \cdot 100$   $j = C, K, G, E$

| | | Percentage distribution $[F_d]_j$ | | $[F_m]_j$ Direct imports $(\delta_1)$ | | $R[F_d]_j$ Induced imports | | $[F_m]_j + R[F_d]_j$ Total imports | |
|---|---|---|---|---|---|---|---|---|---|
| | | 1948 | 1963 | 1948 | 1963 | 1948 | 1963 | 1948 | 1963 |
| A. | Consumption (C) | | | | | | | | |
| 1. | Agriculture | 5.26 | 5.07 | 5.01 | 1.57 | 2.34 | 0.49 | 7.35 | 2.06 |
| 4. | Processed foods | 34.67 | 23.50 | 17.17 | 9.74 | 2.52 | 1.94 | 19.69 | 11.68 |
| 5. | Textiles | 6.56 | 7.18 | 4.58 | 6.23 | 1.21 | 0.59 | 5.80 | 6.82 |
| 6. | Leather | 2.30 | 1.71 | 2.30 | 1.71 | 0.06 | 0.04 | 2.36 | 1.75 |
| 7. | Furniture | 1.26 | 2.48 | 0.39 | 0.97 | 0.84 | 0.35 | 1.22 | 1.33 |
| 10. | Chemical | 3.07 | 3.74 | 2.77 | 2.99 | 1.22 | 1.07 | 3.99 | 4.06 |
| 12. | Petroleum | 0.52 | 2.47 | 0.52 | 0.39 | 1.76 | 0.37 | 2.28 | 0.76 |
| 13. | Metal | 4.66 | 7.76 | 4.41 | 6.84 | 2.30 | 2.15 | 6.71 | 8.99 |
| 15. | Other manuf. | 2.85 | 2.49 | 2.35 | 2.19 | 0.81 | 0.49 | 3.16 | 2.68 |
| 17. | Hotels and res. | 8.84 | 2.68 | 0.0 | 0.0 | 0.0 | 0.03 | 0.0 | 0.03 |
| 22–26 | Services | 26.79 | 35.86 | 1.04 | 6.25 | 2.32 | 1.24 | 3.36 | 7.49 |
| 27. | Totals | 98.45 | 100.00 | 40.95 | 39.48 | 16.88 | 11.81 | 57.84 | 51.28 |
| B. | Capital formation (K) | | | | | | | | |
| 7. | Leather | 0.51 | 1.94 | 0.18 | 0.89 | 6.41 | 4.60 | 6.59 | 5.49 |
| 10. | Chemical | 0.0 | 0.0 | 0.0 | 0.0 | 1.97 | 0.71 | 1.97 | 0.71 |
| 11. | Nonmetal | 0.0 | 0.0 | 0.0 | 0.0 | 1.17 | 1.43 | 1.17 | 1.43 |
| 12. | Petroleum | 0.0 | 0.0 | 0.0 | 0.0 | 1.98 | 0.44 | 1.98 | 0.44 |
| 13. | Metal ind. | 45.61 | 29.72 | 42.78 | 26.65 | 12.16 | 7.61 | 59.94 | 34.26 |
| 14. | Mining | 0.0 | 0.0 | 0.0 | 0.0 | 0.21 | 1.56 | 0.21 | 1.56 |
| 15. | Other manuf. | 0.0 | 0.38 | 0.0 | 0.30 | 0.52 | 0.70 | 0.52 | 1.00 |
| 16. | Construction | 53.33 | 65.69 | 0.0 | 0.0 | 0.0 | 0.0 | 0.0 | 0.0 |
| 27. | Totals | 100.00 | 100.00 | 42.95 | 27.91 | 25.53 | 19.78 | 68.48 | 47.68 |
| C. | Government (G) | | | | | | | | |
| 4. | Processed Foods | | | | | 0.36 | 4.73 | | |
| 7. | Furniture | | | | | 3.49 | 0.14 | | |
| 10. | Chemical | | | | | 1.95 | 1.28 | | |
| 12. | Petroleum | | | | | 1.54 | 0.28 | | |
| 13. | Metal ind. | | | | | 12.12 | 3.00 | | |
| 14. | Mining | | | | | 0.12 | 2.35 | | |
| 15. | Other manuf. | | | | | 1.34 | 0.45 | | |
| 27. | Totals | | | | | 23.52 | 16.71 | | |
| D. | Exports (E) | | | | | | | | |
| 3. | Sugar mill. | 34.08 | 10.71 | | | 0.0 | 0.0 | | |
| 4. | Processed foods | 6.74 | 14.91 | | | 1.10 | 2.27 | | |
| 5. | Textiles | 10.46 | 13.90 | | | 6.43 | 4.94 | | |
| 6. | Leather | 0.0 | 2.84 | | | 0.12 | 0.03 | | |
| 10. | Chemical | 0.0 | 3.45 | | | 2.50 | 2.93 | | |
| 12. | Petroleum | 0.0 | 3.95 | | | 3.64 | 0.52 | | |
| 13. | Metal ind. | 0.0 | 8.00 | | | 3.73 | 5.03 | | |
| 14. | Mining | 0.0 | 0.01 | | | 0.68 | 3.60 | | |
| 15. | Other manuf. | 1.79 | 4.15 | | | 1.46 | 0.92 | | |
| 17. | Hotel and rest. | 1.43 | 6.18 | | | 0.0 | 0.05 | | |
| 21. | Trade | 21.36 | 15.13 | | | 0.0 | 0.04 | | |
| 25. | Transport | 8.54 | 3.57 | | | 3.63 | 0.38 | | |
| 26. | Govt. services | 13.22 | 7.76 | | | 0.0 | 0.61 | | |
| 27. | Totals | 100.00 | 100.00 | | | 26.57 | 25.96 | | |

effects of two sets of forces. Leading to greater imports were (a) an income effect through which American products were viewed as "superior" goods, (b) a price effect resulting from declining shipping costs, and (c) a general shift in preferences toward mainland-style commodities. Working to reduce imports was the establishment of American branch plants which supplied a portion of the expanding local demand in addition to exporting production. Dramatic changes in the sectoral distribution of imports occurred despite the overall stability of import shares. Most notably, direct imports of agricultural goods declined due to greater home production of beef, pork, poultry, and vegetables.[18]

The decline in total leakage due to investment demand (Table 26.4, B) is due primarily to the rise in the share of construction and the decline in the share of metal industries. Despite the decline in total imports (columns 7 and 8) from $68 to $48 per $100 of capital formation, total "openness" is high compared to $18.60 per $100 of capital formation in the United Kingdom.[19]

Imports induced by $100 of government services (Table 26.4, C) declined from $24 to $17, mirroring the transformation of the government from a sluggish bureaucracy for maintaining agricultural infrastructure and limited urban services into a more active interventionist and promotional role. Government activities[20] in 1963 generated imports of processed foods for hospitals and schools, petroleum for government vehicles, metal products for office supplies, and business and transport services for industrial and tourist promotion.

The diversification of the export bundle from sugar cane to manufactures failed to affect the level of induced imports (see Table 26.4, D). This stability was due to the balancing of two opposing effects. The first was a move away from sugar milling, a low import generator, toward more manufactures which were relatively more import-intensive. The second was a reduction in the import needs of manufacturers between 1948 and 1963.

The change in induced imports [Eq. (26–3)] generated by each element of final demand was decomposed into two effects: a demand effect ($F_d$) and an effect of the import and technical structure ($R$). In each case (Table 26.5), the change in the import and technical structure ($R$) resulted in a decline of induced imports, but the

Table 26.5   Induced Imports from Varying Demand and Production Structures

|  | $(D_j)$ $100 of Final Demand | | $(R)$ Import and Technical Structure | |
|---|---|---|---|---|
|  |  |  | 1948 | 1963 |
| 1. | Consumption ($C$) | 1948 | $28.59 | $18.98 |
|  |  | 1963 | 25.34 | 19.51 |
| 2. | Capital Formation ($K$) | 1948 | 44.75 | 27.66 |
|  |  | 1963 | 44.22 | 27.43 |
| 3. | Government ($G$) | 1948 | 23.52 | — |
|  |  | 1963 | — | 16.71 |
| 4. | Exports ($E$) | 1948 | 26.57 | 18.96 |
|  |  | 1963 | 31.39 | 25.96 |
| 5. | Total Final Demand | 1948 | 28.26 | 19.27 |
|  |  | 1963 | 29.89 | 22.88 |

demand effects ($F_d$) on induced imports vary. The changing demand composition is negative for consumption in 1948, neutral for capital formation, and is positive for 1963 consumption and exports and total final demand in both years. In all cases, the import-saving effect of the technical structure was the dominant factor in the decline of induced imports.


## Conclusion

The first fifteen years of export-propelled industrialization resulted in stability of the share of import demand. This stability is seen as the consequence of changes in demand composition, technology, and a "natural" import-replacement, and is hypothesized to be in part the outcome of two offsetting "phases" of export-led industrialization. In the first phase, American branch plants were established on the island and relied on imports of mainland materials. In the next phase, suppliers of the first-wave industries settled on the island, partially closing the leakages created in the first phase and importing only lower-valued, raw materials.

The hypothesis that the absence of protection from mainland industries would necessarily lead to a higher level of import dependence is unsubstantiated. In the case of Puerto Rico, many sectors have undergone considerable import-replacement; others have remained heavily import dependent. Perhaps the stability and growth of the Puerto Rican economy have been achieved by means of its annexation to and integration with the North American economy in the absence of a more autonomous industrialization.


## NOTES

1. The full panoply of promotion and development devices have been described elsewhere. Here, we are concerned only with the skeleton which will assist the reader in examining the empirical material. See D. F. Ross (1966), Ch. VI–VIII.
2. It should be noted that in 1963 Puerto Rico ranked fourth behind Trinidad and Tobago, Hong Kong and Mauritius among low income countries in trade dependence as measured by (exports + imports)/GDP. Data from United Nations (1964).
3. See A. O. Hirschman (1968) and Maria Conceiçao Tavares (1964) for a classic survey of Brazil.
4. Inputs are conventionally summarized as a single row or column vector. In the Puerto Rican studies, the imports are available both by sector of origin and destination. Comparable single-year studies exist for India, the United Kingdom, and Israel and are presented herein.
5. The different cost structure United States firms face in Puerto Rico makes it difficult to compare the efficiency of Puerto Rican with mainland industry.
6. The concept of dependency examined here is narrow in scope and refers exclusively to annual trade flows. Other dimensions of dependency include ownership of capital, price stability of exports, the power relations which determine the type and level of economic activity, imitation of consumption patterns, provision of various kinds of social and private infrastructure, and reliance on foreign sources for technological and product change. See Girvan (1972) and Dos Santos (1968). While data are plentiful on the diversification of products, purchasers, and suppliers in Puerto Rico, almost no information is available on capital stock, its ownership, and the change in ownership patterns.
7. See Chenery and Watanabe (1958), p. 488, for a discussion of this term. The distinction between the

import content of consumer necessities and that of consumer luxuries is made in B. R. Hazari (1967). Chenery develops a measure of import substitution involving a linear projection of imports, by sector of origin, as a function of per capita income. He then measures the difference between projected imports and actual imports. (See Chenery (1960), p. 640.) This aggregate measure fails to capture the differential effects of changing input structure, changing composition of final demand, and changing weights among industries and components of final demand.

8. Algebraically, a dollar's worth of the final product of a sector nets into $v$ worth of value added and $(1 - v)$ worth of imports.

9. Panchumukki (1965), p. 117, argues that the latter is the most important dimension of import dependence.

10. The original 1948 Input-Output Table was constructed by Amos Gosfield in purchaser prices for 65 domestic and 30 foreign exporting sectors. 1963 data were supplied by the Puerto Rican Planning Board, Division of Social Accounts, on the 500 commodity level for domestic, primary, and subsidiary sectors and for imports. The aggregation of this latter table, its reconciliation with national accounts, its alignment with the 1948 table, and the conversion of the earlier data to 1963 prices, were performed by the authors and are described in the Technical Appendix I to Weisskoff et al. (U.S. Department of Labor, 1971).

11. The relative difference of $x$ and $y$ is defined as $2[(x - y)/(x + y)]$ or the difference divided by the mean. The overall difference is defined as: $\sum_i |V_i^2 - V_i^1|$ where $V^2$ and $V^1$ are vectors of distributive shares which sum to unity. The overall difference ranges from zero to two. The overall difference of the change in GDO is 0.39, of the change in value added is 0.33, and of the change in consumption is 0.05.

12. However, the ratio of imports to gross domestic product showed a decline from 0.9202 to 0.6786. The gross output generated by domestic final demand also declined, with a multiplier of 2.27 in 1948 and 1.91 in 1963. A dollar of final demand in 1963 generated, as a result, less interindustry demand than a dollar of final demand in 1948. Since the GDO multipliers rose in 16 sectors and fell in only 10, demand shifted to those sectors with falling GDO multipliers. The fall in imports relative to final demand between 1948 and 1963 was therefore caused by a fall in the total supply necessary to fill the final demand bill of goods.

13. The fall in the import share of agricultural produce was due to the increased domestic production of grains for animal feeds. The existence of a minor share of imported processed sugar in 1963 was traced to packaged sugar cubes, such as used in coffee, from a New Jersey refinery. The increased import share in intermediate government services between 1948 and 1963 reflected the increased importance of postal, insurance, and other specialized federal government services in production.

14. The rise in the share of value added was due, we hypothesize, to the following three factors: (1) the rise in the wage rate relative to the cost of intermediate inputs; (2) the rise in the contribution of fixed capital relative to intermediate inputs and the consequent rise of the profit share relative to current account inputs; (3) tax exemptions on corporate profits which created an incentive for American branch plants to record profits in Puerto Rico rather than the United States through accounting and inventory practices.

15. The constant agricultural share reflects the rise of poultry, fruits, and vegetables. The rising textile share, contrary to the normal behavior of "necessities" may have been the result of the shift to ready-made clothing from bulk cotton fabrics.

16. The decline in the share of expenditures on hotels and restaurants may reflect accounting differences in the two years. In the 1948 table, both processed foods sold to hotels and the total value were recorded as sales to households; in the second year only the mark-up on food was recorded as a sale to households. Aggregation prohibits the separation of the food from mark-up in the earlier year.

17. Barker and Lecumber, op. cit., p. 4; Hazari, B. R., op. cit., p. 167.

18. Note that in 1948, 70 percent of the total household consumption of textiles was directly imported while in 1963, 87 percent of textile consumption was satisfied by direct imports. However, in 1948, 72 percent of domestically produced textiles were exported, and by 1963 the fraction had risen to 91 percent. Elsewhere, we have called this the "criss-cross" effect, with more expensive textiles exported to the United States and cheaper ones imported. The same "criss-cross" effect occurred in leather goods as 100 percent of household consumption was imported in both years, despite the establishment of a leather goods export industry in this period. (See R. Weisskoff and E. Wolff,

"Linkages and Leakages: Industrial Tracking in an Enclave Economy," Economics Dept. Staff Paper No. 11, Iowa State Univ., Apr. 1975.

19. See Barker and Lecumber, op. cit., p. 9.

20. The final demand delivery of "government services," Sector 26, consists of one entry of government service expenditure, and no compositional changes are recorded. Thus, changes in the "basket" of government services must be traced directly to the input structure and related to the performance and programs of the insular government. Part of the decline in imports is due to a shift of government construction from government demand in 1948 to capital formation in 1963. The increase in imports of business services is traced to the treatment of the Commonwealth's promotional and advertising expenditures in New York as imports. Also note that government-induced imports for the United Kingdom total $10.80 compared to $23.52 in 1948 and $16.71 in 1963 for Puerto-Rico. (See Barker and Lecumber, op. cit., p. 9.)

**References**

Barkley, T. S. and J. R. C. Lecumber. 1970. The import content of final expenditure for the U.K. *Bulletin of Oxford University* 32(1).

Chenery, H. B. 1960. Patterns of industrial growth. *American Economic Review* L(4).

Chenery, H. B. and T. Watanabe. 1958. International comparisons of the structure of production. *Econometrica* 26 (October).

Dos Santos, T. 1968. *El neuvo carácter de la dependencia.* Santiago, Chile: Centro de Estudios Socio-Económicos.

Girvan, N. 1972. The development of dependency economics in the Caribbean and Latin America: Review and comparison. Discussion Paper. University of the West Indies.

Hazari, B. R. 1967. The import intensity of consumption in India. *Indian Economic Review*, 2(2) New Series.

Hirschman, A. O. 1968. The political economy of import substituting industrialization. *Quarterly Journal of Economics* 82(1).

Ingram, J. C. 1969. Some implications of the Puerto Rican experience. In *International Finance*, R. N. Cooper (ed.). Harmundsworth, Middlesex, England: Penguin.

Pack, H. 1971. *Structural Change and Economic Policy in Israel.* New Haven: Yale University Press.

Panchamukki, V. R. 1965. Import substitution in relation to technical change and economic growth. *Arthaniti* (July).

Puerto Rico Planning Board. 1964. *External Trade Statistics.* San Juan.

———. 1968. *Income and Product 1967.* San Juan.

Ross, D. F. 1966. *The Long Uphill Path.* San Juan: Talleres Gráficos Interamericana.

Tavares, M. C. 1964. The growth and decline of import substitution in Brazil. *Economic Bulletin for Latin America* 9 (March) pp. 1–60.

Tugwell, R. 1947. *The Stricken Land.* New York: Doubleday.

United Nations. 1964. *International Trade Statistics.* New York.

Weisskoff, R., R. Levy, L. Nisonoff, and E. Wolff. 1971. A multisector simulation model of employment, growth and income distribution in Puerto Rico: A re-evaluation of 'successful' development strategy. U.S. Department of Labor Research Report, July 1971.

# 27

## Energy, Environment, and Economic Growth

ANNE P. CARTER

Recent environmental problems and energy shortages have prompted questions about the future course of economic growth in the United States. Spokesmen for industry and government express concern lest capacity bottlenecks interfere with growth at the 3.5-percent rate that we have sustained over the past quarter century. Others challenge the desirability of the rapid economic growth that we have long taken for granted.

Economic growth is not a simple policy variable. Its future course will depend on many factors: on the availability of natural resources, now only vaguely appraised, both at home and abroad; on the expansion and composition of the labor force; on public and private consumption; on environmental policy; and on the specific techniques of production that will be used. While technology does not uniquely determine our development, it limits the economy's capacity to expand with any given consumption and resources. As resource supplies and environmental standards change, new technologies are developed in response. In principle, these new technologies could increase or decrease the economy's growth potential. Actually, the "first generation" technologies addressed to today's environmental and resource conditions tend to decrease it. This study gives a rough quantitative appraisal of the implications of some specific pollution abatement and energy technologies that tend to reduce the rate of economic growth.

The data for this analysis are drawn from a series of studies that articulate individual abatement techniques and energy processes in a dynamic input-output framework.[1] Three sets of innovations are considered: projected changes in the technologies of generation, transmission, and distribution of electric power over the next ten years; coal gasification; and a broad range of pollution abatement activities that would satisfy announced or projected standards for air, water, and solid waste pollution. While far from exhaustive, this is a fairly broad list of direct technological responses to environmental and resource conditions expected to come on line over the next decade. Their adoption could make important differences in our growth potential and in the relative importance of sectoral contributions to output and to the capital stock. If these changes are introduced, how will the economy of 1980–1985 differ from today's?

This article was published in *The Bell Journal of Economics and Management Science*, Volume 5, Number 2, Autumn, 1974, pp. 587–592. © AT&T Bell Laboratories, Inc. Reprinted by permission.

In a closed dynamic input-output system, the maximum rate of economic growth that can be sustained depends on all sectors' input structures, both on capital and on current account, including the level and composition of final consumption. One important conclusion of this study is that growth potential is more sensitive to changes in the propensity to consume than to increasing costs in the energy sectors. The expansion of the economic system is also constrained implicitly by the availability of labor and specific natural resources. Until very recently, the size of the labor force seemed to be the only major resource constraint that was binding on the U.S. economy. Since the work force was increasing at only one percent per year, continual increases in productivity were necessary to keep the effective labor supply sufficient for 3 to 4 percent growth. Thus, labor-saving progress dominated technical change during the post-war period. The great bulk of industrial innovations—automation, computerization, material and design changes— represented direct or indirect economies of labor, with direct labor-saving predominating.[2] Had per capita consumption remained constant, increasing labor productivity might have freed capacity for increased capital accumulation and accelerated economic expansion. However, increasing productivity was offset by rising consumption per worker and, thus, the average rate of growth was fairly steady. Commoner[3] and others point out that much of the recent increase in labor productivity and in consumption has been achieved through more intensive exploitation of nonhuman resources that once seemed plentiful, particularly energy and the natural environment. Saving labor at the expense of other natural resources may well have hastened the present "crisis" at a time when environmental and energy constraints have also begun to bind.

Just as labor-saving innovations forestall labor shortages, so our environment- and energy-oriented innovations address potential scarcities of natural gas, oil, clean air and water, etc. Precipitators, filters, and treatment plants are designed to protect the air and water; nuclear power and coal gasification are designed to supplement or supplant scarce supplies of clean fossil fuels; higher-voltage transmission and underground distribution lines should convey electric power to consumers with minimal environmental damage. While these new technologies, once installed, do not require much direct labor, they generally require more capital and intermediate inputs than the old technologies that they supersede or augment. These added requirements impose a drag on the growing system.

In sum, two types of influences tend to limit growth in the coming decade. First, increased requirements for delivering a given level of current consumption leave a smaller portion of productive capacity for capital formation. Second, environmental considerations and scarcities of nonhuman resources complicate the technological challenge of increasing the productivity of labor. Innovators must now deal with environmental problems as well as labor efficiency. It may well prove feasible to increase the efficiency of natural resources and labor simultaneously, but the problem is more difficult than that posed by labor alone. Unless effective labor and resource supplies increase in step with capital accumulation, plans to expand capacity cannot be realized.

To the extent that natural resource pressures actually reduce the rate of economic growth, there will be less pressure to increase labor productivity. If growth of industrial capacity were very slow, natural increases in the labor force alone might provide sufficient manpower for the growing stock of equipment. However, we shall show that minor changes in consumption could readily offset the increased intermediate and capital requirements of the new technologies so as to maintain or even increase the long-run growth rate. Thus, it is hard to predict whether the problems of environment and energy supply will tend to relax or to intensify historic pressures for ever-increasing labor productivity.

## Computations

### Analytical Framework

A closed dynamic input-output model is the basis for all the computations:

$$(I - A)X - B\dot{X} = 0 \qquad (27\text{--}1)$$

where $X$ is a vector of total sectoral outputs and $\dot{X}$ is a vector of their time derivatives.

$A$ and $B$ are current account and capital coefficient matrices, respectively. The model is closed by the addition of a household row and column to $A$ and $B$. The household row is a vector of coefficients representing income and indirect tax payments; the household column is a vector of coefficients representing expenditures by households and government plus net exports.

Equation (27–1) says that outputs are allocated to current account uses, $AX$, and to expansion, $B\dot{X}$. Since capital formation is endogenous, the sum of expenditures by the household industry is equal in national accounts terminology to gross national product minus gross private capital formation. If all sectors were to grow at a uniform rate, Eq. (27–1) could be rewritten:

$$(I - A - \lambda B)X = 0 \qquad (27\text{--}2)$$

where $\lambda$ is the uniform or "turnpike" growth rate. Tsukui (1966) has shown that $\lambda$ represents the unique uniform growth rate for the economy consistent with full capacity utilization of all sectors. This turnpike growth rate proves to be the maximum growth rate consistent with the given flow and capital coefficient matrices for a broad range of initial and terminal outputs. Corresponding to the turnpike growth rate, $\lambda$, is a set of output proportions, the positive eigenvector of the matrix $(I - A - \lambda B)$. Thus, $\lambda$ measures the economy's long-run growth potential with the specified input structures $A$ and $B$, while the turnpike output proportions indicate rough norms for the relative importance of sectors on the uniform growth path. Tsukui (see Murakami et al., 1970) and Bródy (1970) have estimated turnpike paths for Japan and the United States. Their findings suggest that both these economies operate reasonably close to the computed paths.

### Computation Procedures

The iterative algorithm of Bródy (1970) was used to compute uniform growth rates and output proportions. We begin with a set of base year output proportions, $X^0$, estimate a trial value of $\lambda$ and reestimate equilibrium output proportions, $X^1$, consistent with the first estimate of $\lambda$. In outline, the computation sequence is

$$\lambda^k = \frac{(1, 1, \ldots, 1)(I - A)X^{k-1}}{(1, 1, \ldots, 1)BX^{k-1}} \tag{27-3}$$

$$X^k = (A + \lambda^k B)X^{k-1}$$

where $k$ is the number of the iteration. All computations converged within seven iterations.[4]

### Comparative Dynamics

Structural changes in matrices $A$ and $B$ will of course bring changes in the turnpike growth rate and output proportions. In this study the economic impacts of structural changes are evaluated in terms of comparative dynamics. We establish benchmark growth rate and output proportions on the basis of capital and current account matrices representing the structure of production in the early 1970s. Successive changes in the base year matrices are introduced to simulate the adoption of new technologies of energy production and stricter standards of pollution abatement. The effects of these structural changes, singly and in combination, on $\lambda$ measure their impacts on the economy's growth potential; changes in computed output proportions represent differential impacts of the specified structural changes on individual sectors.

The effects of changes in intermediate input and capital coefficients differ depending on concomitant changes in the levels and patterns of final consumption. To illustrate the impact of changing consumption, alternate consumption vectors were introduced in combination with the changes in industrial coefficients. The alternative consumption patterns are discussed at the conclusion of this section.

### Implementation

In this portion of the paper we shall outline the base year matrices, the technological variants, and the consumption variants.

*Base Year Matrices.* The 83-order 1970 coefficient matrix of the Interagency Growth Project[5] was chosen as the base year $A$ matrix. This coefficient matrix is the result of a set of 1970 projections made in the mid 1960s and is not based on actual 1970 statistics. The capital matrix is an update of the 1958 capital matrix of the Harvard Economic Research Project to 1970–1975 technologies by the Battelle Memorial Institute.[6] Updating incorporates engineering information and conjecture. The fact that the $A$ and $B$ matrices are both projections should not seriously impair their usefulness in this study, since they serve primarily as general points of reference for analyzing the impact of specific further changes. Both $A$ and $B$ are in 1958 prices. Annual replacement requirements were estimated and added to the $A$ matrix. The current account and capital matrices were augmented to include a "household" row consisting of the sum of income payments and indirect taxes per dollar of output for each industry and a "household" column representing all purchases of final outputs (exclusive of gross capital formation) per dollar of net national income. The turnpike growth rate computed for the base $A$ and $B$ matrices was 3.5 percent per annum.

*Technological Variants.* The base matrices were modified by introducing coefficients that represent the new technologies. Quantitative estimates of new industrial structures are drawn from specialized studies by Istvan, Just, Jenkins, Berlinsky, Kok, and Dorsey.[7] Four major sets of structural changes were considered.

*i.* The first consists of changes in technology for electric power generation, transmission, and distribution. The input coefficients in the base year's flow coefficient matrix were replaced by coefficients for a projected 1980 input structure of the electric power sector. These new coefficients were estimated by Istvan (1972). The corresponding column of the base year capital coefficient matrix was similarly replaced by Istvan's estimates of 1980 capital coefficients. Given Federal Power Commission projections of future process mix, Istvan's projections show that increases in the relative importance of nuclear generation, additions of pollution abatement equipment, extra-high voltage transmission and underground distribution lines will result in greater costs and, in particular, greater capital intensity in electric power delivery. He projects an increase in the total capital coefficient for the electric power industry from 3.2 to 4.8 and a 9-percent increase in real costs on current account (including replacement).

*ii.* The second set of modifications converts the increment in projected over present fossil-fuel consumption from electric utilities to gasified coal. If all additional consumption of fossil fuel took the form of gasified coal, then by 1985 the input coefficients of coal, petroleum, and natural gas into electric utilities would be cut to half of their former values. These coefficients were so cut and the coal gasification industry was then added to the base matrix. Its row has a single entry representing utilities' consumption of gasified coal equivalent in Btu value to the reductions in conventional fossil-fuel consumption. The column represents Just's projected input structure for the Hygas process of coal gasification (Just, 1972). This represents the most speculative of all the new technologies included in this study. Because there is no on-line experience with Hygas or any of the other possible methods, the coefficient estimates are more speculative than estimates for, say, nuclear power generation, cooling towers or sanitary landfill of solid wastes. Environmental standards will require some process of coal gasification and/or liquifaction as reliance on coal increases. This process represents a feasible medium-cost variant of a number of coal gasification alternatives. In all probability, the volume of gasified coal assumed here is far greater than the amount we shall actually produce by 1985. To the extent that gasification capacity is lower, however, costs of alternative environmental protection policies would have to be taken into account.

*iii.* Over the past quarter century electric power input coefficients into all consuming sectors have been increasing steadily at an average rate of roughly 3.5 percent per year (Carter, 1970) and Federal Power Commission projections of electricity consumption implicitly assume that these trends will continue over the next decades (U.S. Federal Power Commission, 1969). Increased electricity consumption is likely to intensify the impact of changing energy technologies. Computations representing the technological changes in (i) and (ii) were repeated with the base year energy row coefficients increased by 40 percent. The adjusted electric power row represents the economy's power consumption propensities a decade hence if present trends continue.

*iv.* To simulate the impact of economy-wide pollution controls, the basic flow and capital coefficient matrices were augmented to include an abatement cost row and a dummy abatement column for each of six types of pollution: particulate air pollution, industrial water pollution, thermal pollution, municipal sewage disposal, strip mining pollution, and municipal solid waste disposal.[8] The levels of abatement activities are calculated to satisfy the abatement standards of the Clean Air Act of 1971 for particulates and requirements for primary treatment (elimination of 85 percent of biochemical oxygen demand and suspended solids) of municipal and industrial waste water.

*Consumption Variants.* In the base matrix consumption proportions are those of 1970 and final consumption grows at the same rate as all other sectors on the turnpike path. Per capita consumption in the future will depend on the rate of growth of the population relative to the rate of growth of total final goods. In the absence of increases in the work force the economy can expand only as fast as increasing per capita productivity permits. Under these conditions, per capita real income increases at the same rate as the economy's uniform growth rate. Should population increase at the same rate as the economy, the rate of increase in per capita income would be zero. Over the next decade the rate of growth of the population will be positive but probably lower than the growth rate of the economy and rising per capita income is generally expected. Consumption patterns can be expected to change as per capita income rises and as product prices change. Because the household sector is large, the structure of consumption has a major influence on growth potential. The impact of changing industrial technologies was computed with each of four different consumption structures.

*i.* The first is that of the base year.

*ii.* The second makes a rough allowance for expected changes due to rising per capita income. Ten years hence, assuming 2.5-percent growth per annum in labor productivity, we can expect real income per capita to be 28 percent higher than it is now.[9] To estimate the effect of this rise in average per capita income on consumption patterns, we assume that spending patterns will remain the same as they are now for any given income level. But as average income goes up, the relative importance of high-income spending patterns will increase. To estimate one possible structure of future consumption we assume that all incomes will increase proportionally as GNP rises. Each income-specific spending pattern is then assigned a weight equal to the proportion of income received by those who earned 28 percent *less* income in the base year. The most striking difference between this projected structure of consumption and that of the base year lies in the personal savings rates. Because people with high incomes save much more than those in lower income groups, the percentage rate of savings under variant (ii) is triple that

Table 27.1  Scenarios Computed in This Study

| Consumption structure | (i) 1970 | | (ii) Income specific consumption structures reweighted for 28 percent increase in real income | (iii) One percent increase in savings over 1970 | (iv) Energy saving changes |
|---|---|---|---|---|---|
| Electricity consumption coefficients | (a) 1970 | (b) 1970 × 1.4 | | | |
| Technology | | | | | |
| 1. Base year | X | X | X | X | X |
| 9. All pollution abatement | X | | X | X | X |
| 10. 1980 electric | X | | | | |
| 11. 1980 electric with coal gasification | X | X | | | |
| 12. All abatement 1980 electric and coal gasification | X | X | X | X | X |

in the base year. History shows, however, that personal savings remained a remarkably stable percentage of the national income, rising only very slowly with time and/or per capita income. Consumers may be ready to save more in the future if environmental improvement requires it.

*iii.* Here the savings rate is allowed to increase by only *one* percentage point over the base pattern (i) in keeping with historical trends. Proportions of final spending on goods and services are kept the same as in variant (ii).

*iv.* Over the next ten years rising energy prices are likely to induce substitution in favor of less energy-intensive products. This variant superimposes energy-saving changes into the base year consumption patterns. Direct consumption of energy is cut by 20 percent while expenditures on other consumer items are increased proportionally so as to maintain the base year savings rate.

Table 27.1 lists the various scenarios that were computed.


# Results


## Effects of Selected Changes on $\lambda$

Table 27.2 shows the effects on growth potential of specific technological shifts, separately and in combination, with the fixed base year consumption structure. Few of these changes affect growth potential by more than a few tenths of a percentage point. However, the cumulative effect of groups of changes is appreciable. Abating all water, air, and solid waste pollution to meet currently

Table 27.2   Long-term Growth Potential ($\lambda$) With New Energy and Abatement Technologies

| Structural change | $\lambda$ percent per annum |
|---|---|
| 1. None (base year structure) pollution abatement | 3.54 |
| 2. Particulate air pollution ( $\simeq 99$ per cent) | 3.44 |
| 3. Industrial bod and suspended solids (primary treatment) | 3.47 |
| 4. Municipal bod and suspended solids (primary treatment) | 3.45 |
| 5. Acid mine drainage | 3.49 |
| 6. Thermal pollution (cooling towers) | 3.52 |
| 7. All water pollution | 3.33 |
| 8. Solid waste disposal | 3.33 |
| 9. All pollution | 3.03 |
| New energy technologies | |
| 10. 1980 electric power technology | 3.32 |
| 11. 10 + coal gasification | 3.06 |
| 12. Energy and pollution control (9 × 11) | 2.59 |

*Source:* See text.

announced goals reduces $\lambda$ from 3.5 to 3.0 percent. Projected changes in electric power generation and distribution technology are introduced separately and also combined with coal gasification. The combined changes lead to a 0.5 percentage point cut in $\lambda$. With pollution controls and changes in electric power generation combined, long-run growth potential is now reduced to 2.6 percent per year.

### *Effects of Identified Changes Combined With Rising Industrial Use of Electric Power*

The impact of electric power technology on the growth rate is magnified greatly when electric power usage expands. Table 27.3 shows how new technologies affect the growth rate when power consumption coefficients are increased. Istvan's projected changes in electric power technology alone reduce growth potential by two-tenths of a percentage point. If electric power consumption coefficients are

Table 27.3   Long-term Growth Potential ($\lambda$) with Varying Energy Technologies and Electricity Use (Percent Per Annum)

| Energy technology | Electricity consumption coefficients | |
|---|---|---|
| | 1970 (a) | 1970 × 1.4 (b) |
| 1. Base year | 3.5 | 2.9 |
| 10. 1980 electric | 3.3 | 2.5 |
| 11. 1980 electric with coal gasification | 3.1 | 2.1 |

*Source:* See text.

Table 27.4  Long-term Growth Potential (λ) with Varying Technologies and Consumption (Percent Per Annum)

| Technology | Consumption Structure | | | |
|---|---|---|---|---|
| | (i) | (ii) | (iii) | (iv) |
| 1. Base year | 3.5 | 5.7 | 3.8 | 3.6 |
| 9. All pollution abatement | 3.0 | 5.1 | 3.3 | 3.1 |
| 12. Electric coal gasification and all pollution abatement | 2.6 | 4.7 | 2.9 | 2.7 |

*Source:* See text.

multiplied by 1.4 in the base matrix, λ is reduced by a 0.6 percentage point. If this increase in electric power consumption is used in conjunction with Istvan's changes, λ is reduced by an 0.8 percentage point. The effects of increased power consumption are even more dramatic—a reduction of 1.0 percentage point—when the increase takes place in a technological scenario that includes coal gasification as well as Istvan's changes. The costs associated with projected increases in electric power usage are larger than those incurred from new developments in electric power technology. Hence, increasing power consumption cuts the growth rate more than the switch to new techniques. The tendency of these two developments to compound each other is striking. Separately, neither has an overwhelming impact on growth potential. The combined effect, 1.4 percentage points, is certainly large enough to be taken seriously.

### Technological Changes Combined with Changes in Consumption

"Households" is the largest sector of our model, comprising all public and private final consumption. At present its level of consumption is roughly five times the volume of gross private domestic investment. Hence, it is no surprise that growth potential is more sensitive to a given percentage change in household spending and saving patterns than to the same percentage change in individual industrial sectors. For each of the four consumption structures discussed in "Computations," Table 27.4 shows computed values of λ assuming all pollution controls, separately and in combination with new electric generating and coal gasification technology.

Table 27.4 confirms the sensitivity of the economy's growth to changes in consumption patterns. Reducing the energy content of final consumption (iv) greatly attenuates the impact of new energy technologies. Were household savings at any given real income level to remain constant as in variant (ii), increasing per capita income would raise savings to a point where 5- or 6-percent growth could be sustained, even in the face of our major "deteriorations" in technology. Greater affluence will mean spending more on entertainment and services, less on food and rent. The effect of such a shift in spending patterns tends to increase growth potential even when savings are restricted to only small percentage point increases over the base year [variant (iii)].

Figure 27.1 shows the trade-offs between consumption and growth with and without pollution abatement and new energy technologies.

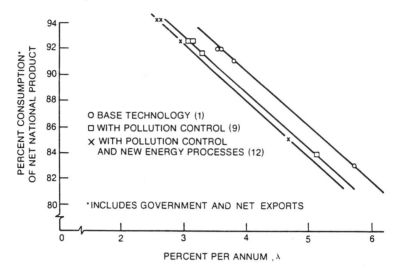

Fig. 27.1. Tradeoffs between consumption and growth with changing technologies.

### Output Proportions

By changing intermediate demand, new technologies alter the relative importance of individual sectors. Higher capital requirements of new energy and pollution control technologies might be expected to increase the relative importance of capital producing sectors. On the other hand, deceleration of economic growth means a smaller proportion of economic resources devoted to capital formation. Because these tendencies offset each other, it is difficult to make a priori judgments about the effects of the new technologies on the relative importance of individual sectors. Our computations give some indication of what they may be.

Figures 27.2 and 27.3 show the proportionate contributions of the new construction and iron and steel sectors to total gross output under various scenarios that we have computed. The proportion of total gross domestic activity is measured on the vertical axis while $\lambda$ is shown on the horizontal axis. The number next to each point indicates a combination of technological and consumption structures identified in Tables 27.2, 27.3, and 27.4.

Table 27.5 shows that the relative importance of new construction in the economy varies from 2.5 to 4.1 for different assumptions. Since new construction is the largest single industrial component of most sectors' capital stocks and also of household capital, it is not surprising that this sector's importance varies directly with the growth rate. However, the relative importance of the iron and steel sector is constant, because steel is used in many consumer products (automobiles, cans, household equipment) as well as in capital goods.

The relative importance of most sectors is, like that of steel, virtually invariant with respect to the changes considered in this study. This is demonstrated in Table

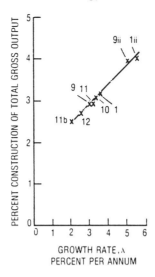

Fig. 27.2. Importance of construction with changing growth rates.

27.5. Utilities increase and decrease in relative importance as electricity input coefficients are varied. Food declines in relative importance as high-income consumption patterns are introduced and the growth rate rises into the neighborhood of 5 percent. By and large, however, the proportions cited for each sector in Table 27.5 are remarkably stable. This tendency is explained by the limited range of the growth rates covered and the modest proportion of total resources devoted to growth under any of the options considered.

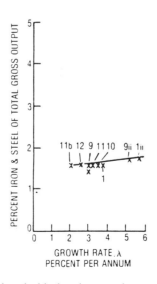

Fig. 27.3. Importance of iron and steel with changing growth rates.

Table 27.5  Sectoral Shares of Total Gross Output with Various Technologies and Consumption Patterns (Percent of Total Gross Output)

|  | | Scenario number[a] | | | | | | | |
|---|---|---|---|---|---|---|---|---|---|
|  | | 11b | 12 | 9 | 11 | 10 | 1 | 9ii | 1ii |
|  | $\lambda =$ | 2.1 | 2.6 | 3.0 | 3.1 | 3.3 | 3.5 | 5.1 | 5.7 |
| Producing sector | | | | | | | | | |
| New construction | | 2.5 | 2.7 | 2.9 | 2.9 | 3.0 | 3.1 | 3.9 | 4.1 |
| Maintenance construction | | 2.2 | 2.1 | 4.4 | 2.1 | 2.1 | 2.2 | 2.1 | 2.1 |
| Food | | 4.4 | 4.3 | 4.4 | 4.4 | 4.3 | 4.4 | 3.8 | 3.8 |
| Iron and steel | | 1.6 | 1.6 | 1.6 | 1.6 | 1.6 | 1.6 | 1.8 | 1.8 |
| Automobiles | | 3.1 | 3.1 | 3.1 | 3.1 | 3.1 | 3.1 | 3.1 | 3.1 |
| Electric utilities | | 2.8 | 2.2 | 2.0 | 2.1 | 2.2 | 2.0 | 1.8 | 1.8 |

[a] Scenarios are listed in Table 27.1.
*Source:* See text.

### *Changing Composition of Capital Stock*

While the composition of annual output remains relatively stable with respect to our changes, a growing proportion of the capital stock is devoted to energy and abatement capital and the average capital-output ratio for the economy rises by as much as 6 percent over the base year economy. Table 27.6 shows that by and large higher capital intensity is characteristic of the lower growth rate scenarios. It also gives the proportion of total capital invested in utilities and in pollution abatement under various sets of assumptions. Utilities accounted for 9.9 percent of total investment in the base year. Initially, they have a much greater capital-output ratio than most other sectors. Increasing investment requirements for utilities and increasing use of electric power add significantly to overall capital requirements.

### Conclusions

The above results reveal the sensitivity of the economy—of its sectoral proportions and growth potential—to prospective structural changes. Taken separately, no one of the technological changes considered in this study exerts a decisive effect on the capital and output proportions of the economy nor on its growth potential. Nevertheless, their combined effect could be substantial. Furthermore, we have dealt with some of the more important but not all of the major adjustments that environmental and resource pressures will induce over the next decades. As the economy gets larger, progressively stricter environmental safeguards will be necessary simply to maintain any given standards of air and water quality. More substances will be recognized as pollutants. Most known abatement technologies involve very sharply increasing costs when the percentage of residual pollution is to be lowered. Significantly higher costs of extraction and refining of all fuels and ores are generally anticipated.

Table 27.6  Economy-wide Capital/Output Ratios and Proportion of Investment in Utilities and Abatement with Varying Structures (Dollars Per Dollars of Total Gross Output)

| Scenario number | Growth rate ($\lambda$) (% per annum) | Capital/output ratio | Percent of total investment in | |
|---|---|---|---|---|
| | | | Utilities | Pollution abatement |
| 11b | 2.1 | 0.84 | 15.3 | —[a] |
| 1b | 2.9 | 0.81 | 12.1 | —[a] |
| 9i | 3.0 | 0.80 | 9.8 | 3.2 |
| 11a | 3.1 | 0.81 | 11.8 | —[a] |
| 1 | 3.5 | 0.79 | 9.9 | —[a] |
| 12ii | 4.7 | 0.82 | 10.8 | 2.5 |
| 9ii | 5.1 | 0.80 | 8.9 | 3.2 |
| 1ii | 5.7 | 0.78 | 9.1 | —[a] |

[a] Pollution abatement costs are not estimated for these scenarios.

*Source:* See text.

On the other hand, adaptive responses to rising energy and pollution costs soften the impact of these resource and environmental constraints. Our computations show, for example, that a 3.5-percent per year increase in electricity consumption coefficients greatly enhances the growth-limiting effects of new energy technologies. Conversely, a modest reduction in electricity coefficients could neutralize their impact. Ayres and Gutmanis (1972) point out that current best practice technologies in many polluting sectors already show substantially lower gross emissions than average technologies now in use. The cumulative impact of small industrial and consumer adaptations to changing energy and environmental cost conditions is hard to anticipate. It could be very large.

Within the framework of this analysis we could study changing economic options as additional technological changes evolve. While the relative contributions of individual sectors do not change markedly in the face of the changes just considered, it would be unrealistic to ignore the lags and potential bottlenecks likely to arise in a sequence of major technological transitions. Leontief (1970) and Istvan (1976) offer a more flexible framework for tracing year-to-year changes in outputs and investment that depart from the turnpike path.

The value of a general equilibrium approach extends beyond the limited area of technological policy to broader economic questions concerning the changing balance of income and tax policy, public spending, prices, and labor productivity. For some years, at least, the economy's growth potential with any given consumption structure will be lower than it used to be. However, this study has shown that a small once-for-all increase in consumer savings or other shifts in private or public final spending patterns can easily offset the drag. Thus, there remains a wide range of trade-offs between faster growth and consumption at any given point. Cross-sectional evidence shows that high-income consumers save a substantially greater portion of their incomes than low-income consumers. Were consumer savings to rise with income over time at even half the rate that they do in cross section, the growth rates of the past 25 years could easily be sustained. Actual future growth will depend on the complex policies that determine real savings for any given technological context.

Fundamentally, the turnpike growth rate measures the rate at which the economy's productive capacity—its capital stock—can be increased. Real growth also requires that the effective labor supply be sufficient to man the growing productive capacity. Increases in labor productivity have contributed more than growth in the labor force to rising output in recent years.[10] Recent studies suggest, however, that the rate of increase of productivity is decelerating.[11] Some of this deceleration may reflect resource pressures. Whatever its explanation, lagging productivity change could also limit growth. Here, as in the case of consumer savings, the economy retains a broad range of options. While both population and productivity increases are decelerating, rising labor force participation can easily offset their effect on growth potential. At a time when women are clamoring for a share in the job market, when youth suffer high rates of unemployment, and when firms experiment with the four-day week, labor shortages need not obstruct growth—not yet anyway.

Pollution abatement might be viewed as a public good (i.e., as a form of consumption), but our conclusions do not rest on this assumption. We can meet environmental standards ånd resource constraints over the next decade and still maintain or even increase present growth rates for conventional goods. This should not be construed as advocacy of rapid economic expansion. The social and environmental evils of bigness are not all captured and dealt with in pollution abatement vectors; nor is it possible to represent all the trade-offs between the present and the remote future, at home and abroad, in a growth model. Despite present constraints, we can still grow. Whether or not we should or want to is another question.

## NOTES

1. See Carter (1976).
2. See Carter (1970).
3. See Commoner et al. (1971).
4. Computations were performed on the PDP-10 computer at the Feldberg Computer Center, Brandeis University, using the PASTIM matrix manipulation program of Richard Drost.
5. See U.S. Department of Labor (1966).
6. See Fisher and Chilton (1971).
7. In Carter (1976).
8. See Leontief (1970).
9. Since the computations are illustrative, the exact increase in income per capita chosen here is somewhat arbitrary. Precise consistency with the computed growth rate could be achieved by an iterative process.
10. See Solow (1957).
11. See Nordhaus (1972) and Almon et al. (1973).

## References

Almon, Jr., C., M. B. Buckler, L. M. Horwitz, and T. C. Reimbold. 1973. *1985 Interindustry Forecasts of the American Economy*, INFORUM Research Report No. 9. College Park, Maryland: University of Maryland, Bureau of Business and Economic Research, August 1973 (preliminary).

Ayres, R. U. and I. Gutmanis. 1972. Technological change, pollution and treatment cost coefficients in input-output analysis. In *Population, Resources, and the Environment*, Vol. 3. Ronald G. Ridker, ed. The Commission on Population Growth and the American Future *Research Reports*. Washington, D.C.: U.S. Government Printing Office.

Bródy, A. 1970. *Proportions, Prices and Planning*. Amsterdam: North-Holland Publishing Company.

Carter, A. P. 1970. *Structural Change in the American Economy*. Cambridge: Harvard University Press.

————. Ed. 1976: *Energy and the Environment: A Structural Analysis*. Hanover, N.H.: University Press of New England.

Commoner, B., M. Corr, and P. J. Stamler. 1971. The causes of pollution. *Environment*, Vol. 13, No. 3 (April): pp. 59–78.

Fisher, W. H. and C. Chilton. 1971. *An Ex Ante Capital Matrix for the United States, 1970–1975*. Columbus: Battelle Memorial Institute.

Istvan, R. 1972. *1890 Inputs for Private Electric Utilities*. Report to the Interagency Growth Project, Harvard Economic Research Project, Cambridge, August 1972, mimeographed.

————. 1976. Interindustry impacts of projected electric utility capital formation. In A. P. Carter, ed., *Energy and the Environment: A Structural Analysis*. Hanover N.H.: University Press of New England.

Just, J. L. 1972. Impacts of new energy technology using generalized input-output analysis. Unpublished Ph.D. dissertation. Cambridge: Massachusetts Institute of Technology.

Leontief, W. W. 1970. The dynamic inverse. In A. P. Carter and A. Bródy, eds., *Contributions to Input-Output Analysis*, Vol. 1, pp. 17–46. Amsterdam: North-Holland Publishing Company.

————. 1970. Environmental repercussions and the economic structure: An input-output approach. *Review of Economics and Statistics* 52, 262–271.

Murakami, Y., K. Tokoyama, and J. Tsukui. 1970. Efficient paths of accumulation and the turnpike of the Japanese economy. In A. P. Carter and A. Bródy, eds., *Contributions to Input-Output Analysis*, Vol. 2, pp. 24–47, Amsterdam: North-Holland Publishing Company.

Nordhaus, W. D. 1972. The recent productivity slowdown. *Brookings Papers on Economic Activity*, Vol. 3.

Solow, R. M. 1957. Technical change and the aggregate production function. *Review of Economics and Statistics* 39, 312–320.

Tsukui, J. 1966. Turnpike theorem in a generalized dynamic input-output system. *Econometrica* 34, 396–407.

U.S. Department of Labor, Bureau of Labor Statistics. 1966. *Projections 1970*—Interindustry Relationships—Potential Demand Employment. Bulletin No. 1536. Washington, D.C.: U.S. Government Printing Office.

U.S. Federal Power Commission, Bureau of Power. 1969. *Trends and Growth Projections of the Electric Power Industry*. Washington, D.C.: U.S. Government Printing Office.

# 28

# Embodied Energy and Economic Valuation

ROBERT COSTANZA

The thesis that available energy both limits and governs the structure of human economies is not new. In 1886, Boltzmann suggested that life is primarily a struggle for available energy. Soddy stated in 1933: "If we have available energy, we may maintain life and produce every material requisite necessary. That is why the flow of energy should be the primary concern of economics" (p. 56).[1] The flow of energy has not been the primary concern of mainstream economists, although the importance of energy to the functioning of economic systems has by now been recognized by almost everyone. The debate now focuses on the nature and details of the energy connection, and the conclusions are critically important to several aspects of national policy. In this article, the earlier input-output analyses of energy-economy linkages are extended by incorporating the energy costs of labor and government services and solar energy inputs.

The flow of energy is the primary concern of what has come to be known as energy analysis.[2-4] An important aspect of energy analysis is the determination of the total (direct and indirect) energy required for the production of economic or environmental goods and services. This total has been termed the embodied energy. For example, the energy embodied in an automobile includes the energy consumed directly in the manufacturing plant plus all the energy consumed indirectly to produce the other inputs to auto manufacturing, such as glass, steel, labor, and capital. A problem immediately apparent from this definition is the choice of procedures for calculating indirect energy requirements. Embodied energy values are thus contingent on methodological considerations.

Input-output (I-O) analysis is well suited to calculating indirect effects in a systematic and all-inclusive accounting framework. Hannon[5] and Herendeen and Bullard[6] adapted this technique to calculate embodied energy. Controversy still exists concerning the relevant system boundaries for such calculations (see note 3).

## System Boundaries

The choice of system boundaries is critical because it determines the distinction between net inputs and internal transactions. Net inputs are considered to be

This article was published in *Science*, Volume 210, December 12, 1980, pp. 1219–1224. © American Association for the Advancement of Science. Reprinted by permission.

independent and exogenously determined, whereas internal transactions are endogenous and interdependent. The net inputs are what economists refer to as primary factors. In the national income accounts, they are "value added." The I-O technique, in essence, distributes a net input vector through a matrix of internal interactions to balance against a net output vector.

Most recent embodied energy calculations based on national I-O tables have employed the standard definitions of economic I-O boundaries (see note 6). With these definitions, the net input (or value added) vector includes labor, government services, capital services, and energy and other natural resources (raw materials). The corresponding financial categories are employee compensation, indirect business taxes, and property-type income. The sum of these net inputs, in dollar units, is the gross national product (GNP). Energy (fossil fuels, nuclear fuels, and solar) is a small component of the GNP in dollar units. This has led several people to conclude that energy is a minor component in economic production, a conclusion that would be accurate if the components of the net input vector as currently defined (GNP) were mutually independent, as is usually assumed.

Most proposals to increase the "energy efficiency" of economic activity are ultimately based on the assumption of mutual independence of primary factors, since increasing energy efficiency entails substituting other primary factors (capital, labor, government services, or other natural resources) for fuel inputs. The question is: Are the components of the net input vector as currently defined really independent? Are the conventional primary factors—capital, labor, natural resources, and government services—free from indirect energy costs? A strong case can be made for the contention that they are not.[7-9] In this article, I present the case for the interdependence of the currently defined primary factors, detail a method for using I-O data to calculate embodied energies so as to take account of this interdependence, and interpret the results.

**Primary Factors**

From a physical perspective, the earth has one principal net input—solar energy. Although very small amounts of meteoric matter also enter the earth's atmosphere, and deep residual heat may continue to drive crustal movement, there is no stream of spacecraft carrying workers, government mandates, and capital structures onto the planet. Thus, practically everything on the earth can be considered to be a direct or indirect product of past and present solar energy. The same cannot be said for the other "primary" factors. Fossil fuels and other natural resources represent millions of years of embodied sunlight. Environmental flows (such as winds, rain, and rivers) represent embodied sunlight of more recent origin. Humans, under this view, are the product of millions of years of solar-powered research and development and are maintained by an agriculture that uses both current sunlight and fossil sunlight. From this perspective, industrial capital is obviously created by the economic process and is not a net (or primary) input.

As Georgescu-Roegen points out: "On paper, one can write a production function any way one likes, without regard to dimensions or to other physical constraints" (p. 97).[10] Doing just this has allowed some economists to ignore critical real interdependences and to conclude, for example, that "There are presently extensive possibilities of substitution between resources and other factors [capital]" (p. 64).[11] Georgescu-Roegen goes on to say: "In actuality, the increase of capital implies an additional depletion of resources." Odum (see note 9) has pointed out that the currently defined primary factors are really interdependent by-products of our one observable net input—solar energy.

How can this interdependence of primary factors be taken into account in an analytical model? In an I-O framework, one can simply expand the boundaries so that the net input to the model coincides with the net input to the real system. In practice, most of the interdependences can be captured by considering households and government to be endogenous sectors. This represents a return to Leontief's concept of a "closed" economic system,[12] with the system boundaries, in this case, placed so that only current solar energy and the energy embodied in fuels and other natural resources enter as a net input.

In a closed Leontief model, households and government are treated like any other sector, with technical coefficients based on the household and government consumption (inputs) used to produce labor and government service outputs. As with standard I-O analysis, this is strictly an accounting of inputs and outputs. The question of whether the current standard of living and level of government spending is good or bad, too high or too low, necessary or wasteful is not and need not be asked in this format.

### Input-Output-Based Energy Accounting

The I-O technique for calculating embodied energy involves defining a set of energy balance equations (one for each sector) and solving the resulting set of simultaneous linear equations for the energy intensity coefficient vector $\varepsilon$, which is the energy required directly and indirectly to produce a unit commodity flow.

Figure 28.1 shows the basic energy balance for a sector, where $x_{ij}$ is the transaction from sector $i$ to sector $j$. $x_j$ is the total output of sector $j$, $\varepsilon_j$ is the embodied energy intensity per unit of $x_j$, and $E_j$ is the external direct energy input to sector $j$. Thus the energy balance for the $j$th sector is

$$E_j = \varepsilon_j x_j - \sum_{i=1}^{n} \varepsilon_i x_{ij} \qquad (28\text{--}1)$$

In matrix notation for all $n$ sectors

$$E = \varepsilon(\hat{x} - x) \qquad (28\text{--}2)$$

where $E$ is a row vector of direct external energy inputs, $\hat{x}$ is a diagonalized vector of gross output flows, and $x$ is the $n$ by $n$ transactions matrix. One can solve for $\varepsilon$ as

$$\varepsilon = E(\hat{x} - x)^{-1} \qquad (28\text{--}3)$$

More detailed expositions of the technique and examples can be found in notes 6, 13, and 14.

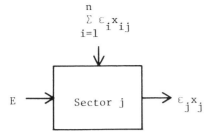

Fig. 28.1. Single-sector energy balance.

The modifications to include labor and government in the model require expanding the transactions matrix $(x)$ to include two more sectors, households and government (Figure 28.2). These sectors receive goods and services from the other sectors in proportion to personal consumption expenditures and government expenditures. They provide services to the other sectors in proportion to employee compensation, indirect business taxes, and some percentage of property-type income. The modifications and approximations are discussed in greater detail in note 14, along with the method employed for estimating solar energy inputs based on land and water use. With household and government sectors considered endogenous, the GNP as currently defined is no longer the net input and output of the model. Personal consumption and government expenditures are now internal transactions, leaving gross capital formation, net inventory change, and net exports as the new net output. Likewise, wages and taxes are internal payments, leaving capital consumption allowances and payments to land and resources as the new net input. To be complete, gross capital formation should include human and government capital formation as well. This implies that some consumption categories in the current model (such as spending on education) would be better handled as capital formation categories. Data on these categories for the U.S. economy have recently been calculated by Kendrick[15] but have not been incorporated here. The effects of this omission will be discussed later.

Input-output methodology does not include capital stocks explicitly, since a static equilibrium is assumed. The flows produced by stocks and the flows necessary to maintain and expand stocks are included (as gross capital formation), so stocks are taken into account implicitly. However, even this picture is somewhat distorted, because by convention gross capital formation is credited to the industries producing the capital, not to those utilizing it. This distortion is correctable and modifications of the I-O model that give a more accurate picture of capital flows have been constructed.[16] These modifications were not included in the results presented here: they have been shown to cause only a 7 to 8 percent change in the average energy intensity.[16]

Solar energy inputs were added to the $E$ vector after correcting for the lower thermodynamic usefulness of direct sunlight in comparison with fossil fuel. Electricity represents an upgraded, more useful form than fossil fuel, requiring, directly and indirectly, about 4 British thermal units (Btu's) of fossil fuel per Btu of electricity to produce (see note 6). Likewise, fossil fuel represents an upgraded, more useful form than the solar energy that produced it. To account for this

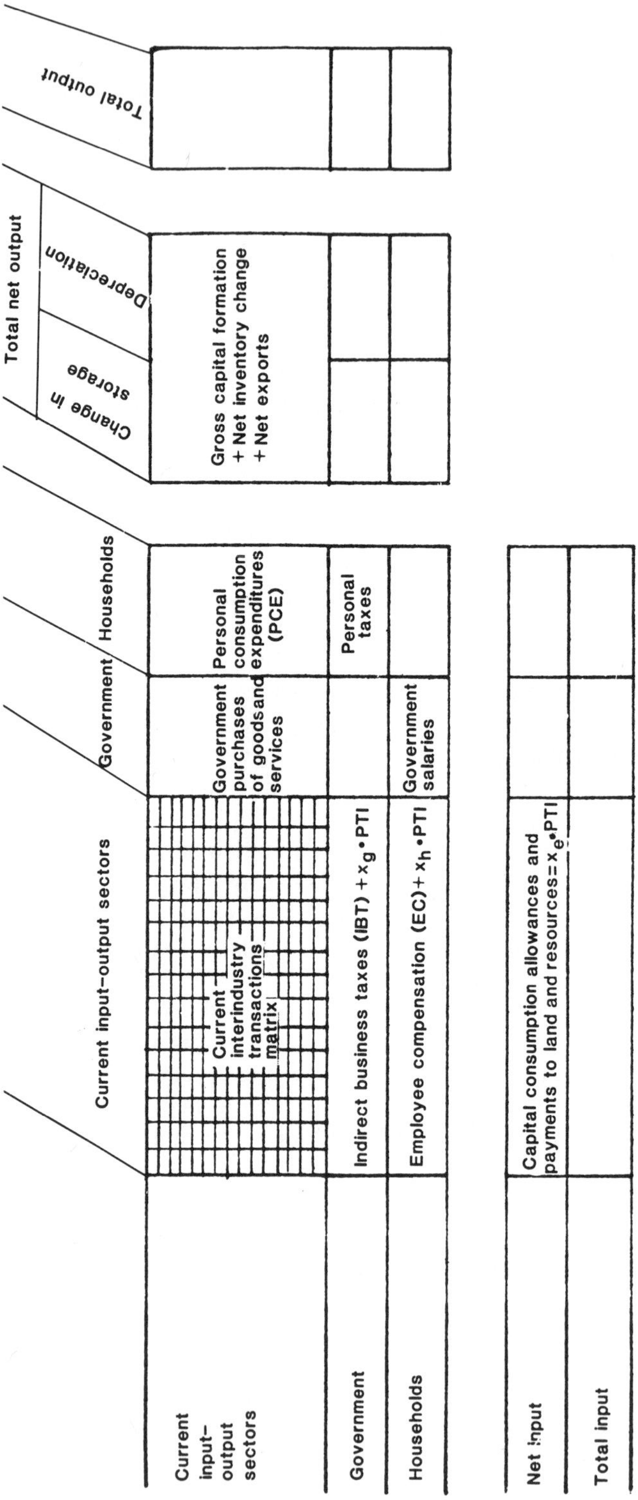

Fig. 28.2. Summary of modifications to the current national input-output conventions. PTI, property-type income.

Table 28.1.   Embodied Energy Intensities for Selected Sectors Evaluated for the 1967 Database

| Sector[a] | Embodied energy intensity (Btu's of fossil fuel per dollar) | | | |
|---|---|---|---|---|
| | A[b] | B[c] | C[d] | D[e] |
| 1.   Coal mining (7) | 5,143,400 | 5,172,000 | 5,455,600 | 5,807,500 |
| 2.   Crude petroleum and natural gas (8) | 2,920,300 | 2,929,200 | 3,188,600 | 3,469,050 |
| 4.   Electric utilities (68.01) | 505,500 | 513,900 | 796,220 | 1,099,950 |
| 6.   Other agricultural products (2) | 81,567 | 775,090 | 381,090 | 1,385,400 |
| 10.   Iron and ferroalloy ores mining (5) | 65,904 | 99,605 | 406,060 | 800,800 |
| 14.   New construction (11) | 54,804 | 230,245 | 389,770 | 913,950 |
| 21.   Apparel (18) | 38,845 | 295,135 | 371,107 | 974,600 |
| 33.   Paints and allied products (30) | 107,100 | 160,680 | 425,290 | 809,300 |
| 73.   Air transportation (65.05) | 122,630 | 143,910 | 452,230 | 814,700 |
| 75.   Transportation services (65.07) | 5,672 | 11,615 | 346,970 | 706,750 |
| 79.   Wholesale and retail trade (69) | 29,302 | 43,265 | 411,490 | 814,350 |
| 91.   Government | | | 717,160 | 1,393,050 |
| 92.   Households | | | 358,350 | 738,050 |

[a] Numbers in parentheses are Bureau of Economic Affairs sector equivalents.

[b] Excluding labor and government energy costs and solar energy inputs.

[c] Including solar energy inputs.

[d] Including labor and government energy costs.

[e] Including both labor and government energy costs and solar energy inputs.

difference in quality, an I-O model of the biosphere showing the complete production relations from sunlight to fossil fuel would be necessary. Since such a model does not yet exist, an approximation based on the conversion of sunlight to tree biomass to electricity in a woodburning power plant was used. This yielded a conversion factor of 2000 Btu's of solar energy per Btu of fossil fuel.[17] The total solar input for the United States was estimated at $103 \times 10^{18}$ Btu's of solar energy per year[18,19] or the functional equivalent of $51.5 \times 10^{15}$ Btu's of fossil fuel per year. Solar energy was assumed to enter the economy through the agriculture, forestry, and fisheries sectors, according to their relative land areas. This distribution is crude and will need to be improved as better data become available. For example, the solar inputs to industrial sectors via the hydrologic cycle in providing water for processing and for carrying away wastes are not properly allocated with this approximation.

## Results of Modifications to System Boundaries

The 90-sector model of energy input and output maintained by the Energy Research Group at the University of Illinois was used to determine the effects of making the above modifications. The model is based on 1967 financial transactions in the U.S. economy. Physical flow I-O data would be preferred for embodied energy calculations, but are not available in the required form at the national level.

Table 28.2. Regression Analysis Results for Direct Plus Indirect Energy Consumption (Embodied Energy) Versus Dollar Value of Output for 92 U.S. Economy Sectors Evaluated for the 1967 Data base

| Alternative[a] | Including energy sectors 1 to 7 | | | Excluding energy sectors 1 to 7 | | |
|---|---|---|---|---|---|---|
| | $R^2$ | $F$ | Significance level of $F$-test | $R^2$ | $F$ | Significance level of $F$-test |
| A | 0.0210 | 1.89 | 0.1729 | 0.5539 | 100.57 | 0.0001 |
| B | 0.0629 | 5.90 | 0.1710 | 0.2042 | 20.78 | 0.0001 |
| C | 0.7809 | 313.73 | 0.0001 | 0.9907 | 8633.95 | 0.0001 |
| D | 0.8535 | 512.74 | 0.0001 | 0.9454 | 1401.31 | 0.0001 |

[a] (A) Excluding labor and government energy costs and solar energy inputs; (B) including solar energy inputs; (C) including labor and government energy costs; and (D) including both labor and government energy costs and solar energy inputs.

Calculations made with financial data are nevertheless useful, because they yield information on the direct and indirect energy required to produce a dollar's worth of each of the commodities in the economy.

Selected embodied energy intensities (in Btu's of fossil fuel per dollar) were calculated for each of four possible alternatives, with the results shown in Table 28.1. The complete calculations may be found in note 13. Alternative A is calculated with the conventional economic I-O boundaries and yields essentially the same results as those of Herendeen and Bullard.[6] Alternative B includes solar energy inputs; alternative C includes labor and government as endogenous sectors; and alternative D includes the modifications of alternatives B and C together.

Figure 28.3 shows frequency plots for the four alternatives, which indicate the reduction in variance of the energy intensities when labor and government are included. Including solar energy did not increase or decrease the variance greatly, but this may be due to the rather crude method used in this study to estimate the distribution of solar energy to the economic sectors. A low variance indicates a more constant relationship from sector to sector of direct-plus-indirect energy consumption and dollar value of output.

The results were put in a regression format to highlight relationships and for significance testing. The calculated energy intensity for each sector (in Btu's per dollar) was multiplied by the sector's dollar output to yield the direct-plus-indirect energy input (in Btu's). This was used as the independent variable, with total dollar output as the dependent variable. Figure 28.4 shows the results for each of the four alternatives. The primary energy sectors (sectors 1 to 7) were found to be outliers, and the regression results excluding them are also presented. The regression lines are indicated in Figure 28.4 along with the $R^2$ values, equations, and $t$ statistics on the parameters (in parentheses below the parameter values). Table 28.2 lists the $R^2$ values, $F$ statistics, and significance levels for each of the alternatives.

The results indicate a significant relationship between embodied energy and dollar output when labor and government are included as endogenous sectors (alternatives C and D). Several problems still exist, but the trend is clear. As more

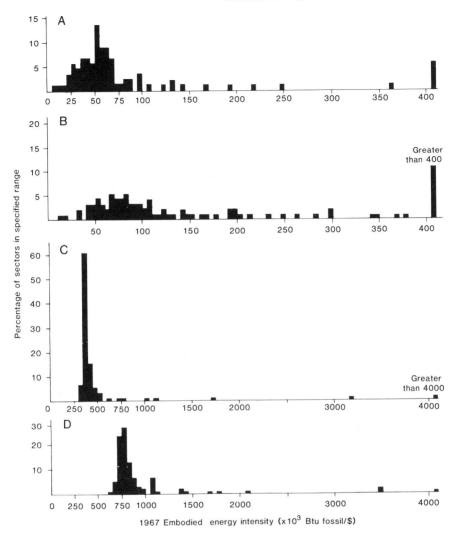

Fig. 28.3. Frequency plots of embodied energy intensity (in $10^3$ Btu's per dollar) by sector for 92 U.S. economy sectors evaluated for the 1967 database (A) excluding labor and government energy costs and solar energy inputs, (B) including solar energy inputs, (C) including labor and government energy costs, and (D) including both labor and government energy costs and solar energy inputs.

of the indirect energy costs are taken into account, the ratio of embodied energy to dollars becomes more nearly constant from sector to sector. The primary input sectors (sectors 1 to 7) are the important exceptions to this rule. Their departure from the regression lines in Figure 28.4, C and D, has several possible interpretations. One interpretation is that the energy intensities for these sectors are high because they represent the points of entry of available energy into the economy. Their degree of departure from the line is an indication of the net energy yield or "energy profit" they provide. In other words, their direct and indirect

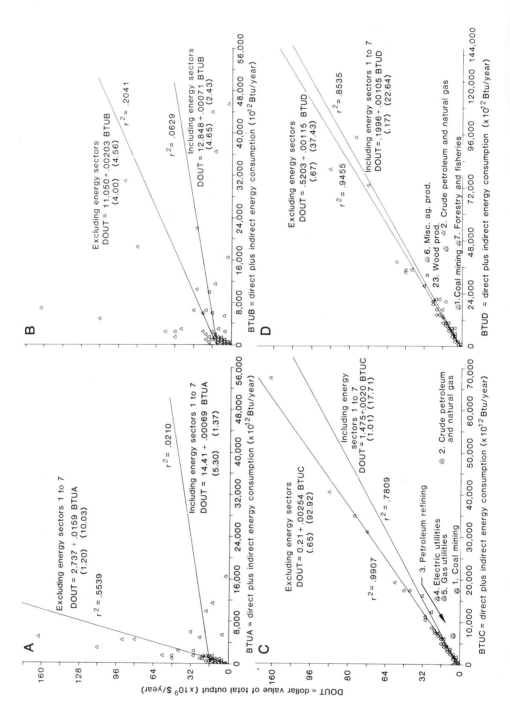

Fig. 28.4. Plots of direct plus indirect consumption (embodied energy) versus dollar value of output for 92 U.S. economy sectors evaluated for the 1967 database (A) excluding labor and government energy costs and solar energy inputs, (B) including solar energy inputs, (C) including labor and government energy costs, and (D) including both labor and government energy costs and solar energy inputs. The primary energy sectors (sectors 1 to 7) were found to be outliers, and regression results excluding them are also presented.

production costs in energy terms are much less than the energy embodied in their outputs, the difference being the amount brought into the economy from outside.

## Ratios of Energy to GNP

Discussions of energy and economics frequently include time series or international comparison plots showing the relation of fossil, nuclear, and hydro energy to GNP or GDP (gross domestic product). The strong historical and international link between these variables is unmistakable. Several authors have suggested that "decoupling" energy and GNP is possible and would allow the economy to continue to grow while decreasing energy consumption.[20,21] Calculations based on the conventional system boundaries (Figures 28.3A and 28.4A) are often used or implied in support of this idea. If the sector-to-sector differences in embodied energy intensities implied in these calculations were real, then it might be possible to simply shift production from high energy intensity sectors to low energy intensity sectors to lower energy use without sacrificing economic activity. This conclusion would follow from the underlying assumption that the currently defined primary factors are independent. But because it takes available energy to produce labor and government services, capital, and other natural resources, the assumption of independence is not warranted. The results presented in Figure 28.3, C and D, and Figure 28.4, C and D, reflect the implications of an attempt to relax the independence assumption and lead to the conclusion that decoupling energy and economic activity by simply shifting production between sectors is not a real possibility. The possibility for large changes in energy efficiency is small since, all things considered, total energy efficiency is fairly constant from sector to sector. Any reductions in direct energy consumption are offset by increases in indirect energy consumption through increased use of labor, land, or capital.

Actually, Figure 28.4, C and D, lead to the conclusion that energy consumption is highly related to gross capital formation plus net inventory change plus net exports because these quantities are the net output from the economy under the revised boundary definition. The GNP includes these quantities, as well as personal consumption and government expenditures. The extent to which gross capital formation, net exports, and net inventory change are separable from the GNP as a whole represents the latitude for decoupling energy and GNP. I suspect that this latitude is small, especially if human and government capital formation are included as part of gross capital formation—as Kendrick (note 15) has suggested—and not as consumption. Inclusion of human and government capital formation would also significantly decrease the mean embodied energy intensity, since it would increase the redefined net output from the economy. For alternative D the mean would drop from $12.20 \times 10^5$ to $1.88 \times 10^5$ Btu's of fossil fuel per 1967 dollar (note 14).

## Double Counting

Slesser[22] commented that including labor costs in embodied energy calculations would involve double counting. This criticism was directed against a specific method of including labor costs, and it remains a valid criticism of that method. If the conventional system boundaries for calculation of embodied energy are used, and then the energy necessary to support labor is simply added on, the same energy has indeed been counted twice. If the energy costs of labor are to be included without double counting, the system boundaries have to be changed. The net output with the revised boundaries does not include the support of labor, which is now an internal transaction (see Figure 28.2). The total energy budget is now allocated to gross capital formation, net inventory change, and net exports. The total energy requirement is equal to the total energy input, and the energy cost of labor is accounted for. However, in any I-O accounting system, gross flows and net flows must be kept straight. One can never add up internal transactions in an I-O table and expect them to equal net inputs or outputs. With the expanded boundaries, net output and input are no longer equal to GNP, but rather to GNP minus labor and government transactions.

### Embodied Energy Theory of Value

Several authors have proposed various forms of an energy theory of value (see notes 1, 8, 23, 24). The idea is summarily dismissed by neoclassical economists[25-27] on the ground that energy is only one of a number of primary inputs to the production process. This dismissal is unwarranted if the traditional primary factors are in reality interdependent. The results presented in this article indicate that if there are interdependences among the currently defined primary factors, then calculated embodied energy values that take this into account show a very good empirical relation to market-determined dollar values. Herendeen[28] has shown that, with the I-O method, the necessary and sufficient condition for the energy intensity vector to be constant is that the value-added vector (in dollars) and the direct-energy input vector (in Btu's) be proportional. If all factors other than energy are moved from the net input vector to the transactions matrix, this proportionality is to be expected.

The question might be asked whether the same thing we have done with energy could not be done with any of the other currently defined primary factors and, thus, support capital, labor, or government service theories of value. The answer is that on paper this could be done. We must look to physical reality to determine which factors are net inputs and which are internal transactions. No one would seriously suggest that labor creates sunlight.

An embodied energy theory of value, thus, makes theoretical sense and is empirically accurate only if the system boundaries are defined in an appropriate way. It is really a cost-of-production theory with all costs carried back to the solar energy necessary directly and indirectly to produce them. The results indicate that there is no inherent conflict between an embodied energy (or energy cost) theory of value and value theories based on utility. The empirical equivalence of these estimates—one from the cost or supply side, and one from the benefit or demand side—supports basic economic principles grounded in optimization which giving them a biophysical basis. The flow of energy *is* the primary concern of economics.

## Conclusions

The results presented here indicate that, with the appropriate perspective and boundaries, market-determined dollar values and embodied energy values are proportional for all but the primary energy sectors. The required perspective is an ecological or "systems" view that considers humans to be a part of, and not apart from, their environment. A few economists have already taken this perspective,[29-31] and the implications for a new ecological economics that links the natural and social sciences are great.[32] The concept of embodied energy may help to provide such a link as an empirically accurate common denominator in ecological and economic systems. With the appropriate boundaries, embodied energy values are accurate indicators of market values where markets exist. Because they are based on physical flows, they may also be used to determine "market values" where markets do not exist—for example, in ecological systems. This is one way of "internalizing" all factors external to the existing market system and solving the natural resource valuation problem. From the ecological perspective, markets can be viewed as an efficient energy allocation device that humans have developed to solve the common problem facing all species—survival.

What does all this imply for national policy? The most important implication is that the physical dimensions of economic activity are not separable from limitations of energy supply. The universally appealing notion of unlimited economic growth with reduced energy consumption must be put firmly to rest beside the equally appealing but impossible idea of perpetual motion. It is easy to think you can get a "free lunch" by looking only at small parts of the system in isolation. When the whole system is analyzed, however, it becomes clear that all you can do is transfer the cost of your lunch to another segment of the system.

These conclusions should not be interpreted as pessimistic. Several authors, notably Daly (note 31), have pointed out the inadequacy of GNP and other yardsticks of physical economic production as measures of social welfare. Indeed, there is nothing inherently appealing about what Boulding (note 30) has called the "cowboy economy," the adolescent phase of rapid, self-conscious, often painful growth. If we are to manage our future wisely, we must be aware of the physical limitations on economic activity and learn to live well within our energy budget.

## NOTES

1. F. Soddy, *Wealth, Virtual Wealth and Debt: The Solution of the Economic Paradox* (Dutton, New York, 1933).
2. M. W. Gilliland, *Science* **189**, 1051 (1975).
3. ———, Ed., *Energy Analysis: A New Public Policy Tool* (Westview Press, Boulder, Colo., 1978).
4. D. E. Gushee, *Energy Accounting as a Policy Analysis Tool* (Congressional Research Service, Washington, D.C., 1976).
5. B. Hannon, *J. Theor. Biol.* **41**, 575 (1973).
6. R. A. Herendeen and C. W. Bullard, *Energy Costs of Goods and Services, 1963 and 1967* (Document 140, Center for Advanced Computation, University of Illinois, Champaign-Urbana, 1974).
7. H. T. Odum, *Ambio* **2**, 220 (1973).

8. ———, in *Ecosystem Modeling in Theory and Practice*, C. A. S. Hall and J. W. Day, Eds. (Wiley, New York, 1977), pp. 173–196.
9. ———, in note 3, pp. 55–87.
10. N. Georgescu-Roegen, in *Scarcity and Growth Reconsidered*, V. K. Smith, Ed. (Johns Hopkins Press, Baltimore, 1979), pp. 95–105.
11. J. E. Stiglitz, in ibid., pp. 36–66.
12. W. W. Leontief, *The Structure of American Economy, 1919, 1929; An Empirical Application of Equilibrium Analysis* (Harvard Univ. Press, Cambridge, Mass., 1941).
13. R. Costanza, *Energy Costs of Goods and Services in 1967 Including Solar Energy Inputs and Labor and Government Service Feedbacks* (Document 262, Center for Advanced Computation, University of Illinois, Champaign-Urbana, 1978).
14. ———, thesis, University of Florida, Gainesville (1979).
15. J. W. Kendrick, *The Formation and Stocks of Total Capital* (National Bureau of Economic Research, New York, 1976).
16. K. Kirkpatrick, *Effect of Including Capital Flows on Energy Coefficients, 1963* (Technical memo **26**, Energy Research Group, University of Illinois, Urbana, 1974).
17. H. T. Odum, F. C. Wang, J. Alexander, M. Gilliland, *Energy Analysis of Environmental Values: A Manual for Estimating Environmental and Societal Values According to Embodied Energies* [Report to the Nuclear Regulatory Commission, contract NRC-04-77-123 (Center for Wetlands, University of Florida, Gainesville, 1978)].
18. T. H. Vonder Haar and V. E. Suomi, *Science* **163**, 667 (1969).
19. M. I. Budyko, in *Climatic Change*, J. Gribben, Ed. (Cambridge Univ. Press, New York, 1978), pp. 85–113.
20. R. Stobaugh and D. Yergin, Eds., *Energy Future: Report of the Energy Project at the Harvard Business School* (Random House, New York, 1979).
21. A. B. Lovins, *Soft Energy Paths: Toward a Durable Peace* (Ballinger, Cambridge, Mass., 1977).
22. M. Slesser, *Science* **196**, 259 (1977).
23. F. Cottrell, *Energy and Society: The Relation Between Energy, Social Change and Economic Development* (McGraw-Hill, New York, 1955).
24. B. Hannon, *Ann. Am. Acad. Polit. Soc. Sci.* **410**, 139 (1973).
25. D. A. Huettner, *Science* **192**, 101 (1976).
26. M. R. Langham and W. W. McPherson, ibid, p. 8.
27. H. M. Peskin, ibid., p. 9.
28. R. A. Herendeen, *On the Concept of Energy Intensity in Ecological Systems* (Document 271, Energy Research Group, University of Illinois, Champaign-Urbana, 1980).
29. N. Georgescu-Roegen, *The Entropy Law and the Economic Process* (Harvard Univ. Press, Cambridge, Mass., 1971).
30. K. E. Boulding, in *Environmental Quality in a Growing Economy*, H. Jarrett, Ed. (Johns Hopkins Press, Baltimore, 1966), pp. 3–14.
31. H. E. Daly, *Steady-State Economics* (Freeman, San Francisco, 1977).
32. E. P. Odum, *Science* **195**, 1289 (1977).

# Index

# About the Editor

Ira Sohn, until 1985, was Senior Research Scientist at the Institute for Economic Analysis (New York University), where, using the Input-Output approach, he carried out studies on global economic modelling.

He is a co-author of *The Future of Nonfuel Minerals in the U.S. and World Economy* (Lexington Books, D.C. Heath and Company: Lexington, MA, 1983) and a co-recipient of the First Russell Ackoff Award (1985) for the best scholarly paper written on the subject of energy, environment, minerals and other resources.

Currently, he is Associate Professor of Finance at Montclair State College in Upper Montclair, New Jersey.